Amphibian Biology

Edited by

Harold Heatwole

Volume 3
Sensory Perception

Co-editor for this volume

Ellen M. Dawley

Published by

Surrey Beatty & Sons

This book is copyright. Apart from any fair dealing for purposes of private study, research, criticism or review, as permitted under the Copyright Act, no part may be reproduced by any process without written permission. Enquiries should be made to the publisher.

Surrey Beatty & Sons Pty Limited is a member of Copyright Agency Limited (CAL) in Australia and is a respondent to overseas collecting societies.

The National Library of Australia
Cataloguing-in-Publication entry:
Amphibian Biology. Volume 3, Sensory Perception.
Bibliography.
Includes index.
ISBN 0 949324 72 8 (v. 3).
ISBN 0 949324 53 1 (set).
1. Amphibians. I. Heatwole, Harold.
597.6

Published February 1998

PRINTED IN AUSTRALIA BY
SURREY BEATTY & SONS PTY LIMITED
43 Rickard Road, Chipping Norton, NSW 2170

Amphibian Biology

Preface to Series

THERE are a number of outstanding treatises of amphibian biology. *Biology of Amphibians* (Duellman and Trueb 1986) is an excellent general work, clearly written and well illustrated, and with a remarkable depth and breadth for a single volume. It will be the standard general reference on amphibians for many years to come and the present generation of herpetologists will consider it the "amphibian bible" much as their predecessors regarded G. K. Noble's (1931) *The Biology of the Amphibia*. However, no single volume fulfils the need for a sequential, monographic treatment of specialized topics. In addition, there are books that treat certain specialized subjects in considerable depth. These include the three volumes of *Physiology of the Amphibia* (Moore 1964, 1974; Lofts 1976), *Frog Neurobiology, a Handbook* (Llinás and Precht 1976) and *The Reproductive Biology of Amphibians* (Taylor and Guttman 1977), but it has now been 29 years since the first of these appeared and 17 years since the last. Many topics are in need of updating. *Environmental Physiology of the Amphibians* (Feder and Burggren 1992) brings some of those topics up to date and treats some new ones. Collectively, these works leave large gaps in physiology unaddressed, and there are no equivalent, recent reviews for most other aspects of amphibian biology.

Recognizing that the discipline of amphibian biology has reached sufficient maturity to warrant detailed, multi-volume review and that such a need has been filled only partly, and with no commitment to continue the process, it was decided to launch the present new series, *Amphibian Biology*. It will not compete with the titles mentioned above. Recently reviewed topics will not be covered in early volumes, but will be reserved for such time as further update is required.

The need for this series was evidenced by the enthusiastic response from potential authors. Of the 64 people contacted with invitations to contribute to the first five volumes, only three declined, and then because of heavy commitments otherwise. Most expressed the view that such a series was long overdue. With this initial encouragement the series was launched.

Amphibian Biology was inspired by *Biology of the Reptilia*, edited by Carl Gans, and is intended as a companion to that series. *Biology of the Reptilia* is a unique, monumental contribution to herpetology and has become the most authoritative single source of information on reptiles that is available. Comprehensive treatments of all aspects of reptilian biology are presented in detail and are exhaustively documented by literature. That series is an ongoing one with topics added as sufficient information comes to hand to warrant a review. It has been, and continues to be invaluable. It is hoped that *Amphibian Biology* will serve herpetologists in the same way and that it will maintain the high standard set by its reptilian counterpart.

Raleigh, North Carolina, USA
January 1993

Harold Heatwole
Series Editor

REFERENCES

Duellman, W. E. and Trueb, L., 1986. Biology of Amphibians. McGraw-Hill Book Company, New York.

Feder, M. E. and Burggren, W. (eds), 1992. Environmental Physiology of the Amphibians. University of Chicago Press, Chicago.

Gans, C. (ed.), 1969–1988+. Biology of the Reptilia. 16+ vols. Academic Press, New York and Alan R. Liss, Inc., New York.

Llinás, R. and Precht, W., 1976. Frog Neurobiology, a Handbook. Springer-Verlag, Berlin.

Lofts, B. (ed.), 1974. Physiology of the Amphibia, Volume 2. Academic Press, New York.

Lofts, B. (ed.), 1976. Physiology of the Amphibia, Volume 3. Academic Press, New York.

Moore, J. A. (ed.), 1964. Physiology of the Amphibia. Academic Press, New York.

Noble, G. K., 1931. The Biology of the Amphibia. McGraw-Hill Book Company, New York.

Taylor, D. H. and Guttman, S. I. (eds). 1977. The Reproductive Biology of Amphibians. Plenum Press, New York.

Preface to Volume 3

THE world is perceived by vertebrates through an array of sensory systems, often with one or two systems predominating. Thus, while auditory and chemosensory systems are a part of the human sensory tool kit, vision so dominates that the perception of a concept often is accompanied by the exclamation "Oh I *see* what you mean". Among amphibians, vision also predominates for most frogs and salamanders but plays a minor, or no role at all, for caecilians. Because of our own visual bias, our perception of different sensory systems in other organisms often are not intuitive. So, while a female salamander's willingness to mate might be increased by pheromones wafted to her from a male, we do not perceive these odours and do not as readily empathize with their effect on her mating behaviour. Finally, certain sensory systems, like the gravistatic and acceleration-sensitive parts of the inner ear or muscle spindles give personal positional information that is mediated nearly subconsciously.

Some sensory systems are not utilized, or are not even present, in all vertebrates. The lateral line system and electroreceptors function only in fishes and aquatic amphibians. Magnetoreception has been found in a few amphibians and a handful of other vertebrates. Amphibians as a group possess the entire array of vertebrate sensory systems and research with these animals has helped reveal a blueprint of basic vertebrate structure and function, as well as some intriguing and unique evolutionary solutions.

The present volume consists of seven chapters that consider the similarities and differences among the sensory systems of the different groups of amphibians, indicates how these systems relate to different life styles, and finally compares them with the systems of other vertebrates.

In Chapter 1, the nasal chemical senses, olfaction and the vomeronasal system, are examined. The nasal chemical senses are perhaps the simplest of the cranial senses with relatively few cell types comprising the sensory epithelium, and with olfactory and vomeronasal receptors both synapsing directly with the telencephalon.

The other major chemical sense, taste, is reviewed in Chapter 2. Taste and smell differ structurally in many fundamental ways, from the morphology of receptor cells to patterns of innervation, but are rather similar physiologically. Although some behavioural information is available about how amphibians may use their sense of smell, taste is more of an unknown.

The complexities of vision are examined in detail in Chapter 3; this system is perhaps, the best known, and extensive tract-tracing and intracellular recording have produced a reasonably clear picture of the neuronal guidance of feeding behaviour. Studies of vision have focused nearly exclusively on its role in feeding. Chapter 3 also touches on extraoptic light perception via the pineal complex.

Chapter 4 reviews the octavolateralis system, including structure and function of the inner and middle ears and of the lateral line. Although audition has been most thoroughly investigated in anurans, this chapter also indicates the unique combinations of adaptations resulting from divergent evolutionary pathways among the different amphibian orders.

The sensations of pressure, touch, temperature, pain, noxious chemicals, stretch, and body position that reside in the skin and muscles are covered in Chapters 5 and 6. Because many amphibians hatch at early developmental stages, separate and sometimes transient sensory pathways develop in the early stages. Thus, embryos and young larvae (Chapter 5) and adults (Chapter 6) are considered separately.

Finally, magnetoreception, covered in Chapter 7, is solely known through behavioural studies of navigation, while the location of receptors and the nature of transduction mechanisms are surely yet to be determined.

Collegeville, Pennsylvania, USA
September 1997

Ellen Dawley
Co-editor

Dedication

DR Robert Capranica is widely considered the founder of the field of amphibian neuroethology. He was born in Los Angeles, California on 29 May 1931. He started his undergraduate education in chemical engineering at the University of California at Los Angeles in 1949. After a three-year stint in a naval aviation group, he returned to California where he received his B.S. degree in electrical engineering from UC Berkeley in 1958. Following his graduation, he joined the technical staff in the Department of Sensory and Perceptual Processes of the Bell Telephone Laboratories. From Bell Labs, he received scholarships to attend New York University where he earned his M.S. degree in electrical engineering in 1960, and the Massachusetts Institute of Technology, where he received his Sc.D. in electrical engineering in 1964.

His doctoral dissertation was published in 1965 by the M.I.T. Press as a monograph entitled: "The evoked vocal response of the bullfrog: A study of communication by sound". In this now classical work, Bob broke new ground in the field of acoustic biology. He pioneered the use of synthetic calls for amphibian playback experiments; he systematically demonstrated that the two principal components of the bullfrog's mating call, namely the high- and low-frequency energy bands, must be simultaneously present to evoke a vocal response from an isolated male bullfrog; and he demonstrated that the mating call of the bullfrog was species-specific.

In 1969 he joined the faculty of Cornell University with a joint appointment in the Section of Neurobiology and Behaviour and in Electrical Engineering. During his years at Cornell, Bob trained more than 16 graduate students and over eleven post-doctoral fellows, many of whom have continued on to make their own contributions in the field of vertebrate neuroethology. Bob's seminal thesis observation of the specificity of the bullfrog's acoustic behaviour led to a series of elegant experiments by Bob and his colleagues culminating in the demonstration that the thalamus is the site in the frog brain of neuronal facilitation responsible for the "detection" of the mating call. He was able to confirm the remarkable specificity of spectral and temporal processing that occurs in the anuran central nervous system and was responsible for elevating natural sound processing by the anuran auditory system to the level of the other vertebrate neuroethological models, e.g., the jamming avoidance response of weakly electric fish, echolocation in bats, sound localization in owls, etc. Some of Bob's major contributions include his multiple studies showing that the auditory systems of different species of frogs have different spectral sensitivities; his seminal studies on acoustic dialects in cricket frog calls with Nevo demonstrating the existence of distinct call types in allopatric populations of *Acris;* his introduction of the concept of the "matched filter" to provide an underlying structure for the coevolution of sender and receiver, and his studies elucidating the remarkable temporal specificity found in the cells of the anuran CNS that are often closely matched with advertisement call features.

Bob cares deeply about his students. All of us remember his famous red pen that would spew forth commentary and constructive criticism (often peppered with "Capranicaisms") into the margins of any manuscript submitted to him. But papers always emerged crisper and clearer after passing through his hi-Q literary filter. He was keenly curious and had a disconcerting way of asking the most fundamental questions: "What path does the energy take?"; "How do the stereocilia move?"; "How does this adaptation benefit the animal?" Basic questions such as these and the animated discussions that followed often led to experiments that produced novel data and insights in vertebrate sensory biology.

Bob is the consummate teacher. Those of us at Cornell fortunate enough to be present in class for the first offering of "Animal Communication" in 1970, experienced the remarkable creative energy put forth by both Bob and Jack Bradbury in developing the ideas for this course. Bob flourished in this environment; challenged by his colleague and by the students, he presented original, difficult material in a manner that was not only rigorous and complete, but also understandable and entertaining. It was clear to all of us that Bob enjoyed that class immensely.

From all of us who have benefited from your wisdom, your insights, your patience and your humour, this volume is dedicated to you, Dr Robert Capranica, that you and your work continue to be a source of inspiration to all students of the frog auditory system.

Peter M. Narins
Los Angeles, California, USA
November 1996

Contributors

BARLOW, Linda, Department of Neurosciences 0201, University of California at San Diego, La Jolla, California 92093-0201, USA. Present address: Department of Biological Sciences, University of Denver, CO 80208, USA.

DAWLEY, Ellen M., P.O. Box 1000, Ursinus College, Collegeville, Pennsylvania 19426-1000, USA.

DICKE, Ursula, Institut für Hirnforschung, Universität Bremen-FB2, Postfach 33 04 40, D-2800 Bremen 33, Germany.

FRITZSCH, Bernd, Department of Biomedical Sciences, Creighton University, Omaha, Nebraska 68178, USA.

HEATWOLE, Harold, Department of Zoology, North Carolina State University, Raleigh, North Carolina 27695-7617, USA.

NEARY, Timothy, Department of Biomedical Sciences, Creighton University, Omaha, Nebraska 68178, USA.

PHILLIPS, John, Department of Zoology, Indiana University, Bloomington, India 47405, USA.

ROBERTS, Alan, School of Biological Sciences, University of Bristol, Woodland Road, Bristol BS8 1UG, United Kingdom.

ROTH, Gerhardt, Institut für Hirnforschung, Universität Bremen-FB2, Postfach 33 04 40, D-2800 Bremen 33, Germany.

WIGGERS, Wolfgang, Institut für Hirnforschung, Universität Bremen-FB2, Postfach 33 04 40, D-2800 Bremen 33, Germany.

Contents

Preface to Series	v
Preface to Volume 3	vii
Dedication	ix
Contributors	x
Chapter 1. Olfaction. Ellen M. Dawley	711
Chapter 2. Taste. Linda Barlow	743
Chapter 3. Vision. Gerhard Roth, Ursula Dicke and Wolfgang Wiggers	783
Chapter 4. The Octavolateralis System of Mechanosensory and Electrosensory Organs. Bernd Fritzsch and Timothy Neary	878
Chapter 5. Skin Sensory System of Amphibian Embryos and Young Larvae. Alan Roberts	923
Chapter 6. Diffuse Cutaneous and Muscular Sensory Systems: Mechanoreception, Thermoreception, Nociception, Chemoreception and Kinesthetic Sense. Harold Heatwole	936
Chapter 7. Magnetoreception. John B. Phillips	954
Index	965

CHAPTER 1

Olfaction

Ellen M. Dawley

I. Introduction
II. Structure and Function
   A. Structure of the Periphery
      1. Receptor Cells
      2. Sustentacular Cells
      3. Basal Cells
      4. Development of the Olfactory/ Vomeronasal Epithelium
      5. *Lamina Propria*
      6. Variation in Nasal Geography and Epithelial Structure
   B. Physiology of Odour Reception, Transduction, and Discrimination
   C. Central Nervous System
      1. Olfactory and Accessory Olfactory Bulbs
      2. Central Projections
   D. Functional Comparisons
      1. Terminal Nerve System
III. Behavioural Studies
   A. Larval Studies
   B. Foraging
   C. Migration and Orientation
   D. Social Interactions
IV. Acknowledgements
V. References

I. INTRODUCTION

MOST vertebrates possess a sense of smell. There is, however, variability in the relative amount of receptor tissue and the architecture of the structures that physically support these tissues. Olfactory chemical receptor cells (to differentiate them from taste receptors) are found in epithelia lining the nasal cavity. In tetrapods, gas- or liquid-borne chemicals enter through the external nares and exit into the oral cavity through internal nares (choanae). In fishes, incurrent and excurrent nares (homologous to external and internal nares) provide a similar one-way flow for liquid-borne chemicals through the nasal cavity. The nasal cavity in tetrapods in subdivided into two or more diverticulae which may be confluent with one another. One of these diverticulae is termed the vomeronasal organ and is thought to be a tetrapod innovation (Parsons 1967; Bertmar 1981; but see Eisthen *et al.* 1994). Vomeronasal receptors appear to be sensitive to heavy odour molecules while the olfactory epithelium is thought to be sensitive to lighter, air borne molecules (Johnson 1985). In anurans and amniotes, the vomeronasal organ is found in a medial position, often, in the case of amniotes, with a separate duct from the oral cavity (Parsons 1967; Bertmar 1981). In caudates and most apodans, the vomeronasal organ (VNO) is usually found in a lateral diverticulum which may remain confluent with the main olfactory chamber (Parsons 1967; Bertmar 1981; Schmidt and Wake 1990). These structural arrangements of the nasal cavities of amniotes and the three orders of amphibians have been used in the debate of the phylogeny of tetrapods.

A great deal is known about the anatomy and physiology of olfactory structures of certain amphibians, i.e., the tiger salamander (*Ambystoma tigrinum*) and the more common species of frogs in the genus *Rana*. These animals often are chosen for a number of characteristics including their relatively large size and the simplicity of the geometry of their nasal cavities, rather than for a specific interest in amphibian olfaction. Many researchers hope to extrapolate their findings to other vertebrates, especially to humans. As a result, a wealth of information about amphibian olfactory structures is also available to non-neurobiologists. The present chapter does not cite every reference to amphibian olfaction, but concentrates especially on studies from the last two decades.

This review is divided into two major parts. The first deals with the anatomy and physiology of olfaction. The second summarizes behavioural studies; these mainly have been carried out by herpetologists. Few studies bridge the gap between the neurobiological basis for olfaction and the behaviour of the animals. Apodans will appear to be neglected by this review, especially in the behavioural section. However, this is not an intentional bias, but rather reflects the paucity of research on these secretive animals.

II. STRUCTURE AND FUNCTION

A. Structure of the Periphery

Reception of odours occurs in the olfactory or vomeronasal mucosae. The mucosa is made up of two distinct layers (Fig. 1), a superficial olfactory or vomeronasal sensory epithelium and a deep *lamina propria* of connective tissue. These two layers are separated by a basement membrane. The epithelium is pseudostratified with three cell types, (1) olfactory or vomeronasal receptor neurons, (2) sustentacular cells, and (3) basal cells. The nuclei of these cells are stratified such that the sustentacular cell nuclei are most superficial, followed by the nuclei of the receptor cells, with the basal cells being deepest (Graziadei 1971; Getchell *et al.* 1984; Dawley and Bass 1988). Because the epithelium is pseudostratified, all three cell types touch the basement membrane. Only the receptor cells and the sustentacular cells extend to the epithelial surface (Rafols and Getchell 1983). Underlying the basal cells is the *lamina propria* which contains blood vessels, olfactory glands (Bowman's glands and deep glands; Getchell *et al.* 1984), and nerve bundles (including olfactory and vomeronasal nerves). Ducts of Bowman's glands pass through the epithelium of the olfactory epithelium to the surface. An acellular mucus layer overlies the epithelium. This mucus layer is particularly thick in amphibians, consisting of a watery upper layer and a viscous deeper layer.

1. Receptor Cells

Olfactory and vomeronasal receptor cells are bipolar neurons, i.e., neurons with a single dendrite and a single axon at opposite ends of the cell body (Fig. 2). The cell body is fusiform, with the long axis perpendicular to the surface of the epithelium. Nuclei are ovoid and show a characteristic "checkerboard" pattern of chromatin (Fig. 1). The cytoplasm contains pigment granules that give the mucosa a characteristic yellowish colour; this pigment is more abundant in superficial neurons (Graziadei and Graziadei 1976). Dendrites project to the epithelial surface where they end in olfactory knobs studded with cilia or microvilli. Although most studies found that olfactory receptor cell dendrites end in cilia and vomeronasal receptor cells end in microvilli, other studies have shown the presence of both microvillar and ciliated olfactory receptor cells (Farbman and Gesteland 1974; Breipohl *et al.* 1982; Eisthen *et al.* 1994). The cilia or microvilli extend into the mucus and contain receptor molecules for odourant detection (Getchell and Getchell 1987). Estimates of the total number of olfactory receptor cells in adult *Rana catesbeiana* range from 19 to 50 million (Burton 1985; Burton *et al.* 1990) and three million in *Rana pipiens* (Daston *et al.* 1990; these are comparable to estimates for other vertebrates, e.g., six million in humans, ten million in pike, 20 million in garfishes and 100 million in rabbits; see references in Burd 1991). In addition, *R. catesbeiana* is estimated to have 295 000 vomeronasal receptor cells (Burton *et al.* 1990). In *R. catesbeiana*, then, there are considerably more olfactory receptors than vomeronasal receptors (at a ratio of 64:1; Burton *et al.* 1990). Axons from the bipolar neurons project without collatorals or synapses with other neurons to the main olfactory or accessory olfactory bulbs (Getchell and Getchell 1987). Thus, there are no evident structures that might allow receptor cells to interact. Vomeronasal and main olfactory axons are virtually identical in appearance (Burton *et al.* 1990), and are among the smallest in the body (i.e., 0.097–0.396 μm in diameter; Burd 1991; Burton and Wentz 1992).

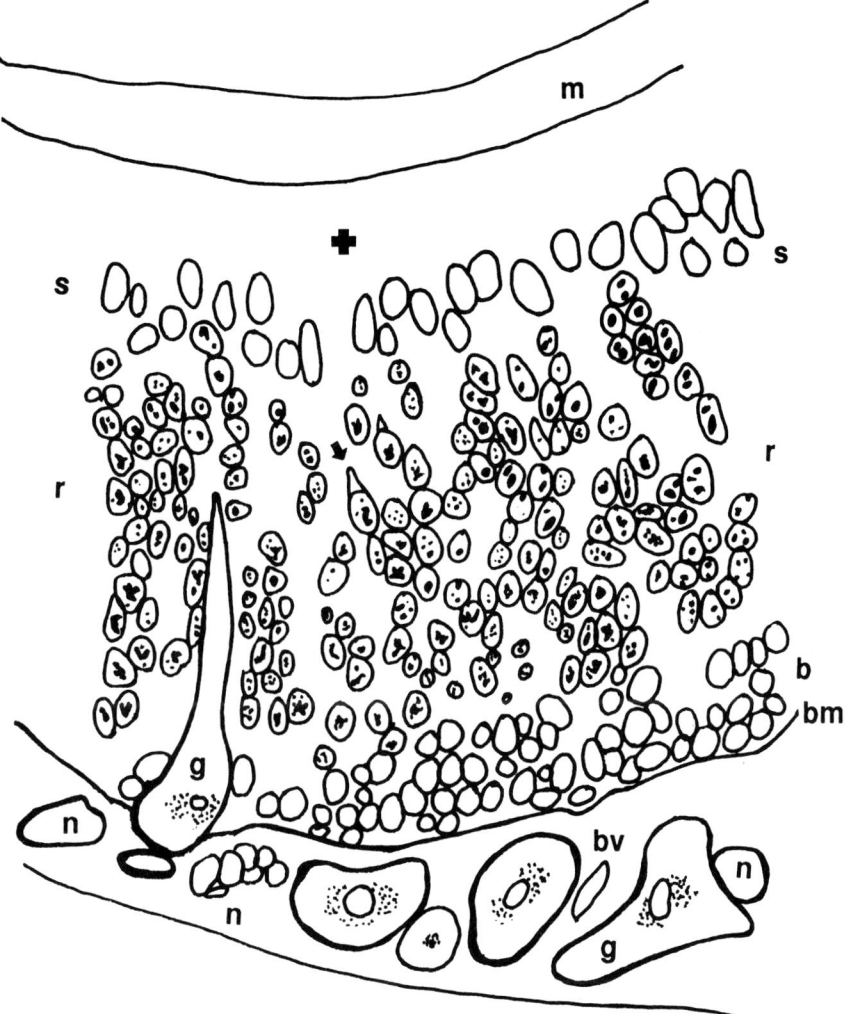

Fig. 1. Camera lucida drawing of the olfactory mucosa of *Plethodon cinereus*. The most superficial structures are at the top of the figure. The sensory epithelium extends from the mucus layer to the basement membrane. The *lamina propria* are those structures deep to the basement membrane. In light microscopic sections, such as this one, only the nuclei of the epithelium are stained. For simplicity, only receptor cell nuclei (in a characteristic "checkerboard" pattern) are shown with chromatin. Basal cells show a similar pattern. On some receptor cells it is possible to see portions of the dendrite as it winds its way to the surface of the epithelium (arrow). The region of the epithelium above all nuclear layers (thick cross) contains receptor cell dendrites and the apical ends of sustentacular cells. Bowman's glands are recognizable because of the presence of secretory granules in gland cells. In some sections, the neck of a Bowman's gland can be partially traced through the sensory epithelium. Vomeronasal mucosa lacks Bowman's glands. b = basal cells; bm = basement membrane; bv = blood vessel; g = Bowman's gland; m = mucus layer; n = nerve; r = receptor cell; s = sustentacular cell.

Among amphibians, the morphology of olfactory and vomeronasal axons and the organization of these axons with glial processes into nerves are similar to that found in other vertebrates. Olfactory and vomeronasal axons are not myelinated. Olfactory or vomeronasal axons are collected together in bundles of 2 to 15 *(Necturus)* to several hundred or even several thousand *(Xenopus* and *Ambystoma)* by the thin processes of ensheathing glial cells that occupy these nerves (Fig. 2; Farbman and Gesteland 1974; Rafols and Getchell 1983; Burd 1991). Vomeronasal nerves usually run alongside either the main trunk or a branch of the olfactory nerve, but remain separate entities *(Rana catesbeiana*, Burton *et al*. 1990; various Plethodontid salamanders, Schmidt *et al*. 1988). Terminal nerve fibers and cell bodies also can be found travelling within the olfactory or vomeronasal nerves (Burd 1991; see section on

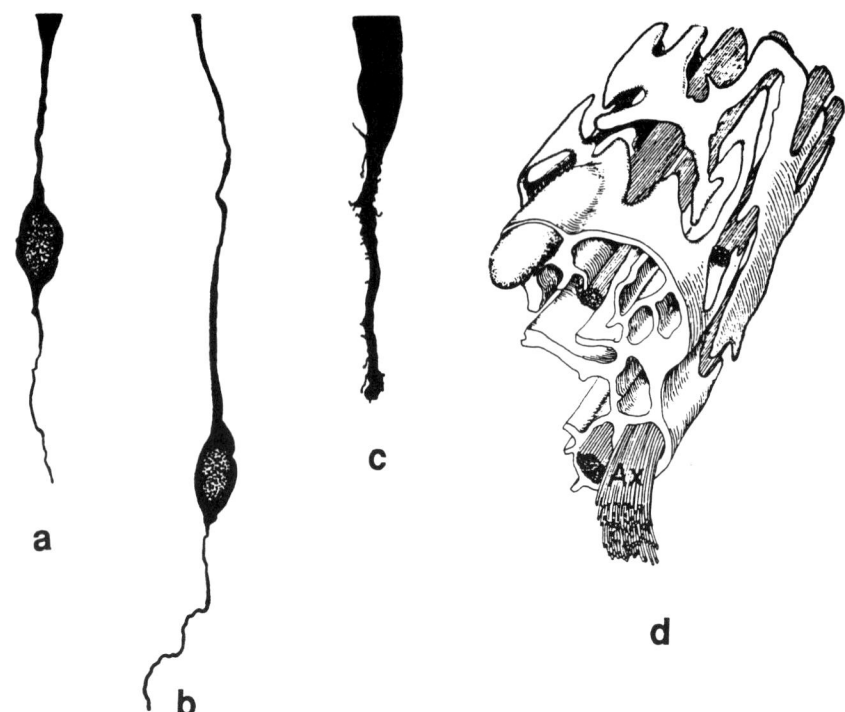

Fig. 2. Camera lucida drawings of **a–b:** olfactory receptor neurons (the cell bodies of a and b are at different depths within the epithelium), **c:** sustentacular cell (cell body is higher within the epithelium than the receptor cells), and a stereographic drawing of **d:** morphological relationships between ensheathing glial cells and axons of receptor cells (ax). After Rafols and Getchell (1983).

terminal nerve for more references). Axons of individual receptor cells show little intermingling along the length of the olfactory nerve so that neighbouring axons tend to be stable along the anterior-posterior axis of the nerve (Daston *et al.* 1990). Thus, axons from receptors in a given region of the olfactory epithelium are segregated within the olfactory nerve, which may allow neighbouring axons to modulate the excitability of each other. This is in contrast to the configuration of axons within the optic nerve where fibers continuously shift their relative positions along a sinuous course (Maturana 1960). Such an arrangement may cut down on interaction between neighbouring cells so that the precise mapping of receptor to CNS target is maintained. Olfactory receptors are not rigidly mapped on to the olfactory bulbs; instead axons near the olfactory bulb are oriented perpendicular or oblique to the plane of the section (Daston *et al.* 1990). This means that the topographical organization of the olfactory epithelium may be maintained in the nerve but not at the surface of the olfactory bulb (see section on reception, transduction, and discrimination).

2. *Sustentacular Cells*

Sustentacular cells are morphologically distinct from the receptor cells (Rafols and Getchell 1983). They are more columnar and lack the apical constriction of a dendrite (Fig. 2). The cell body is elongate and its nucleus is cylindrical rather than ovoid as in receptor cells. The apical ends of the sustentacular cells are studded with microvilli of variable length so that the olfactory mucosa looks like a field of microvilli interrupted by olfactory knobs reaching the external surface (Scalia 1976; Zielinski *et al.* 1988). However, sustentacular cells in the chemosensory epithelium of *Ambystoma mexicanum* either lack surface processes (all of the olfactory epitheliuum and some vomeronasal epithelium) or are ciliated (most of vomeronasal epithelium; Eisthen *et al.* 1994).

The apical region appears to secrete mucus by exocytosis via secretory vesicles lying in close apposition to the surface membranes. Endosome-like vesicles in the apical region also suggest that sustentacular cells endocytose material from the mucous layer (Zielinski *et al.* 1988).

The supranuclear region of the cell, immediately below the apical region, may be a site where secretory material is synthesized and stored because it contains many vesicles and other organelles (Farbman and Gesteland 1974; Zielinski *et al.* 1988). The portion of the cell basal to the soma is constricted into a central stalk or extends cytoplasmic veils that unfurl over the cell bodies of receptor cells (Rafols and Getchell 1983; Zielinski *et al.* 1988). In either case, this basal portion extends through the receptor nuclear layer and the basal cell layer to terminate in an expansion that rests close to the basement membrane and in apposition to structures within the *lamina propria*. These basilar expansions extend over portions of Bowman's glands, parts of the capillary walls, or simply over connective tissue (Rafols and Getchell 1983) and contain ultrastructural features suggesting this region may be metabolically active and may be involved in uptake and transport of material (Zielinski *et al.* 1988).

Many functions have been attributed to sustentacular cells, including many that are glia-like (Graziadei 1971; Farbman and Gesteland 1974; Rafols and Getchell 1983) and include (1) maintenance of an ionic reservoir for electrical activity, (2) uptake and transport of odourants, (3) communication between the surface of the epithelium and secretory and vascular elements in the *lamina propria*, and (4) maintenance of integrity of the epithelium (Graziadei and Graziadei 1976; Getchell 1977; Rafols and Getchell 1983; Masukawa *et al.* 1985). Experimental application of an odourant induces secretory activity from the apical region and increased metabolic and transport activity in the basilar expansions (Getchell *et al.* 1987; Zielinski *et al.* 1988). Sustentacular cells have numerous mucus granules within the apical cytoplasm which are released into the overlying mucus layer and appear to form the deep thick layer of mucus in which the olfactory knobs and proximal parts of olfactory cilia are embedded (Getchell *et al.* 1984a). Glands within the *lamina propria* (Bowman's Glands) also contribute to the mucus layer. Odourants must be absorbed in and diffuse through the mucus layer before interacting with the receptor molecules on the cilia. Thus, different components of the mucus may affect access of the odourant to the receptor sites (Getchell *et al.* 1984). In addition, release of secretory products from the sustentacular cells and Bowman's glands, together with mucociliary transport of odourants toward the internal nares, are the primary extracellular method by which odourants are cleared from the olfactory mucus (Getchell and Getchell 1984; Zielinski *et al.* 1988).

3. Basal Cells

In general, replacement of neurons in post-natal vertebrates is rare. However, olfactory and vomeronasal receptors are vulnerable to damage by exposure to odourants and have a mechanism replacing old receptors. Certain basal cells undergo mitosis and maturation to become functionally mature receptor cells synapsing with neurons in the telencephalon (Graziadei and Monti Graziadei 1978; Simmons and Getchell 1981a,b; Masukawa *et al.* 1985). *Rana pipiens* olfactory receptor neurons, for example, are morphologically mature within 24–28 days (Lidow *et al.* 1987). As the postmitotic receptors mature they migrate towards the surface of the epithelium, eventually to be extruded at the surface after they die (Graziadei and Monti Graziadei 1978; Rafols and Getchell 1983). As receptors mature and migrate to the surface, their morphology changes, especially that of the dendrite (Graziadei and Monti Graziadei 1976; Rafols and Getchell 1983). Receptor cell bodies are located almost throughout the depth of the epithelium; those near the base of the epithelium have long, thin dendrites (Fig. 2; 0.6–3.4 µm in diameter), and those closer to the surface have short, thick ones (1.7–8.5 µm, diameter; *Ambystoma*; Rafols and Getchell 1983).

Two types of basal cells have been identified, basal cells proper that are directly adjacent to the basement membrane, and globose basal cells which lie directly above the basal cells proper (Graziadei and Metcalf 1971). Globose basal cells lack apical dendrites but are most likely an intermediate population of proliferating olfactory/vomeronasal precursor cells. This view is supported by experiments that show that both receptor cells and globose basal cells, but not basal cells proper, express neural cell adhesion molecule (NCAM), a polypeptide of the nervous system (Key and Akeson 1990). NCAM within the olfactory neuroepithelium appears to be found only in the receptor cell lineage.

In the early studies of olfactory receptor birth-dating, researchers thought receptor life spans to be controlled strictly genetically (Graziadei and Monti Graziadei 1978). For example, *Ambystoma tigrinum* receptors appear to have a measurable, finite life span of 102 days (Simmons and Getchell 1981). However, subsequent investigators (e.g., Hinds *et al.* 1984) found sensory neurons to be long-lived and they suggested cell life span to be determined strictly by extrinsic factors related to nutrition, disease, age, hormonal state and/or injury. Not surprisingly, a recent compromise states that both genes and environment regulate olfactory/vomeronasal neurogenesis (Farbman 1990). Continued mitosis in the olfactory/vomeronasal epithelium is probably under genetic control and the longevity of these cells probably is modified by environmental factors. The rate of neurogenesis then can be altered so that the equilibrium between cell division and cell death is maintained.

4. Development of the Olfactory/Vomeronasal Epithelium

The olfactory epithelium develops from paired thickenings of the cranial neural ectoderm called olfactory placodes, plus the overlying non-neural ectoderm (Klein and Graziadei 1983). Some of the cells of the neural ectodermal placodes begin migrating toward the body surface. As they migrate they sprout apical processes which differentiate into dendrites reaching the epithelial free surface. A second process sprouts from the base of these cells and pierces the basal lamina as an axon. Many placodal cells do not migrate and remain at the base of the epithelium as basal cells. As placodal cells migrate to the body surface the non-neural cells superficial to the placode elongate and project processes toward the basal lamina. These cells differentiate into sustentacular cells. Thus, the cells of the mature olfactory epithelium are derived from two different ectodermal cell lines, (1) receptors and basal cells from the neural ectoderm and (2) sustentacular cells from the non-neural ectoderm. As olfactory axons project toward their targets, they impart developmental directives within the telencephalon, inducing hyperplasy and the formation of concentrations of synapses called glomeruli (Stout and Graziadei 1980). Normal telencephalic maturation seems to depend upon innervation by olfactory axons, rather than vice versa. The vomeronasal organ forms as an out-pocketing of the main olfactory chamber (Spaeti 1978).

5. Lamina Propria

The *lamina propria* is found beneath the olfactory or vomeronasal epithelia and their basement membranes and contains several components in addition to connective tissue and connective tissue cells (Fig. 1). These include blood vessels, myelinated and unmyelinated nerve bundles, pigment-containing cells, and, in the olfactory mucosa of terrestrial amphibians, one to three layers of Bowman's glands (Getchell *et al.* 1984). Although the Bowman's glands are within the lamina propria, their ducts extend through the epithelium to reach the surface (see section on *Ambystoma tigrinum*). Aquatic amphibians lack Bowman's glands, as do fish, and therefore such glands may be necessary for reception of air-borne odours. Bowman's glands appear to produce the rather watery secretion overlying the deep thick mucus layer (produced by the sustentacular cells) in which the olfactory knobs and proximal parts of olfactory cilia lie. Bowman's glands also secrete a putative odourant-binding protein (OBP), which may transport odourants through the mucus layer (Getchell and Getchell 1984; Lancet 1988).

6. Variation in Nasal Geography and Epithelial Structure

The main olfactory chamber of apodans can be one of two types (Schmidt and Wake 1990). One type is a simple, dorsoventrally flattened, undivided chamber; olfactory epithelium lines the rostral and medial walls of the cavity (Fig. 3). In the second type, the dorsoventrally flattened main cavity is divided longitudinally into medial and lateral cavities; the rostral walls of the lateral cavity and the entire medial cavity are lined with sensory epithelium. Both types of olfactory chambers are lined with more olfactory epithelium than nonsensory epithelium. The exceptions are two aquatic forms (*Typhlonectes natans* and *Chthonerpeton indistinctum*) in the family Typhlonectidae; their nasal cavities have less sensory than nonsensory epithelium.

The vomeronasal organs of apodans are detached nearly completely from the main olfactory cavities (Jurgens 1970; Badenhorst 1978; Schmidt and Wake 1990). The size and

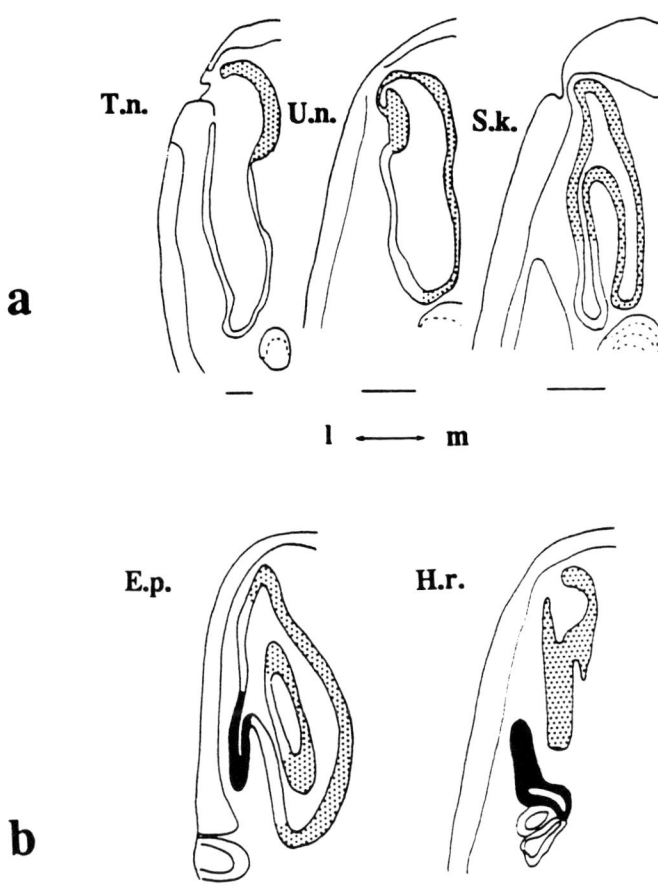

Fig. 3. Line drawings of idealized longitundinal sections of **a:** main olfactory cavities (with sensory epithelium shaded) and **b:** vomeronasal organs (black) of various apodans (after Schmidt and Wake 1988). E.p. = *Epicrionops petersi;* H.r. = *Hypogeophis rostratus;* S.k. = *Scolecomorphus kirkii;* T.n. = *Typholnectes natans;* U.n. = *Ureaotyphlus narayani.* Left is lateral and right is medial. Only one side (left) of the head is shown.

position of the vomeronasal organs in relation to the main cavity varies greatly among species (Fig. 3). The vomeronasal organ is lateral to the main cavity (the general condition in salamanders) in only one species. In other species, its position can be described as (1) lateral with a slight mediolateral extension, (2) "L" shaped, but posterior to the main cavity, (3) "C" shaped and posterior to the main cavity, (4) transverse and posterior to the main cavity, and (5) longitudinal but nestled between the medial and lateral cavities of the main olfactory cavity (see Figure 1 in Schmidt and Wake 1990). The relative size of the VNO also varies a great deal among species, from quite small in two genera in the terrestrial family Caeciliaidae, to quite large in aquatic typholnectids (Schmidt and Wake 1990). Apodan vomeronasal organs are unique because they are connected to tentacles (a pair of protrusible sensory structures located between eye and nostril; Badenhorst 1978; Billo and Wake 1987). Odourants entering the tentacular lumen can reach the vomeronasal organs via tentacular ducts (Schmidt and Wake 1990).

Anuran nasal cavities, which also are flattened dorsoventrally, are subdivided into three chambers by horizontally oriented lamellae (Parsons 1967; Scalia 1976). The largest chamber (principal chamber) is dorsal and ovoid and is bean shaped in cross section at mid-to-posterior levels, where its floor is elevated by a ridge *(eminentia olfactoria).* Olfactory epithelium lines almost all of this cavity. The sensory epithelium covering the *eminentia olfactoria* is surrounded by a perimeter of nonsensory epithelium; axons from these receptors converge to form a ventral division of the olfactory nerve. Axons from receptors at the anterior end, medial wall and roof converge to form the dorsal division of the olfactory nerve. In the aquatic family

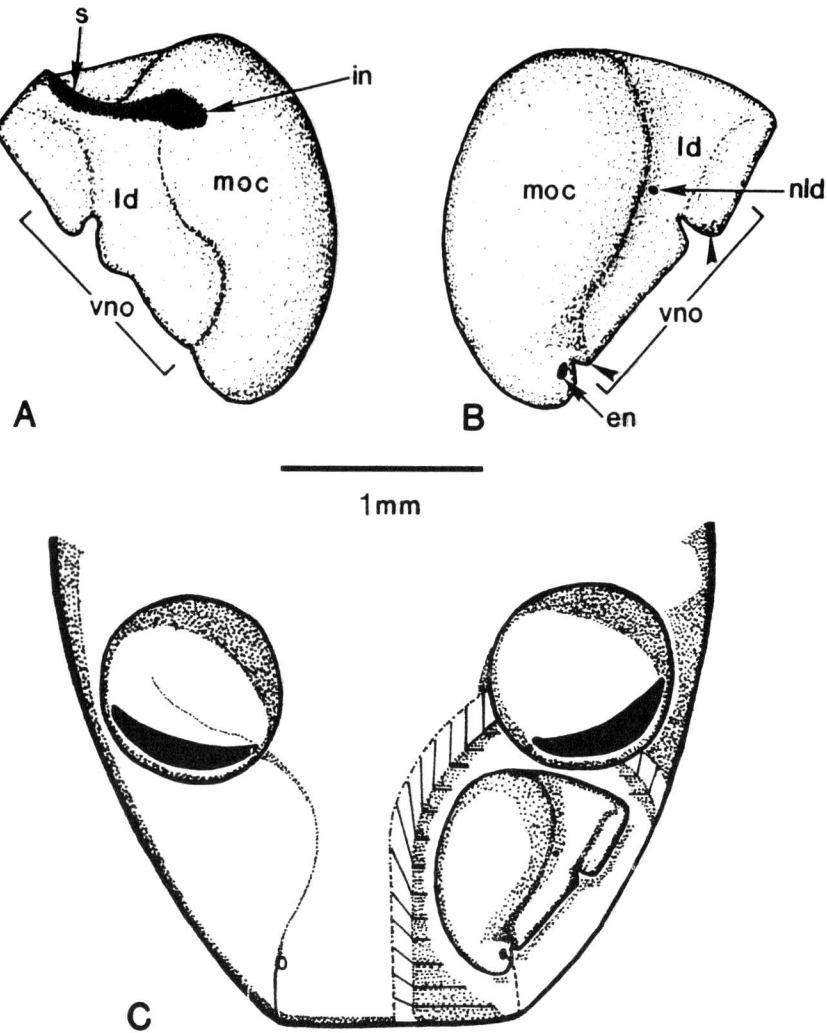

Fig. 4. Graphic reconstructions of the nasal organs of *Plethodon cinereus;* **a:** ventral view; **b:** dorsal view; **c:** nasal organ *in situ* (after Dawley and Bass 1988). en = external nares; in = internal nares; ld = lateral diverticulum; moc = main olfactory chamber; nld = nasolacimal duct; s = slitlike extension of the internal nares; vno = vomeronasal organ.

Pipidae, the principal olfactory chamber is divided into a medial and lateral diverticulum separated by a strip of nonsensory epithelium. A valve-like structure in the external naris closes one side or the other so that the medial diverticulum is open only to air, whereas the lateral diverticulum is open only while under water (Altner 1962). In general, the external nares may enter the principal chamber directly or first through a nonsensory vestibular chamber; the internal nares run from the posterolateral corner of the floor of that chamber to the oral cavity. A middle chamber, ventral to the anterolateral portion of the principal cavity, is lined by nonsensory tissue. An inferior chamber is ventral to the principal chamber and is divided into two main parts, a small medial recess and a larger lateral recess, both of which are dorsoventrally flattened. The lateral recess is nonsensory and the medial recess is lined by vomeronasal epithelium. The inferior chamber is connected to the middle cavity anteriorly and to the principal cavity and internal nares posteriorly. This suggests that chemicals are conveyed from the oral cavity to the medial recess (vomeronasal organ), rather than via the external nares and middle cavity. The relative positions of both the lateral and medial recesses varies a great deal among anuran species (Jurgens 1971).

The main nasal chamber of salamanders generally is oval, flattened dorsoventrally and lined partly by olfactory epithelium and partly by nonsensory (or respiratory) epithelium (Fig. 4).

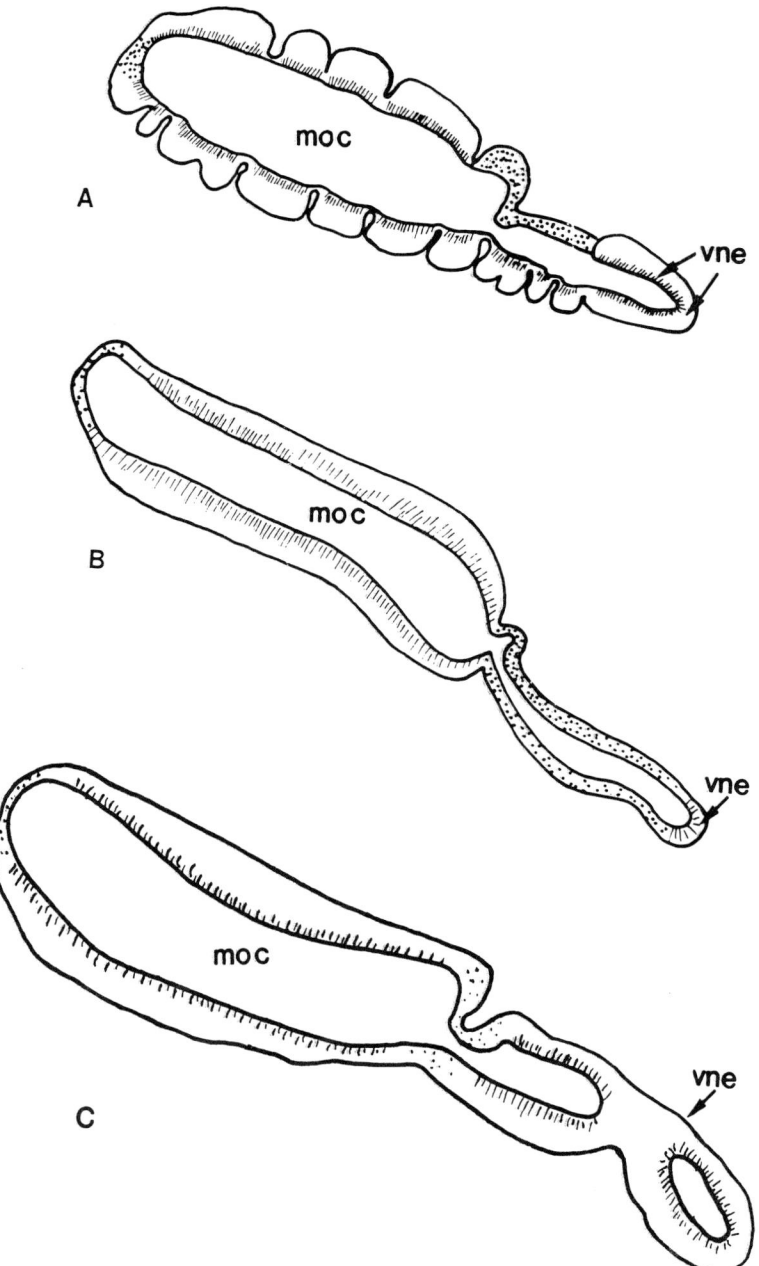

Fig. 5. Line drawings of a representative transverse section of the left nasal organs of **a:** *Triturus alpestris*, **b:** *Ambystoma maculatum* and **c:** *Plethodon cinereus* (after Dawley and Bass 1988). moc = main olfactory chamber; vne = *vomeronasal epithelium*.

In aquatic species, the nasal epithelium is folded with sensory epithelium lining the grooves and nonsensory epithelium covering the ridges (Seydel 1895; Farbman and Gesteland 1975; Graziadei and Monti Graziadei 1976; Eisthen *et al.* 1994). The folds are oriented anteroposteriorly, parallel to the direction of water flow from external to internal nares. In terrestrial species, the sensory epithelium forms a flat sheet on the dorsal and ventral walls of the main nasal chamber (Getchell *et al.* 1984; Arzt *et al.* 1986; Dawley and Bass 1988). In those terrestrial species with aquatic larvae, the larval nasal epithelium again is organized into folds.

The vomeronasal organ of salamanders generally lies in a ventrolateral diverticulum of the main nasal cavity (Fig. 4). Not all of the epithelium lining this diverticulum is sensory. In some *Ambystoma* species (Fig. 5), only a restricted portion of the lateral-most corner is covered with vomeronasal epithelium (Seydel 1895). However, in *Ambystoma mexicanum* the entire lateral diverticulum is lined with vomeronasal sensory epithelium (Eisthen *et al.* 1994). In *Triturus* (Fig. 5), the ventral floor in addition to the lateral corner is sensory (Schuch 1934). The ventrolateral diverticulum of plethodontid salamanders (Fig. 4) is covered nearly completely with vomeronasal epithelium, and it has been suggested that this reflects the importance of the vomeronasal organ for these salamanders (Dawley and Bass 1988). In addition, the presence of specific structures (nasolabial grooves) that direct odourants to the vomeronasal organ during nose-tapping, supports this suggestion (Dawley and Bass 1989). These grooves are not present in salamanders of other families. Some aquatic genera (*Necturus* and *Proteus*) lack a ventrolateral diverticulum, and the vomeronasal epithelium appears to be represented by a lateral row of olfactory buds; others (*Cryptobranchus* and *Amphiuma*) have very reduced lateral diverticulae and vomeronasal organs (Jurgens 1971). The aquatic *Siren* is unique in having a distinctive, inverted "T"-shaped diverticulum housing the vomeronasal organ (Seydel 1895).

Because of the scrutiny of neurobiologists, there are some additional details available concerning the sensory epithelia of *Necturus maculosus* and *A. tigrinum*. The olfactory mucosa of *N. maculosus* and larval *A. tigrinum* is folded into ridges and grooves running antero-posteriorly, as expected of aquatic salamanders. At the epithelial surface, the ridges are lined by a single layer of nonsensory ciliated cells and goblet cells (Farbman and Gesteland 1974; Graziadei and Monti Graziadei 1976). In larval *A. tigrinum* (and, perhaps, *N. maculosus* as well), the basement membrane and *lamina propria* also separate each groove from adjacent grooves (Getchell *et al.* 1984). The grooves are lined by olfactory epithelium that in *N. maculosus* is two to three times the thickness of the olfactory epithelium in most other vertebrates (250–300 μm; Farbman and Gesteland 1974; Graziadei and Monti Graziadei 1976; compared to 80–150 μm in frogs and 100 μm in turtles, Graziadei 1971). Within the mucosa of a single animal there is considerable variation in thickness of the epithelium due to differences in the number but not the size, of receptor or basal cells; there is no variation in number or size of sustentacular cells (Graziadei and Monti Graziadei 1976).

Larval *A. tigrinum* metamorphose into mainly terrestrial adults. Therefore, some direct comparisons can be made between an aquatic and terrestrial morphology (Getchell *et al.* 1984). Although Bowman's glands are found in the basal lamina of the larval stage, no ducts are visible, suggesting the glands are not yet functional. However, because secretory granules are found in gland cells, the cellular machinery for secretion probably is in place. The olfactory epithelium of young salamanders (those lacking gills and fins) has areas of larval-like buds adjacent to regions of adult-type morphology. In addition, Bowman's glands with ducts are found extending through the adult-like epithelium. Thus, at metamorphosis the olfactory epithelium undergoes some reorganization (Getchel *et al.* 1984; Arzt *et al.* 1986). These anatomical changes appear to be accompanied by response characteristics of the olfactory epithelium. Although both larvae and adults respond to all the same odourants (airborne volatiles, volatile solutions, and non-volatile amino acid solutions), magnitudes and threshold values of electro-olfactograms differ. In general, larvae respond more strongly to non-volatile amino acid solutions and adults react more intensely to air-borne volatiles (Arzt *et al.* 1986). These differences in response may be due to changes in either (1) the receptor population or in the mucus overlying the epithelium and, thus, accessibility of odourants to the epithelium, or (2) changes in geography of the epithelium (folded to flat), which would also affect odourant accessibility to the receptor epithelium.

In adult *A. tigrinum*, the ventral olfactory mucosa is thicker than the dorsal one because the receptor cell and basal cell layers are thicker in the epithelium and the glandular layer is thicker in the *lamina propria* (Getchell *et al.* 1984). In these salamanders, there are three layers of olfactory glands in the ventral mucosa (Getchell *et al.* 1984). The superficial layer of Bowman's glands have ducts that extend through the epithelium to the surface. The middle and deep layers of glands do not appear to have ducts (although they simply may be connected to the same ducts as the superficial Bowman's glands), and are separated from the superficial Bowman's glands and each other by olfactory nerve bundles. The three layers of glands have different histological characteristics, staining products, and reactions to nerve section. In addition, both the dorsal and ventral epithelia are thicker anteriorly than posteriorly

(Mackay-Sim and Patel 1984). This gradation largely is due to a greater number of cells, including receptor cells, anteriorly than posteriorly. Paradoxically, despite the thinner posterior epithelium, 3H-thymidine incorporation experiments showed that neurogenesis is more rapid posteriorly than anteriorly (Mackay-Sim and Patel 1984). In addition, electrophysiological studies revealed that the posterior epithelium is more responsive to test odourants than is the anterior epithelium (Mackay-Sim *et al.* 1982). Although this gradient in sensitively might be due to increased density of receptor cell processes (cilia), a scanning electron microscopic (SEM) study of surface morphology did not indicate any regional differences in density (Breihpohl *et al.* 1982). That SEM study did show that on the lateral edge of the ventral surface the epithelium is thrown into almost parallel folds of alternating receptor and non-receptor epithelium.

The only other terrestrial salamander species studied in detail is *Plethodon cinereus*. Like adult *Ambystoma*, the olfactory epithelium of adult *P. cinereus* forms a flat sheet on the dorsal and ventral wall of the main olfactory chamber, with the thickness of those sheets varying in an anterior (thick) to posterior (thin) direction (Dawley and Bass 1988). However, it differs from *Ambystoma* species in that the ventral epithelium is not always thicker than the dorsal epithelium. Anteriorly, the ventral epithelium is thicker than the dorsal one, but more posteriorly the reverse is true.

B. Physiology of Odour Reception, Transduction, and Discrimination

Although the structure of receptor cells is reasonably well understood, exactly how a vertebrate's nervous system distinguishes among the potentially vast number of odours is still largely unknown. Odourants that interact with either main olfactory or vomeronasal receptor cells first must be absorbed in and diffuse through a mucus layer that overlies the receptor mucosa. An extracellular odourant-binding protein (OBP), secreted by Bowman's glands (Getchell and Getchell 1984), may carry odourants through the mucus to the receptor cells (Lancet 1988). Odourants then bind to a putative olfactory receptor molecule, a transmembrane glycoprotein probably located on the cilia or microvilli of the receptor cells (Getchell *et al.* 1984; Lancet *et al.* 1988; Kleene and Gesteland 1991). There are hundreds to thousands of different olfactory receptor molecules, with each perhaps responsive to just one odourant; recent research described a huge family of genes that code for odour-receptor proteins (Buck and Axel 1991). This binding of odourant to receptor results in depolarization of the receptor cell (transduction), sending action potentials to the brain.

The molecular and cellular events of transduction are thought to be the following. The binding of an odourant to its receptor activates a G-protein (G_{olf}), an intracellular protein that converts GTP (guanine triphosphate) into GDP (guanine diphosphate). The G-protein also activates adenylate cyclase, a membrane-bound enzyme, to produce cAMP (cyclic AMP). The cAMP, in turn, opens ion channels in the membrane of receptor cells (Lancet *et al.* 1988; Snyder *et al.* 1988; Kleene and Gesteland 1991). Sodium, potassium, and calcium ions move through these channels to depolarize the receptor cells (Trotier and MacLeod 1983). The membrane depolarization induces an increase in the frequency of spontaneous firing, and the stimulus concentration is encoded in the frequency of firing (high concentrations induce high firing rates; Trotier and MacLeod 1983). However, not all cells fire with application of a particular odourant, and this has lead to speculations about how the brain discriminates among odours.

Ambystoma tigrinum, because of the sheet-like arrangement of its olfactory sensory epithelium, has been one of the organisms of choice for workers interested in the neurophysiology of olfaction (e.g., Mason and Morton 1982; Hamilton and Kauer 1985). One question that has been addressed is whether specific regions of the olfactory epithelium project to specific regions of the main olfactory bulb so that odour quality somehow is encoded in topography. Kauer (1981), using horseradish peroxidase (HRP) tracing, found that a single fascicle from the dorsal epithelium projected to all parts of the glomerular layer, while other fascicles projected to restricted glomerulae. Mackay-Sim and Nathan (1984) found that the anterior half of both the dorsal and the ventral epithelium projected to the ventral half of the main olfactory bulb while the posterior half of both epithelia projected to the dorsal half of the bulb. However, Dubois-Dauphin *et al.* (1981) found *Triton* to be different. In that species, the dorsal olfactory epithelium projected to the anterior bulb while the ventral olfactory epithelium projected to the posterior bulb. In addition, Duncan *et al.* (1990) found that ventral olfactory epithelium of *Rana pipiens* to project to all parts of the glomerular layer of the bulb but more densely to the lateral part; similarly the dorsal epithelium projects most densely to the medial part of the glomerular layer. These studies all show that the projections are broad, and that a point-to-point correspondence does not occur between the locations of receptors in the olfactory epithelium and their terminations in the olfactory bulb. Thus, it is difficult to understand how spatial patterns of activation in the epithelium are conveyed to the bulb to produce any specific discrimination of odours.

On the other hand, it is possible to map areas of the olfactory epithelium that are highly sensitive to certain odourants that are consistent from animal to animal of the same species, although there are cells outside this area that are responsive to the odourant but to a lesser degree (Kauer and Moulton 1974; Mackay-Sim *et al.* 1982; Mackay-Sim and Shaman 1984; Kauer 1991). Different odours produce different patterns of EOG (electro-olfactogram) responses or intracellular recordings across the epithelium (Kubie *et al.* 1980; Mackay-Sim *et al.* 1982) or glomeruli (Lancet *et al.* 1982; Hamilton and Kauer 1988, 1989), which would provide some odour coding. In addition, if different odourants are distributed to different regions of the olfactory epithelium (by turbulence of air flow, shape of nasal cavity, or chromatographic-like interactions of the odour with the mucus covering the surface of the receptor cells), then some encoding of stimulus properties may take place at the receptor epithelium (Kauer 1991). Precisely how the molecular characters of the stimulus are represented is still unknown.

Potential differences among receptor cells also have been shown through examination of the distribution of membrane-bound molecules. Receptor cell membranes are highly specialized, probably possessing unique molecules that are segregated regionally on individual receptors (Key and Akeson 1990a). Carbohydrate containing molecules have been studied extensively using lectins (plant proteins with high affinity for sugar residues). Some lectins bind selectively to olfactory structures. For example, in *Xenopus*, soybean agglutinin (SBA) binds to receptors in the vomeronasal organ and some receptors in the olfactory epithelium (Key and Giorgi 1986). The main olfactory chamber of *Xenopus* and *Pipa* is divided into a medial and lateral diverticulum separated by a strip of nonsensory epithelium; the medial diverticulum is open only to air, while the lateral diverticulum is open only while under water (Altner 1962). Hofmann and Meyer (1991) showed that, other than in the receptors in the vomeronasal organ, only the olfactory receptors in the lateral diverticulum bind with SBA. Thus, in these animals at least, SBA binding seems to delineate two functionally different classes of receptors, those detecting water-borne odours (VNO and lateral diverticulum) and those detecting air-borne ones (medial diverticulum).

Monoclonal antibodies also are being used to distinguish cell types within the olfactory neuroepithelium. Antibodies against frog *(Rana catesbeiana)* olfactory epithelium, for example, are generated by injecting frog olfactory epithelium into mice, fusing spleens from these mice with myeloma cells to create hybridomas, which are then cloned and isotyped for their specific monoclonal antibodies (Key and Akeson 1990a). Tissue sections from bullfrogs are then reacted with these antibodies in order to identify cells, portions of cells or groups of cells that are immunoreactive with specific monoclonal antibodies. So far, monoclonal antibodies have been generated that are immunoreactive with (1) perikarya, axons and dendrites (but not olfactory knobs or cilia) of all olfactory receptor neurons and perikarya of globose basal cells (Key and Akeson 1990a, b), (2) perikarya, axons and dendrites (but not olfactory knobs or cilia) of olfactory receptors and the entire population of basal cell perikarya, (3) cilia and knobs of receptor cells (but not perikarya, axons or dendrites), (4) cilia only, (5) supporting cells and Bowman's glands, and (6) deep glands (Key and Akeson 1991b). A monoclonal antibody directed against the synthetic enzyme responsible for carnosine (a dipeptide found in olfactory and other tissues of vertebrates) has been used to demonstrate diversity of receptors in *Rana pipiens*. Only a specific subset of olfactory receptors was labelled (Crowe and Pixley 1991). These studies indicate that cell types and even specific cell structures have unique membrane

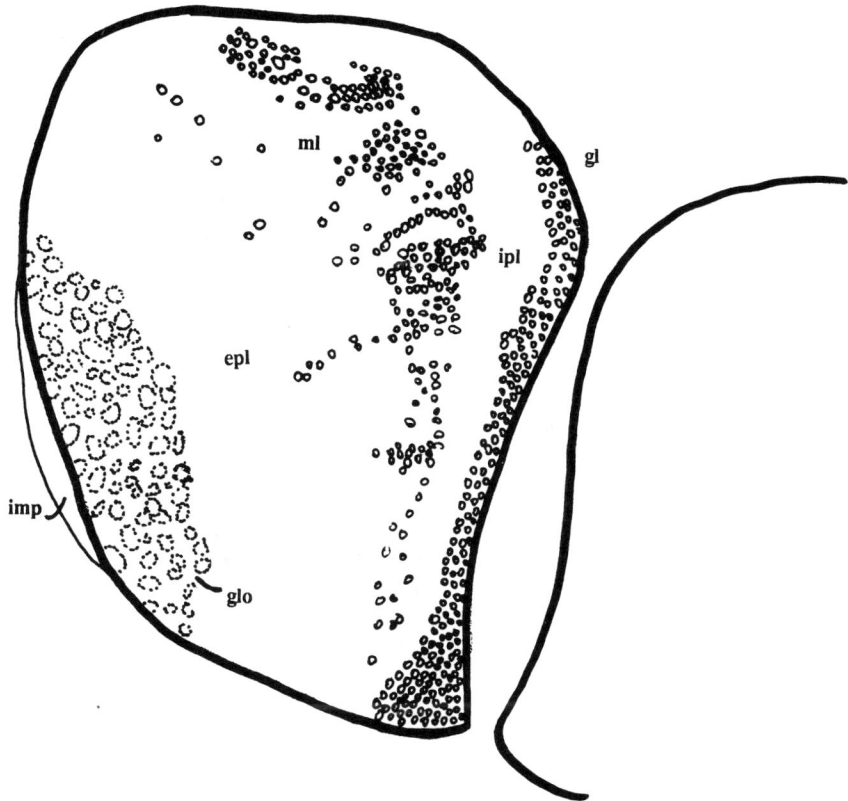

Fig. 6. Camera lucida drawing of a transverse section of the left olfactory bulb of *Plethodon glutinosus.* epl = external plexiform layer; gl = granule cell layer; glo = glomerular layer; imp = implantation cone; ipl = internal plexiform layer; ml = mitral cell layer.

molecules that may be associated with olfactory reception and transduction. This suggests that different chemical signals may be processed by different subsets of cells, but exactly how many different subsets exist is unknown.

C. Central Nervous System

1. Olfactory and Accessory Olfactory Bulbs

Axons of the main olfactory nerves and vomeronasal nerves synapse with first-order relay neurons within the main olfactory and accessory olfactory bulbs, respectively (Schmidt *et al.* 1988; Burton *et al.* 1990). The main olfactory bulbs are paired and form at the anterior end of the telencephalon; the paired accessory olfactory bulbs lie just posteriorly to them (Fig. 7; Hoffman 1963; Northcutt and Kicliter 1980; Burton *et al.* 1990; Schmidt and Wake 1990; Scalia *et al.* 1991a). The olfactory and vomeronasal nerves run parallel to one another, the vomeronasal nerve ventral either to the main trunk or ventral branch of the olfactory nerve. The olfactory nerves travel on a direct, unimpeded route to the main olfactory bulbs while the vomeronasal nerves curve laterally to terminate in the more posterior accessory olfactory bulbs. The bulbs are variably discernible from each other and the rest of the telencephalon depending on the animal (more so in anurans than in apodans or salamanders).

The main olfactory and accessory olfactory bulbs are organized into layers (Fig. 6; Herrick 1910, 1924; Hoffman 1963; Scalia 1976; Schmidt and Wake 1990). The outermost layer of the incoming olfactory or vomeronasal axons is called an implantation cone. In frogs of the genus *Rana* (Scalia 1976) and in two apodan genera *Dermophis* and *Ichthyophis* (Schmidt and Wake 1990), the olfactory implantation cone covers the anterior, ventral, and anterolateral aspects of the olfactory bulb. In salamanders, the olfactory implantation cone occupies a more ventrolateral position (Schmidt *et al.* 1988). The vomeronasal implantation cone covers virtually the entire external aspect of the accessory olfactory bulb in frogs (Scalia 1976); in salamanders and apodans, the shape of this receptive field varies among species (Schmidt *et al.* 1988; Schmidt and Wake 1990). In all three orders of amphibians, the vomeronasal implantation cones and accessory olfactory bulbs are much smaller than their main olfactory counterparts. Below the implantation cone is a layer of glomeruli (synapses between incoming nerves and mitral cells), followed by an external plexiform layer (a fibrous layer crossed by the principal dendrites of mitral cells), then a mitral cell layer (the principal post-synaptic neurons of the olfactory bulb), an internal plexiform layer (containing axons of mitral cells), and finally, a layer of granule cells. Any one of these layers may not be present over the entire bulb; for example, the mitral cell layer in anurans is confined to the anteroventrolateral aspect of the olfactory bulb (Scalia *et al.* 1991). These layers in the accessory olfactory bulbs are separate from those in the main olfactory bulbs (except for the granule cell layer), but are less clearly formed.

The actual synapses between mitral cell dendrites and olfactory or vomeronasal axons occur in roughly spherical areas called glomeruli. Glomeruli contain the axon terminals of the olfactory or vomeronasal nerves and the tassellated ends of the mitral cell dendrites as well as the dendrites and axon terminals of periglomerular cells (Pinching and Powell 1971b). Mitral cells have more than one glomerular dendrite; these may enter into more than one glomerulus. Therefore, these dendritic fibers frequently slant obliquely across the outer plexiform layer, forming a complexly interlacing meshwork (Scalia 1976). Each glomerulus is surrounded by encapsulating periglomerular cells which are interneurons whose dendrites and axons extend into the glomeruli and are thought to function as inhibitory interneurons. Glomeruli are restricted to the anterior, ventral, and lateral portions of the bulb (Scalia 1976; Burton *et al.* 1990; Scalia *et al.* 1991a). Glomerular diameters have been measured in a few species (e.g., *R. pipiens* about 75–100 μm, Scalia 1976; *R. ridibunda* about 20–35 μm, Jiang and Holley 1992). Glomeruli of the main olfactory and accessory olfactory bulbs are very similar in appearance (Burton *et al.* 1990). Differences are seen, however, when comparing adult and larval vomeronasal glomeruli. Glomeruli in adult bullfrogs are more compact, and mitral dendrites within the glomeruli are much more highly branched. However, there are about the same number of vomeronasal receptors in adults and tadpoles. This suggests that as development proceeds, more synaptic junctions between vomeronasal axon terminals and mitral cell dendrites are established. In comparison, there are 29 times more main olfactory receptors in adults than in stage III tadpoles, suggesting that much of the development of olfactory receptors (at least in sheer numbers) occurs after metamorphosis.

The cell bodies of the mitral cells are described by some workers as being organized into a discrete layer (e.g., *Rana pipiens, R. catesbeiana, R. clamitans, Hyla cinerea, Bufo marinus*, Hoffman 1963; *Ambystoma*, Herrick 1910, 1924; Hamilton and Kauer 1988; Eisthen *et al.* 1994; *Salamandra salamandra, Plethodon jordani,* and *Bolitoglossa subpalmata,* Schmidt and Roth 1990). An external plexiform layer separates the glomerular layer from the mitral cell layer; this plexiform layer is crossed by the principal dendrites of the mitral cells extending to the glomeruli and by dendrites of granule cells. The external plexiform layer also is a region of synapse between granule cell dendrites and mitral cell secondary dendrites. Granule cells inhibit mitral cells and thereby play a role in lateral inhibition (that is, mitral cells that are stimulated by odour may be located near cells that are inhibited via granule cells; Hamilton and Kauer 1989). The internal granule layer is densely packed with granule cells and is separated from the mitral cell layer by an internal plexiform layer. The internal plexiform layer provides space for the exiting mitral cell axons (the olfactory tracts).

Other workers describe the mitral cell bodies as being scattered in a broad plexiform zone, contradicting earlier reports (e.g., *R. catesbeiana*, Burton 1990; *R. pipiens*, Scalia *et al.* 1991a). The functional meaning of these two different arrangements is unclear, although mammals have a discrete mitral cell layer (Pinching and Powell 1971a) and a scattered mitral cell arrangement is typical of fish (Kosaka and Hama 1982). In bulbs with this laminar arrangement, the mitral cell layer also includes mitral cell secondary dendrites and the radiating dendrites of deeper granule cells with which they synapse (Scalia *et al.* 1991a). In addition, a superficial plexiform layer is present in the posterior superficial portion of the bulb of *Rana pipiens*, adjacent to the posterior edges of both the glomerular and mitral cell layers. It provides additional space for synapses between mitral and granule cells and for the exiting lateral olfactory tract. The internal plexiform layer also provides space for the exiting lateral olfactory tract.

Amphibian mitral cells may not be a homogeneous group; in mammals, there are several categories of output neurons (types of mitral and tufted cells) with different morphologies and functions. Jiang and Holley (1992) found homologies between two output cell groups in *Rana ridibunda* and mammalian output neurons (superficial/middle tufted cells and deep tufted/mitral cells). These two cell groups seems to be physiologically different. Scalia *et al.* (1991a) also suggest that there are two types of first order neurons; one type may be equivalent to mammalian tufted cells.

To summarize the pathway of incoming olfactory (or vomeronasal) nerves: Axons from the nerve converge on the anterior end of the bulb and spread along the surface as an implantation cone (Hoffman 1963; Northcutt and Kicliter 1980; Burton *et al.* 1990; Scalia *et al.* 1991a). If there are dorsal and ventral rami to the olfactory nerve, the dorsal ramus distributes to the dorsal surface, and the ventral ramus to the ventral surface of the bulb (Northcutt and Kicliter 1980). In this case, the vomeronasal nerve is closely associated with the ventral ramus but still distinct from it. These axons then turn inward to synapse with mitral cell dendrites in glomeruli. Axons of the mitral cells form a secondary olfactory or vomeronasal pathway leaving the main or accessory olfactory bulbs, respectively. The mitral cell bodies and their fibers may form distinct layers or these two layer are intermingled as a plexiform/mitral cell layer. All studies find an internal granule cell layer deepest to all other layers with a narrow internal plexiform layer separating granule cells from mitral cells. Some or all of the initial portions of the exiting mitral cell axons travel within the internal plexiform layer. Granule cell dendrites synapse with mitral cell dendrites and presumably modify the incoming olfactory or vomeronasal nerve signal (Burton 1990; Hamilton and Kauer 1988).

2. Central Projections

Mitral cells of the main olfactory bulb project to other targets within the telencephalon; axons travel posteriorly as lateral and medial olfactory tracts. In anurans such as *Rana pipiens* (Scalia 1976; Scalia *et al.* 1991b), both tracts project ipsilaterally (same side) to several divisions of the telencephalon and then contralaterally to some of these same divisions (Fig. 7). As the lateral olfactory tracts course posteriorly, they synapse widely in the lateral cortex, then continue medially to synapse in the cortical amygdaloid nucleus and nucleus of the hemispheric sulcus (the lateral olfactory tracts are often referred to as the lateral corticohabenular tracts once they turn medially). These tracts are joined by the posterior ends of the medial olfactory tracts (also called the anterior olfactohabenular tracts) where they form a major part of the *stria medullaris thalami* (an area of the diencephalon). These olfactory fibers in the *stria medullaris* cross in the habenular commissure and join the lateral corticohabenular tract on that side to terminate in the contralateral cortical amygdaloid nucleus and the periamygdaloid region in the contralateral lateral cortex. The medial olfactory tracts travel posteriorly over and synapse in the postolfactory eminence, the rostral end of the medial cortex, and the rostral end of the medial septal nucleus, before joining the lateral olfactory tract (or lateral corticohabenular tract) as outlined above.

There are some species-specific variations in the terminations of the medial olfactory tract and of the crossed arm of the olfactory projections. The medial olfactory tract in *Rana esculenta* ends in the *zona limitans* (a cell-free area between the medial cortex and septum of the cerebral hemispheres; Kemali and Guglielmotti 1987) rather than projecting across the postolfactory eminence and other structures as outlined above in both *R. pipiens* and *R. catesbeiana* (Northcutt and Royce 1975). In *R. catesbeiana* the crossed fibers not only enter the contralateral lateral corticohabenular tract, but also the contralateral anterior olfactohabenular tract and run to the olfactory bulb (Northcutt and Royce 1975). In *R. esculenta*, the crossed fibers entered a third tract (the medial corticohabenular tract) and ran forward within the *zona limitans* (Kemali and Guglielmotti 1987).

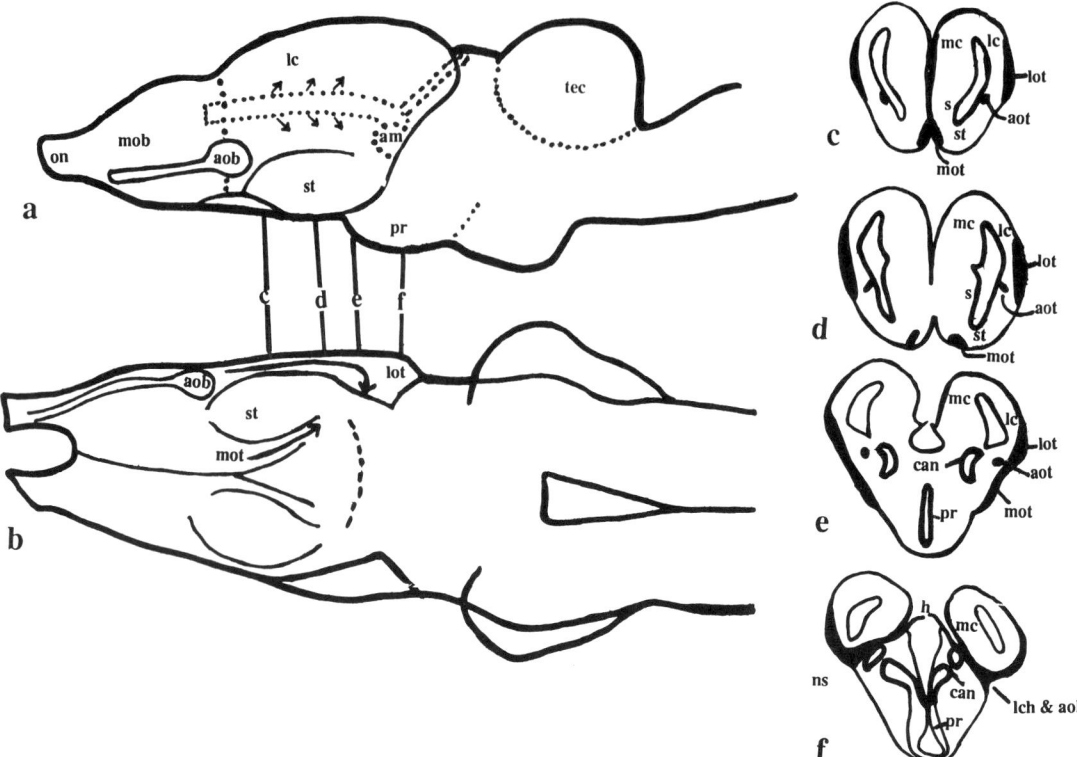

Fig. 7. Brain of a representative amphibian showing central projections of the olfactory and vomeronasal systems (after Scalia 1976 and Schmidt and Roth 1990). Figure 7a is a lateral view, 7b is a ventral view, and c–f represent transverse sections corresponding to the regions lettered in Figure 7a. The lateral olfactory tract is shown as the dotted regions and arrows in Figure 7a and 7b; the superficial part of the medial olfactory tract is shown in Figure 7b. The accessory olfactory tract is deep to the surface and can only be seen in transverse sections (c–f). All three tracts are represented as blackened in areas in c–f. The lateral and medial olfactory tracts join in Figure 7e and turn medially where they are referred to as the lateral corticohabenular and anterior olfactohabenular tracts, respectively. aob = accessory olfactory bulb; aot = anterior olfactohabenular tract; am = amygdala; can = cortical amygdaloid nucleus; h = habenular; lc = lateral cortex; lch = lateral corticohabenular tract; lot = lateral olfactory tract; mc = medial cortex; mob = main olfactory bulb; mot = medial olfactory tract; ns = nucleus of the hemispheric sulcus; on = olfactory nerve; pr = preoptic region; s = septal nucleus; st = striatum; tec = optic tectum.

Mitral cell axons of the accessory olfactory bulb travel as the accessory olfactory tract which project only to the amygdala in *R. pipiens,* synapsing with the medial and cortical amygdaloid nuclei (Scalia *et al.* 1991b). A few fibers continue on to cross in the anterior commissure and terminate in the contralateral amygdala. In the cortical amygdaloid nucleus the main olfactory fibers end on distal dendrites while the accessory olfactory fibers end on proximal dendrites. Thus, the main olfactory and accessory olfactory pathways remain separate from the periphery to secondary central connections (Scalia *et al.* 1991b). In *R. esculenta,* Kemali and Gulglielmotti reported that some accessory olfactory tract fibers project to the posterior part of the contralateral lateral cortex. Northcutt and Royce (1975) suggested a similar accessory olfactory tract projection to the lateral cortex. Both studies suggested some overlap of projections of the main and accessory olfactory tracts; however, Scalia *et al.* (1991b) offered evidence casting doubt on these data.

Salamander central projections are similar to those of anurans (e.g., *Ambystoma tigrinum,* Northcutt and Kicliter 1980; *Salamandra salamandra, Bolitoglossa subpalmata* and *Plethodon jordani,* Schmidt and Roth 1990). The majority of main olfactory fibers form the lateral olfactory tracts, which immediately subdivide into dorsal and ventral lateral olfactory tracts. These tracts project mostly to the lateral cortex and to the *striatum.* Other fibers of the lateral olfactory tracts are joined by the anterior olfactohabenular tract (the proximal portion of the medial olfactory tract) and turn at the amygdala to enter the *stria medullaris.* Fibers do not appear to terminate in the cortical amygdala as they do in frogs. From the *stria medullaris,* fibers cross in the habenula (as in frogs) and enter the contralateral *stria medullaris*. From there, fibers continue into the contralateral anterior olfactohabenular tract and continue anteriorly toward the olfactory bulb. In this way, these salamander species are like *R. catesbeiana* rather than like *R. pipiens*. Fibers do not terminate in the cortical amygdala (as in those two frog species). In *S. salamandra*, some fibers terminate within the habenula itself. The habenula has been proposed to be a center of olfactory processing and involved in courtship; however, such terminals were not found in *B. subpalmata* (Schmidt and Roth 1990).

Main olfactory fibers also form a second tract, called the anterior olfactory habenular tract (Schmidt and Roth 1990) or medial olfactory tract (Northcutt and Kicliter 1980). Most of these fibers cross at the habenular commissure to enter the contralateral hemisphere along with the lateral olfactory tract as described above. Northcutt and Kicliter suggested additional projections into the dorsal portion of the medial cortex and septal nuclei in *A. tigrinum*, making this species like *R. pipiens* and *R. catesbeiana*.

Fibers originating in the accessory olfactory bulb always terminate within the lateral amygdala (Schmidt and Roth 1990). The accessory olfactory tract of *S. salamandra* intermingles with fibers of the lateral olfactory tract. There are two accessory olfactory tracts on both sides of the telencephalon in *P. jordani* and *B. subpalmata*; both tracts terminate in the lateral amygdala without intermingling with main olfactory central projections. In these two species the main olfactory and accessory olfactory pathways remain separate from the periphery to secondary central connections.

D. Functional Comparisons

Some attempts have been made to correlate morphological variation of olfactory and vomeronasal systems with the ecological or evolutionary history of particular species. The major variables that have been examined are size and architecture of the main olfactory or vomeronasal organs and the size and shape of olfactory and vomeronasal nerve projections.

The size of the main olfactory or vomeronasal organs (measuring only those portions of the nasal cavity actually lined with sensory epithelia) has been correlated with both sex and habitat type. For example, male plethodontid salamanders (*Plethodon cinereus* and *Eurycea wilderae*) have significantly larger vomeronasal and main olfactory organs than similar-sized females (Dawley 1992a,b; Dawley and Crowder 1995). Sexual dimorphism at the receptor, neural, or muscular level may be a possible mechanism for generating sexually dimorphic behaviour (Kelley 1988). Because the most obvious behavioural differences between males and females occur during the breeding season, the olfactory and/or vomeronasal system may be involved in locating receptive females and initiating courtship. Dawley and Crowder (1995) found that both male and female *Plethodon cinereus* have larger vomeronasal organs during the summer than at any other time of year. This difference in size is at least partly due to a greater number of receptor cells. They suggested that seasonal fluctuation in vomeronasal cell number may reflect the neurogenesis of specialized receptors used in courtship and mating. Habitat type also seems to affect main olfactory and vomeronasal organ size (refer to the section "Variation in Nasal Geography and Epithelial Structure" for a discussion of the morphology of the sensory epithelia of aquatic versus terrestrial amphibians). Among apodans and plethodontid salamanders, secondarily aquatic species (those with terrestrial ancestors) have reduced olfactory organs while retaining vomeronasal organs as large as or larger than those of their more terrestrial relatives (Schmidt and Wake 1990; Dawley 1992b). This may reflect the importance of the role of the vomeronasal organ in detecting the large, water-borne molecules that may be involved in sex identification and courtship.

There is additional variation in olfactory/vomeronasal system size and morphology that may reflect more subtle ecological or phylogenetic relationships. Schmidt *et al.* (1988) used a modern nerve-tracing technique (injections of horseradish peroxidase) to visualize the projections of the main olfactory nerves on the main olfactory bulb and the vomeronasal nerves on the accessory olfactory bulbs. They compared two species of salamandrid (*Triturus alpestris*, a newt with biphasic development, and *Salamandra salamandra*, a terrestrial, viviparous species) and eight species of plethodontid salamanders (biphasic, semi-aquatic *Desmognathus ochrophaeous, Eurycea bislineata* and direct developing, terrestrial *Ensatina eschscholtzi, Plethodon jordani, Aneides flavipunctatus, Hydromantes italicus, Bolitoglossa subpalmata,* and *Batrachoseps attenuatus*). All 10 species showed similar sizes and shapes of olfactory nerve projections (implantation cones), although there was no attempt at quantification, and the drawings suggest that some species appear to have substantially smaller olfactory fields than do others. However, vomeronasal projection fields varied such that direct-developing species had one or two projection fields (see their Fig. 2) while biphasic developing species had more complex, several-lobed projection fields. Again, quantification of differences were not attempted. The authors concluded that among plethodontid salamanders there may be a reduction of the entire vomeronasal system correlated with terrestrial life.

This conclusion is supported by direct measurement of vomeronasal organ size. The vomeronasal organs of a semiaquatic, biphasic species, *Eurycea wilderae*, are significantly larger than two direct-developing, terrestrial species of about the same size, *Plethodon cinereus* and *Desmognathus wrighti* (Dawley 1992b and in prep.). Other semi-aquatic biphasic species (*D. quadramaculatus* and *E. guttolineata*) also appear to have larger vomeronasal organs than does *P. cinereus*, even after adjusting for differences in body size (Dawley 1992b and in prep.). However, other direct-developing, terrestrial species in the Bolitoglossine lineage (*Chiropterotriton, Bolitoglossa rufescens, B. rostrata, Dendrotriton,* and *Thorius macdougali*) have similar-sized or larger vomeronasal, main olfactory and olfactory/accessory olfactory bulb volumes than those of *E. wilderae*.

1. Terminal Nerve System

Vertebrates may possess another chemosensory system, the terminal nerve (or cranial nerve zero). The terminal nerve (TN) is anatomically and developmentally associated with the olfactory and vomeronasal systems, which has led to the suggestion that it has a chemosensory function as well. There is no hard evidence that the terminal nerve has a sensory function, but may, instead, have an integrative or modulatory function (Wirsig and Getchell 1986). Part of the difficulty in determining whether the TN has a sensory function is because of the diffuse nature of its neurons and the technical difficulty in locating its peripheral projections.

The terminal nerve system in amphibians has been described using classical techniques (e.g., Herrick 1909; McKibben 1911). However, TN neurons contain neuropeptides (e.g., gonadotropin-releasing hormone, GnRH; Schwanzel-Fukuda and Silverman 1980), molluscan cardioexcitatory peptide (FMRFamide; Muske and Moore 1988), or neurotransmitters (e.g., acetylcholine; Wirsig and Getchell 1986). Thus, most of the more recent information about the TN has been the result of immunocytochemical studies of GnRH-immunoreactive, FMRFamide-immunoreactive, and acetylcholinesterase neurons in the nasal cavity and brain. Studies have included salamanders and frogs (*Cynops pyrrhogaster*, Kubo *et al.* 1979; *Ambystoma tigrinum* and *Rana catesbeiana*, Wirsig and Getchell 1986, Dawley *et al.* 1994 and Wirsig-Wiechmann 1993; *Taricha granulosa, Rana pipiens* and *Hyla regila*, Muske and Moore 1988). Salamander TNs also have been studied using horseradish-peroxidase labelling (*Salamandra salamandra, Triturus alpestris, Desmognathus ochrophaeus, Eurycea bislineata, Ensatina eschscholtzi, Plethodon jordani, Aneides flavipunctatus, Hydromantes italicus, Bolitoglossa subpalmata, Batrachoseps attenuatus,* Schmidt *et al.* 1988). In general, these studies show that TN fibers can be found within the olfactory nerve proper, but rarely within the vomeronasal nerve; in salamanders, a group of TN cell bodies can be found caudal to the olfactory bulb or immediately adjacent to the olfactory bulb within the olfactory nerve; in frogs, these cell bodies are found more caudally within the telencephalon (septal and commissural areas). In both salamanders and frogs, central TN projections consist of fibers terminating within the olfactory bulb and fibers that project to the caudal telencephalon (septum and preoptic area) and anterior hypothalamus.

Peripheral projections have been much more difficult to locate using immunocytochemical approaches because of the minute and unpredictable quantities of neurotransmitters or neuropeptides within peripheral processes. However, by using techniques that increase concentrations of the neuropeptide, GnRH, peripherally, Wirsig-Wiechmann (1993) found sparse GnRH-ir fibers within olfactory nerve bundles projecting along ventral and medial areas of the nasal cavities. These terminated within the *lamina propria* of the rostromedial-most region of the ventral olfactory epithelium. Very few fibers terminated within the olfactory or vomeronasal epithelium. This suggests that if there is any interaction between TN and olfactory receptors, it occurs between axons and not between nerve terminals. Few GnRH-ir fibers contact Bowman's glands of the olfactory lamina propria. Some GnRH-ir fibers were found within the trigeminal/facial nerve branches, within palatine ganglia (which innervate the nares constrictor muscle) and some project into this muscle directly. TN fibers also seem to make contact with blood vessels. Thus, the peripheral terminal nerve may modulate and co-ordinate olfactory and vn neurons, glands and smooth muscle of the nasal cavity to facilitate access to and reception of pheromones into the nasal cavity (Wirsig-Wiechmann 1993).

Interest has focused on a link between the TN and reproductive behaviour. Gonadotropin-releasing-hormone in the forebrain causes gonadotropin release from the pituitary, which stimulates activities of gonads. Injection of GnRH induces sexual receptivity and attractivity in female *Xenopus laevis*, apparently independently of the pituitary-gonad axis (Kelley 1982). In male *Taricha granulosa* GnRH injections stimulate sexual behaviours (Moore *et al.* 1982). Concentrations of GnRH-ir within the preoptic area, rostral hypothalamus and infundibulum of *Taricha granulosa* males varies seasonally (Zoeller and Moore 1985). Finally, GnRH-ir concentrations increase significantly within TN areas of the rostral telencephalon in female *T. granulosa* that have been exposed to male mating behaviours (Propper and Moore 1991) and in female *Plethodon cinereus* collected with males during the mating season (Dawley *et al.* 1994). Thus, GnRH is linked with many aspects of vertebrate reproductive cycles and with the TN, suggestive of a link between TN and reproductive behaviour.

IV. BEHAVIOURAL STUDIES

A. Larval Studies

Larvae of some anuran species naturally aggregate in kin groups (e.g., *Bufo americanus*, Waldman and Adler 1979; *Rana cascadae*, Blaustein and O'Hara 1982; *Rana sylvatica*, Waldman 1984; Cornell *et al.* 1989; Fishwild *et al.* 1990; see Blaustein and Waldman 1992 and Blaustein and Walls 1995 for recent reviews). Larval *R. cascadae* (Blaustein and O'Hara 1982a) and *B. americanus* (Waldman 1985) use water-borne chemical cues to discriminate between siblings and non-siblings. *R. cascadae* tadpoles spent more time on the sibling side of a partitioned aquarium (that blocked visual cues but allowed water to flow freely) than on the non-sibling side. In Y-tube tests, *B. americanus* tadpoles preferred sibling water over non-sibling water (water that had flowed past sibling or non-sibling tadpoles, respectively), preferred blank water over non-sibling water, but showed no preference between sibling water and blank water. This suggests that individual tadpoles are avoiding non-sibling conspecifics rather than being attracted to siblings. Both main olfactory and vomeronasal systems are probably functional in *B. americanus* tadpoles, but a well developed lateral line system or free nerve endings in the skin of larvae may also contribute to chemoreception. Blocking tadpole nares with a gelatinous paste, however, removed their ability to discriminate between water-borne cues; this supports the idea that either main olfactory or VNO systems are used in this context. Taste, in contrast, probably is not functioning until metamorphosis is complete. The specific source of tadpole odour (whether from a specific organ, whole body, or a metabolic waste) has not been determined (Waldman 1986).

There are some evolutionary questions associated with sibling recognition in tadpoles: (1) what is the mechanism of recognition and (2) what is the advantage of recognizing siblings? Experiments with *R. cascadae*, *R. sylvatica* and *B. americanus* suggest variable mechanisms that are a combination of genetics and learning (O'Hara and Blaustein 1981; Waldman 1981, 1984, 1986; Blaustein and O'Hara 1982b, 1983). The differences between species as to whether "nature" or "nurture" are more important seem to be correlated with specific early larval ecology; for example, species with more intermingling of clutches have more "hard-wired" and less learned recognition systems (Blaustein and O'Hara 1983; Waldman 1986). Tadpoles that recognize siblings and preferentially school with them may experience increased inclusive fitness compared to tadpoles that school with non-siblings. Distasteful *B. americanus* tadpoles may form a large aposematic group; toads are not fatal if eaten, but toad predators may learn that that group is distasteful and skip over other members of the group (Waldman

1982). Or schooling tadpoles may warn siblings of predators, perhaps through chemicals released from the epidermis when damaged (Waldman 1986; Hews and Blaustein 1985). Schooling siblings may also regulate each others growth, such that small tadpoles may stimulate larger tadpole growth while larger siblings may inhibit small tadpoles. This would have the advantage of assuring that some individuals (those that are already growing most rapidly) would make it to metamorphosis when resources are limiting (Waldman 1982), clearly only remotely an advantage if this altruism is extended to close relatives. Both alarm chemicals and growth regulators become quickly diluted as they diffuse away, thus, schooling of siblings increases the probability that a sibling will encounter these chemicals.

Larvae, alternatively, may aggregate and recognize siblings because they are philopatric, not because there is any advantage to recognizing and associating with kin. The advantage may be in remaining in their familiar (natal) habitat, and kin recognition may be an artifact of habitat selection. Larval *Scaphiopus multiplicatus* show higher growth and survivorship rates when restricted to their natal site than when restricted elsewhere in the same pond (Pfenning 1990). In addition, wild-caught tadpoles preferred water from their natal site over water from other sites within the same pond. Although *S. multiplicatus* tadpoles preferred to associate with unfamiliar siblings over unfamiliar non-siblings, they also preferred (1) unfamiliar non-siblings reared on familiar food over unfamiliar non-siblings reared on unfamiliar food, and (2) water strained through familiar food over unfamiliar siblings reared on familiar food (Pfenning 1990).

Not all species of anurans have tadpoles that school or that school with siblings. For example, *Rana pipiens* and *Pseudacris crucifer* do not school and *Rana pretiosa* and *Hyla regilla* tadpoles aggregate only intermittently; not surprisingly, these tadpoles cannot discriminate between kin and non-kin (Blaustein 1988; O'Hara and Blaustein 1988; Fishwild *et al.* 1990). *Bufo boreas* tadpoles aggregate in large groups that include kin and non-kin, and laboratory experiments show that these tadpoles only preferentially associate with kin if reared solely with kin (O'Hara and Blaustein 1982). Thus, it is possible to identify larval characteristics that correlate aggregates of kin with kin recognition (Waldman 1986; Blaustein 1988). Post-metamorphic *Bufo cognatus* toadlets aggregate in conspecific groups, perhaps as an anti-predator strategy that functions through predator saturation or confusion (Graves *et al.* 1993). Toadlets are attracted to conspecific odours, suggesting that toadlet aggregation is the result of attraction to conspecifics rather than because of environmental features.

Other amphibian larvae use chemical senses to detect predators. *Rana temporaria* tadpoles avoid water taken from aquaria that held predatory fish species (Manteifel 1995). *Eurycea bislineata* larvae avoid water in which a potential predator, the green sunfish, *Lepomis cyanellus*, had been held (Petranka *et al.* 1987). This may explain how *E. bislineata* can coexist in streams inhabited by predators. Similarly, a number of species with larvae inhabiting permanent ponds and streams, thus cohabitating with green sunfish, pike, and other predators, (e.g., *Ambystoma texanum, E. longicauda, E. bislineata, Hyla chrysoscelis;* Kats *et al.* 1988; *A. barbouri,* Sih and Kats 1991; *Rana lessonae* and *Rana esculenta,* Stauffer and Semlitsch 1993) spent more time in refuges or moved more quickly and directly to refuges when exposed to green sunfish-treated water. This behaviour was eliminated when the nares of *A. texanum* were plugged with gelatinous paste (Kats 1988). Other potential predators (crayfish, water snakes, and turtles) did not elicit a refuge seeking response even though *A. texanum* larvae are found in streams with these predators as well as with green sunfish. All of the above amphibian species are palatable; unpalatable larvae in permanent bodies of water (*Rana clamitans, Bufo americanus, Notophthalmus viridescens*) did not respond to green sunfish-conditioned water (Kats *et al.* 1988). However, palatable larvae that live in ephemeral ponds that usually do not contain green sunfish (*Hyla crucifer, Pseudacris triseriata, Rana sylvatica, Ambystoma opacum, A. maculatum* and *A. texanum* population from a temporary pond, as opposed to the population cited above from a permanent stream) also cannot detect chemicals cues from predatory green sunfish through water-borne chemical cues (Petranka *et al.* 1987; Kats *et al.* 1988). Thus, it is possible to predict, based on palatability and habitat, whether or not the larvae of a species will be able to detect water-borne odours of a predator (with some exceptions; e.g., *R. catesbeiana* larvae are unpalatable but can detect sunfish odours). It also often is possible to predict specific anti-predator responses to chemical cues based upon type of predator. *Bufo bufo*, a species found in permanent ponds with fish, will shift to an edge microhabitat when exposed to predator-conditioned water. By contrast, *Bufo calamita* lives in a fish-less habitat and will shift to open water or refuges when exposed to odours of the invertebrate predators that are most abundant along edges (Semlitsch and Gavasso 1992).

B. Foraging

The primary importance that vision plays in the prey catching behaviour of anurans is so well established that this system serves as a model of neuroethological study (e.g., Ewert 1980). Small moving objects elicit (or release) prey-catching; sound and smell are considered of secondary significance. Some early experiments with adult *Bufo americanus* (Risser 1914) suggested that toads do not use chemoreception to locate food, relying probably entirely on vision. When odours of strange chemicals (e.g., oil of clove) were passed directly to the externals nares some reactions were observed (wiping motions; respiratory changes), but odours of prey had no effect. Prey offered in complete darkness were ignored, but when the light was turned on and the prey moved, toads fed upon them. Toad tadpoles, on the other hand, could discriminate between identical cloth-wrapped packets with and without food (dead earthworms, fish, or decomposing meat), nibbling on the packet with food. Tadpoles with vaseline blocked nares could no longer discriminate between these packets.

Since Risser's study, several researchers have sought to show that olfaction could play some role in prey catching. For example, *Rana pipiens* and *Bufo marinus* will approach odour sources more often than a control (non-odour) and the odours will elicit an increased rate of tongue-snapping (Shinn and Dole 1978; Rossi 1983). These studies in no way address whether or not vision is the primary sense used in prey-catching, but they may explain the observations of anurans eating stationary food, such as carrion, rotting vegetation, or faeces (Rossi 1983). In addition, *R. pipiens* can learn to avoid noxious prey using olfactory cues alone, but learn more quickly with visual cues or even more rapidly with a combination of visual and olfactory cues (Sternthal 1974). These results seem to parallel those reported for salamanders (see below).

Among salamanders, both vision and chemosensation guide prey catching behaviours (Reese 1912; Pope 1924; Martin *et al.* 1974; Roth 1976; David and Jaeger 1981; Lindquist and Bachman 1982; Luthardt and Roth 1983). Because salamander species feed under a number of conditions of illumination (e.g., diurnal, nocturnal, cave-dwelling, under and inside cover objects), workers have attempted to tease apart the roles of vision and chemoreception. In general, vision alone or a combination of vision and chemoreception elicited the strongest, quickest reactions. Some studies suggest that prey items are located by vision first and then more specifically identified by olfaction (Martin *et al.* 1974; Linquist and Bachmann 1982). However, when prey are motionless or when visual cues are lacking (as in total darkness or when an extract of a prey item is used), salamanders are able to locate prey using chemical cues alone and chemical cues elicit stronger reactions than tactile ones (Reese 1912; Pope 1924; Martin *et al.* 1974; David and Jaeger 1981; Petranka *et al.* 1987). In addition, salamanders can learn to associate odours with a prey item (Jaeger and David 1981; Luthardt and Roth 1983). These studies included a variety of species (*Ambystoma tigrinum, Eurycea bislineata, Hydromantes italicus, Notophthalmus viridescens, Plethodon cinereus, Salamandra salamandra*) normally feeding under different light conditions.

C. Migration and Orientation

Because many species of anurans return to specific breeding ponds after a period of foraging on land, they probably use a combination of sensory cues, including chemical, to home (Grubb 1970; Dole 1972; Sinsch 1990). Direct evidence that chemical cues are used in orientation to breeding ponds comes from a series of olfactometer tests performed by Martof (1962) and Grubb (1973a, 1973b, 1975, 1976); the olfactometer was a T-shaped wind tunnel with a fan at the base which pulled odour-laden air through the maze. When given a choice between water from the home pond or from other aquatic habitats (permanent or temporary ponds and creeks also used for breeding) or distilled water, at least four species, *Bufo valliceps, Pseudacris clarki, P. strecki* and

Rana utricularia, significantly preferred the odours of their home ponds. If anurans use chemical cues to home, they also must remember home pond odours from one breeding season to the next. *Bufo valliceps* stopped discriminating between water sources two days after capture but when tested 17, 32, and 61 days later they again oriented toward home pond odours if given an injection of gonadotropin 24 hrs before testing (Grubb 1973). *B. valliceps* can be trained in somewhat more complex mazes to orient toward odour cues (Grubb 1976). This suggests that these toads have the chemosensory mechanism available to orient to odours in a natural landscape. Some or all of the natural guiding odours probably come from the breeding body of water (Grubb 1973a,b, 1975; Sinsch 1987, 1990a,b). If anurans return to their natal pond to breed, then odour preferences may develop during embryonic stages. *Rana temporaria* and *R. sylvatica* embryos exposed to odourants subsequently developed preferences for those odours; those preferences were retained after metamorphosis (Hepper and Waldman 1992). Male and female *Dendrobates pumilio*, denizens of tropical forests of Central America, occupy home areas to which they are able to return following displacement. In olfactometer tests, these frogs were able to discriminate the air-borne odours of home aquaria and particularly preferred the odours of a plant clone (a bromeliad) found in their aquaria (Forester and Wisnieski 1991).

Field studies demonstrate that there is variability in the cues used by a particular species for homing, apparently independent of phylogeny. Anosmic toads (*B. boreas*, Tracy and Dole 1969; *B. bufo*, Sinsch 1987; *B. japonicus*, Nishio *et al.* 1988) cannot orient to home; however, *Rana pipiens* (Dole 1972), *B. spinulosus* (Sinsch 1990b), and *Bufo valliceps* (Grubb 1970) continued to home even after being made anosmic (or blind or deaf). When both vision and chemoreception were eliminated, however, *Bufo valliceps* could no longer home (Grubb 1970). Other studies (with *B. bufo*, *B. calamita* and *B. spinulosus*) suggest that chemoreception and magneto-orientation probably supplement each other in different stages of orientation (Sinsch 1987, 1990b; see also Phillips, this volume). Sinsch suggests that species that migrate long distances (like *B. bufo*) rely on olfactory cues for initial orientation, whereas species with short migrations (*B. calamita* and *B. spinulosus*) rely on vision.

Studies of salamander orientation and homing have generally centered on three different life history strategies employed by different species. (1) Ambystomatid salamanders and *Taricha* species (Salamandridae) are terrestrial for most of the year and return to ponds to breed; chemoreception may play a role in homing to breeding ponds and streams. (2) Many plethodontid species remain in a particular home range throughout the warm months of the year, so that mating and egg laying do not occur at a different site than feeding; chemoreception may play a role in home range or territory recognition. In addition, terrestrial or semi-aquatic plethodontid females remain with their nest to care and protect the eggs; females may return to their nest sites using chemical cues. (3) *Notophthalmus viridescens* (Salamandridae) is aquatic as an adult, usually after a terrestrial juvenile stage (red eft), but may overwinter on land or intermittently return to land during the summer; chemoreception may orient adults to a particular pond after these terrestrial excursions.

Almost all studies of salamander homing have been done in the field. Generally, marked individuals are transported some distance from their home range, after ablations of a particular sense, and their ability to return monitored. If the anosmic salamanders cannot home, but can if the olfactory system remains intact, or if some other sense (like vision) is ablated and the animal can still home, then chemoreception is considered to be involved in homing. Such abilities have been shown for *Taricha rivularis* (Twitty 1961; Grant *et al.* 1968), *Desmognathus fuscus* (Barthalmus and Bellis 1972), *Plethodon jordani* (Madison 1972), and *Nothophthalmus viridescens* (Hershey and Forester 1980). In one laboratory study, *Ambystoma maculatum* preferred benthic mud from a home pond to that from a foreign pond, and salamanders were observed nose-tapping and chin-touching, two putative chemosensory sampling behaviours (McGregor and Teska 1989).

Female plethodontids are able to relocate their own nests after displacement (see Forester 1986, for references). Air-borne cues may be important in locating the general vicinity of the nest cavity, although this remains experimentally unproven. However, Forester (1979, 1986) did show that *Desmognathus ochrophaeus* and *D. fuscus* females prefer the odours of their own clutch over those of a conspecific. Females probably are cueing in on egg smells rather than their own odours because they show no preference for strange eggs resting on filter paper marked by that female when the alternative choice was a different set of strange eggs or the female's own eggs, both sets lacking the female marked substrate. Forester suggested that females use air-borne odours to recognize the general area of nest site and then use contact chemoreception (vomeronasal system) to recognize their own clutch.

D. Social Interactions

Interactions among salamanders, whether competitive, predatory, or mating, are mediated by chemical communication as well as by vision. A good deal of research has centered around terrestrial salamanders in the family Plethodontidae. Y-tube olfactometer tests (in which salamanders choose between two air streams first drawn past some odour source) can show what kinds of social odours salamanders perceive. The first such tests were performed by Madison (1975), who gave *P. jordani* a choice between neighbouring and distant same-sex conspecifics before and during the breeding season. Before the breeding season, both males and females preferred neighbours over non-neighbours; during the breeding season they showed no preference. This was interpreted as avoidance of inbreeding (but see below). Later olfactometer tests performed by Dawley (1984a, 1986, 1987) on a number of closely related eastern *Plethodon* species (*P. teyahalee*, *P. glutinosus*, *P. aureolus*, *P. kentucki* and different populations of *P. jordani*) showed that male salamanders prefer female conspecific odours over male odours during breeding and non-breeding seasons while female salamanders are attracted to male conspecific odours only during the breeding season. Similarly, Ovaska (1988) found that male and female *P. vehiculum* (a western species) avoided odours of male conspecifics but not those of females. In addition, when closely related species are sympatric, both males and females are attracted to the odours of opposite sex conspecifics over heterospecifics (Dawley 1984, 1986). However, when salamanders from a hybrid zone between *P. jordani* and *P. glutinosus* were tested, the ability to discriminate between species' odours could not be shown for males of one of the species-pair (Dawley 1987). Thus, salamanders in the genus *Plethodon* probably identify the sex and species of potential mates through chemoreception. Similarly, sympatric *Desmognathus* species may identify potential mates through odours. Male *D. ochrophaeus* and male *D. fuscus* prefer conspecific female odours over the odours of heterospecific females in a Y-maze containing soiled papers from females in the arms of the maze (Uzendoski and Verrell 1993). Male *D. ochrophaeus* and male *D. imitator* likewise prefer conspecific female odours (Verrell 1989). These olfactory discriminatory abilities may account for the lack of interspecific courtship between these two pairs of closely related species.

Plethodon cinereus has been intensively studied, revealing a repertory of competitive and territorial behaviours, in which chemoreception plays a major role (Jaeger 1986). In this species, there is evidence that the vomeronasal system, specifically, is stimulated by conspecific odours. When confronted with substrates previously inhabited by conspecifics, complete with feces, salamanders repeatedly tap and rub their nasolabial grooves (the conduit to the vomeronasal organ) on the substrate and faeces (Tristram 1977; McGavin 1978; Jaeger and Gergits 1979; Jaeger 1981; Simon and Madison 1984; Jaeger *et al.* 1986; Horne and Jaeger 1988; Jaeger and Wise 1991). In these experiments, *P. cinereus* were able to distinguish their own substrates from those of conspecifics (Tristram 1977), familiar conspecific substrates from those of other conspecifics (McGavin 1978; Jaeger 1981), male substrates from female substrates (avoiding male substrates; Jaeger and Gergits 1979), and conspecific substrates from those of sympatric congeners (*P. cinereus* and *P. shenandoah* generally avoided substrates marked by congeners; Jaeger and Gergits 1979). Gravid and non-gravid females were particularly interested in male substrates, nose-tapping more frequently than males on male-marked substrates (Jaeger and Wise 1991). The ability of *P. cinereus* to return to its homesite whether or not the homesite was disturbed has been attributed to its ability to recognize neighouring territorial odours. Jaeger *et al.* (1993) suggested that individual *P. cinereus* may form a cognitive map of neighbouring territorial pheromones and use this map to home. In addition, males apparently find and follow female pheromone trails prior to courtship (Gergits and Jaeger 1990).

Faeces and/or pheromones deposited by glands in or around the cloaca, are a potential source of territorial information for both sexes (Simon and Madison 1984; Jaeger *et al.* 1986; Horne and Jaeger 1988; Simons and Felgenhauer 1992; Jaeger and Gabor 1993; Simons *et al.* 1995). However, because both males and females exhibit territorial behaviour but females lack most of the cloacal glands found in males (Sever 1978), Simons and Felgenhauer (1992) suggested that midventral tail glands are the primary source of territorial pheromones. This is supported by a behaviour called "postcloacal press", which is elicited in animals presented with faecal odours. As the name implies, during this behaviour salamanders seem to press midventral tail glands to mark the substrate. Salamanders often nose-tap each other during interactions (Gergits and Jaeger 1990), and in staged interactions between a "resident" and an "intruding" salamander, 50% of first touches were directed to the postcloacal region (compared to head, anterior trunk, posterior trunk, or distal half of tail) (Jaeger and Gabor 1993). Residents and intruders alike repeatedly assess the postcloacal area and this contact lessens the likelihood of biting. Male and female salamanders are attracted to burrows marked by their own faeces and repelled by burrows marked by same-sex conspecific faeces (Jaeger *et al.* 1986; Horne and Jaeger 1988). Females, however, seem less intimidated by other females' faecal pellets, than by those emanating from males, and exhibit greater "interest" in pellets than do males. Females repeatedly visited pellets, whereas males tended to leave the vicinity of pellets after the initial sampling; females also squashed pellets with their snouts in addition to tapping them with nasolabial grooves (Horne and Jaeger 1988; Jaeger and Wise 1991). Faeces seem to contain information about not only the sex of the individual depositing them, but also, for faeces from males, body size (Mathis 1990) and quality of diet (Walls *et al.* 1989; Jaeger and Wise 1991). Male and female intruders can vary the rate of faecal pellet production, but, apparently, with different functions. Males produce pellets fastest when exposed to female odours, suggesting that males are attracting females through their odours; females produce pellets fastest when exposed to their own odours, suggesting that females are maximizing territorial advertisement (Mathis 1990). Females are attracted more strongly to faecal pellets produced by males on a higher quality diet; however, vision may play a role in assessing dietary composition through faeces (Walls *et al.* 1989; Jaeger and Wise 1991).

Behavioural studies using other species of *Plethodon* and of related genera often do not produce results as clear as those obtained from *P. cinereus*, perhaps reflecting the fact that *P. cinereus* has been more intensely studied than virtually any other salamander species. Two large allopatric species of *Plethodon* found in the mountains of Arkansas and Oklahoma (*P. caddoensis* and *P. ouachitae*) mostly are found singly under objects in woodlands and are aggressive toward conspecifics in laboratory encounters, suggesting that both species are territorial (Antony 1993; Antony and Wicknick 1993). In substrate choice tests, both species spent significantly more time on substrates marked by conspecifics than on their own substrates. Therefore, both species can detect conspecific odours, but they are not repelled by them; both species also investigate faecal pellets and have been observed squashing pellets, pushing and rolling them, and preforming postcloacal presses on them. Ovaska and Davis (1992) looked at the ability of several western plethodontids to recognize faecal pellets from conspecifics and sympatric heterospecifics. *P. vehiculum* and *P. dunni* are able to discriminate faecal pellets of conspecific and heterospecifics. Only *P. dunni* is thought to be territorial, however, and may exclude *P. vehiculum* from the preferred, wetter microhabitats where *P. dunni* is found. On the other hand, *P. vandykei* and *Aneides ferreus* showed no evidence of intraspecific or interspecific odour recognition, although *P. vandykei* can discriminate *P. vehiculum* pellets. *P. vehiculum* is found on Vancouver Island and on mainland western North America. Substrate choice tests show that breeding males and females can discriminate between mainland and island opposite-sex conspecifics, suggesting that pheromones, themselves, have diverged (Ovaska 1989).

Interactions between species of *Desmognathus* also are mediated chemically. Because species of *Desmognathus* vary in size and terrestriality (the largest being most aquatic, and, as size of species decreases, terrestriality increases), predation, as well as competition may play a role in interactions. When *D. quadramaculatus* (the largest species) and sympatric *D. monticola* (the next largest species) were given four simultaneous choices of substrates previously marked by large and small *D. quadramaculatus* and large and small *D. monticola*, all salamanders showed a significant preference for odours from similarly sized conspecifics (Roudebush and Taylor 1987). Both large and small *D. monticola* avoided substrate odours from both sizes of *D. quadramaculatus*, suggesting that *D. monticola* use chemoreception to detect and avoid *D. quadramaculatus*. The reverse was not true; *D. quadramaculatus* neither is attracted to nor repulsed by the odour of *D. monticola*. The source of these odours seems to be cutaneous mucus glands; in a aquatic Y-maze, *D. monticola* avoided the side arm containing skin secretions (Jacobs and Taylor 1992).

Less is known about the involvement of chemoreception in the behaviour of salamanders in other families. Other largely terrestrial salamanders, such as Ambystomatids, generally are not considered to be territorial, although there is potential for competition for burrows and other retreat sites. *Ambystoma maculatum* is the only species shown to exhibit some territorial qualities, such as dispersion in the field, aggressive defense of artificial chambers, and recognition of substrate odours (male residents tended to remain on substrates marked by other salamanders; Ducey and Ritsema 1988). Ambystomatids also may recognize species odours and discriminate between the sexes, although the only experimental evidence to support this is the ability of male *A. jeffersonianum* to discriminate between diploid *A. jeffersonianum* females and triploid *A. platineum* females (Dawley and Dawley 1986). In an aquatic Y-maze, males preferred odours of the diploid females over those of the gynogenetic triploid *A. platineum* females (females that rely on sperm to initiate zygote development without that sperm actually being incorporated in the zygote).

Among Salamandrids (newts), adult *Taricha* return to water to find mates and breed, and chemical cues may guide males to conspecific females. Males are attracted to sponges soaked in water in which conspecific females had been kept, but not sponges soaked in water that contained heterospecific females (Twitty 1955, 1961). Identification of sex and species probably occurs when males nudge females before initiating courtship (Davis and Twitty 1964). Similarly, male *Notophthalmus viridescens* are attracted to odours of conspecific females (Dawley 1984b). When tested in an aquatic Y-maze, males were attracted to water that had contained females but were neither attracted nor repelled by water containing the odours of males. *N. viridescens* engage in mutual sniffing in the preliminary stages of courtship, suggesting mate identification occurs then as well (Verrell 1982). Size also may be assessed using chemical cues; in a Y-maze, males preferred water that contained a large female over water with a smaller female (Verrell 1985). Presumably males would prefer to mate with larger females because they produce more eggs. Similar results were found with a European newt, *Triturus vulgaris*; males were attracted to water in which females had been kept, and showed a trend towards greater attraction to water containing large females than to that containing smaller ones (Verrell 1986). However, in another Y-maze choice experiment, male *N. viridescens* preferred water that contained three small females over water with one large female. This suggests that males use odours to detect and locate females rather than to compare the size of prospective mates (Rowland *et al.* 1990).

Two genera of large, aquatic salamanders in the family Proteidae, *Proteus* and *Necturus*, show some social chemosensory discriminatory abilities. Breeding territorial male *Proteus* can recognize females through direct body contact but not through water-borne cues, although water-borne cues and substrate cues attract both sexes, perhaps facilitating recognition of communal resting places outside the breeding season (Parzefall 1976). *Necturus* also share hiding places with conspecifics and also are attracted to substrate and water borne conspecific odours (Parzefall *et al.* 1980).

In many of the studies cited above, it was not always possible to distinguish between odours used in aggression/territoriality and mate choice, nor was the source of these odours obvious (generalized epidermal glands, special glands, or faeces alone). There are, however, specific chemically mediated behaviours that are involved in courtship and mating, and often specific glandular areas can be identified that produce pheromones. Male delivery of courtship pheromones is known for most species of salamanders (Arnold and Houck 1982). There are several summaries of location and possible functions of courtship glands in salamanders to which the reader is directed for a more complete accounting of references, especially behavioural references (Arnold 1977; Madison 1977; Arnold and Houck 1982; Houck 1986; Houck and Sever 1994). Briefly, Arnold divided the function of courtship glands into those that orient the female to the male and those that serve to enhance female receptivity to courtship. Examples of possible glandular areas in the first category include cloacal papillae in salamandrids and ambystomatids and dorsal bumps found on the tail base of plethodontids. There is little histological confirmation of the glandular nature of these areas. The abdominal gland in *Triturus* (Malacarne *et al.* 1984; see also references in Cedrini and Fasolo 1971) is associated with cloacal papillae and undergoes cyclic fluctuations which are gonadal hormone-dependent. In an electrophysiological study, extracts of male *T. cristatus* accessory glands elicited strong responses in female olfactory bulbs (Cedrini and Fasolo 1971). In an aquatic Y-maze, female *T. cristatus* preferred water inhabited by male-female pairs, particularly courting pairs (Malacarne and Vellano 1987). During the mobile stages of courtship prior to spermatophore deposition, females follow behind males nudging the cloacal papillae (salamandrids or ambystomatids) or resting their snouts on the males' tail base (plethodontids): this orients the female. Sever (1989) confirmed the presence of a cluster of acinar exocrine glands in the skin of the middorsal tail base of male salamanders in the *Eurycea bislineata* complex that are not found in females. These glands produce copious secretions only during the breeding season.

Arnold's second category of glands are those that serve to enhance female receptivity. In aquatic species, these may be cloacal glands, the secretions of which are wafted to the female by tail movements, and/or glands found on the head. Male *Nothophthalmus viridescens* have cheek glands while plethodontid species and *Taricha* males have glands found on the lower jaw or chin; all types hypertrophy during the breeding season and are not found on females. The contents of these head glands are rubbed directly on the female's nares. While these rubbing behaviours have been observed in the courtship sequences of many salamanders (see Arnold 1977 or Madison 1977 for other references), only a few studies have addressed directly the function of these behaviours and associated glands. Verrell (1982), for example, found that male *N. viridescens* perform two basic courtship sequences depending on female's receptivity. Females that are receptive initially are treated only to tail fanning of cloacal glands before males deposit a spermatophore. Females who are initially unresponsive are clasped by the male while he rubs his cheek glands on her snout. Houck and Reagan (1990) present the only quantitative data supporting the hypothesized enhancement function of male pheromones. They ablated the chin glands of male *Desmognathus ochrophaeus* (a plethodontid salamander) before allowing them to court females. Females that had been treated with gland extract before courtship mated 28% sooner with these males than did females who received only a saline control. However, in this species (and several other plethodontids, Arnold 1977) the male circumvents any chemosensory system by rubbing chin gland secretions into scrapes he makes on the female's dorsum with sexually dimorphic premaxillary teeth.

IV. ACKNOWLEDGEMENTS

This work was supported by grants from the Whitehall Foundation and the Howard Hughes Medical Institute. Many thanks to Harold Heatwole for being such a thorough and understanding editor.

V. REFERENCES

Altner, H., 1962. Untersuchungen Über Leistungen und Bau der Nase der südafrikansichen Krallenfrosches *Xenopus laevis* (Daudin, 1803). *Zeitschr. vergl. Physiol.* **45**: 272–306.

Antony, C. D., 1993. Recognition of conspecific odors by *Plethodon caddoensis* and *P. ouachitae*. *Copeia* **1993**: 1028–1033.

Antony, C. D. and Wicknick, J. A., 1993. Aggressive interactions and chemical communication between adult and juvenile salamanders. *J. Herpetol.* **27**: 261–264.

Arnold, S. J., 1977. The evolution of courtship behavior in New World salamanders with some comments on Old World salamanders. Pp. 141–183 *in* "The Reproductive Biology of Amphibians", ed by D. H. Taylor and S. I. Guttman. Plenum Press, New York.

Arnold, S. J. and Houck, L. D., 1982. Courtship pheromones: Evolution by natural and sexual selection. Pp. 173–211 *in* "Biochemical Aspects of Evolutionary Biology", ed by M. H. Nitecki. University of Chicago Press, Chicago.

Arzt, A. H., Silver, W. L., Mason, J. R. and Clark, L., 1986. Olfactory responses of aquatic and terrestrial tiger salamanders to airborne and waterborne stimuli. *J. Comp. Physiol. A* **158**: 479–487.

Badenhorst, A., 1978. The development and the phylogeny of the organ of Jacobson and the tentacular apparatus of *Ichthyophis glutinosus*. *Ann. Univ. Stellenbosch Ser. A2 (Sölogie)* **1**: 1–26.

Barthalmus, G. L. and Bellis, E. D., 1972. Home range, homing and the homing mechanism of the salamanders, *Desmognathus fuscus*. *Copeia* **1972**: 632–642.

Bertmar, G., 1981. Evolution of vomeronasal organs in vertebrates. *Evolution* **35**: 359–366.

Billo, R. and Wake, M. H., 1987. Tentacle development in *Dermophis mexicanus* (Amphibia, Gymnophiona) with an hypothesis of tentacle origin. *J. Morphol.* **192**: 101–111.

Blaustein, A. R., 1988. Ecological correlates and potential functions of kin recognition and kin association in anuran larvae. *Behav. Gen.* **18**: 449–464.

Blaustein, A. R. and O'Hara, R. K., 1982a. Kin recognition cues in *Rana cascadae* tadpoles. *Behav. Neural. Biol.* **36**: 77–87.

Blaustein, A. R. and O'Hara, R. K., 1982b. Kin recognition in *Rana cascadae* tadpoles: Maternal and paternal effects. *Anim. Behav.* **30**: 1151–1157.

Blaustein, A. R. and O'Hara, R. K., 1983. Kin recognition in *Rana cascadae* tadpoles: Effects of rearing with nonsiblings and varying the strength of the stimulus cues. *Behav. Neural. Biol.* **39**: 259–267.

Blaustein, A. R. and Waldman, B., 1992. Kin recognition in anuran amphibians. *Anim. Behav.* **44**: 207–221.

Blaustein, A. R. and Walls, S. C., 1995. Aggregation and kin recognition. Pp. 568–602 *in* "Amphibian Biology", Vol. 2 "Social Behaviour", ed by H. Heatwole and B. K. Sullivan. Surrey Beatty & Sons, Chipping Norton.

Breipohl, W., Moulton, D., Ummels, M. and Matulionis, D., 1982. Spatial pattern of sensory cell terminals in the olfactory sac of the tiger salamander. I. A scanning electron microscope study. *J. Anat.* **134**: 757–769.

Buck, L. and Axel, R., 1991. A novel multigene family may encode odorant receptors: a molecular basis for odour recognition. *Cell* **65**: 175–187.

Burd, G. D., 1991. Development of the olfactory nerve in the African clawed frog, *Xenopus laevis*: I. Normal Development. *J. Comp. Neurol.* **304**: 123–134.

Burton, P. R., 1985. Ultrastructure of the olfactory neuron of the bullfrog: The dendrite and its microtubules. *J. Comp. Neurol.* **242**: 147–160.

Burton, P. R., 1990. Vomeronasal and olfactory nerves of adult and larval bullfrogs: II. Axon terminations and synaptic contacts in the accessory olfactory bulb. *J. Comp. Neurol.* **292**: 624–637.

Burton, P. R., Coogan, M. M. and Borror, C. A., 1990. Vomeronasal and olfactory nerves of adult and larval bullfrogs: I. Axons and the distribution of their glomeruli. *J. Comp. Neurol.* **292**: 614–623.

Burton, P. R. and Wentz, M. A., 1992. Neurofilaments are prominent in bullfrog olfactory axons but are rarely seen in those of the tiger salamander, *Ambystoma tigrinum*. *J. Comp. Neurol.* **317**: 396–406.

Cedrini, L. and Fasolo, A., 1971. Olfactory attractants in sex recognition of the crested newt. An electrophysiological research. *Monit. Zool. Ital. (N.S.)* **5**: 223–229.

Cornell, T. J., Berven, K. A. and Gamboa, G. J., 1989. Kin recognition by tadpoles and froglets of the wood frog *Rana sylvatica*. *Oecologia* **78**: 312–316.

Crowe, M. J. and Pixley, S. K., 1991. Monoclonal antibody to carnosine synthetase identifies a subpopulation of frog olfactory receptor neurons. *Brain Res.* **538**: 147–151.

Daston, M. M., Adamek, G. D. and Gesteland, R. C., 1990. Ultrastructural organization of receptor cell axons in frog olfactory nerve. *Brain Res.* **537**: 69–75.

David, R. S. and Jaeger, R. G., 1981. Prey location through chemical cues by a terrestrial salamander. *Copeia* **1981**: 435–440.

Davis, W. C. and Twitty, V. C., 1964. Courtship behavior and reproductive isolation in the species of *Taricha* (Amphibia, Caudata). *Copeia* **1964**: 602–610.

Dawley, E. M., 1984a. Recognition of individual, sex, and species odours by salamanders in the *Plethodon glutinosus–P. jordani* complex. *Anim. Behav.* **32**: 353–361.

Dawley, E. M., 1984b. Identification of sex through odors by male red-spotted newts, *Nothophthalmus viridescens*. *Herpetologica* **40**: 101–105.

Dawley, E. M., 1986. Behavioral isolating mechanisms in sympatric terrestrial salamanders. *Herpetologica* **42**: 156–164.

Dawley, E. M., 1987. Species discrimination between hybridizing and non-hybridizing terrestrial salamanders. *Copeia* **1987**: 924–931.

Dawley, E. M., 1992a. Sexual dimorphism in a chemosensory system: The role of the vomeronasal organ in salamander reproductive behavior. *Copeia* **1992**: 113–120.

Dawley, E. M., 1992b. Correlation of salamander vomeronasal and main olfactory system with habitat and sex: Behavioral interpretations. Pp. 403–409 *in* "Chemical Signals in Vertebrates", Volume 6, ed by R. Doty and D. Müller-Schwartz. Plenum Press, New York.

Dawley, E. M. and Bass, A. H., 1988. Organization of the vomeronasal organ in a plethodontid salamander. *J. Morph.* **198**: 243–255.

Dawley, E. M. and Bass, A. H., 1989. Chemical access to the vomeronasal organs of a plethodontid salamander. *J. Morph.* **200**: 163–174.

Dawley, E. M. and Crowder, J., 1995. Sexual and seasonal differences in the vomeronasal epithelium of the red-backed salamander (*Plethodon cinereus*). *J. Comp. Neurol.* **359**: 382–390.

Dawley, E. M. and Dawley, R. M., 1986. Species discrimination by chemical cues in a unisexual-bisexual complex of salamanders. *J. Herpetol.* **20**: 114–116.

Dawley, E. M., Crowder, J. and Forlano, P. M., 1994. Sexual and seasonal changes in salamander chemosensory systems. *J. Morphol.* **220**: 338.

Dole, J. W., 1972. The role of olfaction and audition in the orientation of leopard frogs, *Rana pipiens*. *Herpetologica* **28**: 258–260.

Dubois-Dauphin, M., Tribollet, E. and Dreifuss, J. J., 1981. Relations somatotopiques entre la mugueuse olfactive et le bulbe olfactif chez le triton. *Brain Res.* **219**: 269–287.

Ducey, P. K. and Ritsema, P., 1988. Intraspecific aggression and responses to marked substrates in *Ambystoma maculatum* (Caudata, Ambystomatidae). *Copeia* **1988**: 1008–1013.

Duncan, H. J., Nickell, W. T. and Gesteland, R. C., 1990. Organization of projections from olfactory epithelium to olfactory bulb in the frog, *Rana pipiens*. *J. Comp. Neurol.* **299**: 299–311.

Eisthen, H. L., Sengelaub, D. R., Schroeder, D. M. and Alberts, J. R., 1994. Anatomy and forebrain projections of the olfactory and vomeronasal organs in axolotls *(Ambystoma mexicanum)*. *Brain, Behav. Evol.* **44:** 108–124.

Ewert, J-P., 1980. "Neuroethology". Springer-Verlag, Berlin.

Farbman, A. I., 1990. Olfactory neurogenesis: genetic or environmental controls? *Trends Neurosci.* **13:** 362–365.

Farbman, A. I. and Gesteland, R. C., 1974. Fine structure of the olfactory epithelium in the mudpuppy, *Necturus maculosus*. *Am. J. Anat.* **139:** 227–243.

Fishwild, T. G., Schemidt, R. A., Jankens, K. M., Berven, K. A., Gamboa, G. J. and Richards, C. M., 1990. Sibling recognition by larval frogs *(Rana pipiens, R. sylvatica* and *Pseudacris crucifer)*. *J. Herpetol.* **24:** 40–44.

Forester, D. C., 1979. Homing to the nest by female mountain dusky salamanders *(Desmognathus ochrophaeus)* with comments on the sensory modalities essential to clutch recognition. *Herpetologica* **35:** 330–335.

Forester, D. C., 1986. The recognition and use of chemical signals by a nesting salamander. Pp. 205–219 *in* "Chemical Signals in Vertebrates", Volume 6, ed by D. Duvall, D. Müller-Schwartz and R. M. Silverstein. Plenum Press, New York.

Forester, D. C. and Wisnieski, A., 1991. The significance of airborne olfactory cues to the recognition of home area by the dart-poison frog *Dendrobates pumilio*. *Herpetologica* **25:** 502–504.

Gergits, W. F. and Jaeger, R. G., 1990. Field observations of the behavior of the red-backed salamander *(Plethodon cinereus)*: courtship and agonistic interactions. *J. Herpetol.* **24:** 93–95.

Getchell, M. L. and Getchell, T. V., 1984. β-Adrenergic regulation of the secretory granule content of acinar cells in olfactory glands of the salamander. *J. Comp. Physiol. A.* **155:** 435–443.

Getchell, M. L., Rafols, J. A. and Getchell, T. V., 1984. Histological and histochemical studies of the secretory components of the salamander olfactory mucosa: Effects of isoproterenol and olfactory nerve section. *Anat. Rec.* **208:** 553–565.

Getchell, M. L., Zielinski, J. A. and Getchell, T. V., 1987. Odorant stimulation of secretory and neural processes in the salamander olfactory mucosa. *J. Comp. Physio. A.* **160:** 155–168.

Getchell, T. V., 1977. Mechanisms of excitation in vertebrate olfactory receptor and sustentacular cells. Pp. 3–13 *in* "International Symposium on Food Uptake and Chemical Senses", ed by Y. Katsuki, M. Sato, S. F. Takagi and Y. Oomura. University of Tokyo Press, Tokyo.

Getchell, T. V. and Getchell, M. L., 1987. Peripheral mechanisms of olfaction: biochemistry and Neurophysiology. Pp. 91–123 *in* "Neurobiology of Taste and Smell", ed by T. E. Finger. John Wiley and Sons, New York.

Getchell, T. V., Heck, G. L., DeSimone, J. A. and Price, S., 1984. The location of olfactory receptor sites: Inferences from latency measurements. *Biophys. J.* **29:** 397–411.

Grant, D., Anderson, O. and Twitty, V., 1968. Homing orientation by olfaction in newts *(Taricha rivularis)*. *Science* **160:** 1354–1356.

Graves, B. M., Summers, C. H. and Olmstead, K. L., 1993. Sensory mediation of aggregation among postmetamorphic *Bufo cognatus*. *J. Herpetol.* **27:** 315–319.

Graziadei, P. P. C., 1971. The olfactory mucosa of vertebrates. Pp. 29–59 *in* "Handbook of Sensory Physiology", volume 4, ed by L. M. Beidler. Springer-Verlag, Berlin.

Graziadei, P. P. C. and Metcalf, J. F., 1971. Autoradiographic and ultrastructural observations on the frog's olfactory mucosa. *Z. Zellforsch.* **116:** 305–318.

Graziadei, P. P. C. and Monti Graziadei, G. A., 1976. Olfactory epithelium of *Necturus maculosus* and *Ambystoma tigrinum*. *J. Neurocytol.* **5:** 11–32.

Graziadei, P. P. C. and Monti Graziadei, G. A., 1978. The olfactory system: A model for the study of neurogenesis and axon regeneration in mammals. Pp. 131–153 *in* "Neuronal Plasticity", ed by C. W. Cotman. Academic Press, New York.

Grubb, J. C., 1970. Orientation in post-reproductive Mexican toads, *Bufo valliceps*. *Copeia* **1970:** 674–680.

Grubb, J. C., 1973a. Olfactory orientation in breeding Mexican toads, *Bufo valliceps*. *Copeia* **1973:** 490–496.

Grubb, J. C., 1973b. Olfactory orientation in *Bufo woodhousei fowleri, Pseudacris clarki* and *Pseudacris skreckeri*. *Anim. Behav.* **21:** 726–732.

Grubb, J. C., 1975. Olfactory orientation in southern leopard frogs *Rana utricularia*. *Herpetologica* **31:** 219–221.

Grubb, J. C., 1976. Maze orientation by Mexican toads *Bufo valliceps* (Amphibia, Anura, Bufonidae) using olfactory and configurational cues. *J. Herpetol.* **10:** 97–104.

Hamilton, K. A. and Kauer, J. S., 1985. Intracellular potentials of salamander mitral/tufted neurons in response to odour stimulation. *Brain Res.* **338:** 181–185.

Hamilton, K. A. and Kauer, J. S., 1988. Responses of mitral/tufted cells to orthodromic and antidromic electrical stimulation in the olfactory bulb of the tiger salamander. *J. Neurophysiol.* **59:** 1736–1755.

Hamilton, K. A. and Kauer, J. S., 1989. Patterns of intracellular potentials in salamander mitral/tufted cells in response to odor stimulation. *J. Neurophysiol.* **62:** 609–625.

Hepper, P. G. and Waldman, B., 1992. Embryonic olfactory learning in frogs. *Quart. J. Exper. Psychol.* **44B:** 179–197.

Herrick, C. J., 1909. The nervus terminalis (Nerve of Pinkus) in the frog. *J. Comp. Neurol.* **19:** 175–190.

Herrick. C. J., 1910. The morphology of the forebrain in amphibia and reptilia. *J. Comp. Neurol.* **20:** 413–547.

Herrick, C. J., 1924. The amphibian forebrain. I. *Amblystoma*, external form. *J. Comp. Neurol.* **37:** 361–398.

Hershey, J. L. and Forester, D. C., 1980. Sensory orientation in *Notophthalmus v. viridescens* (Amphibia: Salamandridae). *Can. J. Zool.* **58:** 266–276.

Hews, D. K. and Blaustein, A. R., 1985. An investigation of the alarm response in *Bufo boreas* and *Rana cascadae* tadpoles. *Behav. Neural. Biol.* **43**: 47–57.

Hinds, J. W., Hinds, P. L. and McNelly, N. A., 1984. An autoradiographic study of mouse olfactory epithelium: Evidence for long-lived receptors. *Anat. Rec.* **210**: 375–383.

Hoffman, H. H., 1963. The olfactory bulb, accessory olfactory bulb, and hemisphere of some anurans. *J. Comp. Neurol.* **120**: 317–368.

Hofmann, M. H. and Meyer, D. L., 1991. Functional subdivisions of the olfactory system correlate with lectin-binding properties in *Xenopus*. *Brain Res.* **564**: 344–347.

Horne, E. A. and Jaeger, R. G., 1988. Territorial pheromones of female red-backed salamanders. *Ethology* **78**: 143–152.

Houck, L. D., 1986. The evolution of salamander courtship pheromones. Pp. 173–190 *in* "Chemical Signals in Vertebrates", Volume 4, ed by D. Duvall, D. Müller-Schwartze and R. M. Silverstein. Plenum Press, New York.

Houck, L. D. and Reagan, N. L., 1990. Male courtship pheromones increase female receptivity in a plethodontid salamander. *Anim. Behav.* **39**: 729–734.

Houck, L. D. and Sever, D. M., 1994. Role of the skin in reproduction and behaviour. Chapter 10, Pp. 351–381 *in* "Amphibian Behaviour", Volume 1 "The Integument", ed by H. Heatwole, G. T. Barthalmus and A. Y. Heatwole. Surrey Beatty & Sons, Chipping Norton.

Jacobs, A. J. and Taylor, D. H., 1992. Chemical communication between *Desmognathus quadramaculatus* and *Desmognathus monticola*. *J. Herpetol.* **26**: 91–95.

Jaeger, R. G., 1981. Dear enemy recognition and the costs of aggression between salamanders. *Amer. Natur.* **117**: 962.

Jaeger, R. G., 1986. Pheromonal markers as territorial advertisement by terrestrial salamanders. Pp. 191–203 *in* "Chemical Signals in Vertebrates", Volume 4, ed by D. Duvall, D. Müller-Schwartze and D. Silverstein. Plenum Press, New York.

Jaeger, R. G. and Gabor, C. R., 1993. Intraspecific chemical communication by a territorial salamander via the postcloacal gland. *Copeia* **1993**: 1171–1174.

Jaeger, R. G. and Gergits, W. F., 1979. Intra- and interspecific communication in salamanders through chemical signals on the substrate. *Anim. Behav.* **27**: 50–156.

Jaeger, R. G. and Wise, S. E., 1991. A reexamination of the male salamander "sexy faeces hypothesis". *J. Herpetol.* **25**: 370–373.

Jaeger, R. G., Goy, J. M., Tarver, M. and Marquez, C. E., 1986. Salamander territoriality: Pheromonal markers as advertisement by males. *Anim. Behav.* **34**: 860–864.

Jaeger, R. G., Fortune, D., Hill, G., Palen, A. and Risher, G., 1993. Salamander homing behavior and territorial pheromones: alternative hypotheses. *J. Herpetol.* **27**: 236–239.

Jiang, T. and Holley, A., 1992. Morphological variations among output neurons of the olfactory bulb in the frog (*Rana ridibunda*). *J. Comp. Neurol.* **320**: 86–96.

Johnson, R. E., 1985. Olfactory and vomeronasal mechanisms of communication. Pp. 322–346 *in* "Taste, Olfaction, and the Central Nervous System: A Festschrift in Honor of Carl Pfaffmann", ed by D. W. Pfaff. Rockefeller University Press, New York.

Jurgens, J. D., 1971. The morphology of the nasal region of Amphibia and its bearing on the phylogeny of the group. *Ann. Univ. Stell.* **46A2**: 1–146.

Kats, L. B., 1988. The detection of certain predators via olfaction by small-mouthed salamander larvae (*Ambystoma texanum*). *Behav. Neur. Biol.* **50**: 126–131.

Kats, L. B., Petranka, J. W. and Sih, A., 1988. Antipredator defenses and the persistence of amphibian larvae with fishes. *Ecology* **69**: 1865–1870.

Kauer, J. S., 1981. Olfactory receptor cell staining using horseradish peroxidase. *Anat. Rec.* **200**: 331–336.

Kauer, J. S., 1991. Contributions of topography and parallel processing to odor coding in the vertebrate olfactory pathway. *Trends Neurosci.* **14**: 79–85.

Kauer, J. S. and Moulton, D. G., 1974. Responses of olfactory bulb neurones to odour stimulation of small nasal areas in the salamander. *J. Physiol.* **243**: 717–737.

Kelley, D. B., 1982. Female sex behaviours in the South African clawed frog, *Xenopus laevis*: Gonadotropin-releasing, gonadotropic, and steroid hormones. *Horm. Behav.* **16**: 158–174.

Kelley, D. B., 1988. Sexually dimorphic behaviors. *Ann. Rev. Neurosci.* **11**: 225–251.

Kemali, M. and Guglielmotti, V., 1987. A horseradish peroxidase study of the olfactory system of the frog, *Rana esculenta*. *J. Comp. Neurol.* **263**: 400–417.

Key, B. and Akeson, R. A., 1990a. Immunochemical markers for the frog olfactory neuroepithelium. *Dev. Brain Res.* **57**: 103–117.

Key, B. and Akeson, R. A., 1990b. Olfactory neurons express a unique glycosylated form of the neural cell adhesion molecule (N-CAM). *J. Cell Biol.* **110**: 1729–1743.

Key, B. and Akeson, R. A., 1991. Delineation of olfactory pathways in the frog nervous system by unique glycoconjugates and N-CAM glycoforms. *Neuron* **6**: 381–396.

Key, B. and Giorgi, P. P., 1986. Selective binding of soybean agglutinin to the olfactory system of *Xenopus*. *Neurosci.* **18**: 507–515.

Kleene, S. J. and Gesteland, R. C., 1991. Transmembrane currents in frog olfactory cilia. *J. Membrane Biol.* **120**: 75–81.

Klein, S. L. and Graziadei, P. P. C., 1983. The differentiation of the olfactory placode in *Xenopus laevis*: A light and electron microscope study. *J. Comp. Neurol.* **217**: 17–30.

Kosaka, T. and Hama, K., 1982. Structure of the mitral cell in the olfactory bulb of the goldfish (*Carassius auratus*). *J. Comp. Neurol.* **212**: 365–384.

Kubie, J., Mackay-Sim, A. and Moulton, D. G., 1980. Inherent spatial patterning of responses to odourants in the salamander olfactory epithelium. Pp. 163–166 *in* "Olfaction and Taste", Volume 7, ed by H. Van der Starre. Information Retrieval Limited, London.

Kubo, S., Watanabe, K., Ibata, Y. and Sano, Y., 1979. LH-RH neuron system of the newt by immunohistochemical study. *Arch. Histol. Japan* **42**: 235–242.

Lancet, D., 1988. Molecular components of olfactory reception and transduction. Pp. 25–50 *in* "Molecular Neurobiology of the Olfactory System", ed by F. L. Margolis and T.V. Getchell. Plenum Press, New York.

Lancet, D., Lazard, D., Heldman, J., Khen, M. and Nef, P., 1988. Molecular transduction in smell and taste. *Annual Report, Cold Spring Harbor Symposia on Quantitative Biology* **53**: 343–348.

Lidow, M. S., Gesteland, R. C., Shipley, M. T. and Kleene, S. J., 1987. Comparative study to immature and mature olfactory receptor cells in adult frogs. *Dev. Brain Res.* **31**: 243–258.

Lindquist, S. B. and Bachmann, M. D., 1982. The role of visual and olfactory cues in the prey catching behavior of the tiger salamander, *Ambystoma tigrinum*. *Copeia* **1982**: 81–90.

Luthardt, G. and Roth, G., 1983. The interaction of the visual and the olfactory systems in guiding prey catching behaviour in *Salamandra salamandra*. *Behaviour* **83**: 69–79.

Mackay-Sim, A. and Nathan, M. H., 1984. The projection from the olfactory epithelium to the olfactory bulb in the salamander, *Ambystoma tigrinum*. *Anat. Embry.* **170**: 93–97.

Mackay-Sim, A. and Patel, U., 1984. Regional differences in cell density and cell genesis in the olfactory epithelium of the salamander, *Ambystoma tigrinum*. *Exp. Brain Res.* **57**: 99–106.

Mackay-Sim, A. and Shaman, P., 1984. Topographic coding of odorant quality is maintained at different concentrations in the salamander olfactory epithelium. *Brain Res.* **297**: 207–217.

Mackay-Sim, A., Shaman, P. and Moulton, D. G., 1982. Topographic coding of olfactory quality: odorant-specific patterns of epithelial responsivity in the salamander. *J. Neurophysiol.* **48**: 584–596.

Madison, D. M., 1972. Homing orientation in salamanders: A mechanism involving chemical cues. Pp. 485–498 *in* "Animal Orientation and Navigation", ed by S. R. Galler, K. Schmidt-Koenig, G. J. Jacobs and R. E. Belleville. Special Publication 262, NASA, Washington, D.C.

Madison, D. M., 1975. Intraspecific odor preferences between salamanders of the same sex: Dependence on season and proximity of residence. *Can. J. Zool.* **53**: 1356–1361.

Madison, D. M., 1977. Chemical communication in amphibians and reptiles. Pp. 135–168 *in* "Chemical Signals in Vertebrates", ed by D. Müller-Schwartze and M. M. Mozell. Plenum Press, New York.

Malacarne, G., Bottoni, L., Massa, R. and Vellano, C., 1984. The abdominal gland of the crested newt: a possible source of courtship pheromones. Preliminary ethological and biochemical data. *Monit. Zool. Ital. n.s.* **18**: 33–39.

Malacarne, G. and Vellano, C., 1987. Behavioral evidence of a courtship pheromone in the crested newt, *Triturus cristatus carnifex* Laurenti. *Copeia* **1987**: 245–247.

Manteifel, Y., 1995. Chemically-mediated avoidance of predators by *Rana temporaria* tadpoles. *J. Herpetol.* **29**: 461–463.

Martin, J. B., Witherspoon, N. B., Keenleyside, M. H. A., 1974. Analysis of feeding behaviour in the newt, *Notophthalmus viridescens*. *Canad. J. Zool.* **52**: 277–281.

Martof, B. S., 1962. Some observations of the role of olfaction among Salientian Amphibia. *Physiol. Zool.* **35**: 270–272.

Mason, J. R. and Morton, T. H., 1982. Temporary and selective anosmia in tiger salamanders (*Ambystoma tigrinum*) caused by chemical treatment of the olfactory epithelium. *Physiol. and Behav.* **29**: 709–714.

Masukawa, L. M., Hedlund, B. and Shepherd, G. M., 1985. Changes in the electrical properties of olfactory epithelial cells in the tiger salamander after olfactory nerve transection. *J. Neurosci.* **5**: 136–141.

Mathis, A., 1990. Territorial salamanders assess sexual and competitive information using chemical cues. *Anim. Behav.* **40**: 953–962.

Maturana, H. R., 1960. The fine anatomy of the optic nerve of anurans: and electron microscope study. *J. Biophys. Biochem. Cytol.* **7**: 107–135.

McGavin, M., 1978. Recognition of conspecific odors by the salamander *Plethodon cinereus*. *Copeia* **1978**: 356–358.

McGregor, J. H. and Teska, W. R., 1989. Olfaction as an orientation mechanism in migrating *Ambystoma maculatum*. *Copeia* **1989**: 779–781.

McKibben, P. S., 1911. The nervus terminalis in Urodele Amphibians. *J. Comp. Neurol.* **21**: 261–309.

Moore, F. L., Miller, L. J., Spielvoge, S. P., Kubiak, T. and Folkers, K., 1982. Luteinizing hormone-releasing hormone involvement in the reproductive behavior of a male amphibian. *Neuroendocrinology* **35**: 212–216.

Muske, L. E. and Moore, F. L., 1988. The nervus terminalis in amphibians: Anatomy, chemistry and relationship with the hypothalamic gonadotropin-releasing hormone system. *Brain Behav. Evol.* **32**: 141–150.

Nishio, H., Kubokawa, K., Yamanovchi, H., Itoh, M., Kibuchi, M., Funahashi, H., Takizawa, M. and Ishii, S., 1988. The role of olfactory memory for orientation to the breeding pond in *Bufo japonicus*. *Zool. Sci.* **5**: 1330.

Northcutt, R. G. and Kicliter, E., 1980. Organization of the amphibian telencephalon. Pp. 203–255 *in* "Comparative Neurology of the Telencephalon", ed by S. O. E. Ebbesson. Plenum Press, New York.

Northcutt, R. G. and Royce, G. J., 1975. Olfactory bulb projections in the bullfrog *Rana catesbiana*. *J. Morph.* **145**: 251–268.

O'Hara, R. K. and Blaustein, A. R., 1981. An investigation of sibling recognition in *Rana cascadae* tadpoles. *Anim. Behav.* **29**: 1121–1126.

O'Hara, R. K. and Blaustein, A. R., 1982. Kin preference behaviour in *Bufo boreas* tadpoles. *Behav. Ecol. Sociobiol.* **11**: 43–39.

O'Hara, R. K. and Blaustein, A. R., 1988. *Hyla regilla* and *Rana pretiosa* tadpoles fail to display kin recognition behaviour. *Anim. Behav.* **36**: 946–948.

Ovaska, K., 1988. Recognition of conspecific odors by the western red-backed salamander, *Plethodon vehiculum*. *Can. J. Zool.* **66**: 1293–1296.

Ovaska, K., 1989. Pheromonal divergence between populations of the salamander *Plethodon vehiculum* in British Columbia. *Copeia* **1989**: 770–775.

Ovaska, K. and Davis, T. M., 1992. Faecal pellets as burrow markers: intra- and interspecific odour recognition by western plethodontid salamanders. *Anim. Behav.* **43**: 931–939.

Parsons, T. S., 1967. Evolution of the nasal structure in the lower tetrapods. *Amer. Zool.* **7**: 397–413.

Parzefall, J., 1976. Die Rolle der chemischen Information im Verhalten des Grottenolms *Proteus anguinus* Laur. (Proteidae, Urodela). *Z. Tierpsychol.* **42**: 29–49.

Parzefall, J., Durand, J. P. and Rischar, B., 1980. Chemical communication in *Necturus maculatus* and his cave-dwelling relative *Proteus anguinus* (Proteidae, Urodela). *Z. Tierpsychol.* **55**: 133–138.

Petranka, J. W., Kats, L. and Sih, A., 1987. Predator-prey interactions among fish and larval amphibians: Use of chemical cues to detect predatory fish. *Anim. Behav.* **35**: 420–425.

Pfennig, D. W., 1990. "Kin recognition" among spadefoot toad tadpoles: A side-effect of habitat selection? *Evolution* **44**: 785–798.

Pinching, A. J. and Powell, T. P. S., 1971a. The neuron types of the glomerular layer of the olfactory bulb. *J. Cell Sci.* **9**: 305–345.

Pinching, A. J. and Powell, T. P. S., 1971b. The neuropil of the periglomerular region of the olfactory bulb. *J. Cell Sci.* **9**: 379–409.

Pope, P. H., 1924. The life-history of the common water-newt (*Notophthalmus viridescens*), together with observations on the sense of smell. *Ann. Carnegie Mus.* **15**: 305–368.

Propper, C. R. and Moore, F. L., 1991. Effects of courtship on brain gonadotropin hormone-releasing hormone and plasma steroid concentrations in a female amphibian (*Taricha granulosa*). *Gen. Comp. Endocrinol.* **81**: 304–312.

Rafols, J. A. and Getchell, T. V., 1983. Morphological relations between the receptor neurons, sustentacular cells and Schwann cells in the olfactory mucosa of the salamanders. *Anat. Rec.* **206**: 87–101.

Reese, A. M., 1912. Food and chemical reactions of the spotted newt, *Diemctylus viridescens*. *J. Anim. Behav.* **2**: 190–208.

Risser, J., 1914. Olfactory reactions in amphibians. *J. Exp. Zool.* **16**: 617–652.

Rossi, J. V., 1983. The use of olfactory cues by *Bufo marinus*. *J. Herpetol.* **17**: 72–73.

Roth, G., 1976. Experimental analysis of the prey catching behavior of *Hydromantes italicus* Dunn (Amphibia, Plethodontidae). *J. Comp. Physiol.* **109**: 47–58.

Roudabush, R. E. and Taylor, D. H., 1987. Chemical communication between two species of *Desmognathine* salamanders. *Copeia* **1987**: 744–748.

Rowland, W. J., Robb, C. C. and Cortwright, S. A., 1990. Chemically-mediated mate choice in red-spotted newts: Do males select or just detect females? *Anim. Behav.* **37**: 811–813.

Scalia, F., 1976. Structure of the olfactory and accessory olfactory systems. Pp. 213–233 in "Frog Neurobiology: A Handbook", ed by R. Llinas and W. Precht. Springer-Verlag, Berlin.

Scalia, F., Gallousis, G. and Roca, S., 1991a. A note on the organization of the amphibian olfactory bulb. *J. Comp. Neurol.* **305**: 435–442.

Scalia, F., Gallousis, G. and Roca, S., 1991b. Differential projections of the main and accessory olfactory bulb in the frog. *J. Comp. Neurol.* **305**: 443–461.

Schmidt, A., Naujoks-Manteuffel, C. and Roth, G., 1988. Olfactory and vomeronasal projections and the pathway of the nervus terminalis in ten species of salamanders. *Cell Tiss. Res.* **251**: 45–50.

Schmidt, A. and Roth, G., 1990. Central olfactory and vomeronasal pathways in salamanders. *J. Hirnforsch.* **31**: 543–553.

Schmidt, A. and Wake, M. H., 1990. Olfactory and vomeronasal systems of caecilians (Amphibia: Gymnophiona). *J. Morphol.* **205**: 255–268.

Schuch, K., 1934. Das Geruchsorgan von *Triton alpestris*. Eine morphologische, histologische und entwicklungsgeschichtliche Untersuchung. *Zool. Jahrb. Abt. Anat.* **59**: 69–134.

Schwanzel-Fukuda, M. and Silverman, A. J., 1980. The nervus terminalis of the guinea pig: a new luteinizing hormone-releasing hormone (LHRH) neuronal system. *J. Comp. Neurol.* **191**: 213–225.

Semlitsch, R. D. and Gavasso, S., 1992. Behavioural responses of *Bufo bufo* and *Bufo calamita* tadpoles to chemical cues of vertebrate and invertebrate predators. *Ethol., Ecol., Evol.* **4**: 165–173.

Sever, D., 1978. Female cloacal anatomy of *Plethodon cinereus* and *Plethodon dorsalis* (Amphibia, Urodela, Plethodontidae). *J. Herpetol.* **12**: 397–406.

Sever, D. M., 1989. Caudal hedonic glands in salamanders of the *Eurycea bislineata* complex (Amphibia: Plethodontidae). *Herpetologica* **45**: 322–329.

Seydel, O., 1895. Über die Nasenhöhle und das Jacobson'sche Organ der Amphibien. *Morphol. Jahrb.* **23**: 453–543.

Shinn, E. A. and Dole, J. W., 1978. Evidence for a role for olfactory cues in the feeding response of leopard frogs, *Rana pipiens*. *Herpetologica* **34**: 167–172.

Sih, A. and Kats, L. B., 1991. Effects of refuge availability on the responses of salamander larvae to chemical cues from predatory green fishes. *Anim. Behav.* **42**: 330–332.

Simmons, P. A. and Getchell, T. V., 1981a. Physiological activity of newly differentiated olfactory receptor neurons correlated with morphological recovery from olfactory nerve section in salamander. *J. Neurophysiol.* **45**: 529–549.

Simmons, P. A. and Getchell, T. V., 1981b. Neurogenesis in olfactory epithelium: Loss and recovery of transepithelial voltage transients following olfactory nerve section. *J. Neurophysiol.* **45**: 516–528.

Simon, G. S. and Madison, D. M., 1984. Individual recognition in salamanders: Cloacal odours. *Anim. Behav.* **32**: 1017–1020.

Simons, R. R. and Felgenhauer, B. E., 1992. Identifying areas of chemical signal production in the red-backed salamander, *Plethodon cinereus*. *Copeia* **1992**: 776–781.

Simons, R. R., Jaeger, R. G. and Felgenhauer, B. E., 1995. Juvenile terrestrial salamanders have active postcloacal glands. *Copeia* **1995**: 481–483.

Sinsch, U., 1987. Orientation behaviour of toads *(Bufo bufo)* displaced from breeding site. *J. Comp. Physiol. (A)* **161**: 715–727.

Sinsch, U., 1990a. Migration and orientation in anuran amphibians. *Ethol. Ecol. Evol.* **2**: 65–79.

Sinsch, U., 1990b. The orientation behavior of three toad species (genus *Bufo*) displaced from the breeding site. *In* "Biology and Physiology of Amphibians", ed by W. Hanke. Gustav Fischer Verlag, Stuttgart.

Snyder, S. H., Sklar, P. B. and Pevsner, J., 1988. Olfactory receptor mechanisms: Odorant binding protein and adenylate cyclase. Pp. 3–24 *in* "Molecular Neurobiology of the Olfactory System", ed by F. L. Margolis and T. V. Getchell. Plenum Press, New York.

Spaeti, U., 1978. Development of the sensory systems in the larval and metamorphosing European grass frog *(Rana temporaria* L.). *J. Hirnforsch.* **19**: 543–575.

Stauffer, H.-P. and Semlitsch, R. D., 1993. Effects of visual, chemical, and tactile cues of fish on the behavioural responses of tadpoles. *Anim. Behav.* **46**: 355–364.

Sternthal, D. E., 1974. Olfactory and visual cues in the feeding behavior of the leopard frog *(Rana pipiens)*. *Z. Tierpsychol.* **34**: 239–246.

Stout, R. P. and Graziadei, P. P. C., 1980. Influence of the olfactory placode on the development of the brain in *Xenopus laevis* (Daudin). I. Axonal growth and connections of the transplanted olfactory placode. *Neurosci.* **5**: 2175–2186.

Tracy, C. R. and Dole, J. W., 1969. Orientation of displaced California toads, *Bufo boreas*, to their breeding sites. *Copeia* **1969**: 693–700.

Tristram, D. A., 1977. Intraspecific olfactory communication in the terrestrial salamander, *Plethodon cinereus*. *Copeia* **1977**: 597–600.

Trotier, D. and MacLeod, P., 1983. Intracellular recordings from salamander olfactory receptor cells. *Brain Res.* **268**: 225–237.

Twitty, V. C., 1955. Field experiments on the biology and genetic relationships of the Californian species of *Triturus*. *J. Exp. Zool.* **129**: 129–148.

Twitty, V. C., 1961. Experiments on homing behaviour and speciation in *Taricha*. *In* "Vertebrate Speciation", ed by W. F. Blair. University of Texas Press, Austin.

Uzendoski, K. and Verrell, P., 1993. Sexual incompatibility and mate-recognition systems: a study of two species of sympatric salamanders (Plethodontidae). *Anim. Behav.* **46**: 267–278.

Verrell, P., 1982. The sexual behavior of the red-spotted newt, *Notophthalmus viridescens* (Amphibia: Urodela: Salamandridae). *Anim. Behav.* **30**: 1224–1236.

Verrell, P., 1985. Male mate choice for large, fecund females in the red-spotted newts, *Notophthalmus viridescens*: How is sized assessed? *Herpetologica* **41**: 382–386.

Verrell, P., 1986. Male discrimination of larger, more fecund females in the smooth newt, *Triturus vulgaris*. *J. Herp.* **20**: 416–422.

Verrell, P., 1989. An experimental study of the behavoral basis of sexual isolation betyween two sympatric plethodontid salamanders, *Desmognathus imitator* and *D. ochrophaeus*. *Ethology* **80**: 274–282.

Waldman, B., 1981. Sibling recognition in toad tadpoles: The role of experience. *Z. Tierpsychol.* **56**: 341–358.

Waldman, B., 1982. Sibling association among schooling toad tadpoles: Field evidence and implication. *Anim. Behav.* **30**: 700–713.

Waldman, B., 1984. Kin recognition and sibling association among wood frog *(Rana sylvatica)* tadpoles. *Behav. Ecol. Sociobiol.* **14**: 171–180.

Waldman, B., 1985. Olfactory basis of kin recognition in toad tadpoles. *J. Comp. Physiol. A* **156**: 565–577.

Waldman, B., 1986. Chemical ecology of kin recognition in anuran amphibians. Pp. 225–242 *in* "Chemical Signals in Vertebrates", Volume 4, ed by D. Duval, D. Müller-Schwartze and R. M. Silverstein. Plenum Press, New York.

Waldman, B. and Adler, K., 1979. Toad tadpoles associate preferentially with siblings. *Nature* **282**: 611–613.

Walls, S. C., Mathis, A., Jaeger, R. G. and Gergits, W. F., 1989. Male salamanders with high-quality diets have faeces attractive to females. *Anim. Behav.* **38**: 546–547.

Wirsig, C. R. and Getchell, T. V., 1986. Amphibian terminal nerve: Distribution revealed by LHRH and AChE markers. *Brain Res.* **385**: 10–21.

Wirsig-Wiechmann, C. R., 1993. Peripheral projections of nervus terminalis LHRH-containing neurons in the tiger salamander, *Ambystoma tigrinum*. *Cell Tiss. Res.* **273**: 31–40.

Zielinski, B. S., Getchell, M. L. and Getchell, T. V., 1988. Ultrastructural characteristics of sustentacular cells in control and odorant-treated olfactory mucosae of the salamander. *Anat. Rec.* **221**: 769–779.

Zoeller, R. T. and Moore, F. L., 1985. Seasonal changes in luteinizing hormone-releasing hormone concentrations in microdissected brain regions of male roughskinned newts *(Taricha granulosa)*. *Gen. Comp. Endocrinol.* **58**: 222–230.

CHAPTER 2

The Biology of Amphibian Taste

Linda A. Barlow

The valiant never taste of death but once.
WILLIAM SHAKESPEARE, JULIUS CAESAR

I. Introduction
II. Feeding in Amphibians
   A. Diet
   B. Cannibalism
   C. Toxicity
III. Organization of the Taste System: Anatomy and Structure
   A. Urodeles
      1. Distribution of Taste Buds
      2. Morphology and Cytology of Taste Buds
      3. Innervation of Taste Buds
      4. Ganglionic Organization and Central Projections
      5. Modifications of the Taste Periphery in Terrestrial Salamanders
   B. Caecilians
   C. Anurans
      1. Distribution of Taste Organs
      2. Morphology and Cytology of Taste Disks
      3. Innervation of Taste Disks
      4. Ganglionic Organization and Central Projections
      5. Interspecific Variation in Organization of Taste Disks
      6. Taste Receptors in Larval Anurans
IV. Regeneration of the Taste System
V. Embryonic Development of the Taste System
VI. Function of the Taste System
   A. Function of Taste Receptor Cells
      1. Voltage-dependent Properties of Taste Receptor Cells
      2. Transduction of Taste Stimuli
   B. Taste Buds as Peripheral Integrators.
   C. Physiology of the Glossopharyngeal and Facial Nerves
   D. Coding of Taste Information
VII. Summary
VIII. Acknowledgements
IX. References

I. INTRODUCTION

MOST toxic substances produced in nature are known to taste bitter, and, once tasted, are either lethal, or are subsequently avoided by would-be consumers. Thus, a functional taste system allows the detection of these bitter substances and is ultimately essential for the survival both of vertebrate predators and of the creatures that produce these poisons.

The taste system of amphibians, like that of most vertebrates, consists of an array of chemoreceptive organs distributed throughout the mouth and pharynx. The receptor cells within these organs, taste buds, are sensitive to a wide array of chemical substances. They transduce chemical signals into the electrical currency of the nervous system, and synapse chemically upon the neuritic processes of sensory neurons of cranial nerve ganglia. The sensory neurons in turn transmit electrically coded taste information to the central nervous system via projections to specific regions of the hindbrain. This basic anatomical and functional organization is conserved in all vertebrates. However, subtle modifications have occurred throughout evolution in the structure of taste organs and in their distribution.

II. FEEDING IN AMPHIBIANS

A. Diet

The diet of most amphibians is quite generalized, albeit carnivorous (Eycleshymer 1906; Smith and Bragg 1949; Anderson 1968; Sokol 1969; Attar and Maly 1980; Duellman and Trueb 1986). Both frogs and salamanders are primarily visual, opportunistic feeders, whose selection of prey is limited by the gape of the predator (Hamilton 1940; Anderson 1968; Sokol 1969). Thus, diet tends to vary seasonally with prey availability (Donnelly 1991; Parker 1994), as well as ontogenetically (Smith and Bragg 1949; Leff and Bachmann 1986; McWilliams and Bachmann 1989). As amphibians mature and increase in size, they tend to eat fewer but larger prey (Christian 1982; Viertel 1992; Lima and Moreira 1993). The metamorphic transition also imposes many changes in diet (Jenssen 1967; Christian 1982; Parker 1994). Larval anurans are almost exclusively microphagous herbivores (Noble 1954; Duellman and Trueb 1986; Wassersug and Heyer 1988), with some startling exceptions such as the carnivorous larvae of *Lepidobatrachus laevis* (Ruibal and Thomas 1988), *Leptodactylus pentadactylus* (Heyer *et al.* 1975), and *Scaphiopus bombifrons* (Bragg 1960, 1964). After metamorphosis, most anurans become carnivorous (Duellman and Trueb 1986). Larval urodeles, in contrast, initially are carnivorous, and those species that change habitats at metamorphosis, merely switch from aquatic to terrestrial prey (McWilliams and Bachmann 1989; Parker 1994). Although there have been few experimental studies of natural feeding preferences in amphibians (see Leff and Bachmann 1986), laboratory studies have shown that salamanders will initially ingest, and then reject normally palatable food items that have been chemically altered to taste bitter (Bowerman and Kinnamon 1994; Takeuchi *et al.* 1994). Thus, only after visually selected prey have reached the mouth, does the taste system in amphibians function as a toxin detector such that unpalatable and potentially poisonous food is spit out.

Amphibians respond differently to bitter and salty substances. Both mudpuppies *(Necturus maculosus)* and axolotls *(Ambystoma mexicanum)* readily ingest food pellets presented to them in the laboratory. However, if a known bitter compound, such as quinine, is added to the pellet, the food is rejected after it has been mouthed briefly (Bowerman and Kinnamon 1994; Takeuchi *et al.* 1994). Thus, bitter tasting substances are detected readily by salamanders and appear to act as feeding deterrents. In contrast, axolotls may be attracted to salt; when bitter-treated food was further treated with salt, the rejection rate of food pellets declined (Takeuchi *et al.* 1994).

Interestingly, noxious substances do not always deter amphibian predators. Blister beetles produce a substantial quantity of cantharidin, an exceedingly toxic substance. In large part, cantharidin protects these beetles from predation, with the demonstrated exception of predation by adult *Rana catesbeiana* (Kelling *et al.* 1990). Although recordings of the nerves that innervate taste buds demonstrate that the frogs can detect cantharidin, the animals fail to respond to it behaviourally, i.e., the frogs still eat the beetles. Thus, the peripheral components of the taste system are similar to those of other frogs that are deterred by cantharidin in that the substance stimulates the taste periphery appropriately, but the central processing of the bitter signal has been altered such that behavioural modification of feeding does not occur.

B. Cannibalism

Cannibalism is rather common among both anurans and urodeles (Table 1) (for reviews see Polis and Myers 1985; Duellman and Trueb 1986). Many larval salamanders are cannibalistic (Nyman *et al.* 1993). Field studies have revealed that there are distinct cannibalistic morphs (larvae with broader heads and a larger oral gape) that feed primarily on conspecifics, potentially indicating an ability of cannibals to select conspecifics over other food items. Alternatively,

Table 1. Tabular summary of cannibalistic amphibian species compared with known toxicity and/or palatability.

Species	Cannibalism	Toxicity/Palatability
Anurans		
Bufo calimata	Adults eat larvae, density dependent (Heusser 1971)	Toxins in adult skin (Daly and Witkop 1971)
Bufo boreas halophilus	48 mm eat 17 mm (Cunningham 1954)	Toxins in adult skin (Daly and Witkop 1971)
Bombina variegata	Adults eat juveniles (Heusser 1971)	Toxins in adult skin (Daly and Witkop 1971; Lutz 1971)
Scaphiopus bombifrons	Larval cannibal morphs eat larvae (Bragg 1946, 1964, 1965; Pfennig et al. 1993)	Toxins in adult skin (Daly and Witkop 1971; Lutz 1971)
Scaphiopus holbrooki	Larval cannibal morphs eat larvae (Bragg 1946, 1964, 1965)	
Scaphiopus hammondi hammondi	Larval cannibal morphs eat larvae (Bragg 1965)	
Rana pipiens	Juveniles eat juveniles (Kirn 1949)	Serotonin and toxic peptides in adult skin (Daly and Witkop 1971)
Rana palustris		Adults are lethal if ingested (Lutz 1971); serotonin in adult skin (Daly and Witkop 1971)
Rana cyanophlictus	Larvae eat larvae (McCann 1939)	
Rana tigrina	Larvae eat larvae (McCann 1939)	
Rana ridibunda	Adults eat larvae and juveniles (Dushin 1975)	
Rana esculenta	Adults eat juveniles in lab and field (Kalusche 1973; Dushin 1975)	Serotonin and toxic peptides in adult skin (Daly and Witkop 1971)
Rana temporaria	Juveniles and adults eat eggs and juveniles (Heusser 1970; Loman 1979)	
Rana arvalis	Adults eat juveniles (Loman 1979)	
Rana catesbeiana	Adults eat juveniles and eggs (Cohen and Howard 1958; Rose and Rose 1965; Stewart and Sandison 1972)	Toxic peptides in adult skin (Daly and Witkop 1971)
Pyxicephalus adspersus	Adults, juveniles and larvae eat juveniles and larvae (Wager 1965; Grobler 1972)	
Hyla zeteki	Larvae eat eggs (Dunn 1926)	
Hyla brunnea	Larvae eat eggs (Laessle 1961)	
Hyla pseudopuma	Larvae eat eggs and small juveniles (Crump 1983)	
Chacophrys pierottii	Adults and larvae are voracious cannibals of all stages (Cei 1955; Blair 1976)	
Ceratrophrys dorsata	Larvae, juveniles, and adults are cannibalistic (Duellman and Trueb 1986)	
Leptodactylus pentadactylus	Facultative cannibalistic tadpoles (Heyer et al. 1975)	Adult skin is toxic (Daly and Witkop 1971)
Rhinophrynus dorsalis	Larvae eat larvae (Starrett 1960)	
Hymenochirus boettgeri	Larvae eat smaller larvae (Sokol 1962)	
Hoplophryne rogersi	Larvae eat eggs (Dunn 1926)	
Phyllobates terribilis		All dendrobatids (*Phyllobates* and *Dendrobates*) have steroidal alkaloid toxins in adult skin (Daly and Witkop 1971; Myers and Daly 1976, 1983; Daly et al. 1978, 1980, 1990; Myers et al. 1978; Sheridan et al. 1991)
Phyllobates lugubris		
Phyllobates vittatus		
Phyllobates aurotaenia		
Phyllobates bicolor		
Dendrobates histrionicus		
Dendrobates lehmanni		
Dendrobates occultator		
Dendrobates viridis		
Dendrobates pumilio	Larvae eat unfertilized eggs provisioned by mother (Bragg 1965; Brust 1993)	
Dendrobates auratus		Either lethal or rejected by non-indigenous snake predator (Brodie and Tumbarello 1978)
Urodeles		
Ambystoma tigrinum	Adults eat juveniles and eggs (Hamilton 1948; Collins et al. 1980; Collins 1981; Collins and Cheek 1983); larval cannibal morphs eat larvae (Pfennig et al. 1994)	Noxious skin secretions (Duellman and Trueb 1986); eggs eaten by *Notophthalmus viridescens* (Morin 1983); adult skin secretions avoided by rats (Mason et al. 1982)
Ambystoma mexicanum		Eaten by humans (Smith 1969; Luther 1971)
Ambystoma maculatum		Rejected often by shrews (Brodie et al. 1979)
Ambystoma macrodactylum croceum		Rejected by moles (Anderson 1963)
Ambystoma annulatum	Larval cannibalistic morphs eat larvae (Nyman et al. 1993)	

Table 1 — continued.

Species	Cannibalism	Toxicity/Palatability
Urodels — continued		
Cryptobranchus allegansiensis	Adult oophagy (Kaplan and Sherman 1980)	
Dicamptodon ensatus	Adults eat smaller individuals (Anderson 1960)	Noxious and toxic skin secretions in adults of this genus (Habermehl 1971)
Dicamptodon copei	Larvae eat eggs (Kaplan and Sherman 1980)	
Salamandra salamandra	Larval cannibalism of larvae (Malkmus 1975)	Noxious and toxic skin secretions in this genus (Habermehl 1971)
Notophthalmus viridescens	Adults eat eggs and larvae (Morin 1983)	TTX in skin (Mosher et al. 1964); toxic skin secretions (Brodie et al. 1974; Duellman and Trueb 1986); eggs and larvae but not adults are eaten by larval Ambystoma tigrinum (Morin 1983)
Taricha torosa	Adult oophagy (Kaplan and Sherman 1980)	TTX in eggs (Mosher et al. 1964); TTX in skin (Mosher et al. 1964); toxic skin secretions in this genus and species (Habermehl 1971; Brodie et al. 1974)
Desmognathus fuscus fuscus	Adult oophagy (Baldauf 1947)	
Desmognathus ochrophaeus	Adult eat eggs and hatchling larvae (Kaplan and Sherman 1980)	Fully palatable to shrews (Brodie et al. 1979)
Plethodon glutinosus	Adults eat juveniles (Powders 1973)	Toxic and noxious skin secretions in adults of this genus (Habermehl 1971; Duellman and Trueb 1986); adults eaten but rejected often by shrews (Brodie et al. 1979)
Plethodon cinereus	Adults eat eggs and juveniles (Heatwole and Test 1961; Highton and Savage 1961)	Adults eaten but not preferred by shrews (Brodie et al. 1979)
Plethodon dunni	Adults eat juveniles (Altig and Brodie 1971)	
Necturus maculosus	Adult oophagy (Eycleshymer 1906; Kaplan and Sherman 1980)	

cannibalism may arise as a result of opportunistic feeding, when conspecifics represent the majority of available prey in a particular habitat (Nyman et al. 1993). The latter hypothesis is supported by the observation that cannibalism is more prevalent when conspecific larval density is high (Heatwole and Test 1961; Collins and Cheek 1983). Furthermore, cannibalism can be induced in laboratory populations merely by increasing the density of larvae housed together (Collins and Cheek 1983; Morin 1983; Pfennig et al. 1993, 1994). Cannibalism is also prevalent in anurans (Polis and Myers 1985) but because anuran larvae are primarily herbivorous, cannibalistic behaviour in this taxon typically involves adult predation on larvae or small juveniles. However, several anuran species with cannibalistic tadpoles have been reported, e.g., tadpoles of *Chacophrys pierottii* are voracious cannibals (Cei 1955; Blair 1976), as are those of a number of species of *Scaphiopus* (Bragg 1946, 1964, 1965; Pfennig et al. 1993).

Interestingly, sibling conspecifics are less likely to eat one another than unrelated conspecifics. Both *Ambystoma tigrinum* and *Scaphiopus bombifrons* larvae and tadpoles, respectively, are polyphenic, i.e., within a population both cannibalistic and non-cannibalistic morphs will develop. When presented with a group of non-cannibalistic tadpoles, of which some are kin and others are not, a cannibalistic tadpole will selectively prey upon non-kin (Pfennig et al. 1993, 1994). In fact, *A. tigrinum* larvae can distinguish between siblings and cousins, preying selectively on more distantly related larvae! In these kin selection experiments with *A. tigrinum*, occlusion of the nares eliminates the selective ability of the cannibalistic larvae, such that they consume conspecific larvae regardless of relatedness (Pfennig et al. 1994), indicating that kin recognition is mediated by the olfactory or vomeronasal system. However, in the case of *S. bombifrons* tadpoles, kin recognition may be mediated by the gustatory sense. Cannibalistic tadpoles were found to nip at potential prey. After nipping, sibling tadpoles were rejected, whereas non-kin were ingested (Pfennig et al. 1993).

C. Toxicity

In light of the prevalence of cannibalism among urodeles and anurans, it is somewhat paradoxical that in this group of vertebrates the ability to produce extremely toxic substances is also widespread (Noble 1954; Duellman and Trueb 1986). To what degree are these aspects of life history mutually exclusive? For example, is cannibalism less evident in amphibians that produce and store lethal substances in various glands in their skin? Or has immunity to self-produced toxins evolved concomitantly with the ability to produce these lethal compounds, such that cannibalism also occurs in these poisonous species?

A wide variety of amphibians are noxious or toxic. Depending upon the species examined, the skin contains large numbers of granular glands that secrete paralytic neurotoxins, cardiac toxins or other noxious substances (Daly and Witkop 1971; Habermehl 1971; Luther 1971; Lutz 1971; Myers and Daly 1976; Myers et al. 1978; Neuwirth et al. 1979; Duellman and Trueb 1986; Erspamer 1994). The physiological effects of a number of these toxins have been described (Mosher et al. 1964; Hille 1992), and are lethal at exceedingly low concentrations (Mosher et al. 1964; Myers and Daly 1976). Many extremely lethal frogs are also brightly coloured, and their colouration is thought to serve as a general warning of their toxicity (Duellman and Trueb 1986). Thus, it is likely that a would-be predator relies on visual cues rather than on taste to make a decision about ingesting a brightly coloured frog.

A large number of amphibians are not necessarily lethal when consumed, but possess a number of less toxic, noxious substances in the glands of their skin. Many of these animals also taste bitter to humans (Wassersug 1971; Myers et al. 1978; Neuwirth et al. 1979); anecdotally, eating a large toad will make a large dog sick for a day. Amphibians that possess these milder toxins are rejected and subsequently avoided by many other would-be predators (Table 1) (Wassersug 1971; Brodie et al. 1978; Formanowicz and Brodie 1982). Thus, these compounds may function primarily as predation deterrents, and may be tasted by would-be predators before they are rejected (Formanowicz and Brodie 1982; Mason et al. 1982). To what degree they also deter conspecific predation (cannibalism) is not known. Several dendrobatid frog species are immune to the physiological effects of the toxins they produce (Daly et al. 1980); therefore it is unlikely that these toxins would deter cannibalism in these species. However, it is not known how widespread resistance to self-produced toxins is within amphibians. Given that the toxins generally are restricted to glands in the skin, it is possible that this compartmentalization protects the host blood supply from contact with the toxin, obviating the necessity for resistance. Whether toxic species can taste the toxins they produce is also not known.

Table 1 lists both anuran and urodele species that are cannibalistic. The second column contains the form of cannibalism each exhibits, in terms of the life history stage of the consumer and the consumed. Data on the toxicity or palatability of each of these cannibalistic species are contained in the third column. Only one documented case of "cannibalism" exists within the extremely poisonous dendrobatid family; the females of *Dendrobates pumilio* feed their own unfertilized oocytes to their tadpoles (Bragg 1965; Brust 1993). These animals transport their hatchlings from terrestrial nests to accumulations of rain water in the cups of bromeliads, epiphytic plants that are common in the rainforest inhabited by these frogs. Before metamorphosis, the only source of food available to *D. pumilio* tadpoles is the unfertilized eggs provided by the mother. Interestingly, the poison glands in anurans typically do not develop until metamorphosis (Myers and Daly 1976; Duellman and Trueb 1986; Formanowicz and Brodie 1982). Thus, the larvae of this extremely toxic frog consume conspecifics before they become poisonous. In addition, oophagy has been documented in the salamander, *Taricha torosa*, whose eggs are known to contain a potent neurotoxin, tetrodotoxin (TTX), which is the same substance found in its skin and the skin of a number of salamander species (Twitty 1937; Mosher et al. 1964). However, *Taricha* is immune to the effects of its own toxin (Twitty 1937; Mosher et al. 1964), and thus can consume the otherwise poisonous eggs of conspecifics.

These two examples of conspecific oophagy involving toxic species are really exceptions. For the most part, in species in which adults have merely noxious, rather than lethal skin secretions, cannibalism tends to be restricted to adult consumption of juveniles or larvae (Table 1), a finding which may be correlated with the failure of the glands responsible for toxin production to develop until after the metamorphic transition (e.g., Formanowicz and Brodie 1982). For example, *Bombina variegata* adults have toxic skin, and adults have been documented to eat only juveniles (Heuser 1971). Similarly, in species where larval cannibalism is prevalent, such as ambystomid salamanders, larvae have not yet developed the granular glands that produce noxious substances (Formanowicz and Brodie 1982; R. Northcutt, pers. comm.). Once amphibians have metamorphosed and possess mature granular glands, these secretions may deter other predators (Table 1) (Brodie *et al.* 1979; Formanowicz and Brodie 1982). Interestingly, human consumption of adult axolotls *(Ambystoma mexicanum)* has been documented (Smith 1969; Luther 1971). More needs to be known about the identity of these noxious compounds, their biological activity, and whether the species that produce them can sense them, and to what degree these animals are immune to their own toxins.

III. ORGANIZATION OF THE TASTE SYSTEM: ANATOMY AND STRUCTURE

The peripheral taste system of amphibians comprises a field of taste buds distributed within the oropharyngeal cavity. These chemoreceptive organs are innervated by gustatory rami of the facial (VIIth), glossopharyngeal (IXth) and vagal (Xth) cranial nerves. Whereas this basic anatomy is conserved throughout the vertebrates, substantial variation exists in the structure and distribution of taste buds within the oropharyngeal epithelium. For example, the taste buds of terrestrial salamanders and frogs reside upon specialized papillae or elongate bulges, whereas aquatic animals tend to lack papillae and the taste buds are embedded directly in the epithelium.

A. Urodeles

Most studies of the taste system of urodeles have focused on its organization in aquatic salamanders, particularly mudpuppies and axolotls. These animals do not metamorphose, but rather retain their larval form throughout adulthood and reproductive maturity. Because of their phylogenetic position and the simplicity of the organization and morphology of their taste system, aquatic salamanders probably have a taste system closely resembling that of the ancestral condition from which tetrapods in general have evolved. Thus, a description of the taste system of aquatic salamanders will serve to introduce many basic features that have been modified in terrestrial species of urodeles and anurans.

1. Distribution of Taste Buds

In general, taste buds are located on both the palate and floor of the oropharynx in urodeles (Fährmann 1967; Jasinski and Miodonski 1979; Samanen and Bernard 1981a; Toyoshima *et al.* 1987; Northcutt *et al.* 1997). However, only one systematic study has examined the distribution of taste buds in any salamander (Northcutt *et al.* 1997). In adult axolotls, approximately 1 400 taste buds are present in the epithelium lining the roof and floor of the oropharynx. Taste buds are most densely distributed on the lingual surfaces of upper and lower jaws. They are found in slightly lower numbers in the lateral portions of the roof and floor, and are particularly meager in number in the medial portions of the roof and floor (Fig. 1). Given the vast array of tongue morphologies of extant salamanders (Regal 1966; Roth *et al.* 1990), this pattern of taste bud distribution likely varies across taxa in ways that reflect the diverse functional requirements of the tongue for feeding. However, at this juncture, this hypothesis remains a matter of pure conjecture.

2. Morphology and Cytology of Taste Buds

In axolotls (Nagai 1993; Northcutt *et al.* 1997), as well as in another aquatic salamander, the mudpuppy *(Necturus maculosus;* Farbman and Yonkers 1971; Delay and Roper 1988), taste buds are embedded directly in the oropharyngeal epithelium (Fig. 2a). These chemoreceptive organs consist of aggregates of fusiform cells with apical microvillar processes that protrude into the oropharyngeal cavity through a narrow pore in the epithelium (Fig. 2b). There are 40–100 of these elongate cells within a taste bud (Fährmann 1967; Farbman and Yonkers 1971; Reutter and Witt 1993; Northcutt *et al.* 1997), with a ring of serotonergic basal cells surrounding each bud (Fig. 2c; Welton *et al.* 1992; Delay *et al.* 1993; Kim and Roper 1995; Barlow *et al.* 1996). Additionally, a population of basal cells that are not serotonergic occur in taste buds; these ovoid cells may be stem cells (Fährmann 1967; Bigiani and Roper 1993; Delay *et al.* 1994) that continually generate the other cells within each bud throughout the life of the animal (Beidler and Smallman 1965).

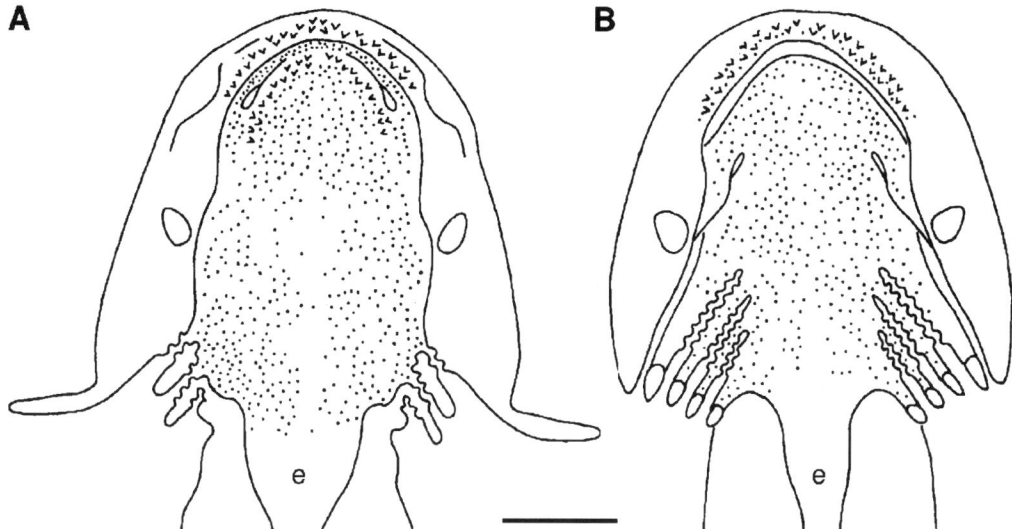

Fig. 1. Distribution of taste buds in a young adult axolotl. Taste buds, as indicated by the black dots, are distributed throughout both the roof (A) and floor (B) of the oropharynx. They are found at highest density in more medial and anterior regions, and become sparse more caudally at the entrance to the esophagus (e). Teeth are represented by small "v"s. Scale bar: 5 mm. Drawing courtesy of Dr. R. Glenn Northcutt.

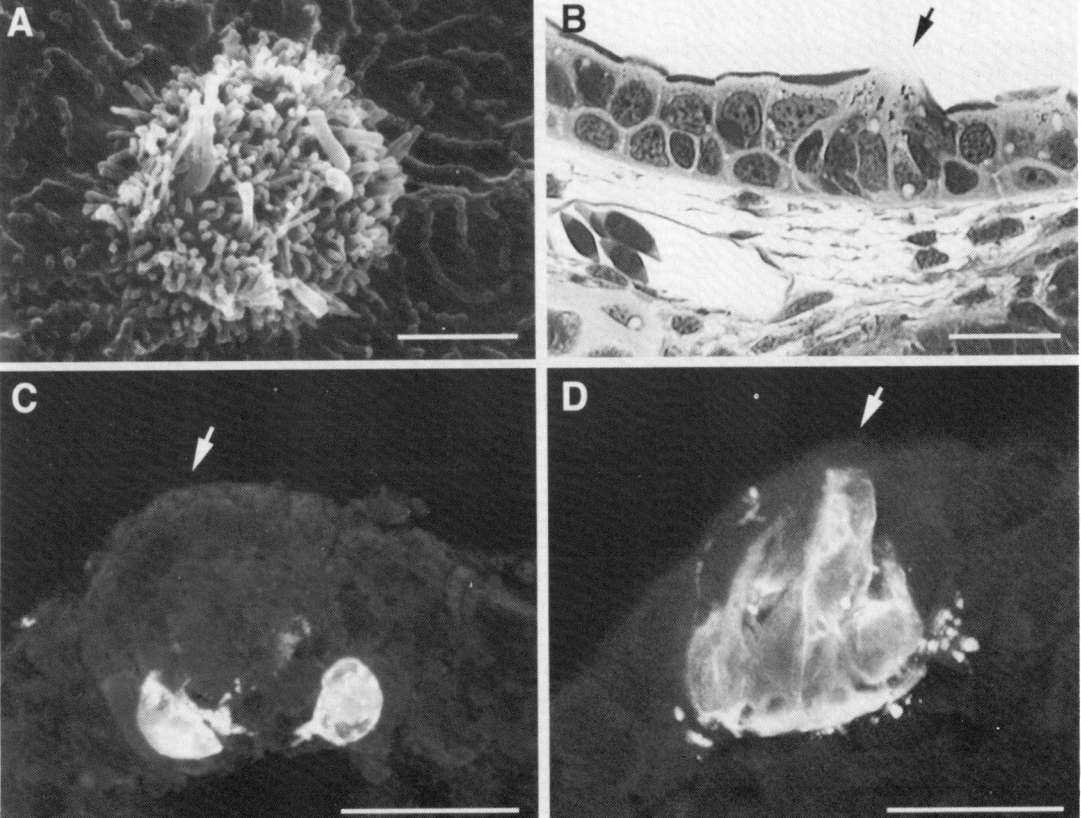

Fig. 2. Normal morphology and immunohistochemistry of taste buds in axolotls. A. Scanning electron micrograph of the apical microvillar processes of a single taste bud. Both large and small microvilli are present, likely representing the apical regions of dark and light cells, respectively, as in mudpuppies (Cummings *et al.* 1987). B. Transverse section through the median plane of an axolotl taste bud, stained with cresyl violet. The apical microvilli protrude through a small taste pore (arrow), while the cell bodies of the fusiform cells are situated within the epithelium. C. Transverse frozen section in which the Merkel-like basal cells are immunoreactive for serotonin, revealing both the position of the cell bodies and the fine centripedally directed processes. Arrow indicates the apex of the taste bud. D. Antiserum against calretinin labels a subset of fusiform cells within each axolotl taste bud. Calretinin-immunostaining is found throughout the cytoplasm of the labelled cells. Arrow indicates taste pore. Scale bars, A: 2 μm; B, C, and D: 40 μm.

The fusiform cells have been categorized further into light and dark cells, based on ultrastructural criteria (Fährmann 1967; Farbman and Yonkers 1971; Toyoshima et al. 1987; Delay and Roper 1988; and see Reutter and Witt 1993 for review). Specifically, dark cells (or A cells) (Fährmann 1967) represent about 50–60% of the cells within a taste bud. They contain dense granular packets or secretory vesicles in the supranuclear and apical regions. Light cells (or B cells) (Fährmann 1967) are less prevalent, and comprise approximately 20–30% of the cells of a taste bud. They lack secretory vesicles, but have a large amount of endoplasmic reticulum and many mitochondria throughout their cytoplasm. Light and dark cells also have different types of apical specializations; dark cells possess long, branched microvilli, whereas light cells have short, unbranched microvilli (Whitear 1976; Cummings et al. 1987; Toyoshima et al. 1987). In *Necturus*, some light cells also terminate in stereocilia (Cummings et al. 1987). Furthermore, Whitear (1976) has shown that only dark cells have extracellular elements likely indicative of secretory material. Serotonergic basal cells make up 10% of the cells of the taste bud. In addition to their serotonergic phenotype, they are recognizable by their elongate processes that travel from their peripherally located cell bodies, along the base of the taste bud, to the more central region of the bud. These cells possess many vesicles, some of which have been shown to contain serotonin (Toyoshima and Shimamura 1987). Basal cells that do not express serotonin and lack radial processes are believed to be stem cells.

There have been numerous attempts to assign specific functions to ultrastructural classes of cells within taste buds. One idea, based on the prevalence of presumed secretory granules, is that the dark cells are support or secretory cells, whereas light cells are actual taste receptors (Fährmann 1967; Farbman and Yonkers 1971; Toyoshima et al. 1987). However, an alternate view is that both light and dark cells within taste buds are receptor cells, but represent discrete phenotypic and functional classes. This latter hypothesis is supported by (1) the presence of synapses, i.e., focal aggregations of synaptic vesicles at membrane thickenings (or active zones) apposed to neuritic membranes in the basal regions of both light and dark cells (Delay and Roper 1988) as well as by (2) electrophysiological data (McPheeters et al. 1994; Mackay-Sim et al. 1996; and see below). Thus, in mudpuppies, both light and dark cells communicate synaptically with afferent nerve fibers, and likely serve a chemoreceptive function, although sensitivity to tactile stimulation cannot be ruled out.

An additional complication of the structural and functional relationship of cells within taste buds is that the population of cells within a bud is constantly turning over throughout the life of the animal. Thus, examination of taste buds at a single point may reveal differences in the ultrastructure of fusiform cells reflecting different stages of maturation of these cells, some of which may not yet possess synapses. It is precisely this type of progression that has been proposed in murine taste buds (Delay et al. 1986). More recent evidence, however, supports the hypothesis that dark and light cells in rat taste buds are discrete cell types rather than stages in a lineage continuum (Pumplin et al. 1996).

Little is known about the histochemistry of cells within amphibian taste buds, other than the well described serotonergic phenotype exhibited by the Merkel-like basal cells. Initially, these basal cells were identified as serotonergic by fluorescence histochemistry (Goosens and Vandenberghe 1974; Hirata and Nada 1975; Toyoshima and Shimamura 1987). These findings have been confirmed subsequently using immunocytochemical methods (Kuramoto 1988; Jain and Roper 1991; Welton et al. 1992; Delay et al. 1993; Kim and Roper 1995; Fig. 2c). Antibodies to the peptide cholecystokinin (CCK) also specifically recognize these basal cells within taste buds (Welton et al. 1992).

A number of functions have been proposed for serotonergic basal cells in taste buds. One hypothesis is that these cells are mechanoreceptors (Toyoshima and Shimamura 1987; Toyoshima 1989a), as has been claimed for cutaneous Merkel cells (for review see Iggo and Findlater 1984). This comparison is based on ultrastructural similarities with cutaneous Merkel cells (e.g., Düring and Andres 1976), and on electrophysiological studies indicating nerves that innervate taste buds respond to mechanical stimulation of the periphery (e.g., Hanamori et al. 1990). Alternatively, serotonergic basal cells may act as local interneurons which modulate taste receptor cell activity. This latter view is in accordance with more recent studies of the function of cutaneous Merkel cells showing that these cells likely modulate the excitability of mechanoreceptive free nerve endings (for reviews see Tachibana 1995; Heatwole 1997). A neuromodulatory role for Merkel-like cells in taste buds is supported by both the large number of synapses from these basal cells on to both taste receptor cells and sensory neurites (Delay and Roper 1988; Delay et al. 1993), as well as by electrophysiological data (Ewald and Roper 1992a, 1992b, 1994; Delay et al. 1997; and see below).

Although some progress has been made in identifying histochemical and immunocytochemical taste receptor cell types in mammalian taste buds (Takeda et al. 1992; Smith et al. 1994; Wong et al. 1994; Kim and Roper 1995; Pumplin et al. 1996), little information is available for amphibian taste receptor cells. Immunoreactive cells within taste buds were not revealed in an extensive immunocytochemical survey of neurotransmitters and neuropeptides, although numerous neuroactive compounds were found in the nerve fibers innervating taste buds (Welton et al. 1992; and tabular summaries therein). Thus, the identity of the neurotransmitter(s) used by amphibian taste receptor cells remains a mystery (for reviews see Roper 1992; Nagai et al. 1996). Interestingly, a subset of fusiform cells within axolotl taste buds are immunoreactive to antisera against a specific calcium-binding protein, calretinin (Fig. 2d) (Barlow et al. 1996; Barlow and Northcutt 1997). This protein is involved in buffering intracellular calcium (Baimbridge et al. 1992), and likely modulates calcium signaling when taste receptor cells are activated by appropriate taste stimuli. However, it is not known whether calretinin immunoreactivity is limited to light or to dark cells, or is found in subsets of both types of receptors.

3. Innervation of Taste Buds

Taste receptor cells and serotonergic basal cells both are innervated by sensory neurites that form a plexus beneath each taste bud (summarized by Reutter and Witt 1993). The general pattern of innervation of taste buds has been described in axolotls (Nagai 1993; Northcutt et al. 1997). The fibers that innervate taste buds, termed special visceral or gustatory fibers, have a distinct pattern of ramification. They travel as unmyelinated fibers from the subepidermal nerve plexus found beneath a given taste bud, into the taste bud where their numerous branches weave among the taste cells (Lu and Roper 1993). The fibers innervating the non-taste epithelium, or general visceral fibers, emerge from the subepidermal plexus, and their varicose branches ramify horizontally among the cuboidal cells of the oropharyngeal epithelium. In contrast, no sub-taste bud plexus exists in *Necturus* (Delay and Roper 1988), although as in axolotls, unmyelinated nerve fibers are intertwined between the cells of taste buds.

In *Necturus*, synaptic contacts occur from both light cells and dark cells on to neurites (Delay and Roper 1988). In addition, serotonergic cells make synapses with light and dark cells, as well as with afferent neurites at the base of taste buds. Because these specialized contacts have accumulations of vesicles on both sides of closely apposed membranes of basal cell-taste cell synapses and basal cell-neurite synapses, these structures may represent bi-directional synapses (Delay and Roper 1988). This synaptic organization is unusual, and may reflect the synaptic plasticity necessary to accommodate the constantly renewing population of taste receptor cells.

The innervation of taste buds has been characterized immunocytochemically by a number of workers. In mudpuppies, taste buds are contacted by nerve fibers that are immunoreactive for vasoactive intestinal peptide (VIP), calcitonin gene-related peptide (CGRP), and Substance P (Welton et al. 1992), as well as for glutamate (Jain and Roper 1991; Lu and Roper 1993) and GABA (Jain and Roper 1991). Whether these immunoreactive nerve fibers represent gustatory afferents, or a combined general visceral/gustatory innervation is unclear (Roper 1992; Nagai et al. 1996). Also unknown is how large a proportion of the innervation of taste buds is not labeled by antibodies to these substances.

In mammals, both Substance P and CGRP immunopositive fibers are associated primarily with the perigemmal innervation that surrounds each taste bud, although sparse intrabud (intragemmal) peptide-immunoreactive fibers are present (Finger 1986; Silverman and Kruger 1990). The perigemmal nerve endings belong to sensory neurons of the trigeminal ganglion and are believed to be somatosensory, i.e., responsive to touch, pain and temperature. The intragemmal, peptide-immunonegative processes belong to the sensory neurons of the VIIth or facial ganglion, and are believed to carry gustatory information (Silverman and Kruger 1990). Thus, in axolotls, the peptide-immunoreactive fibers evident within taste buds may be general visceral (pain, temperature or touch) or a combination of general visceral and gustatory nerve fibers.

The fine, unmyelinated nerve endings found within taste buds belong to sensory neurons with cell bodies within the ganglia of the facial (VIIth), glossopharyngeal (IXth) and vagal (Xth) cranial nerves (Coghill 1902; Nagai and Matsushima 1990; Nagai and Oka 1991; Northcutt et al. 1997). Territories of the branches of these three cranial nerves that innervate taste buds in salamanders have been delineated in axolotls (Nagai and Oka 1991; Northcutt et al. 1997). Taste buds and the epithelium of the rostral 3/5's of the roof and the internal surface of the lower jaw are innervated by branches of the facial or VIIth nerve. The glossopharyngeal or IXth cranial nerve innervates the taste buds and epithelium of more posterior regions of the roof of the oropharynx, as well as the rostral half of the floor (including the anterior tongue) and ventromedial surfaces of the hyoid and first branchial arches. The vagal or Xth nerve branches to innervate the most caudal regions of the roof and floor of the oropharynx where only a small number of taste buds are found, as well as those on the second and third branchial arches. To what degree this pattern of innervation varies among urodeles is unknown, but is of interest given the vast diversity of oropharyngeal morphologies in adult salamanders (Regal 1966; and see below). However, a cursory description of taste bud distribution and innervation in mudpuppies indicates that the pattern in ambystomid salamanders may be consistent with that of proteids (Samanen and Bernard 1981a).

4. Ganglionic Organization and Central Projections

Sensory neurons innervating taste buds have their cell bodies in the ganglia of the VIIth, IXth and Xth cranial nerves, but the degree to which their position within the ganglion reflects their peripheral arborization pattern has not been precisely determined. When the post-trematic or lingual branch of the IXth nerve was retrogradely filled to reveal the distribution of ganglionic cell bodies, a diffuse group of cells were labelled that were localized primarily in the rostral portion of the fused ganglion of IX–X (Nagai and Matsushima 1990). However, additional IXth nerve cell bodies were also found in the caudal region of ganglion IX–X. Retrograde labelling of the Xth nerve trunks produced a population of labelled cells that overlapped in their distribution with those of the IXth nerve. Comparisons of retrograde fills of pre- and post-trematic rami of the IXth nerve also yielded clusters of cell bodies whose distribution overlapped (Nagai and Oka 1991). This rather diffuse organization led these authors to conclude that the postotic ganglionic complex of urodeles is relatively "undifferentiated" (Nagai and Matsushima 1990; Nagai and Oka 1991). However, these results must be interpreted somewhat guardedly because a number of these nerve branches are known to carry mixed fibers, and therefore the diffuse organization of the ganglionic cells may reflect the converging patterns of a number of different types of sensory input.

Like the hindbrain of all vertebrates examined to date, that of urodeles possesses a well defined primary gustatory nucleus, the nucleus of the solitary tract (NTS) (Herrick 1944, 1948; Finger 1987). This nucleus extends the length of the hindbrain from the root of the trigeminus to the more caudal roots of the vagus (Herrick 1948). Gustatory nerve fibers belonging to the VIIth, IXth and Xth nerves enter the solitary tract and both ascend and descend throughout its rostral-caudal extent, where they presumably synapse on neurons that surround the fascicle (Herrick 1948; Nagai and Matsushima 1990; Northcutt et al. 1997). The NTS neurons then project to the secondary gustatory nucleus in the rostral hindbrain. Higher order gustatory nuclei have not yet been identified in amphibians. Given parallels with both mammals and fishes, one might expect that neurons of gustatory nuclei in the hindbrain will project either directly to the thalamus, or indirectly via a nucleus functionally similar to the pontine gustatory nucleus of mammals (Finger 1987).

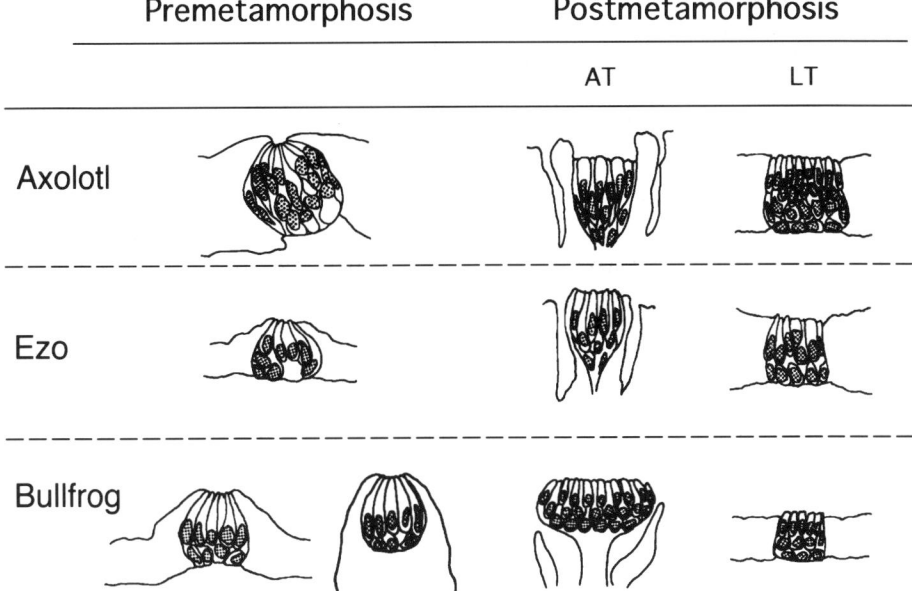

Fig. 3. Summary diagram of pre- and post-metamorphic taste organ morphology in two urodele and one anuran species. In axolotl and Ezo salamander larvae, typical onion-shaped taste buds are present throughout the epithelium of the floor of the oropharyngeal cavity. In bullfrog tadpoles, onion-shaped taste buds are located primarily on finger-like pre-metamorphic papillae, although some taste organs are situated directly within the epithelium. As metamorphosis progresses, taste buds gradually disappear, and taste organs with broad apices that reside in papillae surrounded by deep trenches appear in all three species. In bullfrogs, these organs are termed taste disks. In salamanders, these organs are called taste-disk-like, because their morphology resembles that of anuran taste disks. In all species examined, papillar formation is accentuated in the more anterior region of the tongue (AT: adult tongue) compared to more posteriorly (LT: laval-like tongue). Line drawing courtesy of H. Takeuchi, S. Ido, Y. Kaigawa and T. Nagai, and with permission from Oxford University Press.

5. Modifications of the Taste Periphery in Terrestrial Salamanders

While most studies of the taste system of urodeles pertain to a few aquatic species, the vast majority of urodele species metamorphose and are terrestrial as adults. To what degree does the gustatory system of terrestrial salamanders resemble that of aquatic salamanders? What types of adaptations to terrestrial existence might be encountered in the taste system? Surprisingly few studies address these questions, but the available data suggest that the oropharyngeal epithelium is altered substantially in those species with terrestrial adults. In adult, terrestrial *Salamandra salamandra,* taste buds are organized along ridges that are arranged radially on the tongue (Jasinski and Miodonski 1979). Between the ridges lie discrete grooves that are lined by mucus cells and tubular duct cells that apparently secrete mucus into the grooves. The taste buds themselves are surrounded by rod cells. The extensive distribution of mucus-secreting cells probably serves to maintain an aqueous microenvironment necessary for the proper functioning of the chemoreceptive apices of the taste buds, as well as to promote the clearing of chemical stimuli (Witt 1996).

Only one recent study directly compares the taste buds and attendant oropharyngeal epithelial structures in terrestrial (Ezo; *Hynobius retardatus*) and aquatic (axolotls; *Ambystoma mexicanum*) salamanders (Fig. 3) (Takeuchi *et al.* 1997). Typically, axolotls do not metamorphose and adult taste buds remain as barrel-shaped structures embedded directly in the oropharyngeal epithelium, identical to those found in axolotl larvae (Fig. 2b) (Fährmann 1967; Northcutt *et al.* 1997). However, if induced to metamorphose by treatment with bath-applied thyroxin, taste buds gradually disappear, and new taste organ structures appear in the oropharynx of adult axolotls (Takeuchi *et al.* 1997). These taste organs are aggregates of elongate cells that have a much broader apical receptor region than do those of larval/aquatic adult taste buds (Fig. 3). In fact these metamorphosed taste buds are similar in appearance to the taste disks found in adult frogs (Fig. 3; and see below). These taste disk-like organs in salamanders vary in their morphology, depending upon their location in the oropharynx. In the posterior region of the tongue, the taste disks resemble taste buds with expanded receptor surfaces, but in the tongue's more anterior, adult-like region, the taste disks are surrounded by protrusions of the adjacent epithelium, which in turn are encircled by deep trenches (Fig. 3). These structures are reminiscent of papillae found associated with taste disks in frogs (Jaeger and Hillman 1976; Osculati and Sbarbati 1995; and see below) and with taste buds in other terrestrial tetrapods (for review see Kinnamon 1987). Interestingly, the taste disks of Ezo salamanders, which metamorphose naturally, are intermediate in morphology between those in aquatic and in artificially metamorphosed axolotls (Fig. 3). In both the anterior and posterior tongue, the receptive surfaces of the disks are narrower in metamorphosed Ezo salamanders than in metamorphosed axolotls, and the papillae associated with the anterior taste disks are less extreme, consisting primarily of a slightly elevated papilla surrounded by a trench. However, the expansion of the receptive surface as well as the development of the papillae in metamorphosed animals of both species (and in bullfrogs) suggest that these modifications are likely adaptations to a terrestrial existence and the concomitant switch to chemostimulation in air. It is not known if the organization and types of cells found within taste disks of terrestrial urodeles have been modified, as is the case for terrestrial frogs and toads (Graziadei and DeHan 1971; Düring and Andres 1976; Osculati and Sbarbati 1995; and see below), nor is it known whether the larval taste buds metamorphose directly into taste disk structures, or if the former regress and the latter develop *de novo*.

Another potentially interesting, yet unexplored, aspect of the organization of the taste system is the degree to which the distribution of taste buds within the tongue is correlated with the morphology of tongues, especially in light of the large variation in tongue morphology in urodeles (Regal 1966; Duellman and Trueb 1986). For example, the tongues of aquatic salamanders are non-protrusible, although the tongue of metamorphosed axolotls may transform to a more distensible type (Takeuchi *et al.* 1997), whereas most terrestrial salamanders possess a slightly protrusible tongue. In plethodontids, in contrast, the tongue is highly protrusible (Regal 1966; Lombard and Wake 1977; Thexton *et al.* 1977). Does this variation in tongue morphology and function during feeding correlate with the distribution of taste receptor organs in the tongue and oropharyngeal epithelium in general? This aspect of taste system organization is completely unexplored, despite the large literature devoted to the functional morphology of feeding in amphibians (for review see Roth *et al.* 1990), and the potential correlation between taste bud distribution and tongue function of both lizards (Schwenk 1985) and some fishes (Sibbing *et al.* 1986; Finger 1988).

B. Caecilians

Few studies have examined the taste system of this group of limbless amphibians; there is exactly one short paper describing the distribution and morphology of taste buds in adult *Typhlonectes compressicaudus* and larval *Icthyophis* of an unknown species (Wake and Schwenk 1986). Both *Icthyophis* larvae and *Typhlonectes* adults are aquatic, and the distribution and morphology of their taste buds is similar. The taste buds resemble those of larval axolotls; they are barrel-shaped (compare Fig. 4 with Fig. 2b), and are embedded in the epithelium lining both the roof and the toothed region of the lingual surface of the lower jaw. Surprisingly, taste buds were not found on the non-protrusible tongue of either species. The degree to which this pattern is altered in adult *Icthyophis*, which are terrestrial, remains to be determined, but the authors (Wake and Schwenk 1986) were unable to detect taste buds in any of the terrestrial adults caecilians they examined.

C. Anurans

The taste system of anurans differs substantially from that of aquatic urodeles. Most modern frogs and toads employ a projectile tongue, primarily the lingual flip type, used in feeding on motile prey (Duellman and Trueb 1986). Frogs considered to be more primitive, such as *Bombina orientalis*, possess tongues that are only slightly protrusible, and some paedomorphic frogs, such as the pipid *Xenopus laevis*, lack tongues altogether (Regal and Gans 1976).

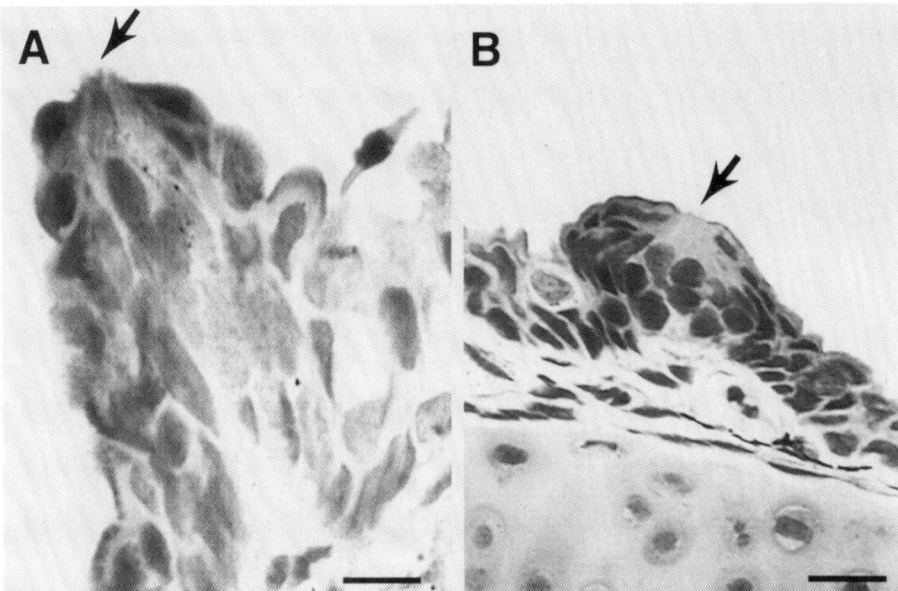

Fig. 4. Taste buds from an aquatic larval caecilian, *Icthyophis kohtaoensis*, stained with cresyl violet. Adults of this species are terrestrial. A. Transverse section through a taste bud located in the rostral, ventrolateral oropharyngeal epithelium. The taste organ is an aggregate of elongate cells with apical microvilli (arrow), and is situated in a small papilla-like structure. B. Transverse section through a taste bud in the caudal, ventrolateral pharyngeal epithelium. This onion-shaped organ is embedded directly in the epithelium, and closely resembles taste buds found in axolotls (see Fig. 2). Arrow indicates apical microvilli. Scale bars in A and B: 20 µm. Sectioned material courtesy of Dr R. Glenn Northcutt.

1. Distribution of Taste Organs

The majority of studies on the taste organs of anurans have been performed on various ranid species. This genus is prevalent throughout the world, and workers from Japan, to Italy, to the United States, and many places in between, have conveniently employed local species to explore the taste system. Ranid frogs, like all terrestrial frogs examined to date, possess an array of specialized taste organs, or taste disks (Fig. 5a), rather than the barrel- or onion-shaped taste buds characteristic of aquatic salamanders. Taste disks appear to be a unique and highly derived taste receptor organ restricted to anurans, and differing greatly from those of all other vertebrates (for review see Reutter and Witt 1993).

2. Morphology and Cytology of Taste Disks

Taste disks are large chemosensory organs containing a few hundred cells, the nuclei of which are arranged in three relatively discrete laminae. The receptive surface of each taste disk is expanded, such that the diameter of the apical and basal regions of the organ are approximately equal (Fig. 5a,c). On the dorsal surface of the tongue, each taste disk sits within a papillar structure. The papillae consist of raised portions of the tongue epithelium surrounded by a trench (Fig. 5a,c). In contrast, taste disks located on the palate (Düring and Andres 1976) and ventral surface of the tongue (Honda *et al.* 1994) are not situated in papillae, but instead are flush with the luminal surface of the epithelium.

Early light microscopists were quick to recognize the stratified organization of the cells within taste disks, describing a superficial, intermediate and deep layer (Fig. 5b) (for reviews see Osculati and Sbarbati 1995; Jaeger and Hillman 1976). It is not surprising that this complex structure also was found to have a more complex cellular organization than that of typical vertebrate taste buds. By the end of the nineteenth century, a number of anatomists had described five discrete cell types within the taste disk of anurans, including cylindric, supporting, rod, wing and basal cells (for review see Jaeger and Hillman 1976). In the last few decades, interest in frog taste organs has been renewed, and several classification schemes have been proposed for the various cell types within taste disks, again primarily based on

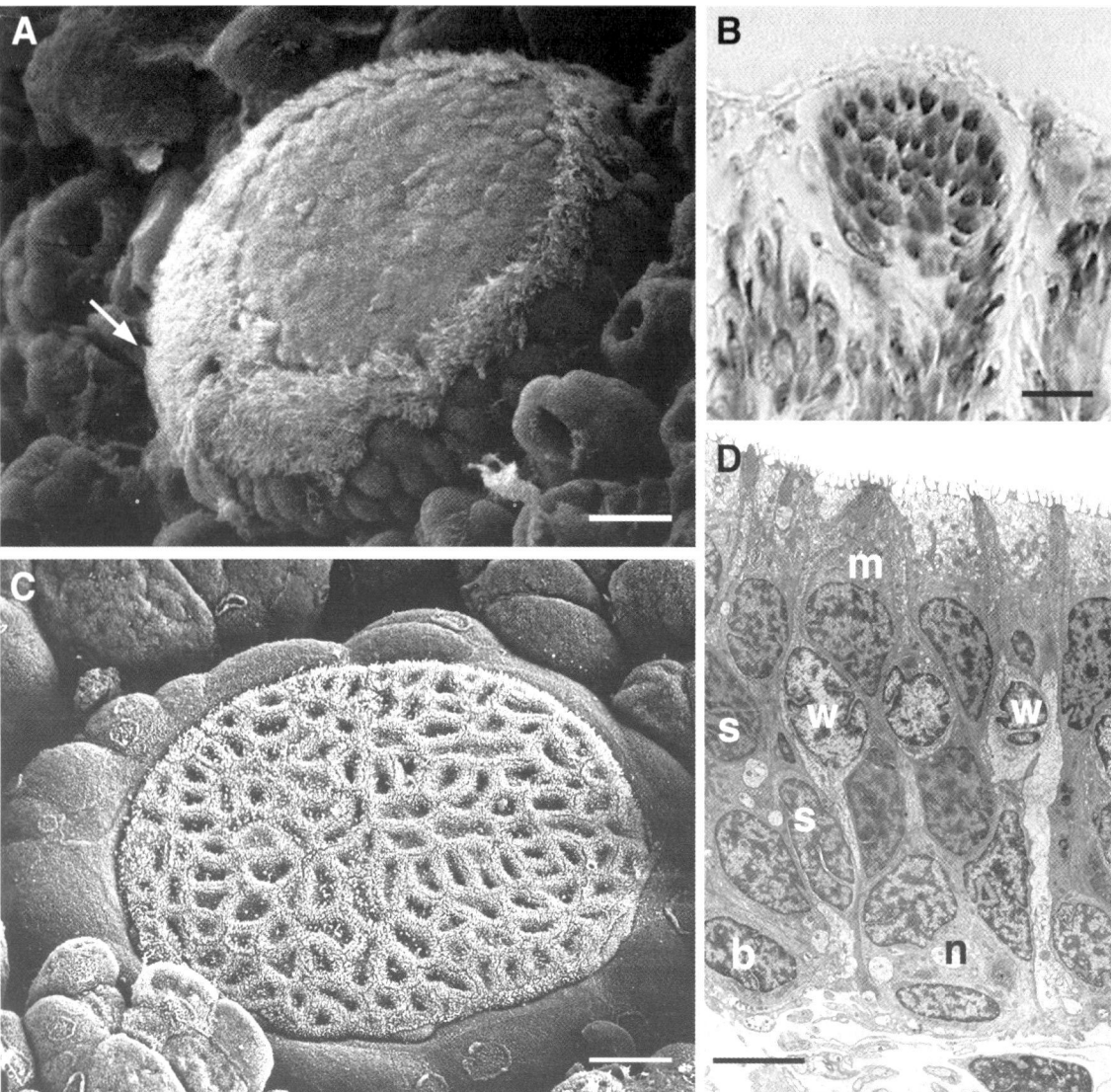

Fig. 5. Morphology of taste disks in several anuran species. A. Scanning electron micrograph of a taste disk of *Rana pipiens* at the apex of a fungiform papilla. The receptive surface is surrounded by a ring of heavily ciliated epithelial cells (arrow). The fungiform papilla is surrounded by numerous filiform papillae. B. Transverse section through a taste disk of *Discoglossus pictus* stained with palmgren silver stain. This disk is from a recently metamorphosed frog (Gossner stage 46), which accounts for the relatively small size of the taste organ. Note the readily evident layering of the cell somata. C. Scanning electron micrograph of a taste disk of *Bombina orientalis* at the apex of a fungiform papilla. Unlike ranid frogs, the taste disks of *B. orientalis* lack the ring of ciliated cells, and are instead, surrounded by a mound-like ring of epithelial cells. D. Transmission electron micrograph of a transverse section through a taste disk of *Bombina orientalis*. Numerous discrete cell types are evident, including mucus cells (m), wing cells (w), Merkel-like basal cells (b) and sensory cells (s). Nerve fiber profiles (n) are present at the base of the taste organ. Scale bars, A: 8 μm, B: 25 μm, C: 20 μm, and D: 5 μm. Micrographs A, C, and D courtesy of Dr Martin Witt. Sectioned material in B courtesy of Dr Gerhard Schlosser.

observations on ranid frogs. None of these classifications has included the traditional cell types, light, dark and basal, found in the taste buds of urodeles and other vertebrates. For the most part, researchers have avoided correlating cell types found in anuran taste disks with those found in the taste buds of other vertebrate species.

Graziadei and DeHan (1971) and DeHan and Graziadei (1971) proposed that cells within each taste disk were either receptor cells or support cells, based on the similar organization of the vertebrate olfactory epithelium (see Dawley 1997). The top third of each disk was proposed to contain support or accessory cells, whereas the receptor cells were suggested to possess apical and basal processes extending from cell bodies located in the basal two-thirds of the disk. The putative support cells were full of secretory granules, and thus were proposed to be involved in mucus secretion. The putative receptor cells had bifurcated, and in some cases trifurcated, apical processes. More recent studies have revealed this categorization to be an oversimplification.

Five to seven categories of cells have now been recognized, based on descriptions of taste disks in *R. temporaria* and *R. esculenta* (for reviews see Düring and Andres 1976; Osculati and Sbarbati 1995). Although a rough consensus on the number of cell types now appears to exist, there are still substantial differences in the organizational schemes put forth by different workers. To some degree, this may be due to the absence of studies in which individual cells have been serially reconstructed in three dimensions (3D). This method, when applied to cells in mammalian (Kinnamon *et al.* 1985, 1988) and urodele (Delay and Roper 1988) taste buds, has allowed the concrete association of basal and apical specializations with specific ultrastructurally defined cell bodies. Without 3D reconstruction of serial thin sections, it not possible to unequivocally identify all components of a cell whose thickness and extent of processes span many sections.

With that caveat in mind, the broad classes of cell types that have been described can be reviewed. In general, these include mucus cells, wing cells, taste receptor cells, basal stem cells, and Merkel cells (Fig. 5d). Some workers also recognize a class of sustentacular cells. Because the details of taste disk cell-type classification have been completely reviewed recently (Osculati and Sbarbati 1995), only an overview of the morphological and ultrastructural features used to categorize the types of cells found within anuran taste disks is presented here. Readers with more specialized, as well as historical interests, are encouraged to consult the original review.

Mucus Cells: Consistent with earlier classifications, mucus cells are localized in the superficial layer of the taste disk. Microvilli cover the luminal surfaces of these cells, and large secretory granules are prevalent throughout the apical cytoplasm (Fig. 5d). These cells are in close apposition with the apical processes of the wing cells, and form tight junctional complexes with them. Mucus cells lack basal processes (DeHan and Graziadei 1971; Graziadei and DeHan 1971; Düring and Andres 1976; Osculati and Sbarbati 1995).

Wing Cells: The cell bodies of wing cells reside beneath the superficial mucus cells in the intermediate layer of the disk, and have apical processes that are branched and flattened (Fig. 5d). These processes extend to the luminal surface via gaps among the hexagonally arranged mucus cells, forming tight junctions with mucus cells as they ascend. At the surface, the processes terminate in ridges of microvilli. Wing cells may have small, unipolar, basal processes, but these do not form synaptic contacts with afferent neurites (Osculati and Sbarbati 1995). von Düring and Andres (1976) also described a cell type similar to the wing cells with its soma in the more basal layer of the taste disk, and large, flattened apical processes interdigitating among the glandular support cells. However, in their observations, the lamellar apical processes wrap the apical processes of the taste receptor cells, and these cells lack basal processes. They termed these cells "satellite support cells".

Sustentacular Cells: The sustentacular cell type has been recognized only by Osculati and Sbarbati (1995) who consider them to be glial-like cells within the taste disks. Sustentacular cell bodies lie within the intermediate layer of cells, and extend processes both apically and basally. While they typically possess only one apical process, sustentacular cells send out several basal processes. The sustentacular apical process wraps the apical process of a taste receptor cell such that it terminates at the luminal surface in a crown of microvilli around the taste cell apex, much in the way described for satellite support cells by von Düring and Andres (1976). The basal processes in some cases wrap neuritic fibers that have invaded the base of the taste disk, but are never seen synapsing upon them. Sustentacular cells are extensively interconnected to one another by desmosomes, and in this manner a subset of them form a meshwork that rings the margin of the taste disk. Interestingly, some, but not all, aspects of both sustentacular and wing cell structure described by Osculati and Sbarbati (1995) are found in a single support cell type (satellite support cells) recognized by von Düring and Andres (1976). 3D reconstruction of each of these putative cell types should help to resolve this dichotomy.

Taste receptor cells: von Düring and Andres (1976) recognized a single taste receptor cell type, of which they could distinguish an immature and a mature form. The mature taste receptor cells are fusiform, with cell bodies located in the disk's more basal portion, and elongate apical processes that protrude from gaps among the mucus cells (Fig. 5d). The apical process of each taste cell is replete with smooth endoplasmic reticulum and bundles of microtubules, and generally terminates as a single microvillar rod that seldom branches. Taste receptor cells also have short basal processes that form synapses with afferent neurites that ramify at the base of the taste disk. Immature taste cells resemble mature taste cells and are in close contact with afferent neurites, but lack basal synaptic contacts.

Osculati and Sbarbati (1995) recognized two categories of mature taste receptor cells: type II and type III cells. The latter closely resemble the mature taste receptor cell type described by von Düring and Andres (1976). These cells are distinctly bipolar, fusiform cells, with the soma residing in the intermediate layer. The apical process is fine and sometimes is forked, although it more often terminates in a single, large microvillus. The basal region of each cell extends into one or more processes that make synapse-like contacts with neurites; these specialized contacts consist of aggregations of cytoplasmic granules or vesicles, but lack associated membranous thickenings. Type II cells were only recognized by Osculati and Sbarbati (1995). These cells tend to be found in greater numbers in taste disks located on the proximal portion of the tongue. This type of cell has a pear-shaped soma located in the intermediate layer, with a single apical process and sometimes a single basal process. The apical process is wrapped by a sustentacular cell, and terminates in microvilli at the receptive surface of the taste disk. The basal portion of type II cells make specialized, non-synaptic contacts with afferent neurites, which in some cases resemble gap junctions. Again, 3D serial reconstructions would help to clarify the issue of numbers of taste receptor cell types and their maturational state.

Basal stem cells: These cells are situated at the base of each taste organ, and cytologically are virtually indistinguishable from the basal stem cells found in the general epithelium of the tongue. They are believed to be mitotically active, and to generate the other cell types found in the taste organ (Düring and Andres 1976; Reutter and Witt 1993). These cells lack the processes found in another class of basal cells, the Merkel-like cells.

Merkel-like basal cells: Between 10 and 20 of these basal cells ring the base of each taste disk. These club-shaped cells are found in the basal layer of each disk (Fig. 5d), and have processes that extend basally toward the taste disk center. Basal cells possess several distinct ultrastructural characteristics, particularly a number of short spine-like protrusions from the cell body, which are reminiscent of cutaneous Merkel cells (Breathnach 1980; Whitear 1989; Tachibana 1995). The large processes of the Merkel-like basal cells form true synapses on afferent nerve fibers, as evidenced by the presence of focal aggregations of cytoplasmic granules in regions where basal cell processes and neuritic fibers are tightly apposed. Implicit in giving these cells within the taste disk the name of Merkel-like cells is the hypothesis that they are comparable to the Merkel cells found in the skin of all vertebrates (Tachibana 1995). It is this class of basal cells that is serotonergic in frogs (Goosens and Vandenberghe 1974; Toyoshima and Shimamura 1987; Kuramoto 1988; Sbarbati *et al.* 1989).

Attempts have been made to assign function to each of these ultrastructural categories of cells found in anuran taste disks. The mucus cells that lie in the apical third of the disk are generally agreed to be responsible for secreting mucus that maintains the relatively hydrated state of the receptive surface (DeHan and Graziadei 1971; Graziadei and DeHan 1971; Düring and Andres 1976; Jaeger and Hillman 1976). Wing cells have been viewed as a receptor cell type (Jaeger and Hillman 1976). Alternatively, they may be glandular cells that produce the specific secretions that are restricted to the microenvironment of the apices of their lamellar processes (Sbarbati *et al.* 1990, 1991). In addition, Osculati and Sbarbati (1995) believed that the wing cells provide structural support to the taste disk. Although von Düring and Andres (1976) did not explicitly recognize a wing cell type, they did describe a satellite cell type with lamellar apical processes that is believed to have a supportive function.

The more basally situated sustentacular cells identified by Raviola and Osculati (1967) and Osculati and Sbarbati (1995) were proposed to function as receptor organ glial cells, in that they insulate basal processes of receptor cells, somewhat like peripheral glial cells in the nervous system. In addition, these cells are thought to contribute to the structural integrity of the taste organ.

There is some controversy concerning the function of the serotonergic basal or Merkel cells; they may be mechanoreceptors (Düring and Andres 1976; Toyoshima and Shimamura 1982), or local paracrine or neuromodulatory cells (Kuramoto 1988; Sbarbati *et al.* 1988), or both (Kuramoto 1988; Osculati and Sbarbati 1995).

In contrast to the rather disparate opinions concerning the precise function of most cell types within taste disks, the fusiform, bipolar cells have been universally recognized as taste receptor cells (DeHan and Graziadei 1971; Graziadei and DeHan 1971; Düring and Andres 1976; Jaeger and Hillman 1976; Osculati and Sbarbati 1995). In the case of the cell types described above, type II cells would likely be recognized as taste receptor cells by virtually all workers in the field (see review by Reutter and Witt 1993), given their elongate apical processes extending to the oropharyngeal lumen, and their extensive synapse-like contacts with afferent neurites. The type III cells recognized by Osculati and Sbarbati (1995) are believed to function as gustatory cells as well, but may simply represent a different stage of taste cell development (Düring and Andres 1976).

As is the case in the taste buds of urodeles, most cell types within anuran taste disks have yet to be associated with particular histochemical or immunocytochemical properties, with the exception of the serotonergic Merkel-like basal cells (Hirata and Nada 1975; Toyoshima and Shimamura 1987; Kuramoto 1988; Sbarbati *et al.* 1989). A subset of basal cells, likely the serotonergic class, also express Neuron Specific Enolase (NSE) (Kuramoto 1988). Although other cells within taste disks are immunoreactive to antibodies against several proteins, this immunostaining has not been correlated with any ultrastructurally identified cell type. In *R. catesbeiana*, a subset of fusiform cells with cell bodies in the upper to middle layers of the disk are recognized by an antibody to NSE (Kuramoto 1988). These NSE-immunopositive cells have one or more apical processes that ascend to the receptive surface of the disk, as well as basal processes directed toward the basal lamina. In addition, a non-overlapping subset of fusiform cells is calbindin-immunoreactive. This calcium-binding protein was initially named Spot 35 Protein, but subsequently was recognized as a member of the large family of calcium-binding proteins and renamed calbindin (e.g., Abe *et al.* 1992). The calbindin-immunoreactive cells have cell bodies in the middle to lower levels of the taste disk, and have single apical and branching basal processes. Fusiform cells immunoreactive to a variety of calcium-binding proteins are also evident in the palatal taste disks of *Xenopus laevis* (Kerschbaum and Hermann 1992). Antibodies against calmodulin, S-100 protein, and parvalbumin, all calcium-binding proteins (Baimbridge *et al.* 1992), specifically labeled elongate cells that were considered to be taste receptor cells (Kerschbaum and Hermann 1992). Merkel-like basal cells were immunopositive for calmodulin and S-100 protein. Interestingly, calbindin, which is present in taste receptor cells in *R. catesbeiana* (Kuramoto 1988), was found only in what were termed support cells in *X. laevis* taste disks, i.e., in cells that lacked basal processes (Kerschbaum and Hermann 1992). These data suggest that there is substantial phylogenetic diversity of the type of calcium-binding proteins employed by different cell types within taste receptor organs. This conclusion is consistent with data on the varied expression of specific calcium-binding proteins in other sensory systems of diverse taxa (Baimbridge *et al.* 1992).

There is ample evidence that the mucus covering the general epithelium of the tongue is decidedly different in histochemical composition compared with the mucus covering the taste receptive surface (Witt and Reutter 1988; Osculati and Sbarbati 1995; Witt 1996). Differences in local mucus histochemistry have been proposed to play an important role in regulating the response of taste receptor cells to sapid stimuli (Witt and Reutter 1988; Osculati and Sbarbati 1995; Witt 1996). Discrete microenvironments are associated with the apices of mucus, wing and taste receptor cell types (Carmignani et al. 1975; Carmignani and Zaccone 1977; Witt and Reutter 1988; Sbarbati et al. 1990; Zancanaro et al. 1990; Sbarbati et al. 1991; Witt 1996). For example, the mucus localized on the apices of wing cells and taste cells is associated with a histochemically identified surfactant-like material that is high in phosphorus and calcium (Sbarbati et al. 1991) and lanthanum (Zancanaro et al. 1990), as well as in lipids (Sbarbati et al. 1990). Because calcium ions are known to be necessary for taste transduction in general, these authors suggested that the high calcium associated with the apices of putative taste receptor cells may be critical for the function of these cells. In addition, lectin histochemistry has revealed that different types of glycoproteins are localized in different cell types within the frog taste disk (Witt and Reutter 1988; Witt 1996). In fact, results of these latter studies indicate that ultrastructurally similar mucus cells may actually represent several biochemically discrete cell types or maturational stages (Witt 1996). Exhaustive reviews of this topic can be found in Osculati and Sbarbati (1995) and Witt (1996).

3. Innervation of Taste Disks

As in urodeles, taste organs in frogs and toads are innervated by sensory neurons of the VIIth, IXth and Xth cranial nerve ganglia (Strong 1895; Jaeger and Hillman 1976; Duellman and Trueb 1986; Schlosser and Roth 1995). Five to ten nerve fibers enter each taste papilla (Jaeger and Hillman 1976), and these fibers form a small subepithelial plexus beneath the sensory epithelium (Sbarbati et al. 1988). Fine, unmyelinated nerve fibers leave the plexus and ramify within the taste organ proper (Düring and Andres 1976; Jaeger and Hillman 1976). These neurites can be divided into two calibers of fibers, thick and thin; the thick fibers contact Merkel-like basal cells exclusively, whereas the thinner neurites travel to innervate cells in more apical strata (Düring and Andres 1976). Synapses can be detected in several cell types within taste disks, including well-organized synapses of both taste receptor cells (Jaeger and Hillman 1976; Toyoshima et al. 1984; Suzuki and Takeda 1989) and Merkel-like basal cells (Sbarbati et al. 1988, 1989) onto neurites. Efferent synapses, in which neurites appear presynaptic to taste receptor cells (Toyoshima et al. 1984), and bi-directional synapses between neurites and taste receptor cells (Zuwala and Jacubowski 1991) have also been reported.

The innervation of the oropharynx in anurans (Strong 1895) is generally similar to that of urodeles (Coghill 1902; Herrick 1948; Samanen and Bernard 1981a; Northcutt et al. 1997). The palatal taste buds are innervated by the palatine ramus of the VIIth nerve (Strong 1895; Suzuki 1966). Taste buds in the lingual surface of the lower jaw are innervated by the mandibular branch of the VIIth nerve, which Strong (1895) termed the hyomandibularis. In amphibians in general, the IXth and Xth cranial nerve ganglia are fused with several of the ganglia of the lateral line system, and the nerves that exit the fused ganglionic complex initially are mixed (Strong 1895; Coghill 1902; Norris 1908; Norris and Buckley 1911; Northcutt 1992; Northcutt and Brändle 1995). Without modern nerve tracing techniques, Strong (1895) was unable to resolve the specific innervation territories of sensory neurons of the IXth and Xth cranial nerves, and was able to determine only that several branches of the postotic ganglionic trunks innervate taste buds within the more posterior regions of the pharynx and in the gill arches (Strong 1895).

Modern electrophysiological investigations have confirmed that the posttrematic branch of the glossopharyngeal (IXth) nerve is primarily responsible for the gustatory and general visceral innervation of the tongue of frogs (Jaeger and Hillman 1976; Ishiko et al. 1979; Sato et al. 1983; Hanamori et al. 1990). Specifically, the lateral and medial branches of the posttrematic branch of the glossopharyngeal nerve innervate the anterior third and posterior two-thirds of the tongue, respectively (Ishiko et al. 1979). In addition, recordings of the palatine ramus of the facial (VIIth) nerve confirm that taste buds of the palate are innervated by this nerve (Suzuki 1966). The innervation territories of each of the branches of the vagal (Xth) nerves known to innervate taste buds in axolotls for example (Northcutt *et al.* 1997), have not been delineated in any anuran species.

Whether the specific fiber types innervating taste disks and their associated papillae are sensitive to touch or to gustatory stimuli has not been determined. This question is of interest because individual neurons within the lingual branch of the IXth nerve respond to both mechanical and gustatory stimulation (Rapuzzi and Casella 1965; Samanen and Bernard 1981b; Hanamori *et al.* 1990). Certainly, the fibers that innervate each sensory papilla represent an immunocytochemically and histochemically heterogeneous population. The nerve bundle that ascends through the core of each taste papilla contains cholinesterase (DeHan and Graziadei 1973), vasoactive intestinal peptide (VIP) (Witt 1995), gastrin-releasing peptide (GRP), calcitonin gene-related peptide (CGRP), peptide histadine isoleucine (PHI), tyrosine hydroxylase (TH) (Kuramoto 1988), and Substance P (Hirata and Kanaseki 1987; Kuramoto 1988). The subepidermal nerve plexus beneath each taste disk is immunoreactive for VIP (Witt 1995), and PHI (Kuramoto 1988). GRP immunoreactivity is also found in fibers in the region of the papilla beneath the taste organ proper (Kuramoto 1988). Substance-P and CGRP immunoreactivity colocalize to a subset of perigemmal fibers that ramify among the ciliated epithelial cells surrounding the taste disk, although a few of these peptide-containing fibers occasionally are found within the sensory epithelium proper (Hirata and Kanaseki 1987; Kuramoto 1988). When detected intragemmally, these Substance-P-immunopositive and CGRP-immunopositive, as well as the VIP-immunoreactive fibers arising from the nerve plexus, appear to contact the serotonergic basal cells (Kuramoto 1988). Given that the perigemmal, and perhaps mechanosensory, fibers innervating taste papillae in mammals can be both Substance-P and CGRP immunoreactive (Finger 1986; Silverman and Kruger 1990), it is possible that this category of peptidergic fibers is also mechanosensory in frog taste papillae. However, the precise immunocytochemical identity of the neurites contacting taste receptor cells is not known.

4. Ganglionic Organization and Central Projections

The cell bodies of neurons that innervate taste organs in anurans are found in the ganglia of the VIIth, IXth and Xth nerves (Strong 1895; Schlosser and Roth 1997a), as is the case in all vertebrates. However, the precise organization of specific sensory neuron types within these mixed sensory ganglia is not known. There are some data on the distribution of sensory neurons that project via specific branches of the IXth nerve (Hanamori and Ishiko 1983). Neurons with axons that travel in the lingual or posttrematic ramus are arranged in two discrete groups, in the distal and the proximal portions of the glossopharyngeal ganglion. This arrangement was also observed when the pretrematic ramus of IX was filled (Hanamori and Ishiko 1983). The presence of two discrete cell groups, one proximal and one distal, suggests that they may have different functions. It has long been assumed that the distal portions of the cranial nerve ganglia contain sensory neurons that subserve gustation, whereas proximally situated neurons are responsible for somato- or general visceral sensation (Strong 1895; Landacre 1907). It is tempting to presume that this may be the organization observed in the study of Hanamori and Ishiko (1983).

Nerve fibers of the VIIth, IXth and Xth nerve project directly to the nucleus of the solitary tract (NTS), located in the dorsal hindbrain (Nieuwenhuys and Opdam 1976; Opdam *et al.* 1976; Hanamori and Ishiko 1983), where they form synapses with neurons surrounding the solitary tract. As in urodeles, sensory afferents in anurans are bifurcated and ascend and descend within this fascicle. Gustatory fibers extend rostrally to the root of the trigeminal ganglion, and caudally to the last vagal root (Nieuwenhuys and Opdam 1976; Opdam *et al.* 1976). Efferents of neurons of the NTS travel within the solitary tract and ultimately form synapses with neurons of the secondary gustatory nucleus, located in the rostral hindbrain (Opdam *et al.* 1976). From there gustatory information likely progresses to thalamic nuclei, as has been demonstrated in mammals and fishes (Finger 1987).

5. Interspecific Variation in the Organization of Taste Disks

Tongue morphology is highly variable in anurans (Regal and Gans 1976). Both ranids and bufonids possess projectile tongues (Gans 1961), whereas other anurans have tongues that are either reduced in their range of movement, e.g., most discoglossids (Regal and Gans 1976), or completely absent as in aquatic pipids (Griffiths 1963; Duellman and Trueb 1986). Does the manner in which the tongue is employed, or its gross morphology, affect the organization or distribution of taste organs? Remarkably, this aspect of taste system anatomy has been relatively ignored. The sparse literature devoted to the taste systems of non-ranid frogs has focused primarily on the ultrastructure of cells within taste disks, and on the morphology of the epithelium covering the tongue. The cellular organization of taste disks appears to be conserved throughout the few anuran taxa that have been examined at the light or electron microscopic levels (*Bombina orientalis*: Witt 1993; *Xenopus laevis*: Toyoshima and Shimamura 1982; Witt and Reutter 1995; *Bufo bufo*: Jasinski 1979); *Calyptocephalella gayi*: Stensaas 1971). Upon more systematic and detailed examination, i.e., 3D reconstruction of serial sections, the observed conservation may fall by the wayside; however, currently it stands in contrast to the wide variety of tongue morphologies and feeding apparati found in diverse anuran species (Regal and Gans 1976).

The general epithelium of the tongue does vary across anuran taxa, but too few studies exist to permit correlations with tongue function. The dorsal surface of the projectile tongue of ranid frogs is covered with filiform papillae (Iwasaki *et al*. 1986; Iwasaki and Wanichanon 1991). Interspersed among the filiform papillae are the fungiform papillae, at the apices of which the taste disks reside. A fringe of ciliated cells surrounds the receptive surface of each taste disk (Fig. 5a) (DeHan and Graziadei 1971; Graziadei and DeHan 1971; Düring and Andres 1976). Both papillar types have copious mucus cells along their lateral walls. The tongue of bufonids, such as *Calyptocephalella gayi* (Stensaas 1971), *Bufo bufo* (Jasinski 1979) and *B. japonicus* (Iwasaki and Kobayashi 1988), resembles that of ranids, although those of *Bufo bufo* and *B. japonicus* lack filiform papillae. Instead, the tongues of these species have elongate epithelial ridges with mucus cells in the grooves between ridges, and taste disks on typical fungiform papillae distributed among the epithelial folds. Unlike the taste disks of ranids that are ringed by ciliated cells, the taste disks of *B. japonicus* are ringed by mound-shaped non-ciliated cells (Iwasaki and Kobayashi 1988). The epithelium of the relatively non-motile tongue of a discoglossid frog *Bombina orientalis* (Witt 1993), is surprisingly similar to the projectile tongue of bufonids and ranids. The dorsal surface is covered with filiform papillae and large numbers of fungiform papillae with taste disks; the latter are particularly prevalent at the anterior portion of the tongue. Like the arrangement of taste disks in *Bufo japonicus,* the receptive surface of the taste disk of *Bombina orientalis* is not ringed by ciliated cells, but instead is surrounded by a circular epithelial mound (Fig. 5c) (Witt 1993). Aquatic pipid frogs, such as *Xenopus laevis*, completely lack a tongue (Regal and Gans 1976), yet possess ample numbers of taste disks situated in the mucosa of both the oropharyngeal floor and palate (Toyoshima and Shimamura 1982; Witt and Reutter 1995). However, fungiform papillae are absent in this species (Witt and Reutter 1995), such that the taste disks are embedded directly in the epithelium, as are palatal taste buds of ranid frogs (Düring and Andres 1976).

One caveat in this litany of tongue structures is that descriptions of species with movable tongues have been based almost exclusively on tongues at rest. The only report on the morphology of the extended or stretched tongue was that by Jasinski (1979) on *Bufo bufo*. Interestingly, the epithelial folds found at rest are flattened in the extended tongue, so the mucus cells between the folds are brought close to the tongue surface, as are the taste disk receptive surfaces, both of which are found below the plane of the epithelium at rest. This finding has implications for the functional morphology of feeding in projectile-tongued frogs (Roth *et al*. 1990; Anderson 1993) in terms of the potential involvement of mucus secretion and gustatory feedback in the ingestion of prey.

To date, taste disk distribution has not been compared systematically in species with very different tongues. Neither has it been correlated with functional morphological studies in which the precise manner of the use of the tongue in capturing prey is known. In ranid and bufonid frogs the distal portion of the tongue is fixed, and the proximal region of the tongue is free. Thus the proximal region of the tongue is forced out of the mouth and flipped during feeding, so that when outside of the mouth the dorsal surface faces downwards (Gans 1961; Regal and Gans 1976; Roth *et al.* 1990). Greatly simplified, the movement of the tongue during capture of prey by these frogs is initiated and controlled by contraction of the musculature of the mandibular and hyoid arches. One study of *R. catesbeiana* revealed that the density of taste disks is actually significantly higher in the unattached proximal portion of the dorsal surface of the tongue (Sato *et al.* 1983). However, another study found taste disks concentrated distally on the tongue of the same ranid species (Jaeger and Hillman 1976). The findings of Sato *et al.* (1983) suggest that taste organs are concentrated in the region of the tongue that, when extruded, comes in immediate contact with the prey. Taste disk density is also highest on the anterior tip of the tongue of *Bombina orientalis* (Witt 1993). This frog has a proximally attached tongue, such that only the very anterior or distal region, with its numerous taste organs, is unattached and is believed to be projected slightly out of the mouth during feeding (Regal and Gans 1976). Thus, the region of the tongue that initially comes into contact with potential prey may also be the region with the highest concentration of taste organs. The implication is that, in these two species at least, tasting of prey may relay information important for feeding success.

6. Taste Receptors in Larval Anurans

Because most anuran tadpoles are filter feeding herbivores, it is not surprising that their taste system is dramatically different from that of carnivorous adults. The mouth and pharynx of tadpoles has been substantially modified throughout evolution; there is no larval tongue, the oropharyngeal epithelium is studded with a multitude of different types of papillae, and the posterior region of the pharynx terminates in branchial baskets or food traps (Wassersug and Rosenberg 1979; Wassersug and Heyer 1988). Although these epithelial structures have been described in a wide variety of anuran species in the context of feeding (Helff and Mellicker 1941a; Kenny 1969; Wassersug and Heyer 1988), they have not been systematically examined for the presence of taste organs. Thus far, taste buds are only known to be localized in the apices of so-called premetamorphic papillae within the oropharynx of ranid tadpoles (Schulze, 1888 as cited in Strong 1895; Nomura *et al.* 1979b; Shiba *et al.* 1980; Zuwala 1991; Zuwala and Jacubowski 1991). These papillae are found on the roof and floor of the pharynx, and are single or multibranched protrusions of the epithelium. Typically, a single taste bud is found at each papillar termination (Zuwala and Jacubowski 1991). The larval taste buds resemble those of aquatic urodeles, in that they consist of aggregates of elongate cells (Zuwala and Jacubowski 1991). Interestingly, however, taste buds in larval *R. temporaria* (Zuwala and Jacubowski 1991) and *R. japonica* (Nomura *et al.* 1979a) lack the Merkel-like basal cells found in both urodele taste buds and adult anuran taste disks.

The fate of larval taste buds at metamorphosis is uncertain. As metamorphosis proceeds, the specialized oropharyngeal feeding apparatus disappears, and there subsequently develops a muscularized, often projectile tongue that is used in predaceous feeding (Helff and Mellicker 1941a, 1941b; Nomura *et al.* 1979a; Hourdry *et al.* 1996). During this transition from larva to adult, the taste buds are replaced by the adult taste disks (Nomura *et al.* 1979a; Shiba *et al.* 1980; Zuwala 1991; Zuwala and Jacubowski 1991). As is the case in urodeles, it is not known if at metamorphosis, taste buds give rise to taste disks directly, or if taste buds degenerate and taste disks then arise *de novo*.

The taste system of frogs with direct-developing embryos has not been examined. In these species, such as the leptodactylid frog *Eleutherodactylus coqui*, the development of larval structures is generally bypassed, and tiny froglets hatch directly from their egg capsules. These animals demonstrate a remarkable plasticity in the relative timing of various developmental events. For example, limb (Ellinson 1994) and retinal (Schlosser and Roth 1997a) development are greatly accelerated in *E. coqui*, whereas a mechanosensory lateral line system fails to develop during embryogenesis (Schlosser *et al.* 1997), and is absent at hatching (Schlosser and Roth 1997b). To what degree does direct development influence the development of taste receptor organs? Are taste buds generated initially and subsequently replaced by taste disks? Or is taste bud formation lost such that only taste disks develop from the very beginning?

From this survey of the organization of the periphery of amphibian taste systems, it is clear that the variation in morphology, relevant both to phylogenetics and to life history strategies, provides a vast arena in which to enquire into the functional organization of the taste system. Using a comparative approach in concert with data from functional morphological studies of feeding behaviour, it may be possible to determine if the peripheral organization of taste organs influences the manner in which taste information is coded by the sensory neurons that innervate them, and to ascertain how gustatory information is then transferred to the central nervous system. In this way, one may be able to begin to construct a working model of how the differing patterns of gustatory organization seen in frogs and salamanders relate to feeding behaviour.

IV. REGENERATION OF THE TASTE SYSTEM

Over a century ago, Vintschgau and Honigschmied (1876) demonstrated that when the lingual nerve supply to the taste buds was cut, taste buds degenerated on the side of the tongue that had been innervated by that nerve branch. Subsequently, this phenomenon has been repeatedly demonstrated using a wide variety of experimental manipulations, including denervation-reinnervation studies and ectopic transplantation (Torrey 1940; Wright 1955, 1964; Guth 1957; Robbins 1967; Fujimoto and Murray 1970). In large part, the results are in agreement: nerve fibers of the cranial nerve ganglia provide trophic support to taste buds located in the oropharyngeal cavity. The majority of studies concerning amphibian taste buds indicate that these chemosensory organs, while still reliant upon neurotrophic support, are less dependent than are mammalian taste buds (Mintz and Stone 1933; Wright 1955, 1964; Poritsky and Singer 1963, 1977; Toyoshima *et al.* 1984; Toyoshima 1989b). For example, in bullfrogs, denervated taste disks did not disappear, but were found to decrease in size due to cell loss and a decrease in cell size within each taste disk over time (Robbins 1967). However, when the adult tongue of a newt was grafted ectopically to a position within the peritoneum, liver, or optic capsule, taste buds initially degenerated in the grafts, but reappeared within a few weeks of grafting (Wright 1955, 1964; Poritsky and Singer 1963, 1977; Toyoshima 1989b). Some workers claim that this reappearance occurs in the absence of innervation, although the degree to which ectopic taste buds were contacted and possibly supported by local, inappropriate nerves was not systematically examined in these studies (Wright 1955, 1964; Toyoshima 1989b). This proved to be critical, as Poritsky and Singer (1963, 1977) found that reappearance of ectopic taste buds in grafts was coincident with innervation of the graft by local nerve fibers. Thus, the necessity of neurotrophic support for maintained differentiation of taste buds remains equivocal in amphibians.

The results of these studies of grafting and regeneration have been generally extrapolated to embryonic development of taste buds *de novo* (Landacre 1907; Torrey 1934; Torrey 1940; Farbman and Yonkers 1971; Farbman 1972); that is, if adult taste buds are maintained by nerves, then it is possible that embryonic taste buds are also induced to differentiate by nerves.

V. EMBRYONIC DEVELOPMENT OF THE TASTE SYSTEM

In virtually all vertebrate embryos, taste buds differentiate late in embryonic development, about the time of birth or hatching (Landacre 1907; Johnston 1910; Cook and Neal 1921; Farbman 1965; Miller and Smith 1988; Belecky and Smith 1990; Farbman and Mbiene 1991; Oakley *et al.* 1991; Zuwala and Jacubowski 1991; Whitehead and Kachele 1994; Barlow and Northcutt 1995). Typically, their appearance in the epithelium of the oral and pharyngeal cavities coincides with or slightly follows innervation of these regions (Landacre 1907; Farbman 1965; Whitehead and Kachele 1994; Barlow *et al.* 1996). This coincidence in timing, and the trophic requirement of adult taste buds for nerve contact, has led to the presumption that embryonic taste buds are induced to form by contact with ingrowing neurites of the cranial nerve ganglion cells (Landacre 1907; Torrey 1940; Farbman 1965; Munger 1977; Hosley *et al.* 1987a, 1987b; Farbman and Mbiene 1991; Oakley 1993a, 1993b; Whitehead and Kachele 1994).

Implicit in the neural induction model is the assumption that taste buds arise directly from the epithelium in which they are found. However, until recently, this had not been conclusively demonstrated in any vertebrate. Taste buds were found to arise directly from the local endoderm of the oropharyngeal epithelium in axolotls (Barlow and Northcutt 1995). Using microsurgical fate-mapping techniques, the migration of neurogenic ectodermal cells was tracked, and these cells did not migrate to the oropharyngeal epithelium and give rise to a distributed field of taste bud progenitor cells, as had been proposed (Farbman and Yonkers 1971; Reutter 1978; Gans and Northcutt 1983). Rather, when the fate of the presumptive cranial endoderm of axolotl embryos (Vogt 1929; Pasteels 1942) was traced with a fluorescent dye, these cells were found to give rise both to the general epithelium of the oropharynx and to the taste buds found in this region (Barlow and Northcutt 1995). These results are consistent with those obtained simultaneously from experiments using transgenic mice (Stone et al. 1995), and with early descriptive observations (Landacre 1907; Johnston 1910; Cook and Neal 1921).

Is differentiation of taste buds from the local epithelium induced by contact with ingrowing nerves? This question has been pursued historically using a paradigm of denervation and regeneration of adult taste buds as a model for embryonic development (e.g., Torrey 1940; Poritsky and Singer 1977). These studies indicated that taste buds require contact with nerves in order to maintain their normal morphology, as discussed above. In contrast, early tests of the neural induction hypothesis in embryos revealed that taste buds in amphibians could definitely develop in the absence of appropriate innervation by cranial nerves (Stone 1932, 1940). When the oropharyngeal region of a salamander embryo was grafted ectopically to the flank of a conspecific host embryo, a site quite distant from the innervation fields of the cranial nerves, the graft developed taste buds. However, local spinal nerve fibers may have supported differentiation and maintenance of these ectopic taste organs. This caveat is particularly critical, given the wide array of nerve fiber types that subsequently have been shown to maintain taste buds in adult animals (Wright 1955, 1964; Zalewski 1973; Kinnman and Aldskogius 1988). In a repeat of Stone's original grafting experiments (Barlow et al. 1996), taste buds did indeed form in the oropharyngeal grafts, and although spinal nerve fibers entered the transplants, nerve fibers were not found in association with the ectopic taste organs. This finding could not rule out the possibility, however, that transient contact by spinal nerve fibers had been sufficient to trigger taste bud differentiation, but simply did not result in the formation of synaptic contacts. In another set of experiments, the oropharyngeal region was dissected out prior to its innervation and placed in isolation in culture. These explants developed taste buds in the complete absence of innervation (Barlow et al. 1996), thus confirming that initial taste bud differentiation is nerve-independent. Recent results from transgenic mouse embryos now tend to support this finding (Mistretta et al. 1996; Fritzsch et al. 1997; Nosrat et al. 1997).

If nerves are not inducing the formation of taste buds, what other embryonic tissues might be involved? Inductive interactions between epithelial and mesenchymal cells are prevalent throughout embryogenesis (Oliver 1980; Hall 1981, 1983; Kollar 1983; Dhouailly 1984; Lumsden 1988; Link et al. 1990; Richman and Tickle 1992; Song et al. 1994). Interactions of this type are responsible for the formation of a number of specialized epithelial structures including teeth, cartilage, feather buds, and hair follicles. Are taste buds also specialized epithelial organs that are induced by contact with cranial mesenchymal cells migrating into the oropharyngeal region during embryogenesis?

This hypothesis was tested by examining the role of two embryonic mesenchymal cell populations: the cranial neural crest and the cranial paraxial mesoderm (Barlow and Northcutt 1997). Both populations of cells initially reside dorsal to the presumptive oropharynx, but move laterally and ventrally to surround it as embryogenesis progresses (Stone 1922; Adelmann 1932; Hörstadius 1950; Jacobson and Meier 1984; Noden 1991; Couly et al. 1992; Northcutt and Brändle 1995). These migrations occur substantially earlier than the innervation and differentiation of taste buds late in embryogenesis (Barlow and Northcutt 1995; Northcutt and Brändle 1995; Barlow et al. 1996). With this timing in mind, a series of grafting and culture experiments were performed using the presumptive oropharyngeal epithelium from late gastrula to late neurula stages (stage 13–19) (Bordzilovskaya et al. 1989). In both transplantation and culture experiments, well developed taste buds were found in the oropharyngeal epithelium, while mesenchymal cell derivatives were not encountered (Barlow and Northcutt 1997). This result is striking; head endoderm taken from embryos that have just finished gastrulating will form taste buds independent of contact with any other embryonic tissue. However, taste bud differentiation does not occur until the time when intact embryos are hatching, approximately 10 days after head endoderm has been specified. This finding raises several intriguing questions: What cellular mechanisms control the intrinsic differentiation of taste buds? How are taste bud progenitor cells determined? When do these embryonic events occur? If taste bud formation is an inherent feature of head endoderm and is independent of contact by nerves, how do sensory neurites find and innervate their taste bud targets? As of yet, no answers have been forthcoming. However, these findings serve as a basis for future experimental approaches to explore the embryogenesis of the taste system.

VI. FUNCTION OF THE TASTE SYSTEM

Because behavioural studies of the taste system in amphibians are rare, electrophysiological studies have been instrumental in shaping an understanding of the function of this system in frogs and salamanders.

A. Function of Taste Receptor Cells

1. Voltage-dependent Properties of Taste Receptor Cells

Early investigations of taste receptor cell function were confounded by problems with intracellular recording techniques, as microelectrode impalements caused cell damage and resulted in very low intracellular resistance. Thus, responses to tastants such as monovalent and divalent cations were passive, and simply reflected the change in the electrochemical gradient across the taste receptor cell membrane imposed by the charge and concentration of the taste stimuli (Kinnamon and Cummings 1992; Roper 1992).

Since the advent of the patch clamp technique, as well as the development of improved methods for intracellular recording, rapid progress has been made toward an understanding of the physiology of taste receptor cells. One of the first important findings was that these receptor cells are electrically excitable, producing action potentials in response to anode break excitation (Kashiwayanagi *et al.* 1983) or current injection (Roper 1983). Subsequently, the voltage-dependent properties of taste receptor cells were characterized in detail (Avenet and Lindemann 1987; Kinnamon and Roper 1987, 1988b; Sugimoto and Teeter 1990; Miyamoto *et al.* 1991; Sugimoto and Teeter 1991). These studies revealed that among cells of the oral epithelium, only taste receptor cells and Merkel-like basal cells possess a host of voltage-dependent currents, reminiscent of those typically found in neurons (Hille 1992). The rapid inward current during the regenerative action potential is carried by a voltage-dependent, TTX-blockable sodium current, while the slower plateau potential is due to a voltage-dependent calcium current. Like neuronal action potentials, the repolarization of taste receptor cells is achieved via an outward delayed rectifier-type potassium current. A number of other potassium currents have now been described for taste cells in amphibians, including an inward rectifier (Kinnamon and Roper 1988b; Sugimoto and Teeter 1990) and a calcium-dependent potassium current (Sugimoto and Teeter 1990). In addition, mudpuppy taste cells possess a calcium-dependent chloride conductance (Taylor and Roper 1994). Interestingly, all of these voltage- and calcium-dependent channels also have a small open probability when the taste receptor cell is at rest, and thus substantially affect the resting membrane potential (Kinnamon and Roper 1988b; Cummings and Kinnamon 1992).

In an early study of mudpuppies, considerable variation was found in the voltage properties of the taste receptor cells (Kinnamon and Roper 1987). These findings have subsequently been refined (Bigiani and Roper 1993), and particular voltage profiles have been assigned to ultrastructurally identifiable cell types within taste buds (Delay *et al.* 1994; McPheeters *et al.* 1994; Bigiani *et al.* 1996; Mackay-Sim *et al.* 1996). Using whole cell patch clamp of enzymatically dissociated taste cells, mature dark cells, with numerous electron-dense granules in their apical cytoplasm, were found to possess all of the voltage-sensitive currents described above, including voltage-dependent inward sodium and calcium, and outward potassium currents (McPheeters *et al.* 1994). This complement of currents was also present in one class of mature light cells, whereas only potassium currents were elicited by a series of voltage steps in a second electrophysiological class of light cells (McPheeters *et al.* 1994). In this study, all three categories of cells were fully differentiated, i.e., with apical processes reaching the taste pore.

Subsequently, differences in electrical properties were correlated with the maturity of taste cells (Mackay-Sim *et al.* 1996); only those cells with apical processes extruding through the taste pore possessed the full range of voltage-sensitive currents. The majority of these fully excitable cells were dark cells (n = 4), although a single excitable light cell also was found. Taste receptor cells with apical processes that did not reach the taste pore had only outward potassium currents, but surprisingly, did not resemble the light cells found in an earlier study (McPheeters *et al.* 1994). Rather they had novel ultrastructural characteristics, never before described in mudpuppy taste buds.

A similar study was performed by Bigiani *et al.* (1996), using a slightly different approach. These workers recorded from individual taste receptor cells using whole-cell patch clamp of cells within an intact lingual slice. In this way, potential artifacts resulting from enzymatic dissociation of cells within taste buds could be avoided (Delay *et al.* 1994; McPheeters *et al.* 1994; Mackay-Sim *et al.* 1996). Again, two electrophysiological classes of fusiform cells were identified: those with voltage-sensitive sodium, calcium and potassium currents, and those with only voltage-dependent potassium currents. Ultrastructural examination revealed that the majority of the former category were found to be dark cells, whereas all of the latter, relatively electrically quiescent cells were light cells. Interestingly, a small number of cells with the full complement of voltage-sensitive currents turned out to be light cells. These findings are for the most part in agreement with those described by McPheeters *et al.* (1994). However, the unique ultrastructural cell type associated with electrical quiescence found by McPheeters *et al.* (1994), was not encountered in lingual slice preparations. Rather, quiescent cells were consistently identified as light cells. These structural data, in concert with differences in the voltage regulation of potassium currents encountered in these cells (Bigiani *et al.* 1996), suggest that enzymatic treatment to produce isolated cells may have altered both the structure and function of these receptor cells (McPheeters *et al.* 1994).

Differences in voltage profiles also have been correlated with discrete classes of basal cells (Bigiani and Roper 1993; Delay *et al.* 1994), and findings from lingual slice and isolated cell preparations are in good agreement here. The serotonergic Merkel-like basal cells have a suite of voltage-dependent currents comparable to those found in the taste receptor cells, consistent with their putative role as either local neuromodulators (Roper 1992) or cellular transducers of mechanical stimulation (DeHan and Graziadei 1971). Non-serotonergic basal cells, in contrast, have only voltage-dependent potassium currents, reflective of the undifferentiated state of a putative stem cell.

2. Transduction of Taste Stimuli

The next step in examining taste receptor cell function was assessment of how specific taste stimuli affect the excitability of these cells. Initially, amphibian taste cells were a particularly useful model system for examining the cellular machinery responsible for the transduction of taste stimuli, primarily because of their large size compared to taste cells of other vertebrates (Kinnamon and Cummings 1992; Kinnamon and Margolskee 1996; Lindemann 1996). However, the transduction cascades that have been identified subsequently for specific sapid molecules vary substantially among vertebrate classes. These differences are likely due to dietary differences among the species examined (Kinnamon and Cummings 1992; Kinnamon and Margolskee 1996). Nonetheless, general principles of taste cell function have arisen from this comparative approach.

Mechanisms of taste transduction fall into two broad functional categories: those in which tastants act directly on channels in the taste cell membrane, and those which act via a G-protein-mediated intracellular cascade (Kinnamon and Cummings 1992; Kinnamon and Margolskee 1996; Stewart *et al.* 1997). The apical microvillar membranes of taste receptor cells are the site of transduction, which results in cellular events that trigger synaptic vesicle release at the basal membrane. Taste stimuli are thus converted at the apical membrane of the taste cell to a change in membrane potential (a graded receptor potential, either depolarizing or hyperpolarizing), or to an increase in intracellular calcium, both of which are thought to influence neurotransmitter release at the taste cell–sensory neuron synapse (Kinnamon and Cummings 1992; Kinnamon and Margolskee 1996; Stewart *et al.* 1997). In general, sour, salty and some bitter substances act directly on ion channels located in the membranes of the apices of taste receptor cells, whereas most bitter, as well as all sweet compounds are thought to bind to unidentified G-protein-coupled membrane receptors also located in the apices of taste cells (Kinnamon and Cummings 1992; Kinnamon and Margolskee 1996). Because several excellent reviews of the function of taste receptor cells have been published recently (Kinnamon and Cummings 1992; Kinnamon and Margolskee 1996; Lindemann 1996; Stewart *et al.* 1997), comments here area restricted to current knowledge about taste transduction in amphibians.

The best studied transduction mechanisms in amphibians are those involved in the perception of salty, sour and bitter substances. Responses to sweet substances, while common in frogs, are generally not found in salamanders (for review see Kinnamon and Margolskee 1996). Salt, sour and bitter stimuli all have been demonstrated to act directly at ion channels, whereas sweet, and some bitter stimuli may be mediated by G-protein-coupled second-messenger cascades.

A. SOUR TASTE

Acidic or sour stimuli block a specific outward, voltage-dependent potassium current in the taste cells of aquatic salamanders, resulting in the depolarization of the receptor cell (Kinnamon *et al.* 1988; Kinnamon and Roper 1988a, 1988b; Sugimoto and Teeter 1991; Cummings and Kinnamon 1992). These sour-blockable potassium channels are located almost exclusively, and in high density, at the apical membrane of taste cells (Kinnamon *et al.* 1988; Roper and McBride 1989; Cummings and Kinnamon 1992). Potassium channels located elsewhere on taste cells are not sensitive to acid block, and stimulation of the taste receptor cell body with protons does not result in a depolarizing receptor potential (Cummings and Kinnamon 1992). In taste cells isolated from bullfrogs, however, acid stimuli such as acetic acid activate a partially calcium-selective channel, resulting in depolarization of the membranes of the taste receptor cells (Okada *et al.* 1987).

Sour taste also may be mediated by regulation of electrical coupling between taste receptor cells (Bigiani and Roper 1994). As many as 20% of the taste cells within mudpuppy taste buds are electrically coupled (West and Bernard 1978; Yang and Roper 1987; Sata *et al.* 1992), and changes in extracellular pH cause changes in intracellular pH (e.g., Moody 1981); these changes are known to affect electrical coupling (Spray and Bennett 1985). Thus, when Bigiani and Roper (1994) lowered intracellular pH, they found that electrical coupling between taste receptor cells was reduced. They therefore proposed a transduction mechanism for sour taste in which acid stimuli may acidify the cytoplasm of taste receptor cells, thus affecting receptor cell coupling, and therefore membrane potential.

B. SALTY TASTE

Monovalent salts appear to be detected directly via apically located ion channels. In frog taste receptor cells, sodium taste is transduced in some taste receptor cells by an amiloride-blockable sodium channel (Avenet and Lindemann 1988). Application of NaCl results in the depolarization of taste receptor cells. This channel is distinct from the TTX-blockable, voltage-sensitive sodium channel found in all taste receptor cells (Kinnamon and Margolskee 1996), and is less selective for sodium. Interestingly, sodium taste stimuli may also activate discrete sodium-dependent potassium channels in bullfrog taste receptor cells (Miyamoto *et al.* 1996). Although salamanders are able to detect sodium salts in behavioural assays, the specific transduction mechanism employed by receptor cells has not yet been elucidated. However, detection of some potassium salts appears to occur via apically located potassium channels in the taste receptor cells of mudpuppies (Kinnamon *et al.* 1988; Kinnamon and Roper 1988b) and bullfrog (Fujiyama *et al.* 1994). In mudpuppy taste receptor cells, chloride ions may also be transduced via a calcium-dependent chloride conductance, although this channel is found both apically and basolaterally (McBride and Roper 1991).

C. BITTER TASTE

In aquatic salamanders, bitter tasting substances, such as quinine, $CaCl_2$ and some potassium salts, act directly at the apical membrane of the taste receptor cell (Kinnamon and Roper 1988b; Bigiani and Roper 1991; Sugimoto and Teeter 1991; Cummings and Kinnamon 1992). Like sour or acidic stimuli, bitter substances block apically located potassium conductances, thereby reducing the outward flux of potassium ions and depolarizing the taste receptor cells. Alternatively, in bullfrog taste cells, quinine may bind specific membrane receptors located in the apices of the taste cells. These receptors directly activate a chloride conductance, resulting in a depolarizing receptor potential (Sato *et al.* 1994). While more complex bitter substances, such as denatonium, are known to act via G-protein-coupled membrane receptors in mammals (Kinnamon and Margolskee 1996), until recently these types of receptor-mediated cascades had not been demonstrated for bitter taste in amphibians. However, in the taste receptor cells of mudpuppies, denatonium has now been shown to activate a G-protein-coupled phospholipase C cascade that results in the IP3-stimulated release of intracellular calcium stores (Ogura *et al.* 1997).

D. SWEET TASTE

Although the ability to detect sweet substances is present in frogs (Kusano and Sato 1957; Suzuki 1966; Sato *et al.* 1995), a link between sweet stimuli and an intracellular transduction cascade has yet to be elucidated. In mammals, sweet stimuli appear to act at membrane receptors that are coupled to G-protein-regulated cyclic nucleotide cascades (Kinnamon and Cummings 1992; Kinnamon and Margolskee 1996). In this light, cAMP and cAMP-dependent protein kinase have been demonstrated to affect membrane conductances in the taste receptor cells of ranid frogs (Avenet and Lindemann 1987; Okada *et al.* 1987; Avenet and Lindemann 1988). However, the possible role of cAMP in sweet taste transduction in anurans remains to be demonstrated. A cGMP cascade also may be involved in the transduction of sweet stimuli in bullfrog taste receptor cells (Kolesnikov and Margolskee 1995).

E. AMINO ACID TASTE

Whole-nerve recordings and behavioural assays indicate that both frogs and salamanders are able to taste a variety of amino acids. Cellular analysis of this response in amphibians, however, has yet to be attempted. Amino acid detection has also been demonstrated in catfish (Michel and Caprio 1991; Caprio 1992; Kohbara *et al.* 1992; Valentincic and Caprio 1994), and substantial progress has been made toward an understanding of its mechanism in these fish (for review see Caprio *et al.* 1993). Both L-arginine and L-proline bind discrete membrane receptor-channels located in the apical microvillar region of taste cells. Binding of L-arginine to the receptor opens the channel, which is generally selective for cations, and results in either an increase or decrease in intracellular calcium, depending upon the cell being monitored (Zviman *et al.* 1996). Thus, even the detection of a single amino acid is likely mediated by a heterogeneous population of receptor types.

From this cursory review of taste transduction mechanisms, it is clear that a single transduction mechanism does not correlate with a specific taste modality, nor do different amphibian species use identical mechanisms to detect the same sapid molecule. In addition, a given taste receptor cell can respond to more than one type of tastant, a phenomenon which has also been well-documented in mammalian taste cells (for reviews see Kinnamon and Margolskee 1996; Lindemann 1996; Stewart *et al.* 1997). However, it is not known if a single taste cell can respond to more than two, or perhaps all, discrete taste stimuli. The response range of single receptor cells has important implications for the peripheral coding of taste stimuli.

B. Taste Buds as Peripheral Integrators

Substantial modulation and integration of taste receptor cell signaling occurs prior to the excitation of gustatory neurites at the taste cell-neurite synapse. This modulation is due both to neurochemical and to electrical signaling among cells within each taste bud.

Thus far, the only demonstrated neurochemical modulator of taste cell activity is serotonin. The organization of each taste bud is complex, consisting of numerous cell types (Fährmann 1967; Farbman and Yonkers 1971) that have been demonstrated ultrastructurally to have extensive synaptic communication among them (DeHan and Graziadei 1971; Delay and Roper 1988; Sbarbati *et al.* 1988). In particular, the serotonergic basal cells that ring the base of each taste bud, synapse upon taste cells, and taste cells, in turn, synapse upon the serotonergic cells. Elegant work in mudpuppies has shown that these synapses between cells within a taste bud have functional consequences. In a series of experiments, Ewald and Roper (1992b, 1994) and Roper (1992) simultaneously impaled adjacent receptor cells and basal cells with microelectrodes. When taste receptor cell apices were stimulated with a potassium salt solution, receptor potentials were recorded in taste receptor cells with a <75 msec delay. Depolarizing potentials also were recorded in adjacent basal cells in response to chemostimulation of the taste bud apex; however, the changes in membrane potential were smaller than those of taste receptor cells. Basal cell potentials were significantly delayed (>75 msec) with respect to the timing of the receptor cell stimulation and receptor potential generation. In addition, low concentrations of taste stimuli or low level current injections evoked receptor potentials in taste receptor cells, but were below the threshold necessary to induce a change in membrane potential of basal cells. These findings indicate that the response of basal cells to the stimulation of taste receptor cells is mediated by a chemical synapse.

Given the ultrastructural data suggesting the existence of bi-directional synapses between taste cells and basal cells (Delay and Roper 1988), Ewald and Roper (1994) tested whether excitation of basal cells could modulate the excitable response of taste receptor cells to taste stimuli. After repeated stimulation of basal cells via current injection, responses of taste receptor cells to taste stimuli were enhanced. Local application of serotonin, the neurotransmitter found in excitable basal cells, also enhanced the excitability of receptor cells (Delay *et al.* 1993, 1994). More recently, Delay and coworkers (1997) have shown that serotonin specifically modulates calcium currents of taste receptor cells. In some taste receptor cells, calcium conductances are enhanced, while in a larger subset of them calcium currents are reduced by serotonin. Thus, it appears that basal cells within the taste buds of mudpuppies serve a neuromodulatory function via serotonergic signaling.

In addition to chemically mediated intercellular interactions, electrical coupling is evident between taste receptor cells in taste organs of both anurans (Sata *et al.* 1992) and urodeles (West and Bernard 1978; Yang and Roper 1987; Bigiani and Roper 1994, 1995). In mudpuppies, an average two or three, but as many as five, taste receptor cells within a single taste bud may be dye-coupled. Whether coupling is restricted to cells belonging to a specific cell type, e.g., light cells, or is a feature of cells at a particular stage of their lineage, is not known. However, coupled taste cells have similar electrical properties (Bigiani and Roper 1995). Specifically, coupled cells have the full complement of voltage-dependent currents found in dark cells and some light cells (McPheeters *et al.* 1994; Bigiani *et al.* 1996). Given that only excitable taste cells are coupled, regulation of electrical coupling between taste receptor cells has been proposed as a means of transduction of sour or acidic stimuli (Bigiani and Roper 1994). Application of acidic stimuli reduces electrical coupling between taste cells and thus increases the membrane resistance of individual taste receptor cells, such that any change in membrane conductance in response to taste stimuli at the cell's apical microvillus would be enhanced (Bigiani and Roper 1995). Alternatively, maintenance of junctional conductances between taste cells may serve to synchronize the responses of taste cells to chemical stimulation (Bigiani and Roper 1995).

C. Physiology of the Facial and Glossopharyngeal Nerves

Most data on the quality of taste stimuli detected by amphibians have come from extracellular recordings of the lingual or posttrematic branch of the glossopharyngeal nerve and the palatine ramus of the facial nerve. These rami innervate the majority of the taste organs on the tongue and palate, respectively, of frogs and salamanders (see above). In both frogs (Pumphrey 1935; Kusano and Sato 1957; Ishiko et al. 1979; Yoshii et al. 1981; Sato et al. 1983; Hanamori et al. 1990) and aquatic salamanders (Samanen and Bernard 1981a, 1981b; McPheeters and Roper 1985; Takeuchi et al. 1994), the posttrematic branch of the IXth nerve is sensitive to bitter, salty and sour stimuli applied to taste organs on the tongue. Responses to sweet stimuli typically are not observed in aquatic salamanders (McPheeters and Roper 1985; Takeuchi et al. 1994), but have been reported (Samanen and Bernard 1981a, 1981b). Sensitivity to sweet substances is present in a number of frog species (Kusano and Sato 1957; Kusano 1960; Sugimoto and Sato 1978; Hanamori et al. 1990; Honda et al. 1994).

Typically, the response of the whole glossopharyngeal nerve increases as the concentration of the taste stimulus is increased, regardless of the tastant employed (Sato 1976; Samanen and Bernard 1981a). However, when individual nerve units are examined, a variety of response profiles are detected (Samanen and Bernard 1981b; Hanamori et al. 1990; Kitada 1990). For example, some fibers respond to only one tastant, whereas the majority of fibers respond to two or more taste stimuli, but with differing sensitivities (Hanamori et al. 1990). Furthermore, some fibers display reduced excitability in response to particular stimuli (Samanen and Bernard 1981b). Interestingly, gustatory fibers also respond to mechanical stimulation of the taste organs (Kusano and Sato 1957; Rapuzzi and Casella 1965; Samanen and Bernard 1981b; Hanamori et al. 1990). Whether this touch sensitivity is mediated by modulation of the activity of taste receptor cells by mechanical stimulation of the Merkel-like basal cells (Ewald and Roper 1992b, 1994), or directly reflects mechanosensitivity of taste receptor cells is not known. However, several classes of fibers within the lingual nerve are sensitive to both touch and taste stimuli.

The main assumption made in all of these studies is that the stimulus categories, sweet, sour, bitter and salt, used to stimulate amphibian taste organs are a reasonable approximation of the types of chemical signals these animals might be encountering during normal feeding. This is admittedly merely an educated guess based on studies of mammals, humans in particular (Lindemann 1996); studies of taste-mediated behaviour in amphibians are sparse (Bowerman and Kinnamon 1994; Takeuchi et al. 1997). Mechanisms of transduction of specific taste stimuli at the cellular level vary substantially across species, and it is believed that these differences have arisen evolutionarily due to differences in diet (Kinnamon and Cummings 1992; Kinnamon and Margolskee 1996). It is also possible that diet may have played an important role in determining the types of chemical stimuli detected by different species; these stimuli may not be restricted necessarily to conventional taste categories.

For example, amino acids are produced by most animals, and may be detected by amphibian predators. Indeed, robust glossopharyngeal nerve responses have been recorded when various amino acids are applied to taste organs on the tongue (Yoshii et al. 1981, 1982; Gordon and Caprio 1985; McPheeters and Roper 1985; Takeuchi et al. 1994). Not surprisingly, species-specific differences exist with respect to the amino acid(s) that can trigger a threshold IXth nerve response. For example, adult *Xenopus laevis* are sensitive to arginine in the range of 0.1–1.0 μM (Yoshii et al. 1982), whereas adult *Rana sphenocephala* have a threshold response of the lingual nerve to arginine of 7.5–10 mM (Gordon and Caprio 1985). In contrast, *R. sphenocephala* produces a robust lingual nerve response to aspartic acid at <10 mM (Gordon and Caprio 1985), while *X. laevis* is completely unresponsive to this amino acid (Yoshii et al. 1982). These differences are not directly attributable to differences in adult habitat, i.e., aquatic versus terrestrial, since the lingual nerve response profiles of *R. sphenocephala* (Gordon and Caprio 1985) and *R. catesbeiana* (Yoshii et al. 1982), both terrestrial frogs, are also quite distinct.

The taste disks of ranid frogs are also sensitive to fresh water (Zotterman 1949; Kusano and Sato 1957; Kusano 1960; Chernetski 1964; Suzuki 1966; Sato 1976; Hanamori et al. 1990; Kitada 1990). Responses by the glossopharyngeal nerve are readily recorded when taste organs adapted to a low salt Ringer's solution are subsequently exposed to de-ionized water. The water response can be eliminated by simply increasing the concentration of NaCl to low levels. Recently, the water response has been investigated using both whole-nerve and intracellular recordings (Okada et al. 1993). What had initially been thought to be a direct response to water seems now more likely to reflect detection of chloride ions; thus, the water response is actually a response to decreased chloride concentration. This ability to detect water, counterbalanced by an ability to detect salts, has important implications for the role of taste in osmoregulation. Zotterman (1949) proposed that water sensitivity was involved in behavioural osmoregulation, i.e., frogs would keep their mouths closed when in water, thereby avoiding becoming hypoosmotic. Additionally, it has been suggested that the interplay between salt and water stimulation may be critical to maintaining osmotic balance by regulating ingestive behaviour (Lindemann 1996).

D. Coding of Taste Information

In amphibians, as in mammals, all regions of the tongue are generally sensitive to all taste stimuli, (Kusano and Sato 1957; Ishiko *et al.* 1979; Samanen and Bernard 1981a), although some regions of the tongue are more sensitive than others to gustatory stimuli (Samanen and Bernard 1981a; Sato *et al.* 1983). Given the relatively homogeneous organization of the taste periphery, how is taste information coded such that the animal can make sense of sapid stimuli?

In the tongue of ranid frogs, a single gustatory neurite branches to innervate seven or eight taste disks (Rapuzzi and Casella 1965; Taglietti *et al.* 1969; Hanamori *et al.* 1990), whereas the nerve endings of approximately nine different gustatory neurons innervate each taste disk (Rapuzzi and Casella 1965). Furthermore, a given taste disk has two to four neurons in common with neighbouring taste disks (Rapuzzi and Casella 1965). Thus, when a single taste disk is stimulated sufficiently to fire an action potential, this electrical event may reach over 40 taste organs via the interconnection of nerve fibers (Rapuzzi and Casella 1965). One can imagine the electrical mêlée produced in the peripheral taste organs and nerve fibers by a taste stimulus that is present over a wide area of the oropharynx. The non-topographical organization of the gustatory periphery is in stark contrast to the maps of the receptive field of touch-sensitive fibers innervating the taste disks; these fibers branch to innervate on average only three taste disks, and there is no overlap of these fibers across neighbouring taste organs (Rapuzzi and Casella 1965; Taglietti *et al.* 1969; Hanamori *et al.* 1990). Thus, it is easy to envisage a somatotopic map of the tongue, but a topographic gustatory map is unlikely.

A functional organization of the taste periphery is reflected in the presence of discrete categories of gustatory fibers innervating the tongue. As many as 14 classes have been identified in bullfrogs which, while able to respond to a variety of taste stimuli, have a "best" stimulus, i.e., the largest electrical response to a particular tastant (Kusano 1960; Hanamori *et al.* 1990). These range from units that are sensitive to a single stimulus type, to those that respond to all stimuli. However, the latter category is relatively rare, making up less than 10% of the fibers. The majority of glossopharyngeal neurites are selectively sensitive to two or three discrete stimuli with a single "best" stimulus. Several of these types of gustatory fibers are also mechanosensitive. A comparable type of organization of "best" taste-stimulus fibers coupled with bimodal sensitivity (i.e., touch and taste) also have been demonstrated in mammals (Frank 1973, 1991; Hanamori *et al.* 1990).

In bullfrogs, six discrete "best" categories of cells are also present in the NTS of the hindbrain (Hanamori *et al.* 1987). The presence of fewer types of NTS cells than types of IXth nerve fibers suggests that there is convergence, as well as integration, of gustatory information at the sensory neuron–NTS synapse. It is not known, however, if a particular category of NTS cell receives information from a specific subset of gustatory afferent types. In mammals, coding of taste information in the hindbrain appears to occur via comparison of the simultaneous inputs of large numbers of fibers that are differentially responding to any given taste stimulus. However, how central processing of taste information occurs in amphibians is unknown.

VII. SUMMARY

The taste system of amphibians consists of an array of peripheral taste buds distributed throughout the oropharyngeal cavity. Bitter, sweet, salty and sour gustatory stimuli are converted to electrochemical signals by taste receptor cells whose activity is modulated by local serotonergic cells within each taste bud. Thus, integrated electrochemical signals likely are achieved in the taste periphery. Taste receptor cells and serotonergic cells synapse chemically upon the fine nerve endings of sensory neurons that have their cell bodies in the ganglia of the VIIth, IXth and Xth cranial nerves. These neurons then transmit gustatory and tactile information to the appropriate central nuclei within the hindbrain.

This general anatomical scenario is consistent across amphibian taxa; however, the morphology of the taste buds themselves varies considerably. In particular, substantial differences exist between terrestrial and aquatic amphibians, perhaps due to specific adaptations for tasting in air and water, respectively. Aquatic salamanders possess relatively simple taste buds, i.e., aggregates of less than 100 fusiform cells embedded directly in the epithelium of the oropharynx, while terrestrial frogs and toads have enlarged, complexly organized taste organs (taste disks) that often are found atop large epithelial papillae. The taste organs of terrestrial salamanders have yet to be examined systematically. In addition, tongue morphology varies widely across amphibian taxa, but how this variation impacts organization of the taste periphery has not been investigated.

In the few species of amphibians that have been examined, the taste system is sensitive to a variety of taste stimuli, including those known to elicit responses in mammals. However, taste stimuli relevant to the natural environment of any amphibian have not been identified. Because most amphibians are primarily visual feeders, the biological importance of taste likely resides in the ability of these animals to detect and reject unpalatable, potentially dangerous prey once it is captured.

VIII. ACKNOWLEDGEMENTS

I thank Chris Braun, Heather Eisthen, Sue Kinnamon, Tom Finger and Glenn Northcutt for careful comments on drafts of this chapter, and Harold Heatwole and Ellen Dawley for their expert editing.

IX. REFERENCES

Abe, H., Watanabe, M. and Kondo, H., 1992. Transient appearance of Ca-binding protein (Spot 35-calbindin) in bronchial epithelial cells, thyroid parafollicular cells and thymic epithelial cells during development of rats. *Histochem.* **97**: 155–160.

Adelmann, H. B., 1932. The development of the prechordal plate and mesoderm of *Amblystoma punctatum*. *J. Morphol.* **54**: 1–67.

Altig, R. and Brodie, E. D., 1971. Foods of *Plethodon larselli*, *Plethodon dunni* and *Ensatina eschscholtzi* in the Columbia River Gorge, Multnomah County, Oregon. *Am. Midl. Nat.* **62**: 226–228.

Anderson, C. W., 1993. The modulation of feeding behavior in response to prey type in the frog *Rana pipiens*. *J. Exp. Biol.* **179**: 1–12.

Anderson, J. D., 1960. Cannibalism in *Dicamptodon ensatus*. *Herpetologica* **16**: 260.

Anderson, J. D., 1963. Reactions of the western mole to skin secretions of *Ambystoma macrodactylum croceum*. *Herpetologica* **19**: 282–284.

Anderson, J. D., 1968. A comparison of the food habits of *Ambystoma macrodactylum sigillatum*, *Ambystoma macrodactylum croceum*, and *Ambystoma tigrinum californiense*. *Herpetologica* **24**: 273–284.

Attar, E. N. and Maly, E. J., 1980. A laboratory study of preferential predation by the newt, *Notophthalmus v. viridescens*. *Can. J. Zool.* **58**: 1712–1717.

Avenet, P. and Lindemann, B., 1987. Patch-clamp study of isolated taste receptor cells of the frog. *J. Memb. Biol.* **97**: 223–240.

Avenet, P. and Lindemann, B., 1988. Amiloride-blockable sodium currents in isolated taste receptor cells. *J. Memb. Biol.* **105**: 245–255.

Baimbridge, K. G., Celio, M. R. and Rogers, J. H., 1992. Calcium-binding proteins in the nervous system. *Trends Neurosci.* **15**: 303–308.

Baldauf, R., 1947. *Desmognathus f. fuscus* eating eggs of its own species. *Copeia* **1947**: 66.

Barlow, L. A. and Northcutt, R. G., 1995. Embryonic origin of amphibian taste buds. *Dev. Biol.* **169**: 273–285.

Barlow, L. A. and Northcutt, R. G., 1997. Taste buds develop autonomously from endoderm without induction by cephalic neural crest or paraxial mesoderm. *Development* **124**: 949–957.

Barlow, L. A., Chien, C.-B. and Northcutt, R. G., 1996. Embryonic taste buds develop in the absence of innervation. *Development* **122**: 1103–1111.

Beidler, L. M. and Smallman, R. L., 1965. Renewal of cells within taste buds. *J. Cell Biol.* **27**: 263–272.

Belecky, T. L. and Smith, D. V., 1990. Postnatal development of palatal and laryngeal taste buds in the hamster. *J. Comp. Neurol.* **293**: 646–654.

Bigiani, A. and Roper, S. D., 1993. Identification of electrophysiologically distinct cell subpopulations in *Necturus* taste buds. *J. Gen. Physiol.* **102**: 143–170.

Bigiani, A. and Roper, S. D., 1994. Reduction of electrical coupling between *Necturus* taste receptor cells, a possible role in acid taste. *Neurosci. Letters* **176**: 212–216.

Bigiani, A. and Roper, S. D., 1995. Estimation of the junctional resistance between electrically coupled receptor cells in *Necturus* taste buds. *J. Gen. Physiol.* **106**: 705–725.

Bigiani, A. R. and Roper, S. D., 1991. Mediation of responses to calcium in taste cells by modulation of a potassium conductance. *Science* **252**: 126–128.

Bigiani, A., Kim, D.-J. and Roper, S. D., 1996. Membrane properties and cell ultrastructure of taste receptor cells in *Necturus* lingual slices. *J. Neurophysiol.* **75**: 1944–1956.

Blair, F. W., 1976. Adaptations of anurans to equivalent desert scrub of North and South America. Pp. 197–222 *in* Evolution of Desert Biota, ed by D. W. Goodall. Texas Press, Austin.

Bordzilovskaya, N. P., Dettlaff, T. A., Duhon, S. T. and Malacinski, G. M., 1989. Developmental-stage series of Axolotl embryos. Pp. 201–219 *in* Developmental Biology of the Axolotl, ed by J. B. Armstrong and G. M. Malacinski. Oxford University Press, Oxford.

Bowerman, A. G. and Kinnamon, S. C., 1994. The significance of apical K^+ channels in mudpuppy feeding behavior. *Chem. Senses* **19**: 303–315.

Bragg, A. N., 1946. Aggregation with cannibalism in tadpoles of *Scaphiopus bombifrons* with some general remarks on the probable evolutionary significance of such phenomena. *Herpetologica* **3**: 89–97.

Bragg, A. N., 1960. Experimental observations on the feeding of spadefoot tadpoles. *SW Nat.* **5**: 201–207.

Bragg, A. N., 1964. Further study of predation and cannibalism in Spadefoot tadpoles. *Herpetologica* **20**: 12–24.

Bragg, A. N., 1965. Gnomes of the Night: The Spadefoot Tadpoles. U. Penn. Press, Philadelphia.

Breathnach, A. S., 1980. The mammalian and avian Merkel cell. Pp. 283–292 *in* The Skin of Vertebrates, ed by R. I. C. Spearman and P. A. Riley. Academic Press, London.

Brodie, E. D. and Tumbarello, M. S., 1978. The antipredator functions of *Dendrobates auratus* (Amphibia, Anura, Dendrobatidae) skin secretion in regard to a snake predator *(Thamnophis). J. Herpetol.* **12**: 264–265.

Brodie, E. D., Hensel, J. L. and Johnson, J. A., 1974. Toxicity of the urodele amphibians *Taricha, Notophthalmus, Cynops* and *Paramesotriton* (Salamandridae). *Copeia* **1974**: 506–511.

Brodie, E. D., Nowak, R. T. and Harvey, W. R., 1979. The effectiveness of antipredator secretions and behavior of selected salamanders against shrews. *Copeia* **1979**: 270–274.

Brodie, E. D. J., Formanowicz, D. R. and Brodie, E. D. I., 1978. The development of noxiousness of *Bufo americanus* tadpoles to aquatic insect predators. *Herpetologica* **34**: 302–306.

Brust, D. G., 1993. Maternal brood care by *Dendrobates pumilio:* a frog that feeds its young. *J. Herpetol.* **27**: 96–98.

Caprio, J., 1992. Peripheral mechanisms of signal processing for amino acid taste. Pp. 331–353 *in* The Science of Food Regulation. Food Intake, Taste, Nutrient Partitioning, and Energy Expenditure, ed by G. A. Bray and D. H. Ryan. Louisiana State Univ. Press, Baton Rouge.

Caprio, J., Brand, J. G., Teeter, J. H., Valentincic, T., Kalinoski, D. L., Kohbara, J., Kumazawa, T. and Wegert, S., 1993. The taste system of the channel catfish: from biophysics to behavior. *Trends Neurosci.* **16**: 192–197.

Carmignani, M. P. A. and Zaccone, G., 1977. Histochemical studies on the tongue of anuran amphibians-II. Comparative morphochemical study of the taste buds and the lingual glands in *Bufo viridis* Laurenti and *Rana graeca* Boulenger with particular reference to the mucosaccharide histochemistry. *Cell. Molec. Biol.* **22**: 203–217.

Carmignani, M. P. A., Zaccone, G. and Cannata, F., 1975. Histochemical studies on the tongue of anuran amphibian: I. Mucopolysaccharide histochemistry of the papillae and the lingual glands in *Hyla arborea* L., *Rana esculenta* L., and *Bufo vulgaris* Laur. *Ann. Histochim.* **20**: 47–65.

Cei, J., 1955. Chacoan batrachians in central Argentina. *Copeia* **1955**: 291–293.

Chernetski, K. E., 1964. Sympathetic enhancement of peripheral sensory input in the frog. *J. Neurophysiol.* **24**: 493–515.

Christian, K. A., 1982. Changes in the food niche during postmetamorphic ontogeny of the frog *Pseudacris triseriata. Copeia* **1982**: 73–80.

Coghill, G. E., 1902. The cranial nerves of *Ambylstoma tigrinum. J. Comp. Neurol.* **12**: 207–289.

Cohen, N. and Howard, W., 1958. Bullfrog food and growth at the San Joaquin experimental range, California. *Copeia* **1958**: 223–224.

Collins, J. P., 1981. Distribution, habitats, and life history variation in the tiger salamander, *Ambystoma tigrinum,* in east, central and south-west Arizona. *Copeia* **1981**: 666–675.

Collins, J. P. and Cheek, J. E., 1983. Effect of food and density on development of typical and cannibalistic salamander larvae in *Ambystoma tigrinum nebulosum. Am. Zool.* **23**: 77–84.

Collins, J. P., Mitton, J. B. and Pierce, B. A., 1980. *Ambystoma tigrinum:* a multispecies conglomerate? *Copeia* **1980**: 938–941.

Cook, M. H. and Neal, H. V., 1921. Are the taste buds of elasmobranchs endodermal in origin? *J. Comp. Neurol.* **33**: 45–63.

Couly, G. F., Coltey, P. M. and Le Douarin, N. M., 1992. The developmental fate of the cephalic mesoderm in quail-chick chimeras. *Development* **114**: 1–15.

Crump, M. L., 1983. Opportunistic cannibalism by amphibian larvae in temporary aquatic environments. *Am. Nat.* **121**: 281–287.

Cummings, T. A. and Kinnamon, S. C., 1992. Apical K^+ channels in *Necturus* taste cells. Modulation by intracellular factors and taste stimuli. *J. Gen. Physiol.* **99**: 591–613.

Cummings, T., Delay, R. J. and Roper, S. D., 1987. Ultrastructure of apical specializations of the taste cells in the mudpuppy, *Necturus maculosus. J. Comp. Neurol.* **261**: 604–615.

Cunningham, J., 1954. A case of cannibalism in the toad *Bufo boreas halophilus. Herpetologica* **10**: 166.

Daly, J. W. and Witkop, B., 1971. Chemistry and pharmacology of frog venoms. Pp. 497–520 *in* Venomous Animals and their Venoms, ed by W. Bücherl and E. E. Buckley. Academic Press, New York.

Daly, J. W., Brown, G. B., Mensah-Dwumah, M. and Myers, C. W., 1978. Classification of skin alkaloids from neotropical poison-dart frogs (Dendrobatidae). *Toxicon* **16**: 163–188.

Daly, J. W., Myers, C. W., Warnick, J. E. and Albuquerque, E. X., 1980. Levels of batrachotoxin and lack of sensitivity to its action in poison-dart frogs *(Phyllobates)*. *Science* **208**: 1383–1385.

Daly, J. W., Gusovsky, F., Mcneal, E. T., Secunda, S., Bell, M., Creveling, C. R., Nishizawa, Y., Overman, L. E., Sharp, M. J. and Rossignol, D. P., 1990. Pumiliotoxin alkaloids: A new class of sodium channel agents. *Biochem. Pharmacol.* **40**: 315–326.

Dawley, E., 1997. Olfaction. Pp. 711–742 *in* Sensory Perception, Vol. 3 of Amphibian Biology, ed by H. Heatwole and E. Dawley. Surrey Beatty & Sons, Chipping Norton.

DeHan, R. and Graziadei, P. P. C., 1971. Functional anatomy of frog's taste organs. *Experientia* **27**: 823–826.

DeHan, R. S. and Graziadei, P. P. C., 1973. The innervation of frog's taste organ. A histochemical study. *Life Sci.* **13**: 1435–1449.

Delay, R. J. and Roper, S. D., 1988. Ultrastructure of taste cells and synapses in the mudpuppy *Necturus maculosus*. *J. Comp. Neurol.* **277**: 268–280.

Delay, R. J., Kinnamon, J. C. and Roper, S. D., 1986. Ultrastructure of mouse vallate taste buds: II. cell types and cell lineage. *J. Comp. Neurol.* **253**: 242–252.

Delay, R. J., Kinnamon, S. C. and Roper, S. D., 1997. Serotonin modulates voltage-dependent calcium current in *Necturus* taste cells. *J. Neurophysiol.* **77**: 2515–2524.

Delay, R. J., Mackay-Sim, A. and Roper, S. D., 1994. Membrane properties of two types of basal cells in *Necturus* taste buds. *J. Neurosci.* **14**: 6132–6143.

Delay, R. J., Taylor, R. and Roper, S. D., 1993. Merkel-like basal cells in *Necturus* taste buds contain serotonin. *J. Comp. Neurol.* **335**: 606–613.

Dhouailly, D., 1984. Specification of feather and scale patterns. Pp. 581–602 *in* Pattern Formation, ed by G. M. Malacinski and S. V. Bryant. Macmillan Publishing Co., New York.

Donnelly, M. A., 1991. Feeding patterns of the strawberry poison frog, *Dendrobates pumilio* (Anura: Dendrobatidae). *Copeia* **1991**: 723–730.

Duellman, W. E. and Trueb, L, 1986. The Biology of the Amphibia. McGraw-Hill Book Co., New York.

Dunn, E. R., 1926. The frogs of Jamaica. *Proc. Boston Soc. Nat. Hist.* **38**: 111–130.

Düring, M. V. and Andres, K. H., 1976. The ultrastructure of taste and touch receptors of the frog's taste organ. *Cell Tiss. Res.* **165**: 185–198.

Dushin, A. I., 1975. Diet of two frog species in fishery ponds of the Mordovian (ASSR). *Soviet J. Ecol.* **5**: 87–90.

Ellinson, R. P., 1994. Leg development in a frog without a tadpole *(Eleuterodactylus coqui)*. *J. Exp. Zool.* **270**: 202–210.

Erspamer, V., 1994. Bioactive secretions of the amphibian integument. Pp. 178–350 *in* The Integument, Vol. 1 of Amphibian Biology, ed by H. Heatwole, G. Barthalmus and A. Heatwole. Surrey Beatty & Sons, Chipping Norton.

Ewald, D. A. and Roper, S. D., 1992a. Basal cells modulate receptor cell function in *Necturus* taste buds by a serotonergic mechanism. *Soc. Neurosci. Abs.* **18**: 596.

Ewald, D. A. and Roper, S. D., 1992b. Intercellular signaling in *Necturus* taste buds: Chemical excitation of receptor cells elicits responses in basal cells. *J. Neurophysiol.* **67**: 1316–1324.

Ewald, D. A. and Roper, S. D., 1994. Bidirectional synaptic transmission in *Necturus* taste buds. *J. Neurosci.* **14**: 3791–3804.

Eycleshymer, A. C., 1906. The habits of *Necturus maculosus*. *Am. Nat.* **60**: 123–136.

Fährmann, W., 1967. Licht- und electronenmikroskopische Untersuchungen an der Geschmacksknospe des neotenen Axolotls *(Siredon mexicanum* Shaw). *Zeit. Microsk. Anat. Forsch.* **77**: 117–152.

Farbman, A. I., 1965. Electron microscope study of the developing taste bud in rat fungiform papilla. *Dev. Biol.* **11**: 110–135.

Farbman, A. I., 1972. The taste bud: a model system for developmental studies. Pp. 109–123 *in* Developmental Aspects of Oral Biology, ed by H. C. Slavkin and L. A. Bavetta. Academic Press, New York.

Farbman, A. I. and Mbiene, J.-P., 1991. Early development and innervation of taste bud-bearing papillae on the rat tongue. *J. Comp. Neurol.* **304**: 172–186.

Farbman, A. L. and Yonkers, J. D., 1971. Fine structure of the taste bud in the mudpuppy, *Necturus maculosus*. *Am. J. Anat.* **12**: 328–350.

Finger, T. E., 1986. Peptide immunohistochemistry demonstrates multiple classes of perigemmal nerve fibers in the circumvallate papilla of the rat. *Chem. Senses* **11**: 135–144.

Finger, T. E., 1987. Gustatory nuclei and pathways in the central nervous system. Pp. 331–354 *in* Neurobiology of Taste and Smell, ed by T. E. Finger and W. L. Silver. John Wiley and Sons, New York.

Finger, T. E., 1988. Sensorimotor mapping and oropharyngeal reflexes in goldfish, *Carassius auratus*. *Brain Behav. Evol.* **31**: 17–24.

Formanowicz, D. R. and Brodie, E. D., 1982. Relative palatabilities of members of a larval amphibian community. *Copeia* **1982**: 91–97.

Frank, M. E., 1973. An analysis of hamster afferent taste nerve response functions. *J. Gen. Physiol.* **61**: 588–618.

Frank, M. E., 1991. Taste-responsive neurons of the glossopharyngeal nerve of the rat. *J. Neurophysiol.* **65**: 1452–1463.

Fritzsch, B., Sarai, P. A., Barbacid, M. and Silos-Santiago, I., 1997. Mice lacking the neurotrophin receptor trkB lose their specific afferent innervation but do develop taste buds. *Intl. J. Dev. Neurosci.* **15**: 563–576.

Fujimoto, S. and Murray, R. G., 1970. Fine structure of degeneration and regeneration in denervated rabbit vallate taste buds. *Anat. Rec.* **168**: 393–414.

Fujiyama, R., Miyamoto, T. and Sato, T., 1994. Differential distribution of two Ca^{++}-dependent and -independent K^+ channels throughout the receptive and basolateral membrane of bullfrog taste cells. *Pfluegers Arch.* **429**: 285–290.

Gans, C., 1961. The bullfrog and its prey. A look at the biomechanics of jumping. *Nat. Hist.* **52**: 26–37.

Gans, C. and Northcutt, R. G., 1983. Neural crest and the origin of vertebrates: a new head. *Science* **220**: 268–274.

Goosens, N. and Vandenberghe, M.-P., 1974. The basal cells in the papillae fungiformes of the tongue of the common frog, *Rana temporaria*. L. *Arch. Histol. Jap.* **36**: 173–179.

Gordon, K. D. and Caprio, J., 1985. Taste responses to amino acids in the Southern leopard frog, *Rana sphenocephala*. *Comp. Biochem. Physiol.* **81A**: 525–530.

Graziadei, P. P. C. and DeHan, R. S., 1971. The ultrastructure of frogs' taste organs. *Acta anat.* **80**: 563–603.

Griffiths, L. G., 1963. The phylogeny of the Salientia. *Biol. Rev.* **38**: 241–292.

Grobler, J., 1972. Observations on the amphibian *Pyxicephalus adspersus* Tschudi in Rhodesia. *Arnoldia* **6**: 1–4.

Guth, L., 1957. The effects of glossopharyngeal nerve transection on the circumvallate papilla of the rat. *Anat. Rec.* **128**: 715–731.

Habermehl, G., 1971. Toxicology, pharmacology, chemistry and biochemistry of salamander venom. Pp. 569–587 *in* Venomous Animals and their Venoms, ed by W. Bücherl and E. E. Buckley. Academic Press, New York.

Hall, B. K., 1981. The induction of neural crest-derived cartilage and bone by embryonic epithelia: an analysis of the mode of action of an epithelial-mesenchymal interaction. *J. Embryol. Exp. Morph.* **64**: 305–320.

Hall, B. K., 1983. Epithelial-mesenchymal interactions in cartilage and bone development. Pp. 189–214 *in* Epithelial-Mesenchymal Interactions in Development, ed by R. H. Sawyer and J. F. Fallon. Praeger Publishers, New York.

Hamilton, R., 1948. The egg-laying process in the tiger salamander. *Copeia* **1948**: 212–213.

Hamilton, W. J., 1940. The feeding habits of larval newts with reference to availability and predilection of food items. *Ecology* **21**: 351–356.

Hanamori, T. and Ishiko, N., 1983. Intraganglionic distribution of the primary afferent neurons in the frog glossopharyngeal nerve and its transganglionic projection to the rhombencephalon studied by the HRP method. *Brain Res.* **260**: 191–199.

Hanamori, T., Hirota, K. and Ishiko, N., 1990. Receptive fields and gustatory responsiveness of frog glossopharyngeal nerve. A single fiber analysis. *J. Gen. Physiol.* **95**: 1159–1182.

Hanamori, T., Ishiko, N. and Smith, D. V., 1987. Multimodal responses of taste neurons in the frog nucleus tractus solitarius. *Brain Res. Bull.* **18**: 87–97.

Heatwole, H., 1997. Diffuse cutaneous and muscular sensory systems: Mechanoreception, thermoreception, nociception, chemoreception and kinesthetic sense. Pp. 937–953 *in* Sensory Perception, Vol. 3 of Amphibian Biology, ed by H. Heatwole and E. Dawley. Surrey Beatty & Sons, Chipping Norton.

Heatwole, H. and Test, F., 1961. Cannibalism in the salamander, *Plethodon cinereus*. *Herpetologica* **17**: 143.

Helff, O. M. and Mellicker, M. C., 1941a. Studies on amphibian metamorphosis. XIX. Development of the tongue in *Rana sylvatica*, including the histogenesis of "premetamorphic" and filiform papillae and the mucous glands. *Am. J. Anat.* **68**: 339–368.

Helff, O. M. and Mellicker, M. C., 1941b. Studies on amphibian metamorphosis. XX. Development of the fungiform papillae of the tongue in *Rana sylvatica*. *Am. J. Anat.* **68**: 371–394.

Herrick, C. J., 1944. The fasciculus solitarius and its connections in amphibians and fishes. *J. Comp. Neurol.* **81**: 307–331.

Herrick, C. J., 1948. The Brain of the Tiger Salamander. University of Chicago Press, Chicago.

Heusser, H., 1970. Spawn eating by tadpoles as possible cause of specific biotype preferences and short breeding times in European anurans (Amphibia, Anura). *Oecologia* **4**: 83–88.

Heusser, H., 1971. Laich Räubern und kannibalismus bei sympatrischen anurans Kaulquappen. *Experientia* **27**: 474–475.

Heyer, W. R., McDiarmid, R. W. and Weigmann, D. L., 1975. Tadpoles, predation and pond habitats in the tropics. *Biotropica* **7**: 100–111.

Highton, R. and Savage, J., 1961. Functions of brooding behavior in female red-backed salamander, *Plethodon cinereus*. *Copeia* **1961**: 95–98.

Hille, B., 1992. Ionic Channels of Excitable Membranes. Sinauer, Sunderland, MA.

Hirata, K. and Kanaseki, T., 1987. Substance P-like immunoreactive fibers in the frog taste organs. *Experientia* **43**: 386–388.

Hirata, K. and Nada, O., 1975. A monoamine in the gustatory cell of the frog's taste organ. *Cell Tiss. Res.* **159**: 101–108.

Honda, E., Toyoshima, K., Hirakawa, T., Nakamura, S. and Nakahara, S., 1994. Structure and physiological properties of the taste organs on the ventral side of frog tongue (*Rana catesbeiana*). *Chem. Senses* **19**: 231–238.

Hörstadius, S., 1950. The Neural Crest: Its Properties and Derivatives in the Light of Experimental Research. Oxford University Press, Oxford.

Hosley, M. A., Hughes, S. E. and Oakley, B., 1987a. Neural induction of taste buds. *J. Comp. Neurol.* **260**: 224–232.

Hosley, M. A., Hughes, S. E., Morton, L. L. and Oakley, B., 1987b. A sensitive period for the neural induction of taste buds. *J. Neurosci.* **7**: 2075–2080.

Hourdry, J., L'Hermite, A. and Ferrand, R., 1996. Changes in the digestive tract and feeding behavior of anuran amphibians during metamorphosis. *Physiol. Zool.* **69**: 219–251.

Iggo, A. and Findlater, G. S., 1984. A review of Merkel cell mechanisms. Pp. 117–131 *in* Sensory Receptor Mechanisms, ed by W. Hamann and A. Iggo. World Scientific Publ., Singapore.

Ishiko, N., Hanamori, T. and Murayama, N., 1979. Frog's tongue receptive areas: neural organization and gustatory function. *Experientia* **35**: 773–774.

Iwasaki, S. and Kobayashi, K., 1988. Fine structure of the dorsal tongue surface in the Japanese toad, *Bufo japonicus* (Anura, Bufonidae). *Zool. Sci.* **5**: 331–336.

Iwasaki, S. and Wanichanon, C., 1991. Fine structure of the dorsal lingual epithelium of the frog, *Rana rugosa*. *Tissue Cell* **23**: 385–391.

Iwasaki, S., Miyata, K. and Kobayashi, K., 1986. Studies on the fine structure of the lingual dorsal surface in the frog, *Rana nigromaculata*. *Zool. Sci.* **3**: 265–272.

Jacobson, A. G. and Meier, S., 1984. Morphogenesis of the head of a newt: mesodermal segments, neuromeres, and distribution of neural crest. *Dev. Biol.* **106**: 181–193.

Jaeger, C. B. and Hillman, D. E., 1976. Morphology of gustatory organs. Pp. 588–606 *in* Frog Neurobiology: A Handbook, ed by R. Llinas and W. Precht. Springer-Verlag, Berlin.

Jain, S. and Roper, S. D., 1991. Immunocytochemistry of gamma-aminobutyric acid, glutamate, serotonin, and histamine in *Necturus* taste buds. *J. Comp. Neurol.* **307**: 675–682.

Jasinski, A., 1979. Light and scanning microscopy of the tongue and its gustatory organs in the common toad, *Bufo bufo* (L.). *Z. mikrosk.-anat. Forsch. Leipzig* **93**: 465–476.

Jasinski, A. and Miodonski, A., 1979. Light and scanning microscopy of the taste organs and vascularization of the tongue of the spotted salamander, *Salamandra salamandra* (L.). *Z. Microsk.-Anat. Forsch.* **93**: 780–792.

Jenssen, T. A., 1967. Food habits of the green frog, *Rana clamitans*, before and during metamorphosis. *Herpetologica* **29**: 66–72.

Johnston, J. B., 1910. The limit between the ectoderm and entoderm in the mouth, and the origin of taste buds. I. Amphibians. *Am. J. Anat.* **10**: 41–67.

Kalusche, D., 1973. Kaulquappen als Beute von Wasserfröschen. *Salamandra* **9**: 164–165.

Kaplan, R. and Sherman, P., 1980. Intraspecific oophagy in California newts. *J. Herpetol.* **14**: 183–185.

Kashiwayanagi, M., Miyake, M. and Kurihara, K., 1983. Voltage-dependent Ca^{2+} channel and Na^+ channel in frog taste cells. *Am. J. Physiol.* **244**: C82–C88.

Kelling, S. T., Halpern, B. P. and Eisner, T., 1990. Gustatory sensitivity of an anuran to cantharidin. *Experientia* **46**: 763–764.

Kenny, J. S., 1969. Pharyngeal mucous secreting epithelia of anuran larvae. *Acta Zool.* **50**: 143–153.

Kerschbaum, H. H. and Hermann, A., 1992. Calcium-binding proteins in chemoreceptors of *Xenopus laevis*. *Tiss. Cell* **24**: 719–724.

Kim, D.-J. and Roper, S. D., 1995. Localization of serotonin in taste buds: a comparative study in four vertebrates. *J. Comp. Neurol.* **353**: 363–370.

Kinnamon, J. C., 1987. Organization and innervation of taste buds. Pp. 277–297 *in* Neurobiology of Taste and Smell, ed by T. E. Finger and W. L. Silver. Wiley, New York.

Kinnamon, J. C., Sherman, T. A. and Roper, S. D., 1988. Ultrastructure of mouse vallate taste buds: III. Patterns of synaptic connectivity. *J. Comp. Neurol.* **270**: 1–10.

Kinnamon, J. C., Taylor, B. J., Delay, R. J. and Roper, S. D., 1985. Ultrastructure of mouse vallate taste buds. I. Taste cells and their associated synapses. *J. Comp. Neurol.* **235**: 48–60.

Kinnamon, S. C. and Cummings, T. A., 1992. Chemosensory transduction mechanisms in taste. *Ann. Rev. Neurosci.* **54**: 715–731.

Kinnamon, S. C. and Margolskee, R. F., 1996. Mechanisms of taste transduction. *Curr. Opin. Neurobiol.* **6**: 506–513.

Kinnamon, S. C. and Roper, S. D., 1987. Passive and active membrane properties of mud puppy taste receptor cells. *J. Physiol.* **383**: 601–614.

Kinnamon, S. C. and Roper, S. D., 1988a. Evidence for a role of voltage-sensitive apical K^+ channels in sour and salt taste transduction. *Chem. Senses* **13**: 115–121.

Kinnamon, S. C. and Roper, S. D., 1988b. Membrane properties of isolated mudpuppy taste cells. *J. Gen. Physiol.* **91**: 351–371.

Kinnamon, S. C., Dionne, V. E. and Beam, K. G., 1988. Apical localization of K^+ channels in taste cells provides the basis for sour taste transduction. *Proc. Natl. Acad. Sci. USA* **85**: 7023–7027.

Kinnman, I. and Aldskogius, H., 1988. Collateral reinnervation of taste buds after chronic sensory denervation: A morphological study. *J. Comp. Neurol.* **270**: 569–574.

Kirn, A., 1949. Cannibalism among *Rana pipiens berlandieri* and possibly by *Rana catesbeiana*, near Somerset, Texas. *Herpetologica* **5**: 84.

Kitada, Y., 1990. Taste responses to electrolytes in the frog glossopharyngeal nerve: enhancement by Ni^{2+} ions. *Dent. Japan* **27**: 41–44.

Kohbara, J., Michel, W. and Caprio, J., 1992. Responses of single facial taste fibers in the channel catfish, *Ictalurus punctatus*, to amino acids. *J. Neurophysiol.* **68**: 1012–1026.

Kolesnikov, S. S. and Margolskee, R. F., 1995. A cyclic-nucleotide-suppressible conductance activated by transducin in taste cells. *Nature* **376**: 85–88.

Kollar, E. J., 1983. Epithelial-mesenchymal interactions in the mammalian integument: Tooth development as a model for instructive induction. Pp. 27–49 *in* Epithelial-Mesenchymal Interactions in Development, ed by R. H. Sawyer and J. F. Fallon Praeger Publ., New York.

Kuramoto, H., 1988. An immunohistochemical study of cellular and nervous elements in the taste organ of the bullfrog, *Rana catesbeiana*. *Arch. Histol. Cytol.* **51**: 205–221.

Kusano, K., 1960. Analysis of the single unit activity of gustatory receptors in the frog tongue. *Jap. J. Physiol.* **10**: 620–633.

Kusano, K. and Sato, M., 1957. Properties of fungiform papillae in frog's tongue. *Jap. J. Physiol.* **7**: 324–338.

Laessle, A. M., 1961. A micro-limnological study of Jamaican bromeliads. *Ecology* **42**: 499–517.

Landacre, F. L., 1907. On the place of origin and method of distribution of taste buds in *Ameirus melas*. *J. Comp. Neurol.* **17**: 1–66.

Leff, L. G. and Bachmann, M. D., 1986. Ontogenetic changes in the predatory behavior of larval tiger salamanders *(Ambystoma tigrinum)*. *Can. J. Zool.* **64**: 1337–1344.

Lima, A. P. and Moreira, G., 1993. Effects of prey size and foraging mode on the ontogenetic change in feeding niche of *Colostethus stepheni* (Anura: Dendrobatidae). *Oecologia* **95**: 93–102.

Lindemann, B., 1996. Taste reception. *Physiol. Rev.* **76**: 719–766.

Link, R. E., Paus, R., Stenn, K. S., Kuklinska, E. and Moellmann, G., 1990. Epithelial growth by rat vibrissae follicles *in vitro* requires mesenchymal contact via native extracellular matrix. *J. Invest. Dermatol.* **95**: 202–207.

Loman, J., 1979. Food, feeding rates, and prey-size selection in juvenile and adult frogs, *Rana arvalis* and *R. temporaria. Ekol. Pol.* **27**: 581–601.

Lombard, R. E. and Wake, D. B., 1977. Tongue evolution in the lungless salamanders, family Plethodontidae. II. Function and evolutionary diversity. *J. Morph.* **153**: 39–80.

Lu, K.-S. and Roper, S. D., 1993. Electron microscopic immunocytochemistry of glutamate-containing nerve fibers in the taste bud of mud puppy *(Necturus maculosus). Microsc. Res. Tech.* **26**: 225–230.

Lumsden, A. G. S., 1988. Spatial organization of the epithelium and the role of neural crest cells in the initiation of the mammalian tooth germ. *Development* **103**: 155–169.

Luther, W., 1971. Distribution, biology and classification of salamanders. Pp. 557–568 *in* Venomous Animals and their Venoms, ed by W. Bücherl and E. E. Buckley. Academic Press, New York.

Lutz, B., 1971. Venomous toads and frogs. Pp. 423–474 *in* Venomous Animals and their Venoms, ed by W. Bücherl and E. E. Buckley. Academic Press, New York.

Mackay-Sim, A., Delay, R. J., Roper, S. D. and Kinnamon, S. C., 1996. Development of voltage-dependent currents in taste receptor cells. *J. Comp. Neurol.* **365**: 278–288.

Malkmus, R., 1975. Kannibalismus bei der Larvae des Feuersalamanders. *Nachrichten Naturw. Mus. Aschaffenb.* **82**: 39–43.

Mason, F. R., Rabin, M. D. and Stevens, D. A., 1982. Conditioned taste aversions: skin secretions used for defense by tiger salamanders, *Ambystoma tigrinum. Copeia* **1982**: 667–671.

McBride, D. W. and Roper, S. D., 1991. Ca^{2+}-dependent chloride conductance in *Necturus* taste cells. *J. Membrane Biol.* **124**: 85–93.

McCann, C., 1939. Biology of frogs in the genus *Rana. J. Bombay Nat. Hist. Soc.* **36**: 152–180.

McPheeters, M. and Roper, S. D., 1985. Amiloride does not block taste transduction in the mudpuppy, *Necturus maculosus. Chem. Senses* **10**: 341–352.

McPheeters, M., Barber, A. J., Kinnamon, S. C. and Kinnamon, J. C., 1994. Electrophysiological and morphological properties of light and dark cells isolated from mudpuppy taste buds. *J. Comp. Neurol.* **346**: 601–612.

McWilliams, S. R. and Bachmann, M., 1989. Foraging ecology and prey preference of pond-form larval small-mouthed salamanders, *Ambystoma texanum. Copeia* **1989**: 948–961.

Michel, W. and Caprio, J., 1991. Responses of single facial taste fibers in the sea catfish, *Arius felis*, to amino acids. *J. Neurophysiol.* **66**: 247–260.

Miller, I. J. and Smith, D. V., 1988. Proliferation of taste buds in the foliate and vallate papillae of postnatal hamsters. *Growth. Dev. Aging* **52**: 123–131.

Mintz, B. and Stone, L. S., 1933. Transplantation of taste organs in adult *Triturus viridescens. Proc. Soc. Exp. Biol. Med.* **30**: 1080–1082.

Mistretta, C. M., Goosens, K., Farinas, I. and Reichardt, L. F., 1996. BDNF depletion alters gustatory papilla and taste bud size and number in postnatal mouse. *Soc. Neurosci. Abst.* **22**: 991.

Miyamoto, T., Okada, Y. and Sato, T., 1991. Voltage-gated membrane current of isolated bullfrog taste cells. *Zool. Sci.* **8**: 835–845.

Miyamoto, T., Fujiyama, R., Okada, Y. and Sato, T., 1996. Properties of Na^+-dependent K^+ conductance in the apical membrane of frog taste cells. *Brain Res.* **715**: 79–85.

Moody, W. J., 1981. The ionic mechanism of intracellular pH regulation in crayfish neurones. *J. Physiol.* **316**: 293–308.

Morin, P. J., 1983. Competitive and predatory interactions in natural and experimental populations of *Notophthalmus viridescens dorsalis* and *Ambystoma tigrinum. Copeia* **1983**: 628–639.

Mosher, H. S., Fuhrman, F. A., Buchwald, H. D. and Fisher, H. G., 1964. Tarichatoxin-tetrodotoxin: a potent neurotoxin. *Science* **144**: 1100–1110.

Munger, B. L., 1977. Neural-epithelial interactions in sensory receptors. *J. Investigative Dermatology* **69**: 27–40.

Myers, C. W. and Daly, J. W., 1976. Preliminary evaluation of skin toxins and vocalizations in taxonomic and evolutionary studies of poison-dart frogs (Dendrobatidae). *Bull. Am. Mus. Nat. Hist.* **157**: 177–262.

Myers, C. W. and Daly, J. W., 1983. Dart-poison frogs. *Sci. Amer.* **248**: 120–133.

Myers, C. W., Daly, J. W. and Malkin, B., 1978. A dangerously toxic new frog *(Phyllobates)* used by Embera Indians of Western Colombia, with discussion of blowgun fabrication and dart poisoning. *Bull. Am. Mus. Nat. Hist.* **161**: 307–365.

Nagai, T., 1993. Transcellular labeling by DiI demonstrates the glossopharyngeal innervation of taste buds in the lingual epithelium of the Axolotl. *J. Comp. Neurol.* **331**: 122–133.

Nagai, T. and Matsushima, T., 1990. Morphology and distribution of the glossopharyngeal nerve afferent and efferent neurons in the Mexican salamander, Axolotl: A cobaltic-lysine study. *J. Comp. Neurol.* **302**: 473–484.

Nagai, T. and Oka, Y., 1991. The glossopharyngeal nerve of the axolotl labeled with carbocyanine dye (diI). *Neurosci. Lett.* **131**: 125–128.

Nagai, T., Kim, D.-J., Delay, R. J. and Roper, S. D., 1996. Neuromodulation of transduction and signal processing in the end organs of taste. *Chem. Senses* **21**: 353–365.

Neuwirth, M., Daly, J. W., Myers, C. W. and Tice, L. W., 1979. Morphology of the granular secretory glands in skin of poison-dart frogs (Dendrobatidae). *Tissue Cell.* **11**: 755–771.

Nieuwenhuys, R. and Opdam, P., 1976. Structure of the brain stem. Pp. 811–855 *in* Frog Neurobiology: A Handbook ed by R. Llinas and W. Precht. Springer-Verlag, Berlin.

Noble, G. K., 1954. The Biology of the Amphibia. Dover Publications, Inc., New York.

Noden, D. M., 1991. Vertebrate craniofacial development: The relation between ontogenetic process and morphological outcome. *Brain Behav. Evol.* **38**: 190–225.

Nomura, S., Shiba, Y., Muneoka, Y. and Kanno, Y., 1979a. Developmental changes of premetamorphic and fungiform papillae of the frog *(Rana japonica)* during metamorphosis: a scanning electron microscopy. *Hiroshima J. Med. Sci.* **28**: 79–86.

Nomura, S., Shiba, Y., Muneoka, Y. and Kanno, Y., 1979b. A scanning and transmission electron microscope study of the premetamorphic papillae: possible chemoreceptive organs in the oral cavity of an anuran tadpole *(Rana japonica). Arch. Histol. Jap.* **42**: 507–516.

Norris, H. W., 1908. The cranial nerves of *Amphiuma means. J. Comp. Neurol. Psychol.* **18**: 527–555.

Norris, H. W. and Buckley, M., 1911. The peripheral distribution of the cranial nerves of *Necturus maculatus. Proc. Iowa Acad. Arts and Sci.* **18**: 131–135.

Northcutt, R. G., 1992. Distribution and innervation of lateral line organs in the axolotl. *J. Comp. Neurol.* **323**: 1–29.

Northcutt, R. G. and Brändle, K., 1995. Development of branchiomeric and lateral line nerves in the axolotl. *J. Comp. Neurol.* **355**: 427–454.

Northcutt, R. G., Barlow, L. A., Braun, C. B. and Catania, K. C., 1997. Distribution and innervation of taste buds in the axolotl. *Brain Behav. Evol.*, in press.

Nosrat, C. A., Blomlöf, J., ElShamy, W. M., Ernfors, P. and Olson, L., 1997. Lingual deficits in BDNF and NT3 mutant mice leading to gustatory and somatosensory disturbances, respectively. *Development* **124**: 1333–1342.

Nyman, S., Wilkinson, R. F. and Hutcherson, J. E., 1993. Cannibalism and size relations in a cohort of larval ringed salamanders *(Ambystoma annulatum). J. Herpetol.* **27**: 78–84.

Oakley, B., 1993a. Control mechanisms in taste bud development. Pp. 105–125 *in* Mechanisms of Taste Transduction ed by S. A. Simon and S. D. Roper. CRC Press, Boca Raton.

Oakley, B., 1993b. The gustatory competence of the lingual epithelium requires neonatal innervation. *Dev. Brain Res.* **72**: 259–264.

Oakley, B., LaBelle, D. E., Riley, R. A., Wilson, K. and Wu, L.-H., 1991. The rate and locus of development of rat vallate taste buds. *Dev. Brain Res.* **58**: 215–221.

Ogura, T., Mackay-Sim, A. and Kinnamon, S. C., 1997. Bitter taste transduction of denatonium in the mudpuppy *Necturus maculosus. J. Neurosci.* **17**: 3580–3587.

Okada, Y., Miyamoto, T. and Sato, T., 1987. Depolarization induced by injection of cyclic nucleotides into frog taste cell. *Biochim. Biophys. Acta* **904**: 187–190.

Okada, Y., Miyamoto, T. and Sato, T., 1993. The ionic basis of the receptor potential of frog taste cells induced by water stimuli. *J. Exp. Biol.* **174**: 1–17.

Oliver, R. F., 1980. Local interactions in mammalian hair growth. Pp. 199–210 *in* The Skin of Vertebrates, ed by R. I. C. Spearman and P. A. Riley. Academic Press, London.

Opdam, P., Kemali, M. and Nieuwenhuys, R., 1976. Topological analysis of the brain stem of the frogs *Rana esculenta* and *Rana catesbeiana. J. Comp. Neurol.* **165**: 307–332.

Osculati, F. and Sbarbati, A., 1995. The frog taste disc: a prototype of the vertebrate gustatory organ. *Prog. Neurobiol.* **46**: 351–399.

Parker, M. S., 1994. Feeding ecology of stream-dwelling Pacific giant salamander larvae *(Dicamptodon tenebrosus). Copeia* **1994**: 705–718.

Pasteels, J., 1942. New observations concerning the maps of presumptive areas of the young amphibian gastrula. *(Amblystoma* and *Discoglossus). J. Exp. Zool.* **89**: 255–281.

Pfennig, D. W., Reeve, H. K. and Sherman, P. W., 1993. Kin recognition and cannibalism in spadefoot toad tadpoles. *Anim. Behav.* **46**: 87–94.

Pfennig, D. W., Sherman, P. W. and Collins, J. P., 1994. Kin recognition and cannibalism in polyphenic salamanders. *Behav. Ecol.* **5**: 225–232.

Polis, G. A. and Myers, C. A., 1985. A survey of intraspecific predation among reptiles and amphibians. *J. Herpetol.* **19**: 99–107.

Poritsky, R. and Singer, M., 1963. The fate of taste buds in tongue transplants to the orbit in the urodele *Triturus. J. Exp. Zool.* **153**: 211–218.

Poritsky, R. and Singer, M., 1977. Intraperitoneal transplants of taste buds in the newt. *Anat. Rec.* **188**: 219–228.

Powders, V., 1973. Cannibalism by the slimy salamander, *Plethodon glutinosus*, in eastern Tennessee. *J. Herpetol.* **7**: 139–140.

Pumphrey, R. J., 1935. Nerve impulses from receptors in the mouth of the frog. *J. Cell. Comp. Physiol.* **6**: 457–467.

Pumplin, D. W., Yu, C. and Smith, D. V., 1996. Light and dark cells of rat vallate taste buds are morphologically distinct cell types. *J. Comp. Neurol.* **378**: 389–410.

Rapuzzi, G. and Casella, C., 1965. Innervation of the fungiform papillae in the frog tongue. *J. Neurophysiol.* **28**: 154–165.

Raviola, E. and Osculati, F., 1967. La fine struttura dei recettori gustativi della lingua di rana. *Rc. Inst. Lomb. Sci. Lett.* **101**: 599–601.

Regal, P. J., 1966. Feeding specializations and the classification of terrestrial salamanders. *Evolution* **20**: 392–407.

Regal, P. J. and Gans, C., 1976. Functional aspects of the evolution of frog tongues. *Evolution* **30**: 718–734.

Reutter, K., 1978. Taste organ in the bullhead (Teleostei). *Adv. Anat. Embryol. Cell Biol.* **55**: 1–98.

Reutter, K. and Witt, M., 1993. Morphology of vertebrate taste organs and their nerve supply. Pp. 29–82 *in* Mechanisms of Taste Transduction ed by S. A. Simon and S. D. Roper. CRC Press, Boca Raton.

Richman, J. M. and Tickle, C., 1992. Epithelial-mesenchymal interactions in the outgrowth of limb buds and facial primordia in chick embryos. *Dev. Biol.* **154**: 299–308.

Robbins, N., 1967. The role of the nerve in maintenance of frog taste buds. *Exp. Neurol.* **17**: 364–380.

Roper, S. D., 1983. Regenerative impulses in taste cells. *Science* **220**: 1311–1312.

Roper, S. D., 1992. The microphysiology of peripheral taste organs. *J. Neurosci.* **12**: 1127–1134.

Roper, S. D. and McBride, D. W., 1989. Distribution of ion channels on taste cells and its relationship to chemosensory transduction. *J. Membrane Biol.* **109**: 29–39.

Rose, S. M. and Rose, F. C., 1965. The control of growth and reproduction in freshwater organisms by specific products. *Intl. Verein. Theor. Angew. Limnol.* **13**: 21–35.

Roth, G., Nishikawa, K. C., Wake, D. B., Dicke, U. and Matsushima, T., 1990. Mechanics and neuromorphology of feeding in amphibians. *Nether. J. Zool.* **40**: 115–135.

Ruibal, R. and Thomas, E., 1988. The obligate carnivorous larvae of the frog, *Lepidobatrachus laevis* (Leptodactylidae). *Copeia* **1988**: 591–604.

Samanen, D. W. and Bernard, R. A., 1981a. Response properties of the glossopharyngeal taste system of the mud puppy *(Necturus maculosus)*. I. General organization and whole nerve responses. *J. Comp. Physiol.* **143**: 143–150.

Samanen, D. W. and Bernard, R. A., 1981b. Response properties of the glossopharyngeal taste system of the mud puppy *(Necturus maculosus)*. II. Responses of individual first-order neurons. *J. Comp. Physiol.* **143**: 151–158.

Sata, O., Okada, Y., Miyamoto, T. and Sato, T., 1992. Dye-coupling among frog *(Rana catesbeiana)* taste disk cells. *Comp. Biochem. Physiol.* **103**: 99–103.

Sato, M., 1976. Physiology of the gustatory system. Pp. 576–587 *in* Frog Neurobiology: A Handbook ed by R. Llinas and W. Precht. Springer-Verlag, Berlin.

Sato, T., Okada, Y. and Miyamoto, T., 1994. Receptor potential of the frog taste cell in response to bitter stimuli. *Physiol. and Behav.* **56**: 1133–1139.

Sato, T., Okada, Y. and Miyamoto, T., 1995. Molecular mechanisms of gustatory transductions in frog taste cells. *Prog. Neurobiol.* **46**: 239–287.

Sato, T., Ohkusa, M., Okada, Y. and Sasaki, M., 1983. Topographical difference in taste organ density and its sensitivity of frog tongue. *Comp. Biochem. Physiol.* **76A**: 233–239.

Sbarbati, A., Ceresi, E. and Accordini, C., 1991. Surfactant-like material on the chemoreceptorial surface of the frog's taste organ: an ultrastructural and electron spectroscopic imaging study. *J. Struct. Biol.* **107**: 128–135.

Sbarbati, A., Zancanaro, C., Franceschini, F. and Osculati, F., 1989. Basal cells of the frog's taste organ: fluorescence histochemistry with the serotonin analog 5,7-dihydroxytryptamine in supravital conditions. *Bas. Appl. Histochem.* **33**: 289–297.

Sbarbati, A., Franceschini, F., Zancanaro, C., Cecchini, T., Ciaroni, S. and Osculati, F., 1988. The fine morphology of the basal cell in the frog's taste organ. *J. Submicrosc. Cytol. Pathol.* **20**: 73–79.

Sbarbati, A., Zancanaro, C., Franceschini, F., Balercia, G., Morroni, M. and Osculati, F., 1990. Characterization of different microenvironments at the surface of the frog's taste organ. *Am. J. Anat.* **188**: 199–211.

Schlosser, G. and Roth, G., 1995. Distribution of cranial and rostral spinal nerves in tadpoles of the frog *Discoglossus pictus* (Discoglossidae). *J. Morphol.* **226**: 189–212.

Schlosser, G. and Roth, G., 1997a. Development of the retina is altered in the directly developing frog *Eleutherodactylus coqui* (Leptodactylidae). *Neurosci. Letters* **224**: 153–156.

Schlosser, G. and Roth, G., 1997b. Evolution of nerve development in frogs. II. Modified development of the peripheral nervous system in the direct-developing frog *Eleutherodactylus coqui* (Leptodactylidae). *Brain Behav. Evol.* **50**: 94–128.

Schlosser, G., Kintner, C. and Northcutt, R. G., 1997. Loss of lateral line placodes in directly developing frogs. *Dev. Biol.* **186**: 270.

Schwenk, K., 1985. Occurrence, distribution and functional significance of taste buds in lizards. *Copeia* **1985**: 91–101.

Sheridan, R. E., Deshpande, S. S., Lebeda, F. J. and Adler, M., 1991. The effects of pumiliotoxin B on sodium currents in guinea pig hippocampal neurons. *Brain Res.* **556**: 53–60.

Shiba, Y., Sumomogi, H., Nomura, S., Muneoka, Y. and Kanno, Y., 1980. Oral chemoreceptor organs of bullfrog tadpoles during metamorphosis. *Develop. Growth Differ.* **22**: 209–217.

Sibbing, F. A., Osse, J. W. M. and Terlouw, A., 1986. Food handling in the carp *(Cyprinus carpio)*: its movement patterns, mechanisms, and limitations. *J. Zool.* **210**: 161–203.

Silverman, J. D. and Kruger, L., 1990. Analysis of taste bud innervation based on glycoconjugate and peptide neuronal markers. *J. Comp. Neurol.* **292**: 575–584.

Smith, C. C. and Bragg, A. N., 1949. Observations on the ecology and natural history of Anura. VII. Food and feeding habits of the common species of toads in Oklahoma. *Ecology* **30**: 333–348.

Smith, D. V., Klevitsky, R., Akeson, R. A. and Shipley, M. T., 1994. Taste bud expression of human blood group antigens. *J. Comp. Neurol.* **343**: 130–142.

Smith, H. M., 1969. The Mexican Axolotl: some misconceptions and problems. *BioScience* **19**: 593–597.

Sokol, O. M., 1962. The tadpoles of *Hymenochirus boettgeri*. *Copeia* **1962**: 272–284.

Sokol, O. M., 1969. Feeding in the pipid frog *Hymenochirus boettgeri* (Tornier). *Herpetologica* **25**: 9–24.

Song, H. K., Carver, W. E. and Sawyer, R. H., 1994. Pattern formation in chick feather development: distribution of β1-integrin in normal and scaleless embryos. *Dev. Dynamics* **200**: 129–143.

Spray, D. C. and Bennett, M. V. L., 1985. Physiology and pharmacology of gap junctions. *Ann. Rev. Physiol.* **47**: 281–303.

Starrett, P., 1960. Descriptions of tadpoles of middle American frogs. *Misc. Pub., Mus. Zool., Univ. Mich.* **110**: 1–39.

Stensaas, L. J., 1971. The fine structure of fungiform papillae and epithelium of the tongue of a South American toad, *Calyptocephalella gayi*. *Am. J. Anat.* **131:** 443–462.

Stewart, M. and Sandison, P., 1972. Comparative food habits of sympatric mink frogs, bullfrogs and green frogs. *J. Herpetol.* **6:** 241–244.

Stewart, R. E., DeSimone, J. A. and Hill, D. L., 1997. New perspectives in gustatory physiology: transduction, development, and plasticity. *Am. J. Physiol.* **272:** C1–C26.

Stone, L. M., Finger, T. E., Tam, P. P. L. and Tan, S.-S., 1995. Taste receptor cells arise from local epithelium, not neurogenic ectoderm. *Proc. Natl. Acad. Sci. USA* **92:** 1916–1920.

Stone, L. S., 1922. Experiments on the development of the cranial ganglia and the lateral line sense organs in *Amblystoma punctatum*. *J. Exp. Zool.* **35:** 421–496.

Stone, L. S., 1932. Independence of taste organs with respect to their nerve fibers demonstrated in living salamanders. *Proc. Soc. Exp. Biol. Med.* **30:** 1256–1257.

Stone, L. S., 1940. The origin and development of taste organs of salamanders observed in the living condition. *J. Exp. Zool.* **83:** 481–506.

Strong, O. S., 1895. The cranial nerves of amphibia. *J. Morphol.* **10:** 101–222.

Sugimoto, K. and Sato, T., 1978. Depression of frog gustatory neural responses to quinine-HCl after adaptation of the tongue to various taste stimuli. *Experientia* **34:** 196–197.

Sugimoto, K. and Teeter, J. H., 1990. Voltage-dependent ionic currents in taste receptor cells of the larval tiger salamander. *J. Gen. Physiol.* **96:** 809–834.

Sugimoto, K. and Teeter, J. H., 1991. Stimulus-induced currents in isolated taste receptor cells of the larval tiger salamander. *Chem. Senses* **16:** 109–122.

Suzuki, N., 1966. Taste mechanism in the frog's palate. *Zool. Mag.* **75:** 239–246.

Suzuki, Y. and Takeda, M., 1989. Filaments in the cells of frog taste organ. *Zool. Sci.* **6:** 487–497.

Tachibana, T., 1995. The Merkel cell: recent findings and unresolved problems. *Arch. Histol. Cytol.* **58:** 379–396.

Taglietti, V., Casella, C. and Ferrari, E., 1969. Interactions between taste receptors in the frog tongue. *Pflügers Arch.* **312:** 139–148.

Takeda, M., Suzuki, Y., Obara, N. and Nagai, Y., 1992. Neural cell adhesion molecule of taste buds. *J. Electron Microsc.* **41:** 375–380.

Takeuchi, H., Mauda, T. and Nagai, T., 1994. Electrophysiological and behavioral studies of taste discrimination in the axolotl *(Ambystoma mexicanum)*. *Physiol. Behav.* **56:** 121–127.

Takeuchi, H., Ido, S., Kaigawa, Y. and Nagai, T., 1997. Taste disk is induced the lingual epithelium of salamanders during metamorphosis. *Chem. Senses* **22:** in press.

Taylor, R. and Roper, S. D., 1994. Ca^{2+}-dependent Cl^- conductance in taste cells from *Necturus*. *J. Neurophysiol.* **72:** 475–478.

Thexton, A. J., Wake, D. B. and Wake, M. H., 1977. Tongue function in the salamander *Bolitoglossa occidentalis*. *Arch. Oral Biol.* **22:** 361–366.

Torrey, T. W., 1934. The relation of taste buds to their nerve fibers. *J. Comp. Neurol.* **59:** 203–220.

Torrey, T. W., 1940. The influence of nerve fibers upon taste buds during embryonic development. *Proc. Natl. Acad. Sci. USA* **26:** 627–634.

Toyoshima, K., 1989a. Chemoreceptive and mechanoreceptive paraneurons in the tongue. *Arch. Histol. Cytol.* **52:** 383–388.

Toyoshima, K., 1989b. Fine structural and histochemical study of lingual taste organs of *Rana catesbeiana* (Anura: Ranidae) transplanted to liver. *J. Morphol.* **200:** 29–36.

Toyoshima, K. and Shimamura, A., 1982. Comparative study of ultrastructures of the lateral-line organs and the palatal taste organs in the African clawed toad, *Xenopus laevis*. *Anat. Rec.* **204:** 371–381.

Toyoshima, K. and Shimamura, A., 1987. Monoamine-containing basal cells in the taste buds of the newt *Triturus pyrrhogaster*. *Archs Oral Biol.* **32:** 619–621.

Toyoshima, K., Miyamoto, K. and Shimamura, A., 1987. Fine structure of taste buds in the tongue, palatal mucosa and gill arch of the axolotl, *Ambystoma mexicanum*. *Okajimas Folia Anat. Jap.* **64:** 99–110.

Toyoshima, K., Honda, E., Nakahara, S. and Shimamura, A., 1984. Ultrastructural and histochemical changes in the frog taste organ following denervation. *Arch. Histol. Jap.* **47:** 31–42.

Twitty, V. C., 1937. Experiments on the phenomenon of paralysis produced by a toxin occurring in *Triturus* embryos. *J. Exp. Zool.* **76:** 67–104.

Valentincic, T. B. and Caprio, J., 1994. Consummatory feeding behavior to amino acids in intact and anosmic channel catfish *Ictalurus punctatus*. *Physiol. Behav.* **55:** 857–863.

Viertel, B., 1992. Functional response of suspension feeding anuran larvae to different particle sizes at low concentrations. *Hydrobiologia* **234:** 151–173.

Vintschgau, M. and Honigschmied, J., 1876. Nervus glossopharyngeus und Schmeckbecher. *Arch. f. d. gesam. Physiol.* **14:** 443–448.

Vogt, W., 1929. Gestaltungsanalyse am Amphibienkeim mit örtlicher Vitalfärbung. II. Gastrulation und Mesodermbildung bei Urodelen und Anuren. *Wilhelm Roux' Arch. Entwicklungmech. Org.* **120:** 385–706.

Wager, V. A., 1965. The Frogs of South Africa. Purnell and Sons Pty. Ltd., Cape Town.

Wake, M. H. and Schwenk, K., 1986. A preliminary report on the morphology and distribution of taste buds in Gymnophones, with comparison to other amphibians. *J. Herpetol.* **20:** 254–256.

Wassersug, R. J., 1971. On the comparative palatability of some dry-season tadpoles from Costa Rica. *Am. Midl. Nat.* **86:** 101–109.

Wassersug, R. J. and Heyer, W. R., 1988. A survey of internal oral features of Leptodactyloid larvae (Amphibia: Anura). *Smithsonian Contrib. Zool.* **457:** 1–99.

Wassersug, R. A. and Rosenberg, K., 1979. Surface anatomy of branchial food traps of tadpoles: a comparative study. *J. Morphol.* **159**: 393–426.

Welton, J., Taylor, R., Porter, A. J. and Roper, S. D., 1992. Immunocytochemical survey of putative neurotransmitters in taste buds from *Necturus maculosus*. *J. Comp. Neurol.* **324**: 509–521.

West, C. H. K. and Bernard, R. A., 1978. Intracellular characteristics and responses of taste bud and lingual cells of the mudpuppy. *J. Gen. Physiol.* **72**: 305–326.

Whitear, M., 1976. Apical secretion from taste bud and other epithelial cells in amphibians. *Cell Tiss. Res.* **172**: 389–404.

Whitear, M., 1989. Merkel cells in lower vertebrates. *Arch. Histol. Cytol.* **52**: 415–422.

Whitehead, M. C. and Kachele, D. L., 1994. Development of fungiform papillae, taste buds, and their innervation in the hamster. *J. Comp. Neurol.* **340**: 515–530.

Witt, M., 1993. Ultrastructure of the taste disc in the red-bellied toad *Bombina orientalis* (Discoglossidae, Salientia). *Cell Tissue Res.* **272**: 59–70.

Witt, M., 1995. Distribution of vasoactive intestinal peptide-like immunoreactivity in the taste organs of teleost fish and frog. *Histochem. J.* **27**: 161–165.

Witt, M., 1996. Carbohydrate histochemistry of vertebrate taste organs. *Prog. Histochem. Cytochem.* **30**: 172.

Witt, M. and Reutter, K., 1988. Lectin histochemistry on mucous substances of the taste buds and adjacent epithelia of different vertebrates. *Histochem.* **88**: 453.

Witt, M. and Reutter, K., 1995. Ultrastructure of the taste disk of the African clawed frog, *Xenopus laevis*. *Chem. Senses* **19**: 433.

Wong, L., Oakley, B., Lawton, A. and Shiba, Y., 1994. Keratin 19-like immunoreactivity in receptor cells of mammalian taste buds. *Chem. Senses* **19**: 251–264.

Wright, M. R., 1955. Persistence of taste organs in tongue transplants of *Triturus v. viridescens*. *J. Exp. Zool.* **129**: 357–373.

Wright, M. R., 1964. Taste organs in tongue-to-liver grafts in the newt, *Triturus v. viridescens*. *J. Exp. Zool.* **156**: 377–390.

Yang, J. and Roper, S. D., 1987. Dye-coupling in taste buds in the mudpuppy, *Necturus maculosus*. *J. Neurosci.* **7**: 3561–3565.

Yoshii, K., Kobatake, Y. and Kurihara, K., 1981. Selective enhancement and suppression of frog gustatory responses to amino acids. *J. Gen. Physiol.* **77**: 373–385.

Yoshii, K., Yoshii, C., Kobatake, Y. and Kurihara, K., 1982. High sensitivity of *Xenopus* gustatory receptors to amino acids and bitter substances. *Am. J. Physiol.* **243**: R42–R48.

Zalewski, A. A., 1973. Regeneration of taste buds in tongue grafts after reinnervation by neurons in transplanted lumbar sensory ganglia. *Exp. Neurol.* **40**: 161–169.

Zancanaro, C., Sbarbati, A., Franceschini, F., Balercia, G. and Osculati, F., 1990. The chemoreceptor surface of the taste disc in the frog, *Rana esculenta*. An ultrastructural study with lanthanum nitrate. *Histochem. J.* **22**: 480–486.

Zotterman, Y., 1949. The response of the frog's taste fibres to the application of pure water. *Acta Physiol. Scand.* **18**: 181–189.

Zuwala, K., 1991. Developmental changes in the structure of mucous membrane in the oral cavity and taste organs in tadpoles of the frog, *Rana temporaria* (SEM). *Acta Biol. Cracov.* **33**: 59–74.

Zuwala, K. and Jacubowski, M., 1991. Development of taste organs in *Rana temporaria*. *Anat. Embryol.* **184**: 363–369.

Zviman, M. M., Restrepo, D. and Teeter, J. H., 1996. Single taste stimuli elicit either increases or decreases in intracellular calcium in isolated catfish taste cells. *J. Memb. Biol.* **149**: 81–88.

CHAPTER 3

Vision

Gerhard Roth, Ursula Dicke and Wolfgang Wiggers

I. Introduction
II. Visual Behaviour
   A. Feeding Behaviour
      1. Estimation of Distance and Localization of Objects
      2. Prey Recognition
      3. Experimental Analysis of Prey Recognition
   B. Optomotor Behaviour
III. Morphology and Function of the Eye, Retina and Optic Nerve
   A. Anatomy of the Dioptric Apparatus
   B. The Visual Field
   C. Eye Movement
   D. Accommodation Mechanisms
   E. Morphology, Cytoarchitecture and Synaptology of the Retina
      1. Photoreceptors
      2. Interneurons
      3. Retinal Ganglion Cells
      4. Müller Cells
   F. Optic Nerve
   G. Visual Acuity
   H. Eye Size and its Effect on Retinal Structures
IV. Anatomy of the Visual System
   A. Gross Anatomy of the Amphibian Brain
      1. Telencephalon
      2. Diencephalon
      3. Mesencephalon
      4. Cerebellum
      5. *Medulla Oblongata* and Cervical Spinal Cord
   B. Morphology of Visual Centres
      1. Visual Afferents to the Brain
      2. Topic Organization of Retinal Projections to the Diencephalon and Tectum
      3. Organization and Cytoarchitecture of the Optic Tectum
      4. Central Visual and Visuomotor Pathways
V. Neurophysiology of Vision
   A. Methods
   B. Retinal Cells
      1. Photoreceptors
      2. Horizontal Cells
      3. Bipolar Cells
      4. Amacrine Cells
      5. Retinal Ganglion Cells
   C. Tectum
      1. Topic Organization of Retinal Afferents in the Tectum
      2. Topic Organization and Receptive Field Sizes of Neurons
      3. Response Properties of Tectal Cells
      4. Disparity-Sensitive Tectal Neurons
   D. Neurons of the *Nucleus Isthmi*
   E. Response Properties of Isthmic Neurons
   F. Thalamic and Pretectal Neurons
      1. Anurans
      2. Salamanders
   G. Lesion Experiments in Visual Centres
VI. Extraoptic Light Perception
VII. Neuronal Mechanisms Underlying Visually Guided Behaviour: An Overview
VIII. References
Appendix: Abbreviations Used

I. INTRODUCTION

AMPHIBIANS possess a wide variety of sensory mechanisms that they use in orienting within their environment and in detecting prey, enemies, conspecifics, mating partners and obstacles. Modalities include vision, audition, olfaction, vibration sense, mechanoreception and electroreception. For most frogs and salamanders, vision is the predominant sense used in guidance of behaviour, except for mating behaviour in anurans. In the third order of Amphibia, the Gymnophiona or caecilians, vision plays a minor or no role. These animals have a reduced visual system (Wake 1986), and their behaviour mostly is guided by chemosensation.

The present chapter provides a description of visually guided behaviour, an account of the morphology and physiology of the eye, retina and visual centres, and elucidates some ideas about the neural mechanisms underlying visually guided behaviour. Emphasis is laid on feeding because this type of behaviour and its neuroanatomical and neurophysiological bases have been studied most intensely in amphibians.

The visual system of frogs and salamanders is one of the best studied sensory systems in the animal kingdom, although it is still far from being fully understood. It is impossible to report all relevant data here. Additional information is available in various books and monographs such as "Frog Neurobiology" (Llinas and Precht 1976) and "The Amphibian Visual System" (Fite 1976) (both outdated to some extent) and "Visual Behavior in Salamanders" by Roth (1987) and in recent review articles (cf. Ewert 1984, 1989, dealing mostly with *Bufo*).

II. VISUAL BEHAVIOUR

A. Feeding Behaviour

Prey-capture in frogs and salamanders shows the following sequence: In most cases, the first reaction to a prey object is a turning movement that permits the frog or salamander to fixate the object binocularly. In contrast to frogs that turn the whole body, salamanders perform the initial orientation with their heads alone.

Most frogs approach the prey by jumping, combined with a forward lunge (Fig. 1). Grobstein *et al.* (1985) showed in *Rana pipiens* that all responses to prey at distances up to 6 cm are "snaps" and those at distances of 12 cm or greater are all "hops". At intermediate distances, responses can either be snaps or hops, with the probability of "hops" increasing with prey distance. The distance at which frogs switch from predominantly snapping to predominantly hopping increases with body size.

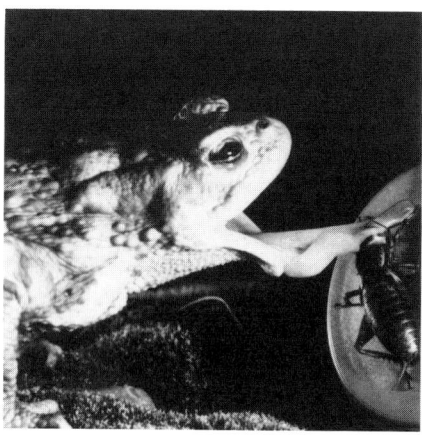

Fig. 1. Prey capture in the toad *Bufo bufo*. Photo courtesy of W. Grunwald.

Salamanders and some anurans (especially toads) approach their prey by walking. Many (e.g., bufonids and salamandrids) actively search for prey ("hunter strategy"), whereas others (e.g., ranids and bolitoglossine salamanders) wait until the prey comes close ("ambush strategy") (Fig. 2). Some frog and salamander species are known to switch from one strategy to another depending on prey density or prey type (salamanders: Jaeger 1972; Roth 1987; Maglia and Pyles 1995; frogs: Deban and Nishikawa 1992).

After reaching the snapping distance, amphibians usually fixate the prey binocularly for a shorter or longer period depending on the prey's attributes (e.g., movement intensity). By snapping, the prey finally is seized and taken into the mouth. Smaller prey are captured by the protrusible tongue; larger prey may be seized by the jaws as well.

Amphibians with a short, unspecialized tongue tend to snap only in a more or less frontal direction. For that reason they have to direct their heads directly toward the prey. In contrast, tongue-projecting salamanders (bolitoglossines), as well as some frogs, can project their tongues laterally (Roth 1987). Angles of 45° and more between tongue-tip and axis of the head have been observed in bolitoglossine salamanders.

Fig. 2. Feeding action in the tongue-projecting salamander *Hydromantes italicus*.

1. Estimation of Distance and Localization of Objects

Accurate localization of objects, including estimation of distance, is a necessary prerequisite for prey-capture in the context of approach behaviour and release of snapping as well as the estimation of absolute size and velocity of objects (see below). Distance can be estimated by a variety of ways: (1) eye vergence; (2) lens accommodation; (3) motion parallax; (4) stereoscopic vision based on retinal image disparity or binocular triangulation; (5) indirect cues such as texture gradients, overlapping of objects, knowledge about absolute object size, etc.

In amphibians, no eye vergence (i.e., inward or outward movement of the eye axes during fixation) takes place, although amphibians show stabilizing eye-reflex movements (cf. Grüsser and Grüsser-Cornehls 1976; Dieringer 1986, 1987). Amphibians, lacking eye vergence and having no fovea, do not achieve foveal stereoscopic vision.

The two main mechanisms for depth perception and localization of objects apparently are lens accommodation and the evaluation of retinal disparities. In amphibians, lens accommodation to a nearby object is achieved by forward movement of the lens by the *protractor lentis* muscle, of which frogs possess two; salamanders have only one ventrally situated muscle. Frogs and some salamanders make use of lens accommodation during fixation of nearby prey objects (Ingle 1972; Werner 1983), and they monitor the accommodative state of the eyes (Ingle 1972; Douglas *et al.* 1986). However, many salamanders, particularly plethodontids, have large lenses and a very small *protractor lentis* which is situated exactly ventral of the lens and seems rather ineffective in moving the lens forward. In amphibians, binocular estimation of distance is possible without accommodation, but at the same time binocularity (presence of two functional eyes) is not necessary for good depth perception (Schneider 1954; Ingle 1968, 1972, 1973, 1976). Collett (1977), Lock and Collett (1979), Jordan *et al.* (1980), and Douglas *et al.* (1986) showed by different procedures that monocular toads, for example, are almost as effective in estimation of distance to objects as are binocular ones. These authors also demonstrated that binocular toads have depth perception independent of accommodation and that they make use of binocular triangulation on the basis of retinal disparity (Collett 1977).

Motion parallax seems to play no, or only a minor, role in depth perception, because most amphibians do not exhibit the jerky head movement necessary for such a mechanism, when they approach or fixate a prey item. Exceptions are toads in which such movements sometimes can be observed.

Binocular and monocular depth perception were studied by Luthardt-Laimer (1983) in *Salamandra salamandra*. She compared three groups: (1) binocular animals; (2) chronically monocular animals (with one eye excised a year before); (3) reversibly monocularized animals (with one eye covered). The animals were tested with live prey and prey dummies, i.e., a black square (8 × 8 mm) moved at 0.5 cm/s at various distances from the salamander.

Binocular salamanders snapped at live prey with a success rate of about 40%. The chronically monocular salamanders were only slightly inferior (37%), but the reversibly monocularized animals were significantly less effective (26%). In the dummy experiments, all salamanders approached the stimulus at an angle leading ahead of the moving object. This angle turned out to be correlated strictly with the distance between stimulus and animal: the more distant the stimulus the larger the lead angle.

Binocular and chronically monocular salamanders did not differ with respect to their lead angle; in both groups it increased with increasing distance between subject and stimulus. However, in reversibly monocularized animals the lead angle did not significantly increase with increasing distance. This means that these animals could not adapt their approach behaviour with regard to the distance of the stimulus.

Wiggers and Roth studied depth perception in binocular and monocular tongue-projecting salamanders, *Bolitoglossa subpalmata* and *Hydromantes italicus* (Wiggers and Roth, unpubl. data; cf. Roth 1987). With one eye covered, often these salamanders showed a conspicuous approach behaviour in front of a live housefly fixed on a needle (Fig. 3). The salamander approached the prey up to snapping distance, then bent the body away from the prey toward the side of the seeing eye, but compensated this bending by turning the head back toward the prey such that the head and body formed angles between 60° and 90°. In such a position, the animals moved around the prey in a circular route 2–3 cm in diameter. In about half of the experiments, the salamanders eventually snapped at the fly; in the other cases they turned away, or the experiment was terminated after many minutes. A similar bending behaviour has been observed in other monocular amphibians, e.g., toads (Podufal 1971).

Fig. 3. Approach behaviour of monocular *Bolitoglossa subpalmata* to pinned-down, live cricket. The right eye of the salamander is covered. For further explanation see text. From Roth (1987).

In contrast to the findings in *Salamandra*, there was no qualitative difference between chronically and reversibly monocularized bolitoglossine salamanders. Under both experimental conditions, both types of behaviour could be observed in the same animal, i.e., a "normal" direct approach and body curvature. However, in both chronically and reversibly monocularized animals there was a strict inverse relationship between the speed of approach and the amount of body curvature; the faster the salamander approached the prey, the less prominent was the bending of the body. Thus, at least in bolitoglossine salamanders, monocularization does not lead to any substantial impairment of depth perception. When highly motivated, both chronically and reversibly monocularized animals behave much like binocular ones.

In summary, two depth perception mechanisms exist, one operating under conditions of binocular vision, where retinal disparity is used, and one under conditions of monocular vision based essentially on eye accommodation.

Amphibians are (necessarily) able to estimate the absolute size of an object. Size-constancy during turning and snapping responses was studied in the frog *Rana pipiens* by Ingle (1968) and in the toad *Bufo bufo* by Ewert and Gebauer (1973). The toad exhibited size-constancy up to a minimal distance of 23 cm. From the fact that visual neurons respond only to angular size of objects, it can be concluded that absolute size is calculated on the basis of angular size and distance to object.

2. Prey Recognition

In the past, the study of visual prey recognition in amphibians suffered from two incorrect assumptions. The first was that amphibians are highly indiscriminate with respect to prey object; i.e., they feed on everything that moves and is not too small to detect and not too large to swallow. From this it was concluded that amphibians cannot discriminate between different kinds of prey due to their poor capacity to recognize objects. While many amphibians in many instances "take what they can get", particularly when there are no choices, there are well-documented cases in which frogs and salamanders make subtle distinctions between different kinds of prey with respect to their size, configuration, movement pattern, nutritive value, digestibility and potential harmfulness.

The second incorrect assumption was that capture of prey by amphibians is a stereotyped behaviour or "fixed action pattern" (FAP, in the sense of Lorenz and Tinbergen; cf. Tinbergen 1951; for a critique: Roth and Dicke 1994) which is triggered by an invariable "sign stimulus" interacting with "feature detectors" in the visual system which then (as "command elements") release the feeding response (for this concept, see Ewert 1968, 1989). However, prey capture is not stereotyped, because amphibians can regulate their tongue protrusion with respect to their feeding strategy (see above) and to distance and velocity of prey (Maglia and Pyles 1995). Neither are there invariable properties of prey that act like sign stimuli.

An important source of information about relevant properties of prey objects is what amphibians eat and prefer under natural conditions. It is reported that some anurans, particularly large ones like *Ceratophrys ornata*, *Pyxicephalus adspersus* or *Rana catesbeiana*, are voracious eaters and commonly feed on large prey such as rats and mice, turtles, snakes and even birds, and the same is true for large salamanders like *Dicamptodon ensatus* (Duellman and Trueb 1986). These reports probably are the basis for the myth that amphibians are completely unselective feeders. There are only a few studies that compare the temporal (e.g., seasonal) fluctuation of the whole spectrum of possible prey in a given habitat with the array of prey actually eaten. In many species of frogs and salamanders, the abundance of food items found in the stomach is correlated roughly with the relative prey abundance in the habitat, and indeed, some rather indifferent feeders seem to exist. On the other hand, there are specialists like the frogs *Breviceps verrucosus* or *Scaphiopus couchii*, with a feeding activity that is timed to the swarming of termites. Most amphibians, however, have a relatively broad prey spectrum but with a certain preference, if there are choices.

For many amphibians, *size* seems to be an important prey parameter. This is seen in ontogenetic changes in the diets of frogs and salamanders. As individuals become larger, they tend to include in their diet increasingly larger prey, as well as a greater diversity of prey. Likewise, many authors found an intraspecific as well as an interspecific correlation between size of the body or head of the predator on the one hand and size of prey on the other. Thus, larger amphibians tend to eat larger prey, although most of them continue to eat small prey (for salamanders see Lynch 1985; Jaeger 1972; Maiorana 1978a,b).

Besides seasonal and geographic variation, differences in the feeding mechanism and feeding "strategy" may strongly influence the spectrum of prey. Sites (1978) studied the prey-spectra of the sympatric plethodontid salamanders, *Desmognathus fuscus* and *Eurycea longicauda*. *Desmognathus* has a "primitive", albeit fast, flipping tongue, adopts a "hunting" strategy and eats mostly earthworms and millipedes, while *Eurycea* has a free, projectile tongue, is an "ambush" feeder and eats mostly collembolans, i.e., very small and fugitive prey. Sites (1978) found no correlation between salamander size and prey size.

The latter is also true for salamanders of the plethodontid tribe Bolitoglossini (comprising about half of all salamander species) which all have free, projectile tongues with a reach of up to two-thirds of snout-vent length. The mean prey length of various neotropical bolitoglossines (e.g., *Bolitoglossa subpalmata*, *B. rostrata*, *Dendrotriton bromeliacia*, with snout to vent lengths between 64 and 17 mm) varied between 3.9 and 1.8 mm (Roth 1987). Terrestrial individuals of one species *(B. subpalmata)* tended to eat larger prey (2.9 mm) than did arboreal ones (1.8 mm). The stomachs of arboreal bolitoglossines contained mostly ants, dipterans, collembolans and mites, whereas those of terrestrial ones contained mostly coleopterans. Again, there was no correlation between predator size and prey size; minimum prey size was 0.17 × 0.08 mm.

In general, Bolitoglossini tend to eat much smaller prey than do other salamanders except those non-bolitoglossine plethodontid salamanders with free tongues like the Hemidactyliini (e.g., *Eurycea;* see above). Tongue-projecting salamanders almost exclusively use their projectile tongues during feeding without use of their jaws, and they have difficulty in feeding on large or elongate prey (Roth 1987). In captivity, relatively large specimens of *Hydromantes genei* often prefer *Drosophila* to larger but still moderately sized prey like young crickets (Roth, unpubl. obs). These studies reveal that many salamander species do not maximize prey size.

The study of Freed (1982) on the feeding habits of the treefrog *Hyla cinerea* reveals other criteria for prey selection. In its natural habitat, *Hyla cinerea* eats mostly larvae of the lepidopteran *Spodoptera* and of the coleopteran *Chauliognathus*, but in the laboratory strongly prefers houseflies *(Musca domestica)*, which are not natural prey. Freed found that prey selection is correlated neither with mass nor size of prey, but with its activity, e.g., *movement pattern*. In crickets, grooming behaviour elicited few feeding responses, but crawling was an effective stimulus. The most preferred prey were crawling spiders and flies.

Most informative with respect to prey discrimination abilities are experiments in which amphibians modify their prey preferences on the basis of experience. It is well known that anurans learn to avoid noxious or unpalatable prey like bumblebees, honey bees or wasps. The bolitoglossine salamander *Hydromantes* avoided *Calliphora erythrocephala* after one to very few trials. The salamanders spat these flies out and subsequently neglected them, whereas they specialized on *Musca domestica* (Roth, unpubl. obs). In a series of experiments with the salamander *Plethodon cinereus*, Jaeger and co-workers demonstrated that these salamanders can change their feeding habits according to individual experience with prey. For example, the salamanders specialized on large species of *Drosophila* at high prey densities by increasingly "ignoring" a smaller species. Also, they changed their feeding strategy from "hunting" to "ambush". When raised with a variety of different *Drosophila*, *Plethodon* specialized on the larger and more profitable prey. The authors could exclude any inborn prey preference, because fruit flies are almost completely absent from the natural diet of *Plethodon cinereus* (Jaeger and Barnard 1981; Jaeger and Rubin 1982; Jaeger *et al.* 1982).

Other studies on amphibian prey preferences reveal further discrimination capabilities probably based on learning. The Californian plethodontid salamander, *Batrachoseps attenuatus*, prefers sminthurid collembolans to oribatid mites, a taxon of equal size but having an exoskeleton that is harder to digest. This certainly requires the ability to distinguish prey by criteria other than size.

Important information about crucial properties of prey is gained from consideration of the conditions of illumination under which prey are captured in nature. Many amphibians are active during darkness, leaving their hiding places after sunset and returning before sunrise. They are able to use vision at very low illumination levels, e.g., on rainy nights or underground. The salamander *Hydromantes italicus* is able to detect prey by vision at an illumination level of 10^{-3} lx (Roth 1976). Himstedt (1982) demonstrated feeding reactions of *Salamandra salamandra* at $10^{-5} - 10^{-6}$ lx, which corresponds to a dark, rainy night. The same is reported for the toad *Bufo bufo* (Larsen and Pedersen 1982).

The question arises as to which visual cues can still be detected under such conditions. Given the fact that visual acuity is poor in darkness and object-background contrast is generally low, details of the anatomy of the prey can hardly be recognized. Also, precise estimates of body size are difficult. What remains is motion and movement pattern. Thus, at illumination levels at which most salamanders and many frogs are active, *movement pattern* seems to be the most important visual cue.

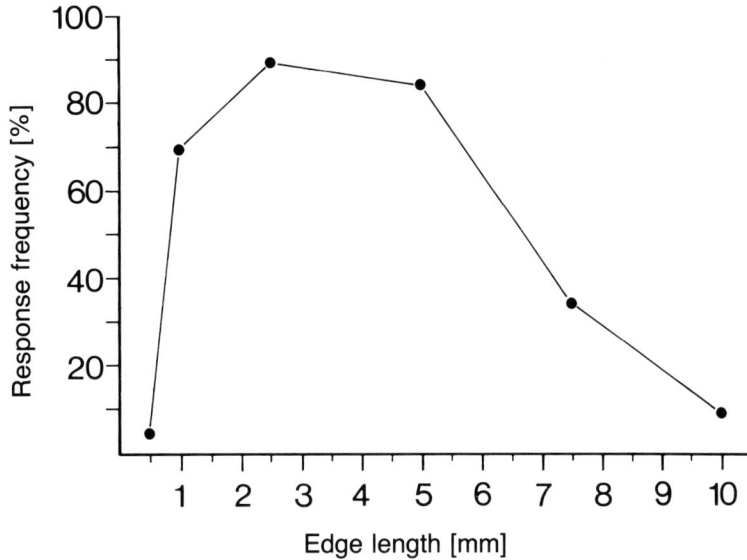

Fig. 4. Dependence of prey-catching responses on stimulus size in the salamander *Hydromantes italicus*. Black squares moving in front of the animal at a velocity of 3 cm/s were used as stimuli. Abscissa indicates edge length of the stimuli. After Roth (1976).

3. Experimental Analysis of Prey Recognition

Grüsser and Grüsser-Cornehls (1968) were the first to measure experimentally the responses of amphibians to visual stimuli. They conducted experiments in the field, i.e., on the banks of lake Neusiedler in Austria, but used standardized visual objects. This has the great advantage of avoiding any influence of captivity or artificial test conditions (territory, moisture, etc.) on the animals. However, the range and reliability of systematic variation of visual stimuli are limited as is the control of boundary conditions. In the laboratory, one can test the relative contribution of visual parameters to the overall attractiveness of a prey object. The most important of these are *size, shape, contrast, configuration, velocity,* and *movement pattern*. Laboratory experiments for determining stimulus preferences usually are carried out using pieces of black cardboard or plastic of different sizes and shapes, moved at different velocities and movement patterns on a white background or a background consisting of a fine-grained, random black and white configuration ("Julesz pattern"). Recently, the development of high-resolution computer screens and sophisticated programmes, has made possible the creation of such configurations of stimuli and provides opportunity for quick and nearly unlimited changes in the visual parameters. However, as will be discussed later, such use of dummies might produce artifacts.

Size and Shape: In laboratory experiments, all amphibians tested exhibit size preferences that are not identical with maximum size of devourable prey. Among salamanders, *Hydromantes italicus* responded to stimuli within a size range of 0.5–10 mm in length with a maximum between 1 and 5 mm (Roth 1976) (Fig. 4). Preferences for even smaller prey dummies were found in other medium-sized plethodontid salamanders like *Bolitoglossa subpalmata*, *Plethodon cinereus* and *P. jordani*; these salamanders preferred dummies with a length of 1 and 2.5 mm and responded well to dummies that were 0.5 mm long (Roth 1987). The large-sized *Salamandra salamandra*, in contrast, responded to stimuli within a size range of 2–32 mm in length (maximum: 16 mm); prey dummies 32 mm long still elicited snapping in half of the tests (Luthardt 1981). Finkenstädt and Ewert (1983a) found response maxima at stimulus sizes between 10 and 40 mm (Fig. 5).

Grüsser and Grüsser-Cornehls (1968), in their field experiments on visual behaviour in *Rana esculenta* using black, round objects moved at different velocities in front of the frogs, found an optimal prey size of 8.5° at a distance of 15 cm. This equals an absolute size of about 22 mm in diameter. Objects 15–20° (39–52 mm) in diameter elicited either snapping or

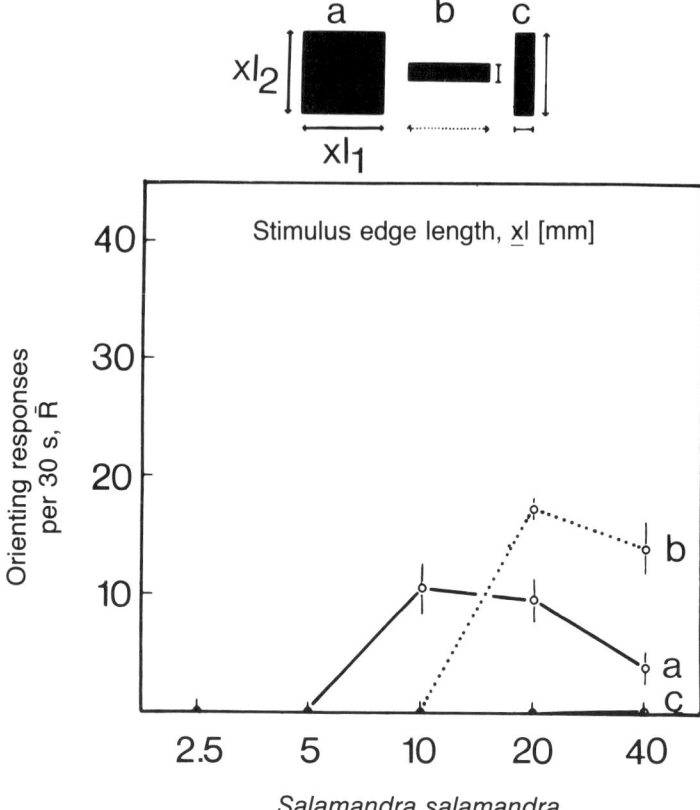

Fig. 5. Size preferences in orienting behaviour in *Salamandra salamandra* in response to black configurational stimuli moving in front of a white background at a velocity of 10°/s. Starting with a 2.5 × 2.5 mm square, edge length of the stimuli was increased parallel (b), and perpendicular (c) to the direction of movement or in both directions (a). After Finkenstädt and Ewert (1983).

avoidance but objects larger than 20° only avoidance. In experiments by Ewert (1968), the orienting response of *Bufo bufo* (not snapping, as in the above experiments) was tested with dummies ranging in length from 2.8 to 44.8 mm which, at a stimulus distance of 8 cm, equals a range in angular size from 2° to 32°. For square stimuli, a maximum was found between 5.6 and 11.2 mm (4° and 8°, respectively); for stimuli oriented with their long axis parallel to the direction of movement ("wormlike" stimuli), the rate of orienting responses increased up to a size of 2.8 × 22.4 mm (2° × 16°) and remained constant for even more elongate dummies. Using stimuli oriented with their long axis perpendicular to the direction of movement ("antiwormlike" stimuli) of increasing length, the rate of orienting responses continuously decreased from 2.8 to 44.6 mm (2°–32°). In a later study using the same procedure, Ewert and co-workers found a maximum at 3.5 × 28 mm for "wormlike" stimuli, with a decrease at 3.5 × 56 mm, and a maximum of 3.5 × 7 mm for "antiwormlike" stimuli. Thus, the worm-antiworm discrimination has an effect at length to width ratios greater than 2:1.

It was concluded from these studies that toads have a preference for "wormlike" prey objects and that stimuli become less attractive the more "antiwormlike" they are. Squared stimuli would, therefore, have an intermediate efficiency. This "worm preference" was shown to be independent of the direction of movement. The question arises whether other amphibians show the same "worm" preference and whether this preference is constant except as modified by direction of movement, e.g., velocity and movement pattern. Among anurans, different species of the genera *Bufo*, *Hyla* and *Rana* and the species *Bombina variegata* and *Alytes obstetricans* are reported to exhibit such a "worm preference", although it seems less pronounced in *Rana temporaria* (Ewert 1984). A number of authors report a preference for compact prey objects by *R. temporaria* and *R. esculenta*.

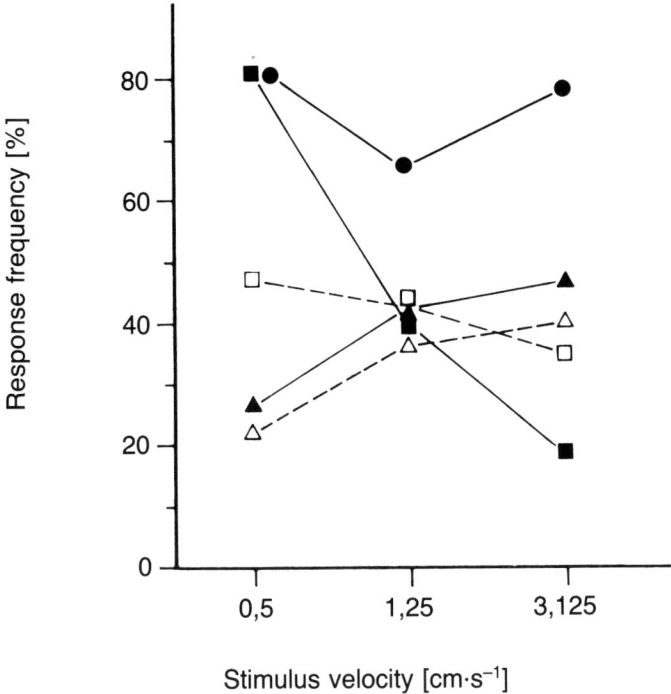

Fig. 6. Dependence of prey-catching behaviour of *Salamandra salamandra* on stimulus orientation and velocity. Stimuli were a square (16 × 16 mm, filled circles) and rectangles (4 × 16 mm) with orientation parallel (filled squares), diagonal with tip down (open squares), diagonal with tip up (open triangles), and perpendicular (filled triangles) to the direction of movement. From Roth (1987).

Among salamanders, *S. salamandra* shows a similar "worm preference", at least at low velocities (0.5–1 cm/s) when tested with 4 × 16 mm or 2.5 × 15 mm "wormlike" and "antiwormlike" stimuli (Luthardt and Roth 1979; Finkenstädt and Ewert 1983a). However, in both series of experiments, square stimuli (16 × 16 and 15 × 15 mm) proved to be nearly equally effective. Furthermore, Luthardt and Roth found that at a velocity of 1.3 cm/s, the effectiveness of the "wormlike" stimulus strongly decreased, while that of the "antiwormlike" stimulus increased such that both stimuli were roughly equally effective (Fig. 6). At still higher velocity (3.1 cm/s), the "antiwormlike" stimulus was now more than twice as effective as the "wormlike" one. Furthermore, diagonally oriented stimuli were intermediate in effectiveness: at low velocity, a diagonal stimulus inclining downwards in the direction of movement proved to be superior to a rectangle oriented in the opposite direction, whereas at high velocity the opposite was the case (Fig. 6). The former diagonal stimulus was apparently "interpreted" by the salamander as being somehow "horizontal" and the latter as being more nearly vertical. At medium and high velocity, the square was by far the most preferred stimulus.

Such a "preference inversion" for "wormlike" and "antiwormlike" stimuli was not found by Finkenstädt and Ewert (1983a) (Fig. 5). In their experiments, the responses to a "wormlike" and a square stimulus (2.5 × 15 mm and 15 × 15 mm, respectively) strongly increased with increasing velocity (4, 10 and 20°/s), but the "antiwormlike" stimulus elicited no response. Particularly the latter result contradicts other experiments carried out with *S. salamandra* using the same set of stimuli (e.g., Stenner 1976; cf. Roth 1987). Furthermore, *Plethodon jordani* and *P. cinereus*, exhibit a "preference inversion" with increasing velocity similar to that found in *Salamandra* (Roth 1987). Thus, in some salamanders there seems to be a tendency for elongate stimuli oriented parallel to the direction of movement to decrease in effectiveness with increasing velocity, but the opposite seems to be the case for elongate stimuli oriented perpendicular to the direction of movement. The reasons for such a velocity dependence of configural stimuli are unclear. It may be that the vertically oriented stimulus is interpreted by the amphibian as the leading edge of a compact object. This view is corroborated by the fact that salamanders confronted with larger squares often snapped at the horizontal edge when the stimulus was moving slowly, and at the vertical edge when it was moving rapidly (Luthardt 1981).

Velocity: In most cases, movement is a necessary prerequisite for prey recognition in amphibians. Motionless prey objects are usually ignored, but prey-catching responses to stationary prey objects or dummies may happen under specific conditions (see below). Most amphibians respond to relatively slow movement of prey. Himstedt (1967) found the effective range of stimulus velocities to be 0.05–2 cm/s (1–45°/s) for the salamander *Triturus vulgaris* and 0.5–2 cm/s (0.5–40°/s) for *S. salamandra*. For the latter species, Luthardt (1981) found a range of 1.2–6 cm/s , with an optimum at 1 cm/s. For the bolitoglossine salamander *Hydromantes italicus*, Roth found a range of 0.05–6 cm/s (0.24–172°/s); velocities between 0.5 and 2.5 cm/s (4.8–72°/s) elicited most responses (Roth 1976). Small neotropical bolitoglossines, e.g., *Bolitoglossa rufescens*, still respond to velocities of 10 cm/s or more, which at snapping distance equals an angular velocity of about 300°/s. These salamanders are able to catch flies on the wing (Roth 1987).

Freed (1982) found that the treefrog *Hyla cinerea* attacks houseflies that have an average flight velocity of 94 cm/s. The angular velocity of *Musca* in flight 18 cm away from a treefrog equals 300°/s which is about the same response shown by small Bolitoglossini. In *Bufo bufo*, Ewert (1968) found an increase in orienting responses toward a 2.8 × 22.4 mm "wormlike" stimulus when its velocity increased from 2 to 60°/s (0.35–10.5 cm/s). At velocities beyond 100°/s (17.4 cm/s), the response rate decreased.

Movement Pattern: The way in which a prey moves seems to be very important in prey recognition. However, the movement pattern of prey seems to interact in a complex manner with velocity. In the salamander *Hydromantes italicus,* continuous movement of a 4 × 4 mm prey dummy proved to be more effective at low velocity (0.5 cm/s), whereas the opposite was the case at medium (1.25 cm/s) and high (3.1 cm/s) velocities (Fig. 7). At these velocities, a saw-tooth movement pattern of 1 Hz was most effective in eliciting snapping (Roth 1978).

More complex results were obtained in experiments with *S. salamandra;* 4 × 16 mm rectangles oriented parallel ("wormlike") or perpendicular ("antiwormlike") to the direction of movement were moved either continuously or stepwise at 0.25–8 Hz at velocities of 0.5, 1.3 and 3.1 cm/s (Fig. 8). In the case of the "wormlike" stimulus, continuous and stepwise movement proved to be equally effective at low velocities, but at medium and high ones continuous movement was much better than stepwise movement except at 8 Hz; stepwise movement at 1 Hz was particularly "disliked". In the case of the "antiwormlike" stimulus, again continuous and stepwise movement proved to be equally effective at low velocities, but at medium and high velocities, stepwise movement was now much better than continuous movement except at 0.25, 0.5 and 8 Hz; stepwise movement at 4 Hz was particularly "liked" (Luthardt and Roth 1979).

These results suggest that salamanders associate an elongate, "wormlike" shape of a stimulus with slow and smooth movement, and that they associate a compact shape (plus vertical bars as leading edges of compact objects) with fast and/or jerky movement. In contrast, a jerky movement of "wormlike" objects are disliked.

Recent experiments with salamanders (*Hydromantes genei, H. italicus* and *Plethodon jordani*) and frogs (*Bombina bombina* and *Discoglossus pictus*) clearly demonstrated that a prey dummy increases in attractiveness, as its movements become more irregular (Roth, unpubl. data). This is in accordance with the observations on *Hyla cinerea* and *Rana esculenta* mentioned above.

Responses to Stationary Prey: According to common belief, an amphibian will starve to death sitting in a heap of dead flies, because it is incapable of recognizing stationary objects (e.g., Lettvin *et al.* 1959). At least for toads and salamanders, this view is incorrect. It is true that amphibians usually do not pay attention to non-moving prey; they stop pursuing a moving prey and often turn away from it as soon as it remains motionless for more than 30 s. In other cases, they stare at the prey and snap at it as soon as it moves again.

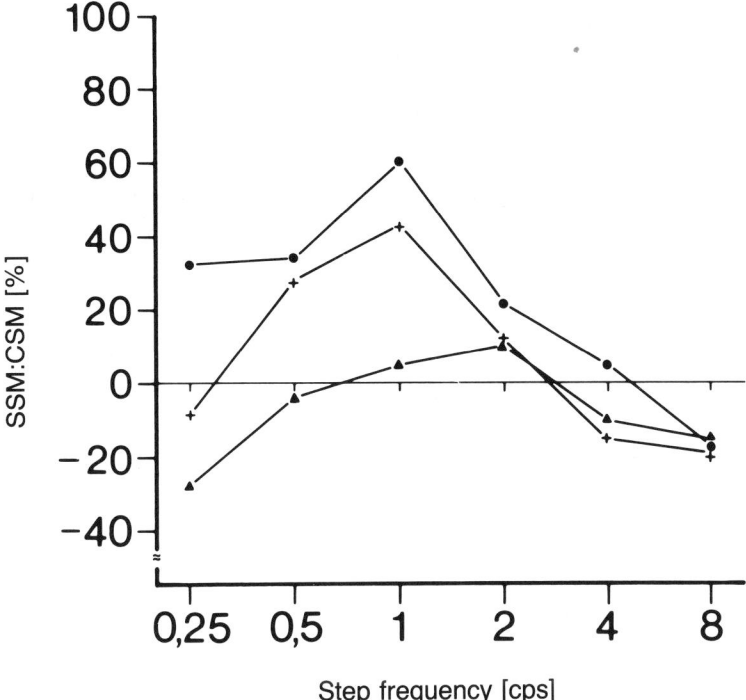

Fig. 7. Comparison between prey-catching responses of *Hydromantes italicus* to stepwise stimulus movement (ssm) and to continuous stimulus movement (csm) at different basic velocities and different step frequencies. The stimulus used was a black square measuring 4 × 4 mm. The curves show how much (in per cent of the respective small value) ssm is more efficient than csm (positive value) or vice versa (negative values). Dots: 0.5 cm/s; crosses: 1.25 cm/s; triangles: 3.1 cm/s. From Roth (1978).

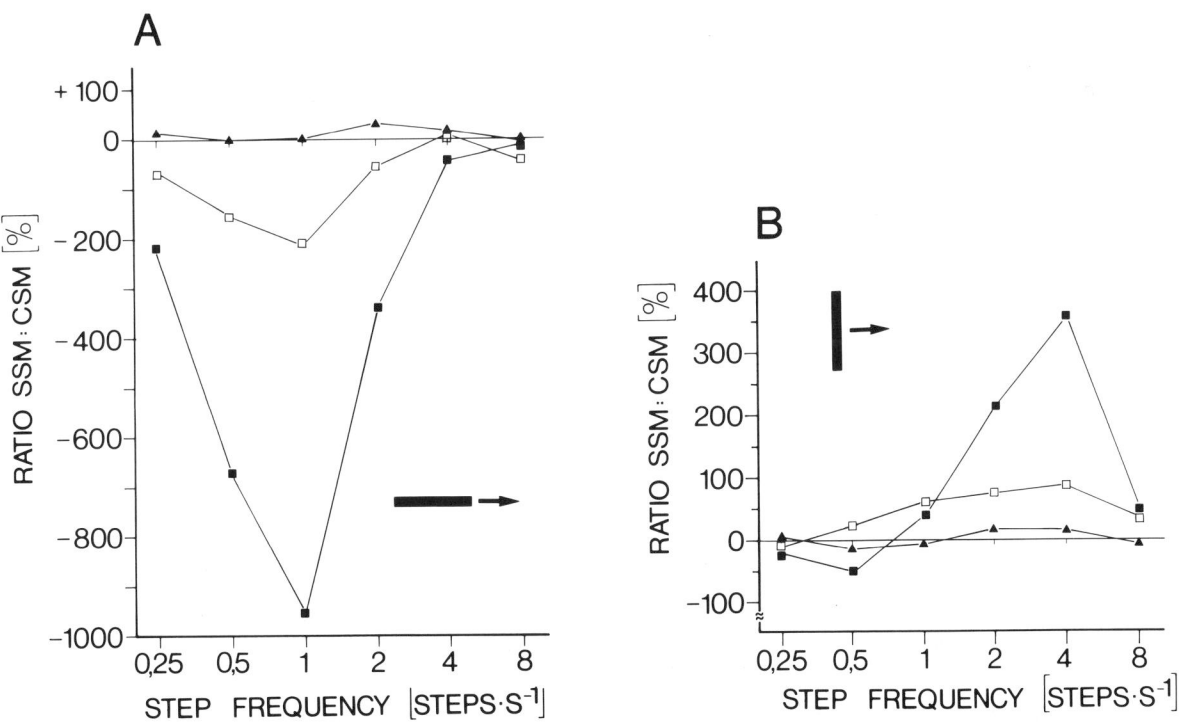

Fig. 8. Comparison between prey-catching responses of *Salamandra salamandra* to stepwise and to continuous stimulus movement. Diagrammatic representation of data as in Figure 7. Stimuli were rectangles measuring 4 × 16 mm oriented parallel (A) and perpendicular (B) to the direction of movement. Solid squares: 0.5 cm/s; open squares: 1.3 cm/s; triangles: 3.1 cm/s. From Luthardt and Roth (1979a).

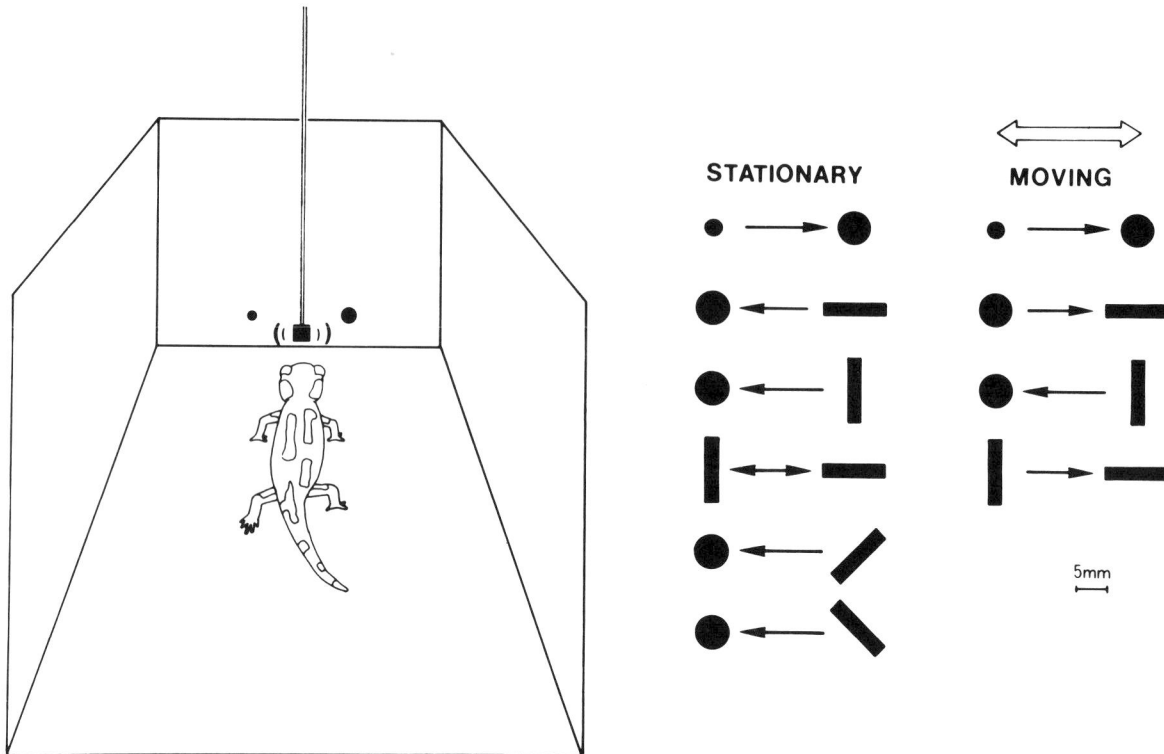

Fig. 9. Responses of *Salamandra salamandra* to stationary patterns. Left: a salamander follows a moving prey dummy to the wall where stationary patterns are fixed. Right: Comparisons of prey-catching responses to stationary and moving patterns. Arrows point toward the pattern preferred in a choice experiment. From Himstedt *et al.* (1978).

However, one easily can show that stationary prey do not simply "disappear" from vision, as has often been argued on the incorrect assumption that amphibians lack involuntary eye movements (see below). Rather, non-moving objects are usually not "interpreted" by amphibians as prey. If one lures salamanders or frogs toward stationary prey stimuli by means of a moving dummy and withdraws the dummy, when the animals are close to the stationary prey, they will snap precisely at this kind of prey (Fig. 9). In such experiments with *S. salamandra*, Himstedt and co-workers demonstrated that configuration preferences are quite different. While in tests with moving stimuli a horizontally oriented dummy was preferred to a small circle and this stimulus to a vertically oriented dummy, in tests with stationary stimuli, the small circle was now more attractive than either the horizontal or vertical dummy, and the vertical stimulus proved to be more effective than the horizontal one (Himstedt *et al.* 1980).

Roth and Wiggers (1983) investigated the response of the toad *Bufo bufo* to stationary stimuli using square prey dummies (Fig. 10). The toads significantly preferred squares with edge lengths of 2–8 mm to both smaller and larger objects. When confronted with squares (10 × 10 mm) and rectangles (5 × 20 mm) oriented either horizontally or vertically and situated in the vertical plane (on the walls of the test container), the square was preferred to the two rectangles, and the horizontal rectangle to the vertical one. When the stimuli were located on the floor of the container, the square was preferred to the rectangle oriented perpendicular to the direction of the animal's movement, and the rectangle with parallel orientation was preferred to the one oriented perpendicularly. The former rectangle had the same efficacy as the square. When one stimulus was presented on the floor and another on the wall, the stimulus on the wall was always preferred, regardless of its shape or orientation.

Finally, a stationary, vertically oriented rectangle 5 mm wide and of decreasing length was compared to a 20 × 5 mm horizontal rectangle. The stimuli were located on the wall. The vertical rectangle was significantly less effective than the reference stimulus at lengths of 20 and 15 mm; it had the same effectiveness at a length of 12.5 mm and was better at lengths of 10 and 5 mm (forming a 5 × 5 mm square in the latter case).

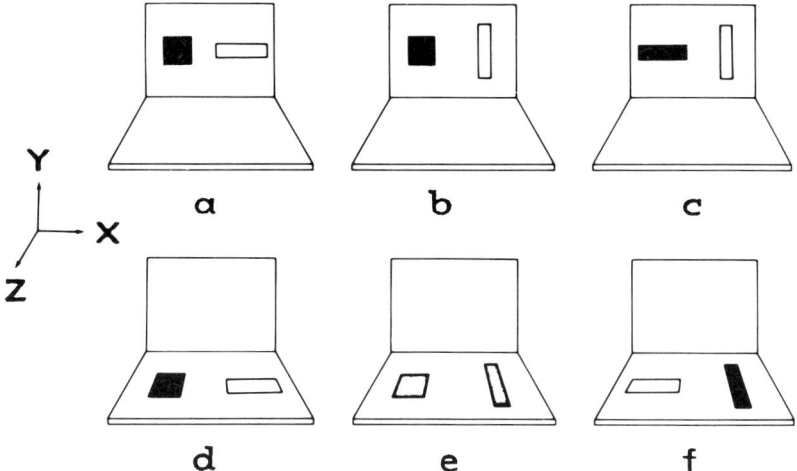

Fig. 10. Responses of the toad *Bufo bufo* to stationary prey stimuli. Comparison between the effect of a square (10 × 10 mm) and a rectangle (5 × 20 mm) oriented parallel to the X, Y and Z axes. Dark filling of the stimuli indicates significant preference ($p < 0.05$). **a–c** presentation in the frontal-vertical (X–Y) plane. **d–f** presentation in the horizontal (X–Z) plane. From Roth and Wiggers (1983).

Burghagen and Ewert (1983) conducted similar experiments with *Bufo*, although in this case the stimuli were located exclusively on the floor. The toads snapped at small dots and at rectangles oriented parallel to the animal's direction of movement, whereas rectangles oriented perpendicular to it elicited only few orienting responses, but no snapping. These findings are in good agreement with those of Roth and Wiggers (1983).

These experiments indicate that in the absence of movement, the toad as well as *S. salamandra* exhibit a preference for small, compact stimuli rather than for horizontal and vertical rectangles. While *Salamandra* prefers a vertical rectangle to a horizontal one, the opposite is true for *Bufo*. In the latter species, vertically oriented rectangles seem to have a negative influence on prey capture. However, this is the case only for stimuli located in the vertical plane. When the stimuli are located in the horizontal plane (on the floor), the orientation relative to the approaching animal comes into play and the rectangle parallel to the direction of movement of the animal is preferred to the one oriented perpendicularly.

It is important to note that neither in the experiments with stationary stimuli (Roth 1976) nor in those of Himstedt *et al.* (1980) did a "vacuum activity" (snapping in the direction of the luring stimulus after its disappearance) occur, such as described for the toad by Hinsche (1935). The subjects either turned away from both stimuli or approached one of them. However, the times elapsed between removal of the luring stimulus and the first approach movements were relatively long with a range between 12 and 87 s. Furthermore, when standing in front of one stimulus, the subjects waited another 8 to 85 s until they either snapped at it or turned away.

In conclusion, these experiments demonstrate convincingly that both anurans and urodeles are able to recognize stationary prey differing in size, shape and orientation. Possible mechanisms underlying recognition of stationary objects are (1) respiratory eye movements, (2) tremorlike eye movements, and (3) small jerky eye movements. These are described in greater detail below.

B. Optomotor Behaviour

In the context of gaze stabilization during self-induced motion or passive drift within the environment (cf. Collewijn 1981), vertebrates perform optomotor responses by movement of the eyes relative to the head (most mammals), or by head movement, often combined with a certain degree of eye movement. These compensatory movements consist of a smooth pursuit and a saccade directed in the opposite direction. Optokinetic eye movements in frogs were studied by Dieringer (1986, 1987). Passive displacement (table movement under light) is

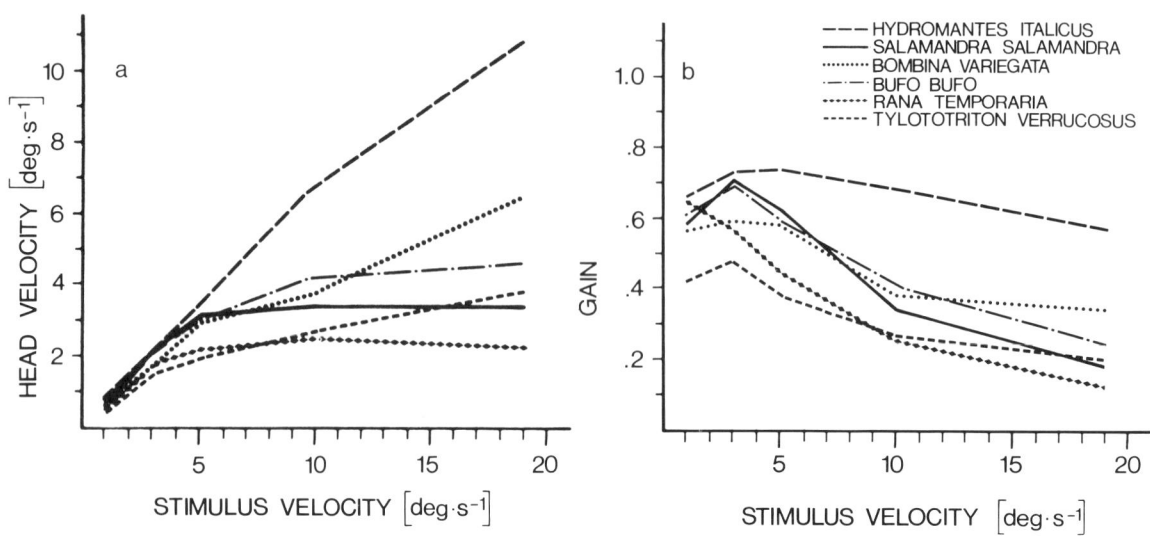

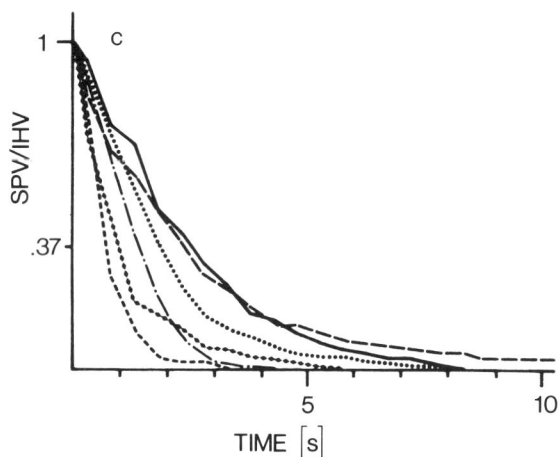

Fig. 11. Optokinetic behaviour in five amphibian species. **a** Relationship between slow phase head velocity and stimulus velocity. **b** Relationship between compensation gain (head velocity/stimulus velocity) and stimulus velocity. **c** Time-course of optokinetic afternystagmus (OKAN) at stimulus velocity of 10°/s. Ordinate indicates ratio of slow phase head velocity (SPV) versus initial head velocity (IHV), at which OKAN started when the light was switched off. From Manteuffel et al. (1986).

compensated with a gain (ratio between head or eye velocity and stimulus velocity) between 0.55 and 0.85, depending on stimulus amplitude. At small amplitudes, gaze was stabilized exclusively by compensatory eye movements; at larger amplitudes, compensatory head movements contributed gaze stabilization by up to 80%.

Salamanders show only optokinetic head movements. The optokinetic response to a continuously moving, structured large-field stimulus consists of a slow compensatory pursuit phase of the head in the direction of stimulus movement, interrupted by occasional saccades in the opposite direction, which reset the head back into a more or less medial position. During slow pursuit movement the head accelerates for a few seconds, until maximum velocity is reached. After a saccade, there is again a build-up phase.

In a comparative study, Manteuffel et al. (1986; see also Manteuffel 1989) measured the optokinetic nystagmus of three salamander species, *S. salamandra*, *Tylototriton verrucosus*, and *Hydromantes italicus*, and of three anuran species, *Bufo bufo*, *Rana temporaria*, and *Bombina variegata* (Fig. 11). The subjects were placed in an optomotor drum, with its inner surface covered with a black-and-white random-dot pattern (average period of 6°) and moved at different velocities.

All species tested showed a clear head nystagmus. *Hydromantes*, *Bombina*, and *Tylototriton* showed an increase in head velocity with increasing stimulus velocity up to a maximum velocity of 19°/s. There were differences in the slopes of the increment of head velocity. The gain was 0.48 in *Hydromantes*, 0.32 in *Bombina*, and only 0.12 in *Tylototriton*. The other amphibian species showed saturation of head velocity with increasing stimulus velocity.

Saturation was reached at 9.5°/s stimulus velocity with a head velocity of 4°/s in *Bufo*, at 5°/s stimulus velocity with a head velocity of 3°/s in *Salamandra*, and at 5°/s stimulus velocity with a head velocity of 2.2°/s in *Rana*.

Of all species tested, *Hydromantes italicus* showed the best compensation over the entire range of the tested velocities. Compensation of *Salamandra* was very similar to that of *Bufo*, showing a maximum at 3°/s stimulus velocity and a pronounced decrease toward higher stimulus velocities. *Bombina* showed a slightly weaker compensation; *Rana* showed good compensation at low stimulus velocities, but relatively weak compensation at higher velocities. The salamander *Tylototriton* showed the weakest compensation of all species tested.

III. MORPHOLOGY AND FUNCTION OF THE EYE, RETINA AND OPTIC NERVE

A. Anatomy of the Dioptric Apparatus

Except for caecilians and troglobitic salamanders, most amphibians have good vision. The eye of amphibians does not differ substantially in gross morphology from that of other vertebrates (Fig. 12). Its dioptric apparatus is characterized by a large cornea, which occupies nearly ⅓ of the eye bulb surface. Behind the *camera anterior bulbi* and the iris, the *camera posterior bulbi* contains the lens. The cornea is the major refractive structure in the dioptric system of terrestrial animals, because of the high difference in density between the cornea and the surrounding air. Terrestrial animals tend to have a more curved cornea and a more flattened lens, while aquatic animals have rather flat corneae and nearly spherical lenses; the refractive index of the surrounding water is similar to that of the cornea and the lens becomes the most important refractive element.

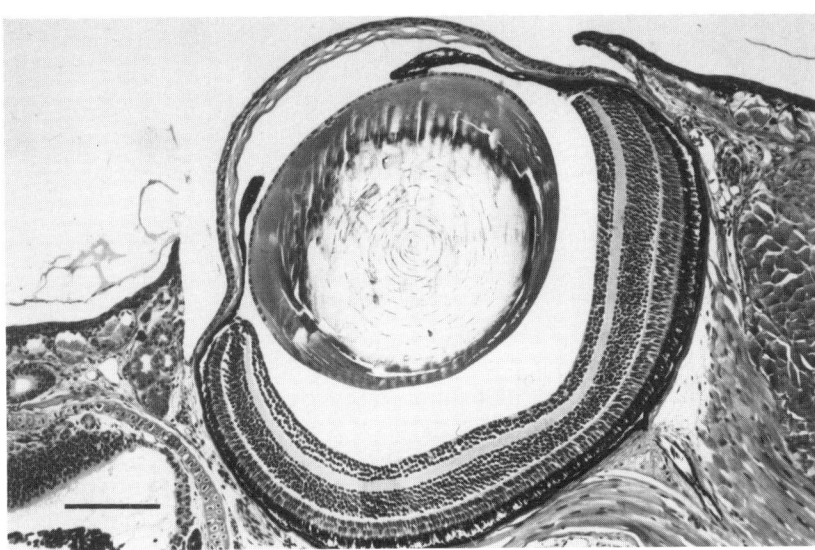

Fig. 12. Photomicrograph of a horizontal section through the eye of *Bolitoglossa subpalmata*. Bar: 500 μm. Courtesy of R. Linke.

The relatively thin vitreous body of the amphibian eye is of minor optical importance. Because of the large lens and the thick retina, the distance between cornea and lens as well as the distance between lens and retina, is small in relation to retinal thickness. In miniaturized salamanders, most of the space of the inner eye is filled with the large and nearly round lens and the thick retina (Linke *et al.* 1986). In some plethodontid salamanders the lens accounts for up to 25% (*Hydromantes, Desmognathus*) and the retina for up to 50% (*Thorius*) of the eye volume.

B. The Visual Field

Amphibians possess the widest visual fields among vertebrates (Schneider 1954; Fite 1973). Especially in anurans, the visual field covers nearly 360° because of the large cornea and protruded position of the eyes. Differences among anuran species exist primarily in the frontality of the eyes and shape and extent of the binocular visual field. A binocular visual

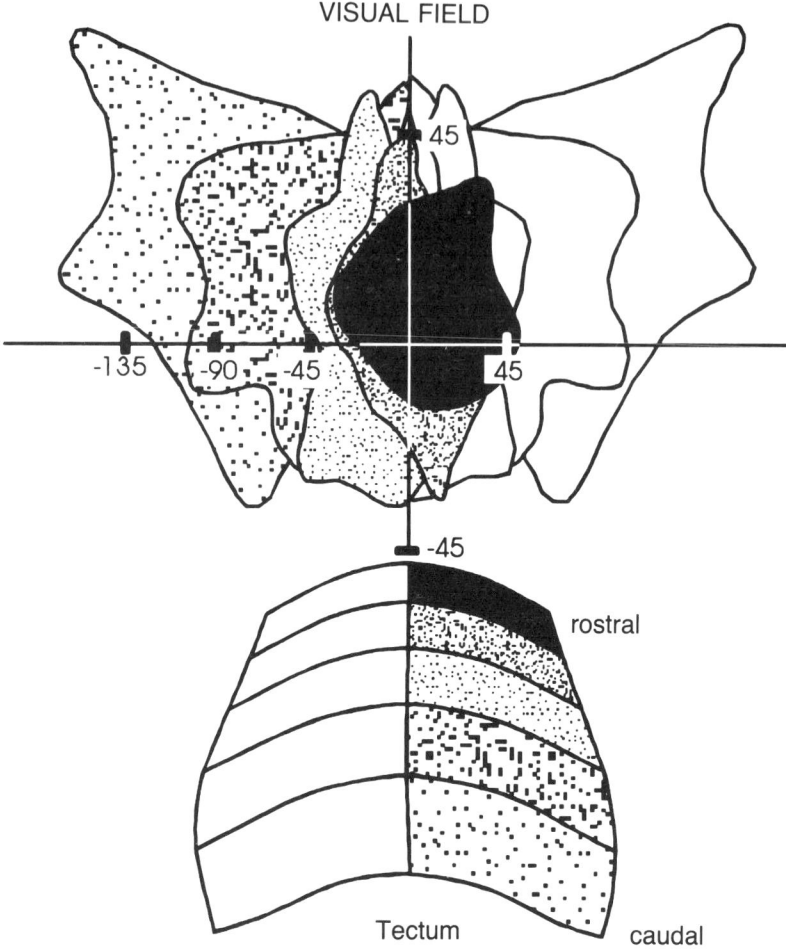

Fig. 13. Representation of the visual hemifield field plus central visual field in the contralateral tectal hemisphere in tongue-projecting salamanders. The central visual field is represented in the rostral tectum of both hemispheres and the caudolateral part of the hemifield in the contralateral caudal tectum. Note that the central visual field (±45°) covers about half of the tectal surface.

field is a necessary prerequisite for stereoscopic depth perception. *Rana esculenta* and *Rana temporaria* have a larger upper and dorsal binocular field (Schneider 1954), while the toad *Bufo bufo* has a larger lower anterior binocular field (Fite 1973). This correlates with differences in the ecological conditions of prey catching behaviour in these species (the ranid frogs are ambush feeders and mostly catch objects above the horizon, while the toad is a "hunter" of objects moving on the ground). According to Schneider, some anurans also have a posterior binocular field (e.g., 24° in *Rana esculenta*). However, it is difficult to estimate the precise extension of the monocular visual fields and their binocular overlap, because electrophysiological, optical, anatomical and behavioural methods often yield different results.

Salamanders, especially plethodontids have more frontalized eyes than do frogs and therefore have a larger binocular visual field (Linke *et al.* 1986; Wiggers 1991). Based on the frontality of the eyes and the maximal aperture, the plethodontid salamander *Hydromantes italicus* has a visual field of about 280°. A frontal region of 90° is perceived binocularly (Fig. 13). Electrophysiological investigations in the same species showed that the receptive fields of tectal neurons cover an area of 300° in horizontal direction. A frontal area extending over 90° is represented binocularly onto both tectal hemispheres. Vertically, the visual field extends from 34° below to 60° above the horizon. Some of the lateral receptive fields have sizes up to 180° in diameter. Thus, it is fair to assume that the field of visual perception covers nearly 360° (Wiggers 1991). The eyes of *Salamandra salamandra* are more laterally located and have a binocular field of only 65° (Werner 1983).

C. Eye Movement

As in most vertebrates, the amphibian eye bulb can be moved horizontally and vertically by six eye muscles. An additional pair of muscles, the *retractor bulbi* and its antagonist, the *levator bulbi*, move the eye during the eye-withdrawal reflex, when the cornea is touched or when the animal swallows.

Eye movement occurs during optomotor behaviour in frogs (see above), and during the vestibulo-ocular reflex in salamanders and frogs (Grüsser and Grüsser-Cornehls 1976; Dieringer 1987; Manteuffel 1989). Voluntary eye movements occur in larval salamanders and larval frogs, but for unknown reasons they are absent in adult amphibians. However, small "involuntary" eye movements occur. Schipperheyn (1965) described "respiratory" eye movements which resulted from pressure changes exerted on the eyeballs during ventilation movements of the throat. In the frog, these eye movements are vertically oriented. Tremor-like eye movements were long denied to occur in amphibians (Autrum 1959), but were demonstrated in *Salamandra salamandra* by Manteuffel *et al.* (1977). These authors used a laser beam reflected from a mirror on to the cornea. These rapid eye movements have an amplitude of approximately 12 minutes of arc, which is sufficient to displace the retinal image over several receptors. This latter mechanism alone could be sufficient for the perception of stationary objects. In *Rana temporaria*, Grüsser and Grüsser-Cornehls (1976) described microscopic short saccades at intervals of 0.5 to 2 minutes, mostly in a horizontal direction.

D. Accommodation Mechanisms

In contrast to amniotes, where accommodation is based on changes in the shape of the lens, the anuran eye is accommodated by forward and backward movement of the lens by means of two muscles, the dorsal and ventral *protractor lentis*. In salamanders this is achieved by one ventral *protractor lentis*, which brings the lens into an outwardly tilted position. The overall effect of accommodation in amphibians is unclear. Although monocular frogs can measure distances precisely in the range from 5 to 20 cm and should be able to accommodate over a range of about 15 diopters (Collett and Harkness 1982), Grüsser and Grüsser-Cornehls (1976) reported the effect of accommodation to be less than 5 diopters; Douglas *et al.* (1986) observed a maximal refractive change of 10 diopters in frogs. Furthermore, particularly in urodeles, the shift of the lens does not seem to be very accurate and is unlikely to be the dominant mechanism for the excellent depth perception present in bolitoglossine salamanders. A relaxation of the *protractor lentis* muscle by application of atropin causes toads (Jordan *et al.* 1980) and frogs (Douglas *et al.* 1986) to undershoot their prey, and an overcontraction of the muscle by application of miotic to overshoot. As already mentioned, frogs exhibit *no* substantial loss of depth perception during binocular vision after bilateral section of the oculomotor nerve. This indicates that even in frogs, accommodation is only *one* of the mechanisms of distance estimation and is mainly used for monocular vision (Douglas 1986).

E. Morphology, Cytoarchitecture and Synaptology of the Retina

The amphibian retina shows the typical five-layered structure of vertebrates (Figs 14, 15). The outer nuclear layer (ONL), the inner nuclear layer (INL) and the layer of retinal ganglion cells (RGC) are separated by two fiber layers: the outer plexiform layer (OPL), which is thin to very thin (in salamanders), and the much thicker inner plexiform layer (IPL). The two plexiform layers are the main site of synaptic contacts between the five major types of retinal cells. The outer nuclear layer contains the inner segments of the photoreceptors and their nuclei.

In most frogs and salamanders, the ONL is formed by two layers of cell bodies. The rod nuclei are aligned at the distal side of the ONL, and the cone nuclei are more proximal (Gordon 1976). This is opposite to the situation in most other vertebrates, in which the rod nuclei are vitread to the cone somata.

In the inner nuclear layer, amacrine cells are concentrated at the vitread side, bipolar cells in the middle, and horizontal cells at the sclerad side. There are also a few displaced

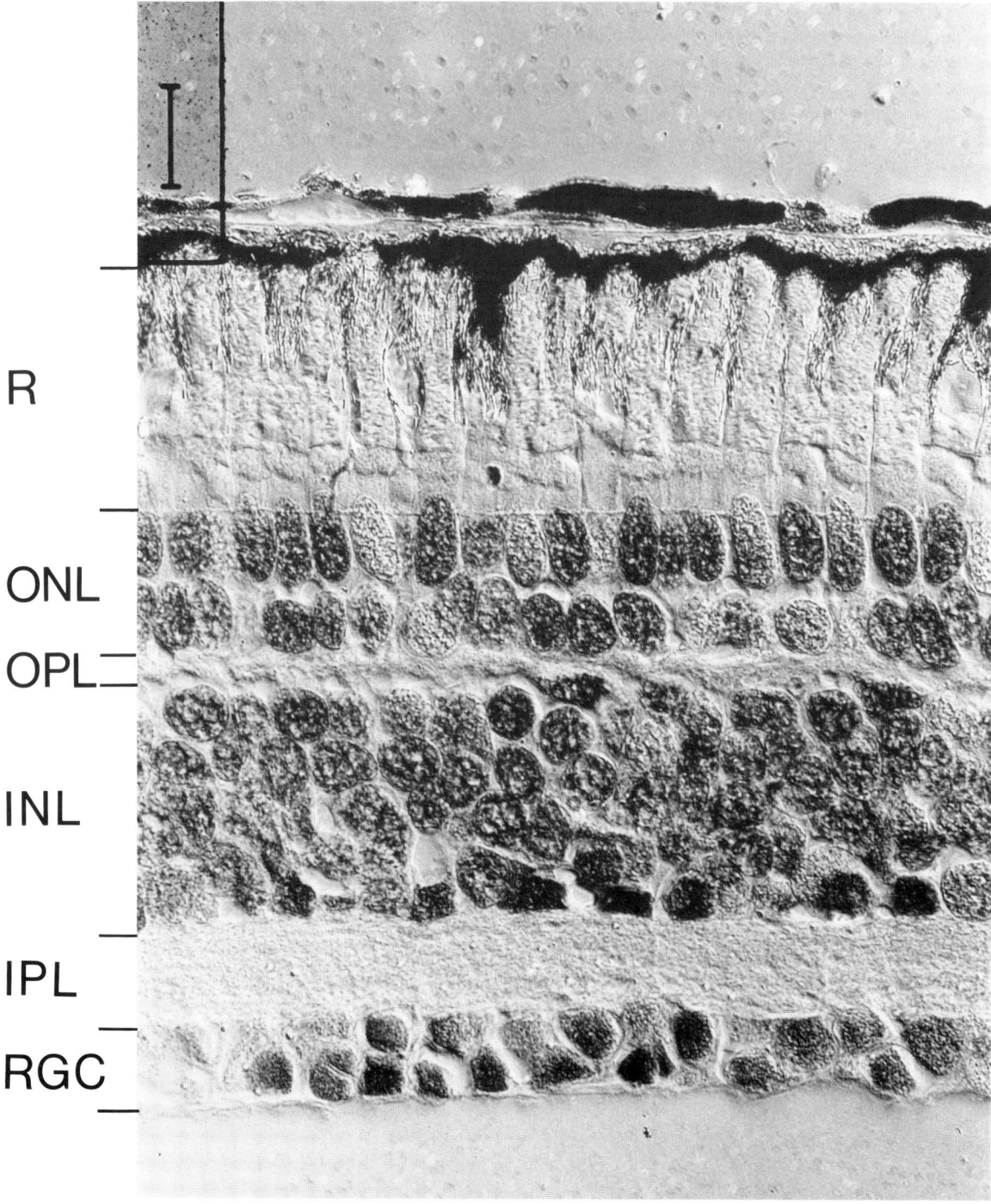

Fig. 14. Transverse section through the retina of *Hydromantes shastae*. RGC = retinal ganglion cell layer; IPL = inner plexiform layer; INL = inner nuclear layer, OPL = outer plexiform layer, ONL = outer nuclear layer, R = receptor outer segments. Bar: 20 μm.

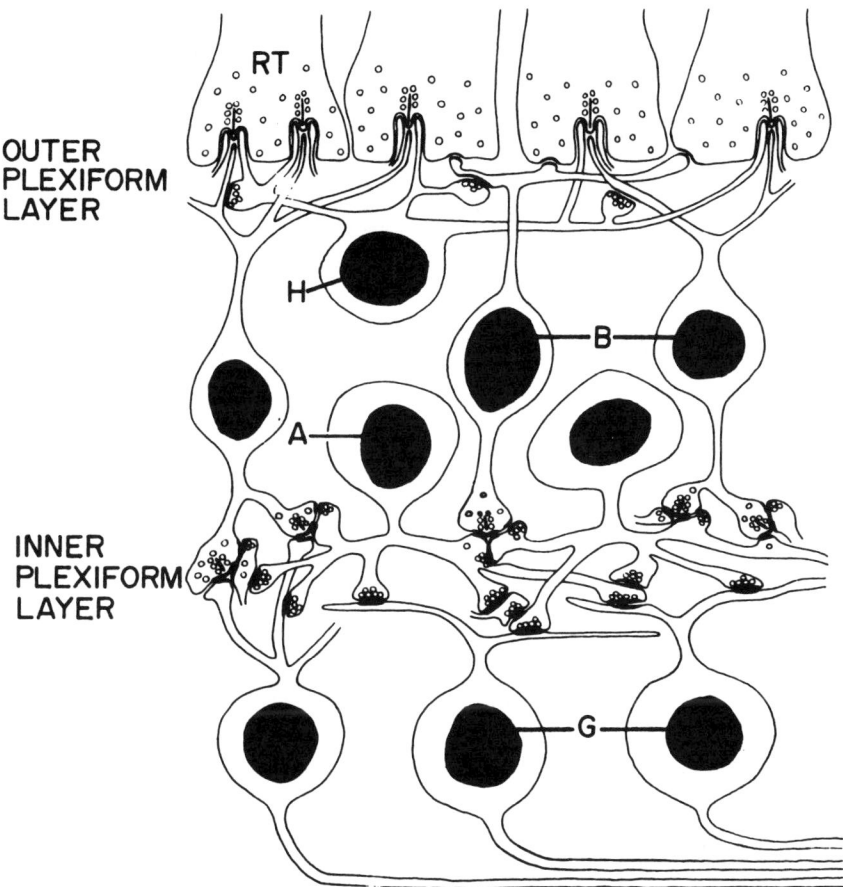

Fig. 15. Summary diagram of synaptic contacts in the amphibian retina. A = amacrine cells; B = bipolar cells; G = ganglion cells; H = horizontal cells; RT = receptor terminals. From Dowling and Werblin (1969).

retinal ganglion cells at the vitread side of the INL. In most amphibians, the RGC layer consists of more than one row of retinal ganglion cells. The axons of the RGC form bundles, which meet in the optic disk and constitute the optic nerve.

Although amphibians have no specialized intraretinal structure like the fovea of primates or birds, some frogs exhibit an uneven distribution of all cell types within the retina. In *Hyla raniceps* (Bousfield and Pessoa 1980), *Heleioporus eyrei* (Dunlop and Beazley 1981) and *Bufo marinus* (Nguyen and Straznicky 1989), a streak of high cell density exists in the RGC layer along the naso-temporal meridian of the retina. The same holds for the inner nuclear layer (Zhu *et al.* 1990) and the outer nuclear layer (Zhang and Straznicky 1991). The increase in cell density is not comparable to the situation in and around the *fovea centralis* in primates, but to the visual streak in the reptilian retina (Wong 1989). In salamanders, no difference in intraretinal cell density has been found so far.

1. Photoreceptors

Photoreceptors of amphibians, like those of other vertebrates, exhibit two main compartments (Fig. 16). The most distal part is the outer segment (OS), the photosensitive region of the receptor. A thin eccentric cilium connects the OS to the inner segment (IS). The nucleus of the receptor cell is located proximal to the IS. The most vitread structure in rods is the spherule, (in cones, the pedicle), which is the synaptic terminal of the receptor.

A. RODS

The largest photoreceptors in the amphibian retina and the most frequent receptors in nocturnal species are the rods, named after the cylindrical shape of their outer segment. There is great interspecific as well as intraretinal variation in size of the rod outer segments.

In the toad, *Bufo marinus*, their length varies from 28 μm in the ventral peripheral retina to 89 μm in the dorsal retina (Zhang and Straznicky 1991). In the salamander, *Plethodon cinereus*, the average length is about 40 μm (Braekevelt 1992), which is remarkably long considering the size of the eye. The outer segment is the photosensitive region of the receptor. It is densely filled with free-floating disks formed by a double-layered membrane. In terrestrial amphibians, these densely stacked disks contain the photoreactive pigment rhodopsin. Some aquatic species like *Necturus* have porphyropsin instead. The disks are constantly renewed by invaginations of the plasma membrane at the basal surface of the OS. Later, these move upwards in the outer segment. At the opposite end (the tip of the rod) they are absorbed and phagocytized by cells of the pigment epithelium. The inner segment is the synthetic center of the receptor cell but also contains several structures that allow waveguiding within the receptor. The ellipsoid is a convex-concave or plane-concave structure with a higher refractive index than that of the surrounding cytoplasm and is located near the apex. It is filled with densely packed mitochondria and often contains additional transparent oil droplets of a high refractive index (Baylor and Fettiplace 1975). It is still uncertain whether these structures are involved in optical processes.

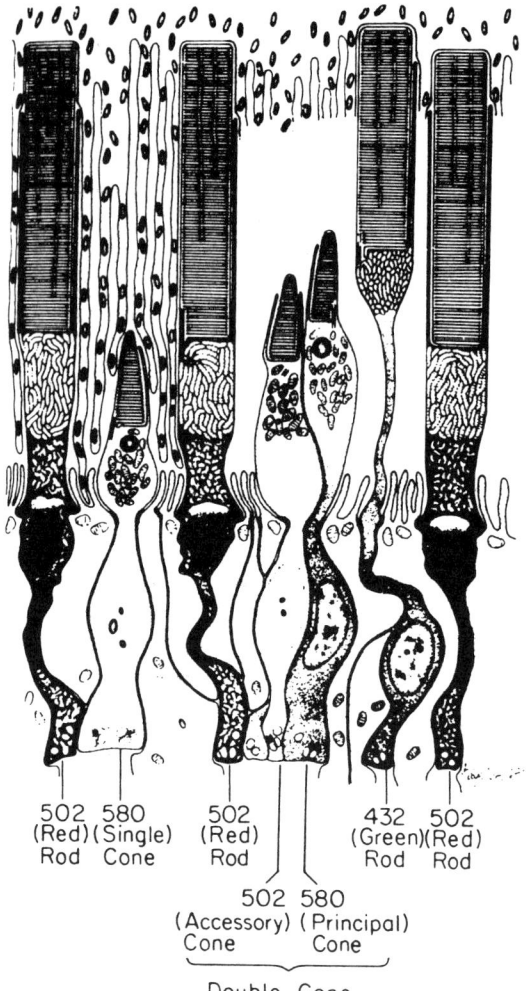

Fig. 16. Schematic drawing of photoreceptors in the frog retina *(Rana pipiens)*. Three red rods, one single cone, one double cone and one green rod are shown. From Donner and Reuter (1976).

As a reaction to light and darkness, photoreceptors can change their position. The myoid, an area between ellipsoid and nucleus, can be lengthened or shortened. It is unclear whether these photomechanical reactions occur in all amphibian species and whether these movements have an effect on light regulation, sensitivity or visual acuity. It has been observed in some frogs (Arey 1916), but not in *Plethodon cinereus* (Braekevelt 1992).

The synaptic terminal of the rods, the spherule, carries gap junctions and a typical additional invagination with a single synaptic ribbon at its bottom. There are several lateral expansions from the terminal (telodendra).

B. CONES

Most diurnal amphibians have a cone-dominated retina. In contrast to the rods, the outer segment of the cone is much shorter and cone-shaped. The pigment disks are formed by infoldings of the plasma membrane. Like the disks of rods, they are also renewed but in a different fashion. Furthermore, there is an additional refractive structure in the ellipsoid of most cones, a droplet of transparent oil.

The synaptic terminal of cones, the pedicle, is a flat, synaptic structure with some typical ribbon synapses at the bottom of its invaginations, additional conventional synapses and gap junctions to adjacent cones and rods.

Further differentiation of photoreceptors: In most amphibians, the two basic types of photoreceptors are differentiated further. Beside the typical red rod, there is a "green rod" which has a different pigment, a much shorter outer segment and an elongated myoid. There is also a second type of cone, the so called "double cone", which is composed of a small ordinary cone and an additional element, an embryonic rod. While the outer segments of this double-receptor are separated and contain different pigments, the inner segments are fused. Similar double receptors are found in most vertebrates.

D. VISUAL PIGMENTS

The amphibian retina differs from the typical primate one, which has a three-colour cone system and a brightness-sensitive rod system. The colour of a freshly excized frog retina is pink due to the absorbance of the photoreactive pigment in the most frequent receptor, the "red rod". The "red rod" visual pigment in frogs (as in other vertebrates) is a rhodopsin (in aquatic amphibians porphyropsin), which has its maximum absorbance in the green spectrum at about 502 nm. The green rods contain a pigment with an absorbance at 432 nm, which is similar to the blue-sensitive cones in other vertebrates. The cone pigment is most sensitive to yellow light at about 575–580 nm. The principal component of a double cone carries the typical cone pigment, while the pigment of the accessory component is the green-sensitive rhodopsin of the "red rod".

In frogs, there are different absorbance spectra in adults and tadpoles. There is an alteration of pigments during metamorphosis, such that the aquatic stages are more sensitive to longer wavelengths. The same holds for some salamanders, in which aquatic stages, larvae as well as adults, are more sensitive to the yellow spectrum (Himstedt 1973a,b).

E. CONTROL OF LIGHT INTENSITY LEVEL

The amphibian eye has to be active over a large range of light intensities. While some frogs catch their prey in bright sunlight, others leave their hiding-place only at dark night (see above) and, therefore, need highly sensitive photoreceptors. In vertebrates, several mechanisms exist for the control of light influx. Most terrestrial species have a very efficient pupillary light reflex, released directly by the decay of photopigments in the *sphincter pupillae*. In aquatic animals, the pupillary reflex plays only a minor role. Furthermore, the sensitivity of the retina can be changed by switching between receptor systems or by changing the sensitivity threshold (Dowling 1977). In the newt, *Notophthalmus viridescens*, diurnal changes occur in the morphology of the photoreceptor synaptic terminal (Ball 1983).

In most amphibians, distinct retinomotoric movement has been observed. In a light-adapted retina, the myoids of the single cones and the principal components of the double cones are shortened, while the red and green rod myoids are elongated somewhat (Donner and Reuter 1976). The opposite occurs in a dark-adapted retina. In addition, there is a striking movement of the melanin granules of the pigment epithelium. In darkness, the granules are concentrated at the sclerad side. When exposed to light, they move into epithelial extensions, which surround the receptors' outer segments.

2. Interneurons

The inner nuclear layer (INL) is the region of interneurons. Due to their shape and function, they are described as horizontal, bipolar and amacrine cells.

Most of the horizontal cells are located on the distal side of the INL. Their soma is round and often flattened at the distal side. Their dendrites extend from the apical side and spread out laterally into the OPL, where they make postsynaptic contacts primarily with the receptor terminals (invaginating synapses) and presynaptic contacts with bipolar cells (regular synapses). In frogs, there are two morphologically distinct classes of horizontal cells. While the so called H2 cells have no discernible axons, H1 cells possess an axon that runs within the OPL and arborizes about a hundred microns away from the soma (Stephan and Weiler 1981). Another classification of large inner cells and small outer cells depends on their location within the INL and the size of the soma (Ramon y Cajal 1892). The same classes of HC can be found in salamanders (Linke and Roth 1989). Furthermore, there is evidence of strong electrical coupling between horizontal cells via gap junctions (Skrzypek 1984). They are thought to act as a single pathway laterally transmitting information in the retina.

In contrast to horizontal cells, the main pathway of bipolar cells runs in a radial direction, i.e., they connect receptors and horizontal cells on the distal side with the amacrine and ganglion cells on the proximal side of the INL. Bipolar cells possess a so-called Landolt club. In *Triturus* (Hendrickson 1966) and *Salamandra* (Linke, unpubl. results), this specific apical fiber is a process that is normally thicker than the other apical dendrites and runs in the longitudinal plane of the bipolar cell. It extends through the OPL and the cell bodies of the photoreceptors and ends at the outer limiting membrane. The terminals of this fiber normally have a bulbous swelling.

Several classes of bipolar cells have been described in amphibians (Werblin and Dowling 1969; Wong-Riley 1974). Except for variation in the pattern of arborization, which may spread in the sclerad or vitread side of the IPL and is smaller than in the horizontal cells, bipolar cells differ in their synaptic contacts with the receptors in the OPL (conventional or ribbon synapses) and in their contacts in the IPL. The striking presynaptic structure of the bipolar cell is the ribbon synapse and usually is connected with two postsynaptic elements, collectively called a dyad. In frogs, most of the ribbon synapses are connected to amacrine cells, and only few make simultaneous contacts with amacrine and ganglion cells. This is opposite to the situation in primates. Another difference between amphibians and primates is the number of conventional synapses in the IPL. In frogs, there are ten times more conventional synapses than ribbon synapses (Gordon 1976). The presynaptic element in these conventional synapses usually belongs to an amacrine cell. The postsynaptic element can be a second amacrine cell, a bipolar cell or a ganglion cell.

Amacrine cells are true interneurons, because they lack an axon. Most of their somata are located in the proximal third of the INL, but some are displaced amacrine cells located in the retinal ganglion cell layer. The number of displaced cells is surprisingly high in urodeles (Ball and Dickson 1983). Electrical coupling is very extensive among amacrine cells. Intracellular application of the substance Biocytin into a single amacrine cell often resulted in the staining of more than 150 amacrine cells, because Biocytin passes gap junctions. In plethodontid salamanders, about half of these coupled cells are "displaced" amacrines and are located in the RGC layer. There are two morphologically distinct types of amacrine cells, with several subclasses in amphibians (Werblin and Dowling 1969; Wong-Riley 1974). The arborization pattern in the IPL of the first type of amacrine cells is mono- or multilayered, while the other type arborizes in a more diffuse manner.

3. Retinal Ganglion Cells

Together with the receptors, ganglion cells are the most studied retinal neurons. In amphibians, there is a wide variation in number, size and type of arborization. In frogs, the diameter of the soma varies from 7 to 20 μm, and their dendrites may spread over an area of up to 600 μm. Such large dendritic trees are likewise common in plethodontids. Since different types of amacrine and bipolar cells branch at different depths of the IPL, the dendritic branching pattern of ganglion cells points to the type of cells, from which they receive synaptic input and may in this way influence their physiological function. There are several morphological types with regard to the size and shape of the soma, the dendritic branching pattern and depth of arborization in the strata of the IPL.

Using the same criteria, the number of RGC types is larger in anurans, than it is in salamanders. Despite some arbitrariness in the morphological classification, it varies among three major types (with 12 subtypes based on arborization) in *Xenopus laevis* (Straznicky and Straznicky 1988) and seven types in *Rana pipiens* (Frank and Hollyfield 1987). In plethodontid salamanders, only four types have been described (Linke and Roth 1989) (Figs 17, 18). The following brief description of RGCs is based on the papers of Frank and Hollyfield (1987) and Linke and Roth (1989).

Frog RGC classes 1 and 2 are characterized by a large dendritic field, which extends over several hundred microns and is symmetric in shape in class 1 and asymmetric in class 2. The arborization is restricted to the sclerad stratum of the IPL. Most dendrites are covered with spine-like processes. The somata are aligned close to the IPL and often give rise to two thick primary dendrites. These classes are comparable to class 4 in salamanders where RGC with symmetric and asymmetric dendritic fields were not distinguished.

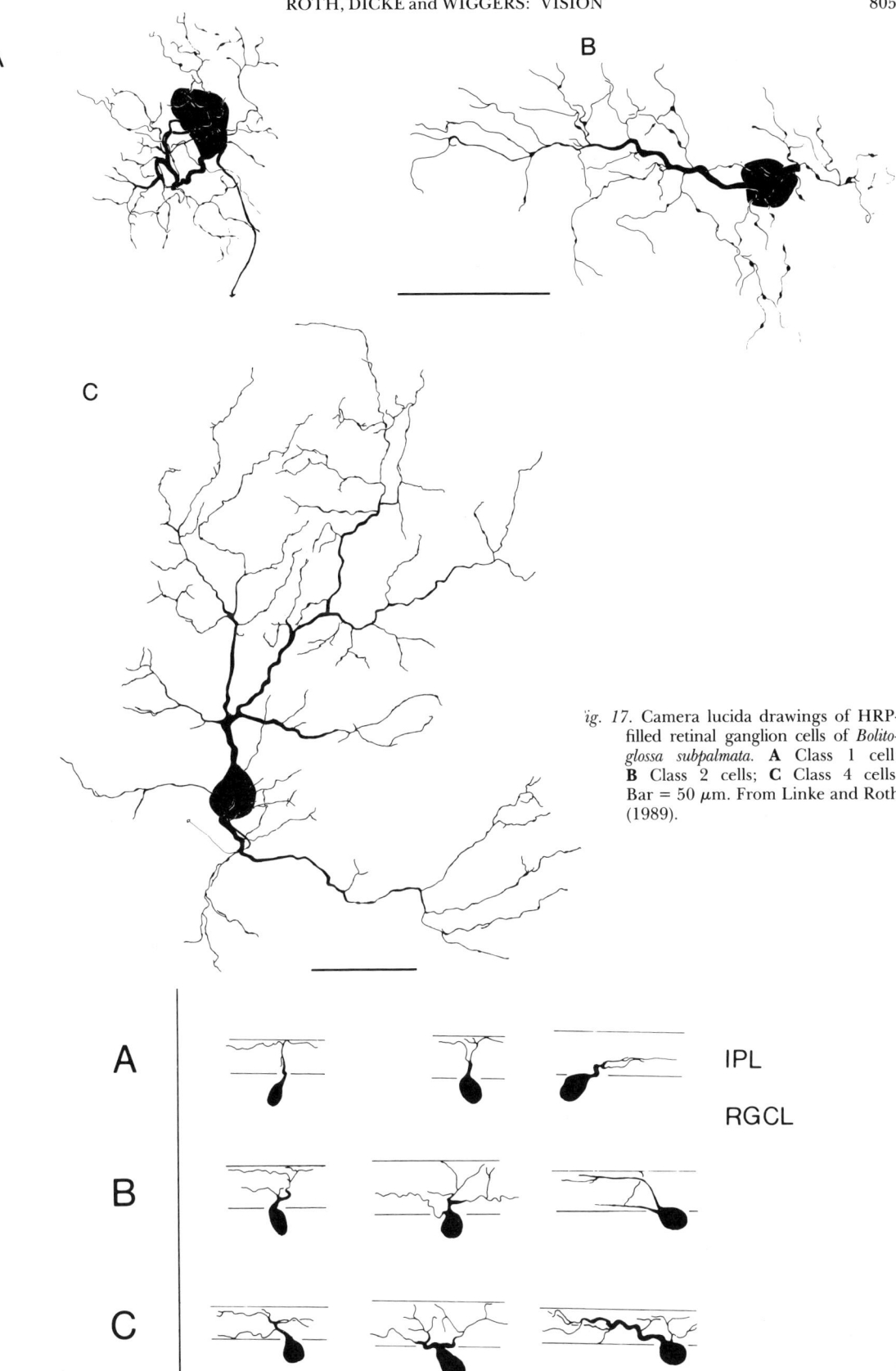

Fig. 17. Camera lucida drawings of HRP-filled retinal ganglion cells of *Bolitoglossa subpalmata*. **A** Class 1 cell; **B** Class 2 cells; **C** Class 4 cells. Bar = 50 μm. From Linke and Roth (1989).

Fig. 18. Lamination patterns of retinal ganglion cell dendrites in the inner plexiform layer in plethodontid salamanders. **A** Monostratified cells; **B** bistratified cells; **C** diffuse cells. IPL = inner plexiform layer; RGCL = retinal ganglion cell layer. From Linke and Roth (1989).

Class 3 RGC in frogs have medium sized dendritic fields arborizing in different laminae of the IPL (bistratified, tristratified and diffuse). Their somata are located more distal to the nerve fiber layer, and one (monochotomous) or two (dichotomous) thick primary processes extend from the soma. This type is comparable to class 2 RGC in salamanders, although in that group the stratification pattern is not as distinct as it is in frogs.

Class 4 RGC in frogs is characterized by a sparsely branched dendritic tree that spreads out in the vitread level of the IPL. This type is comparable to class 3 RGC in salamanders.

Class 5 RGC in frogs consists of displaced ganglion cells with an arborization in both the IPL and the OPL. This type has not been found in salamanders.

The somata of the class 6 and 7 RGC in frogs are located at the most superficial part of the RGC layer. They have only one primary dendrite, which is elongated (class 6) or short (class 7) and a small circular to oval dendritic field confined to the vitread portion of the IPL. These two types are comparable to class 1 RGC in salamanders, where the length of the primary dendrite was not used as a distinguishing feature. For the numbers of RGC see Section F below (Optic nerve).

4. Müller Cells

Another retinal cell type is the Müller cell. These cells are glial cells that extend from the outer limiting membrane to the layer of ganglion cells and are visible as vertical structures in the plexiform layers. Beside supporting functions, they may act as a reservoir of potassium, because there are light-induced changes of the intracellular K^+ potential.

F. Optic Nerve

The optic nerve is constituted by the bundled axons of retinal ganglion cells and neuroglia. Myelinated and unmyelinated fibers are grouped closely together in fascicles. Throughout the length of the nerve to the *chiasma opticum*, the separate fascicles shift their relative position, and individual fibers pass from one fascicle to another. The number of axons and their activity represents the output of the retina, before it is modified in the afferent layers of the tectum or other visual centres.

Among amphibians, the highest number of optic nerve axons (and thus of RGC) are found in anurans; Maturana (1959) counted 470 000 fibers in *Rana pipiens* and 320 000 fibers in *Bufo americanus*. The lowest number presently known among anurans is in *Xenopus laevis* with 68 000–80 000 fibers (Dunlop and Beazley 1984). However, the Australian paedomorphic frog, *Arenophryne*, has the lowest number of central visual cells among frogs (see below), and therefore may be expected also to have the lowest number of RGC and optic nerve fibers. On average, urodeles have 5–10 times fewer optic fibers. They range from a minimum of 26 000 in *Batrachoseps attenuatus* (Linke and Roth 1990) to 75 000 in the salamandrid *Notophthalmus viridescens* (Ball and Dickson 1983). Fritzsch (in Roth 1987) reported 53 000 fibers in *Salamandra salamandra*, and Gruberg (1972) found a nearly equal number (50 000) in *Ambystoma tigrinum*. In plethodontid salamanders, despite their excellent visual abilities, the number of axons is astonishingly low. Linke and Roth (1990) found a range from 26 000 in *Batrachoseps attenuatus* to about 50 000 in *Plethodon jordani*. These numbers are the lowest found among vertebrates with an elaborate visual system.

In contrast to teleosts, birds and mammals, where most optical axons possess a myelin sheet, the percentage of myelination in adult amphibians is low (Maturana 1959; Wilson 1971). *Xenopus laevis* has the highest percentage of myelination with 11% (Wilson 1971; Dunlop and Beazley 1984), while the lowest percentage is found in the plethodontid salamander, *Batrachoseps attenuatus*, which has less than 1% myelinated fibers or even no myelin sheets at all (Linke and Roth 1990).

G. Visual Acuity

The acuity of the amphibian visual system depends on two main factors: (1) the *optical resolution* which is determined by the aperture and the optical quality of the dioptric apparatus, i.e., the cornea and lens, and (2) the *anatomical resolution*, which is determined by the diameter and packing density of the receptor outer segments. Because of the small absolute size and the round lens of most amphibian eyes, the optical quality of the dioptric apparatus is relatively poor. In addition, there is a high chromatic aberration in optical systems with round lenses and short focal length, and in order to maintain maximum resolution, the aperture has to be as large as possible, as is found in most amphibian eyes. In contrast to the optical resolution, the anatomical resolution is not necessarily reduced by the small size of the eye, provided the size of photoreceptors is reduced proportionally. Therefore, miniaturized salamanders with small eyes have also very small receptor diameters, and the angular distance between the receptors is minimized (Linke et al. 1986).

The anatomical resolution power was calculated for some salamanders on the basis of the rod diameter (Linke et al. 1986). It ranged from 0.32° in *Eurycea bislineata* to 0.5° in *Thorius narisovalis* and *Desmognathus ochrophaeus*. Thus, an object of 0.5 mm edge length should be detectable at a distance between 57 and 82 mm. The optical resolution, based on the aperture, is even higher and the same object should be detectable at a distance of about 78 cm in *Hydromantes*. Behavioural studies in *Salamandra salamandra* yielded a range of visual acuity between 8' (aquatic stage) and 12' (terrestrial stage) (Himstedt 1967). Neurons in the optic tectum of the plethodontid salamander, *Batrachoseps attenuatus*, showed a clear response to objects of about 13' (0.23°). On the basis of anatomical, behavioural and electrophysiological investigations, it is fair to assume a visual acuity ranging from 8' to 15' in salamanders. Birukow (1937) found a resolution of about 7' in *Rana temporaria*. However, on the basis of minimal stimulus size that elicits responses in tectal cells of class T 5 (see below), Grüsser and Grüsser-Cornehls (1976) determined what they called "functional visual acuity", with a value of 0.1–0.2'. Generally, due to their larger eyes and smaller receptor diameters, most frogs seem to have a higher visual acuity than do most salamanders, which are characterized by smaller eyes and larger (sometimes much larger) cells.

H. Eye Size and its Effect on Retinal Structures

While most frogs have large eyes and small cells relative to body size, most salamanders have relatively small eyes and at the same time large cells (Linke et al. 1986). This results in an amazingly low number of visual cells, including photoreceptors, particularly in bolitoglossine salamanders, which at the same time have the most precise depth perception. In order to compensate for the resulting low resolution power of the retina, the number of retinal ganglion cells relative to the photoreceptors in bolitoglossines is disproportionally increased (with 3–6 rows of RGC), although the absolute number of RGC is very low in comparison with other amphibians (see above). Due to the high relative number of RGC, the relation to the receptors is roughly 1:1 (Linke et al. 1986), which is similar to the *fovea centralis* of most mammals. Thus, in these animals the whole retina is a functional fovea. This "foveal" compensatory process is increased further by the fact that most tectal cells are devoted to frontal vision (see below).

IV. ANATOMY OF THE VISUAL SYSTEM

A. Gross Anatomy of the Amphibian Brain

Following is given a brief description of the brain of frogs and salamanders, with emphasis of the parts involved in vision and visually guided behaviour (Figs 19, 20 for salamanders; Figs 21, 22 for frogs).

The brain of amphibians is divided into five major parts: (1) telencephalon; (2) diencephalon; (3) mesencephalon; (4) cerebellum; (5) *medulla oblongata*. Generally, the brain of frogs exhibits a degree of morphological differentiation similar to that found in most other vertebrates, with a number of laminated structures (e.g., *tectum mesencephali*, *torus semicircularis*) and morphologically distinct cell nuclei, often found in a migrated position inside the white matter. By contrast, in the salamander brain, the neurons are situated mostly within the periventricular layer; few migrated nuclei are found. This situation long has been considered primitive, but extensive neuroanatomical studies and phylogenetic analysis have demonstrated convincingly that it is the product of secondary simplification within the context of paedomorphosis (Roth et al. 1993).

1. Telencephalon

In anurans, the two hemispheres of the telencephalon are clearly separated from each other as two tubular formations. In urodeles, the separation of the hemispheres is much less distinct and indicated externally only by a shallow groove. The cerebral hemispheres join the *telencephalon impar* at the level of the *foramen interventriculare*. The telencephalon is divided into: (1) olfactory bulbs; (2) pallial structures: (dorsal, lateral and medial pallium, and *amygdala pars lateralis*) and (3) subpallial structures (*nuclei septi*, *amygdala pars medialis*, and *striatum*).

The medial pallium represents the dorsomedial wall of the hemisphere. It is considered to be homologous to the mammalian hippocampus, it projects outside the telencephalon to the *nucleus praeopticus*, the ventral thalamus, and the hypothalamus via the medial forebrain bundle.

The dorsal pallium includes the pallial parts dorsal and caudal to the most rostral extent of the medial pallium. It receives both olfactory fibers from the olfactory bulb and visual, auditory and somatosensory afferents from the thalamus through the medial forebrain bundle. It may be regarded as homologous to the mammalian neocortex. The dorsal pallium does not project outside the telencephalon.

The lateral pallium, a homologue of the mammalian olfactory (piriform) cortex, projects to the dorsal and medial pallium, the striatum, and the septum as well as to the contralateral lateral pallium. Its projections, however, do not leave the telencephalon except when crossing to the contralateral side of the telencephalon in the habenula (Schmidt and Roth 1990).

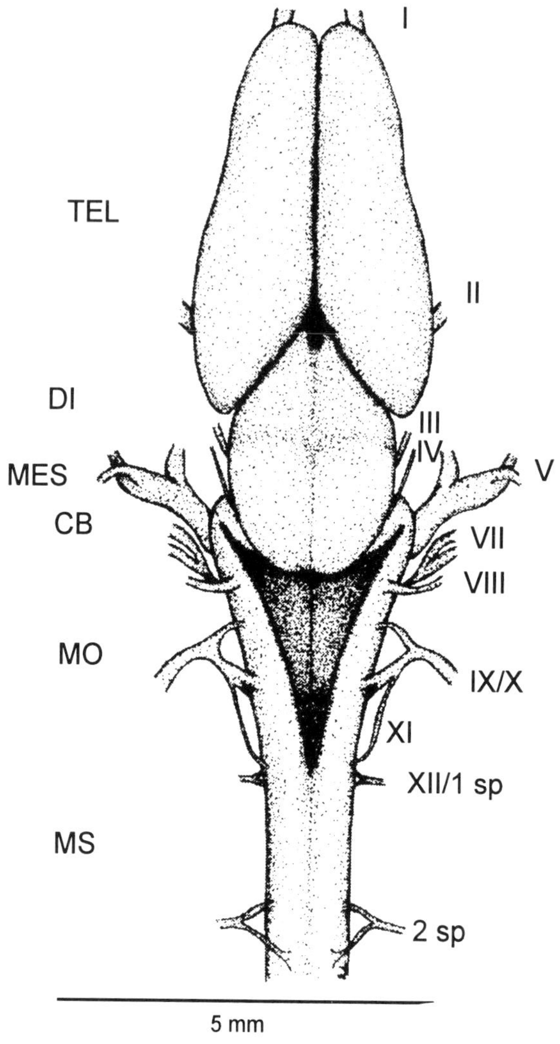

Fig. 19. Dorsal view of the brain of *Hydromantes italicus*. CB = cerebellum; DI = diencephalon; MES = mesencephalon; MO = medulla oblongata; MS = medulla spinalis; TEL = telencephalon; I–XII = cranial nerves 1–12; 1 sp = 1st spinal nerve; 2 sp = second spinal nerve.

The amygdala consists of two distinct cellular nuclei, the *pars lateralis* of pallial origin, and the *pars medialis* of subpallial origin (Källén 1951; Northcutt and Kicliter 1980). Only the medial part seems to be homologous to the mammalian amygdala. The septal nuclear complex occupies the ventromedial wall of the cerebral hemispheres below the medial pallium between the olfactory bulb and the *lamina terminalis*. Its chief efferent pathway is the *fasciculus medialis telencephali*, or "medial forebrain bundle".

The striatum occupies the ventrolateral wall of the telencephalic hemisphere in a position immediately caudal to the accessory olfactory bulb. The striatum receives both olfactory and visual input and seems to be a major co-ordination centre in the telencephalon. The chief efferent pathway is the *fasciculus lateralis telencephali*, or "lateral forebrain bundle".

2. Diencephalon

The diencephalon of amphibians is divided into four parts: (1) epithalamus, (2) dorsal thalamus, (3) ventral thalamus, (4) hypothalamus.

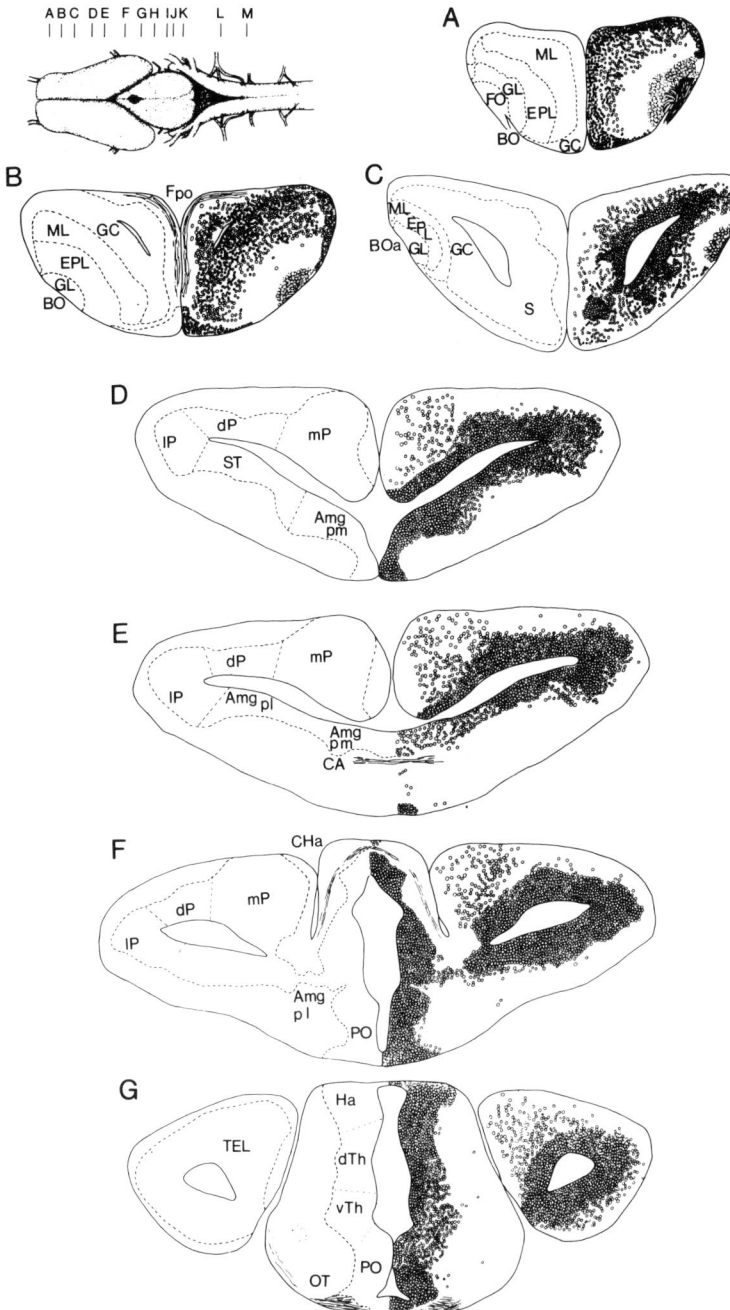

Fig. 20. Drawings of transverse sections through the brain of *Bolitoglossa subpalmata*. **A–B:** telencephalon at the level of the primary olfactory bulb. **C:** telencephalon at the level of the accessory olfactory bulb. **D:** mid-telencephalon. **E:** caudal telencephalon at the level of the commissura anterior. **F:** caudal telencephalon/rostral diencephalon at the level of the commissura habenulae. **G:** mid-diencephalon. **H:** pretectum/rostral mesencephalon at the level of the commissura posterior. **I:** mid-mesencephalon/hypothalamus at the level of the commissura tuberculi posterioris. **J:** mesencephalon and beginning medulla oblongata (auricles). **K:** caudal mesencephalon/cerebellum/rostral medulla oblongata at the level of 5th nerve root. **L:** mid-medulla oblongata at the entrance of 10th nerve roots. **M:** rostral cervical spinal cord. A = *auricula cerebelli*, Amg = *amygdala*, Amgpl = *amygdala pars lateralis*, Amgpm = amygdala pars medialis, BO = *bulbus olfactorius*, CA = *commissura anterior*, CC = *corpus cerebelli*, Cha = *commissura habenulae*, CP = *commissura posterior*, Ctub = *commissura tuberculi posterioris*, dP = dorsal pallium, dT = dorsal tegmentum, dTh = dorsal thalamus, EPL = external plexiform layer, flm = *fasciculus longitudinalis medialis*, FO = *fila olfactoria*, Fpo = *fasciculus postolfactorius*, GC = layer of granule cells, GL = glomeruli, Ha = habenula, Hy = hypothalamus, Hyp = hypophysis, lmsp = *lemniscus spinalis*, lP = lateral pallium, ML = layer of mitral cells, mP = medial pallium, PO = *nucleus praeopticus*, OT = optic tract, S = septum, TEL = telencephalon, TO = *tectum opticum*, vTh = *thalamus ventralis*, ventral thalamus. From Roth (1987). See **H–M** on page 810.

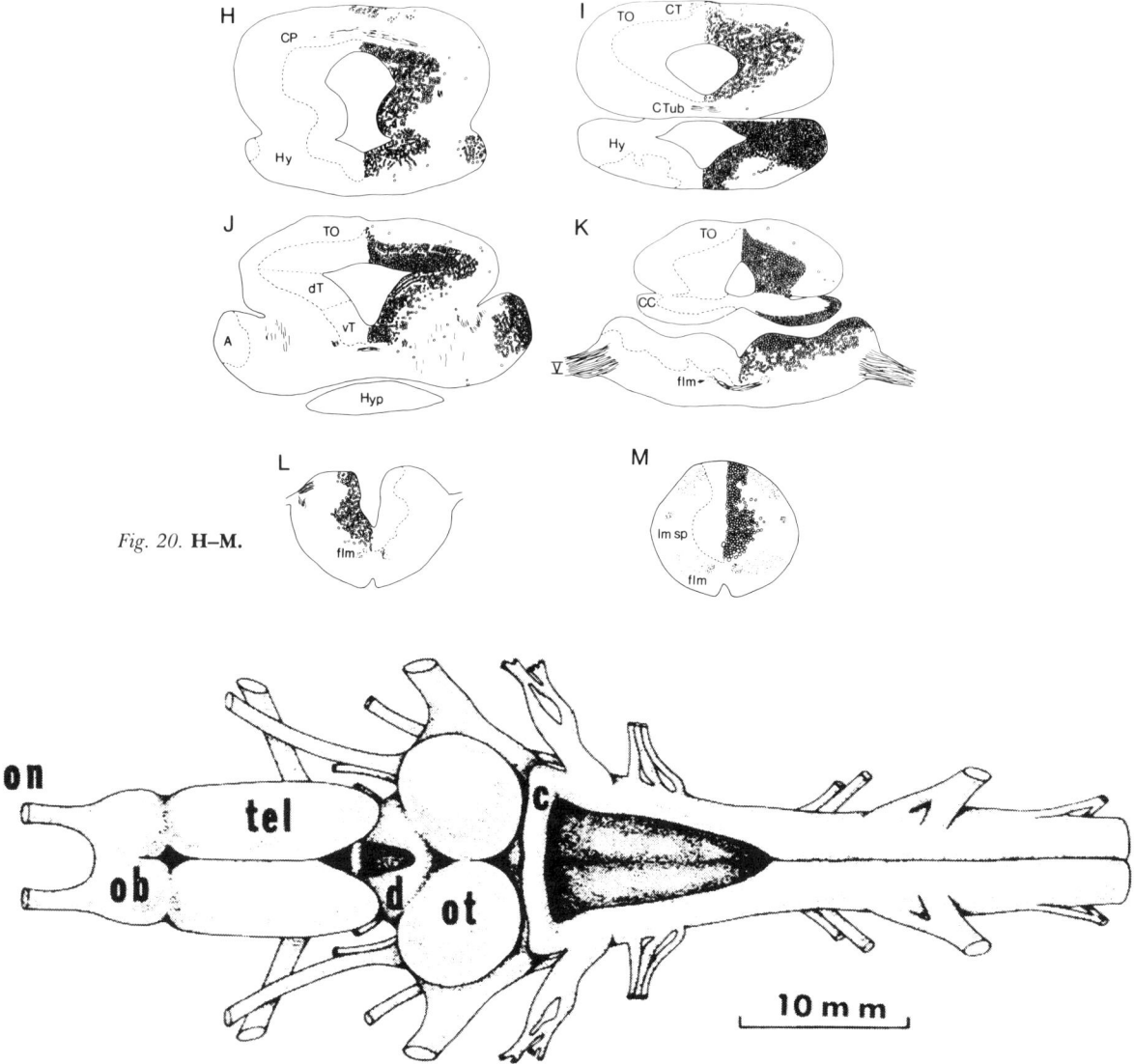

Fig. 20. **H–M.**

Fig. 21. Dorsal view of the brain of the frog *Rana catesbeiana*. c = cerebellum, d = diencephalon, ob = olfactory bulb; on = olfactory nerve, ot = optic tectum, tel = telencephalon. From Northcutt and Kicliter (1980).

The epithalamus represents the dorsal part of the diencephalon. Its anterior part is occupied by the habenular nuclei including the *commissura habenulae*. In the posterior direction, between the *commissura habenulae* and the *commissura posterior*, the *pars intercalaris diencephali* (or *thalami*) is situated. In salamanders, the dorsal thalamus is separated from the epithalamus by the *sulcus dorsalis* (Wicht and Himstedt 1988). The ventral thalamus is separated from the dorsal thalamus by the *sulcus medialis*, and from the hypothalamus by the *sulcus ventralis*.

The hypothalamus is divided by the *sulcus hypothalamicus* into a dorsal and ventral part. The dorsal part is connected with the ventral thalamus and the rostral ventral tegmentum, the ventral part with the *nucleus praeopticus*.

The praetectum of salamanders extends around the fibers of the *commissura posterior* rostrally adjacent to the *pars intercalaris thalami* and descends ventrocaudally. Dorsally, it consists of the *nucleus praetectalis*, which is divided into the *n. praetectalis profundus*, situated within the periventricular grey matter, and the *n. praetectalis superficialis*, embedded within the white matter, forming one of the few migrated nuclei in the salamander brain. Ventrally and more caudally, the *n. Darkschewitsch* is found. Ventrally, the praetectum extends to the level, at which the *pars ventralis thalami* and the mesencephalic tegmentum merge.

In contrast to salamanders, the anuran diencephalon and praetectum shows a substantial morphological differentiation of the periventricular grey matter into separate nuclei and lamination within these nuclei. In the dorsal thalamus of *Rana catesbeiana* (Neary and Northcutt 1983), three periventricular nuclei and one migrated nucleus can be distinguished. In the periventricular zone the anterior, central, and posterior nucleus are found, of which the central and the posterior nucleus show compact lamination. The nucleus migrated from the grey into the white substance in the lateral nucleus, which is further divided into an anterior, posterodorsal, and posteroventral portion. The ventral thalamus also shows both periventricular and migrated nuclei. Periventricularly, the ventromedial nucleus is found, which is replaced rostrally by the thalamic eminence, and caudally by the tuberculum. The migrated nuclei are the ventrolateral nucleus, the *nucleus Bellonci*, situated immediately below the corresponding neuropil, and the superficial nucleus situated ventromedial of the *corpus* (= neuropil) *geniculatum thalamicum*.

The praetectum of *Rana catesbeiana* consists of three nuclei: (1) the *nucleus lentiformis mesencephali*, which is situated laterally to the lateral nucleus of the dorsal thalamus in a vertical orientation, (2) the nucleus of the *commissura posterior* capping this commissure and (3) the pretectal grey dorsal of the *commissura posterior* as a medial continuation of tectal laminae 2 and 4.

The hypothalamus is divided into a preoptic and an infundibular part. The preoptic part is divided further into an anterior and posterior part; the latter contains the magnocellular preoptic nucleus and the suprachiasmatic nucleus. The infundibular hypothalamus possesses two periventricular nuclei, the dorsal and the ventral hypothalamic nucleus.

The diencephalon of other ranid frogs (e.g., *Rana pipiens*) seems to be less differentiated. According to Neary (1975) and Wilczynski and Northcutt (1977), the dorsal thalamus again is divided into an anterior, central, and posterior nucleus and a migrated lateral nucleus, in which the pretectal areas, the posteromedial and posterolateral pretectal nuclei of Trachtenberg and Ingle (1974) are included in the posterior and lateral nucleus.

3. Mesencephalon

The mesencephalon of amphibians is divided into the tectum (often called *"tectum opticum"*, which is incorrect, because it also includes non-visual functions), a subtectal zone and the tegmentum.

Whereas in anurans the tectum is multilayered, in salamanders the separation of the inner layer of cell somata and the outer layer of fibers and dendrites is the only well-defined lamination. Only a few cells can be found in the outer layer. The morphology and cytoarchitecture of the amphibian tectum is described in greater detail below.

The anuran subtectum includes an important auditory zone, called the *torus semicircularis*, (considered homologous to the *colliculus inferior* of amniotes). It consists of three nuclei, (1) the *nucleus laminaris*, represented by a laminated zone beneath and parallel to the ventricular surface and directly continuous with the periventricular zone of the tectum; (2) the *nucleus principalis* situated internal to the *nucleus laminaris*; and (3) the *nucleus magnocellularis*, consisting of clusters of large cells and dorsally and medially bounded by the principal nucleus.

In salamanders, the auditory subtectal center corresponding to the anuran *torus semicircularis* is embedded completely in the periventricular grey matter and shows no, or very little, lamination. It can be identified only by tracing experiments and neurophysiological recordings, as is the case for all mesencephalic nuclei except for the *nucleus opticus tegmenti* (Naujoks-Manteuffel and Manteuffel 1988).

In salamanders, the dorsal tegmentum includes the *nucleus dorsalis tegmenti pars anterior* and *pars posterior* (Naujoks-Manteuffel and Manteuffel 1988). The ventral tegmentum is a conglomerate of several nuclei. The *nucleus fasciculi longitudinalis medialis* is found anteriorly and the *nucleus ventralis tegmenti* (*pars anterior* and *pars posterior*) posteriorly. Ventral to these nuclei, in a rostrocaudal sequence are the *nucleus tuberculi posterioris*, the *nucleus ruber*, the nucleus of the oculomotor nerve (third cranial nerve), the nucleus of the trochlear nerve (fourth cranial nerve) and the *nucleus interpeduncularis*. Lateral to the *nucleus fasciculi longitudinalis* and the *nucleus tuberculi posterioris*, is found the *nucleus opticus tegmenti*, which, together with the *nucleus praetectalis superficialis*, is another migrated nucleus in the salamander brain. The *tegmentum isthmi* is separated from the dorsal tegmentum by the *sulcus isthmi*. Ventrally, it contains the nucleus of the trochlear nerve. At the dorsocaudal end of the dorsal tegmentum, immediately rostral to the *corpus cerebelli*, is found the *nucleus isthmi* (Wiggers and Roth 1991).

The *tegmentum mesencephali* of anurans likewise consists of a series of tegmental nuclei constituting the so-called central tegmental grey (Potter 1969; Opdam and Nieuwenhuys 1976), the *nucleus isthmi*, the *nucleus profundus mesencephali* and the *nucleus opticus tegmenti*. The tegmental grey contains anterodorsal and anteroventral, posterodorsal and posteroventral tegmental nuclei. The large-celled *nucleus profundus mesencephali* is situated in the lateral tegmentum inside the white matter, i.e., in a migrated position. Another migrated nucleus is the *nucleus opticus tegmenti*, which — as in salamanders — is the end station of the basal optic tract. In addition, there is the nucleus of the oculomotor nerve, which extends through most of the length of the tegmentum, as well as the nucleus of the trochlear nerve, which is situated in the rostral part of the isthmic tegmentum.

In contrast to salamanders, the *nucleus isthmi* is well developed in frogs and clearly distinguishable on morphological grounds. It consists of a cortex-like band of small, closely-packed cells, with a ventrolateral interruption surrounding a central area containing numerous fibers.

4. Cerebellum

In salamanders the cerebellum is small and simply organized (Herrick 1948; Larsell 1967). It consists of three parts, (1) the median *corpus cerebelli*; (2) the *auricula cerebelli*, which are enlargements of the sensory zones of the medulla oblongata (Larsell 1967); and (3) a *nucleus cerebelli*, situated ventral to the *corpus cerebelli*. The last was homologized by Herrick (1948) to the deep cerebellar nuclei of mammals. A molecular and a granular layer adjacent to the ependymal layer can be distinguished. The Purkinje cells at the boundary between these two layers are rather irregularly arranged. Granular and stellate cells can be distinguished, but it is unclear whether basket and Golgi cells are present.

▶

Fig. 22. Cross sections through the brain of the frogs *Rana catesbeiana* (**A–E**) and *Rana esculenta* (**F–I**). **A**: mid-telencephalon, **B**: caudal telencephalon-rostral diencephalon, **C**: mid-diencephalon, **D**: caudal diencephalon, **E**: Rostral mesencephalon, **F**: caudal mesencephalon, **G**: rostral medulla oblongata, **H**: Mid medulla oblongata, **I**: caudal medulla oblongata. **A**, **B** from Northcutt (1980); **C–E** from Neary and Northcutt (1983); **F–I** from Nieuwenhuys and Opdam (1976). A = anterior thalamic nucleus, a,pl = *amygdala, pars lateralis*, AD = anterodorsal tegmental nucleus, Ad = *nucleus anterodorsalis tegmenti mesencephali*, Av = *nucleus anteroventralis tegmenti mesencephali*, B = neuropil of Bellonci, bn = bed nucleus of pallial commissure, BON = basal optic nucleus, C = central thalamic nucleus, CO = subcommissural organ, Cp = *corpus geniculatum thalamicum*, dp = dorsal pallium, en = entopeduncular nucleus, fa = *fibrae arcuatae*, flm = *fasciculus longitudinalis medialis*, fsol = *fasciculus solitarius*, ft = *fasciculi tegmentales*, Hd = dorsal habenular nucleus, Hv = ventral habenular nucleus, LH = lateral hypothalamic nucleus, Lpd = lateral thalamic nucleus, posterodorsal division, lp,dp = lateral pallium, *pars dorsalis*, Lpv = lateral thalamic nucleus, posteroventral division, ls = lateral septal nucleus, lsp = *lemniscus spinalis*, Mg = magnocellular preoptic nucleus, mp = medial pallium, ms = medial septal nucleus, na = *nucleus accumbens*, NB = nucleus of Bellonci, NMLF = nucleus of the medial longitudinal fasciculus, NPM = *nucleus profundus mesencephali*, Npv = nucleus of the periventricular organ, n VIII = stato-acusticus, n IX = *nervus glossopharyngeus*, n XII = *nervus hypoglossus*, OC = optic chiasm, Ols = *oliva superior*, on = optic nerve, optl = *tractus opticus marginalis lateralis*, P = posterior thalamic nucleus, pd = *pars dorsalis*, pg = preoptic periventricular grey, PtG = pretectal grey, PtrG = pretoral grey, pv = *pars ventralis*, Ri = *nucleus reticularis inferior*, Ris = *nucleus reticularis isthmi*, Rm = *nucleus reticularis medius*, SC = suprachiasmatic nucleus, slH = sulcus limitans of His, smi = *sulcus medianus inferior*, sms = *sulcus medianus superior*, spcv = *tractus spinocerebellaris ventralis*, st,pd = *striatum, pars dorsalis*, tbsp = *tractus tectobulbaris et spinalis*, tect = *tectum mesencephali*, Tel = telencephalon, TeO = optic tectum, Tl = *nucleus laminaris tori semicircularis*, Tmc = *nucleus magnocellularis tori semicircularis*, TP = posterior tuberculum, Tp = *nucleus princeps tori semicircularis*, tr Vds = *tractus descendens nervi trigemini*, tr VIIIds = *tractus descendens nervi octavi*, VH = ventral hypothalamic nucleus, VIIIc = *nucleus caudalis nervi octavi*, VIIId = *nucleus dorsalis nervi octavi*, VIIIv = *nucleus ventralis nervi octavi*, Vld = ventrolateral thalamic nucleus, dorsal part, Vlv = ventrolateral thalamic nucleus, ventral part, VM = ventromedial thalamic nucleus, vm = *ventriculus mesencephali*, Vs = superficial ventral thalamic nucleus, XII = *nucleus nervi hypoglossi*, IV = *nucleus nervi trochlearis*, Ixm = *nucleus motorius nervi glossopharyngei*, III = oculomotor nucleus, 6 = tectal lamina six. See **F–I** on page 814.

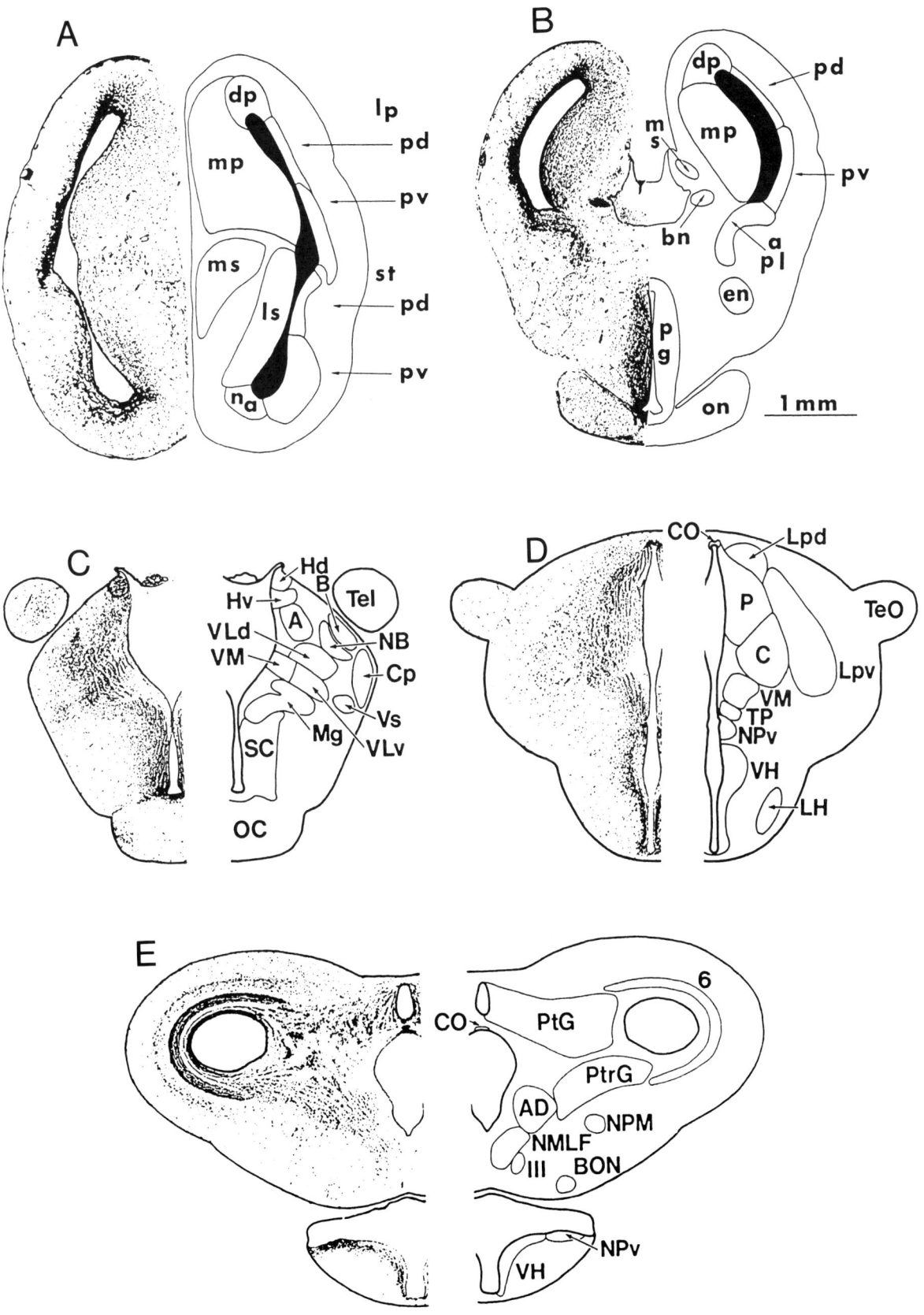

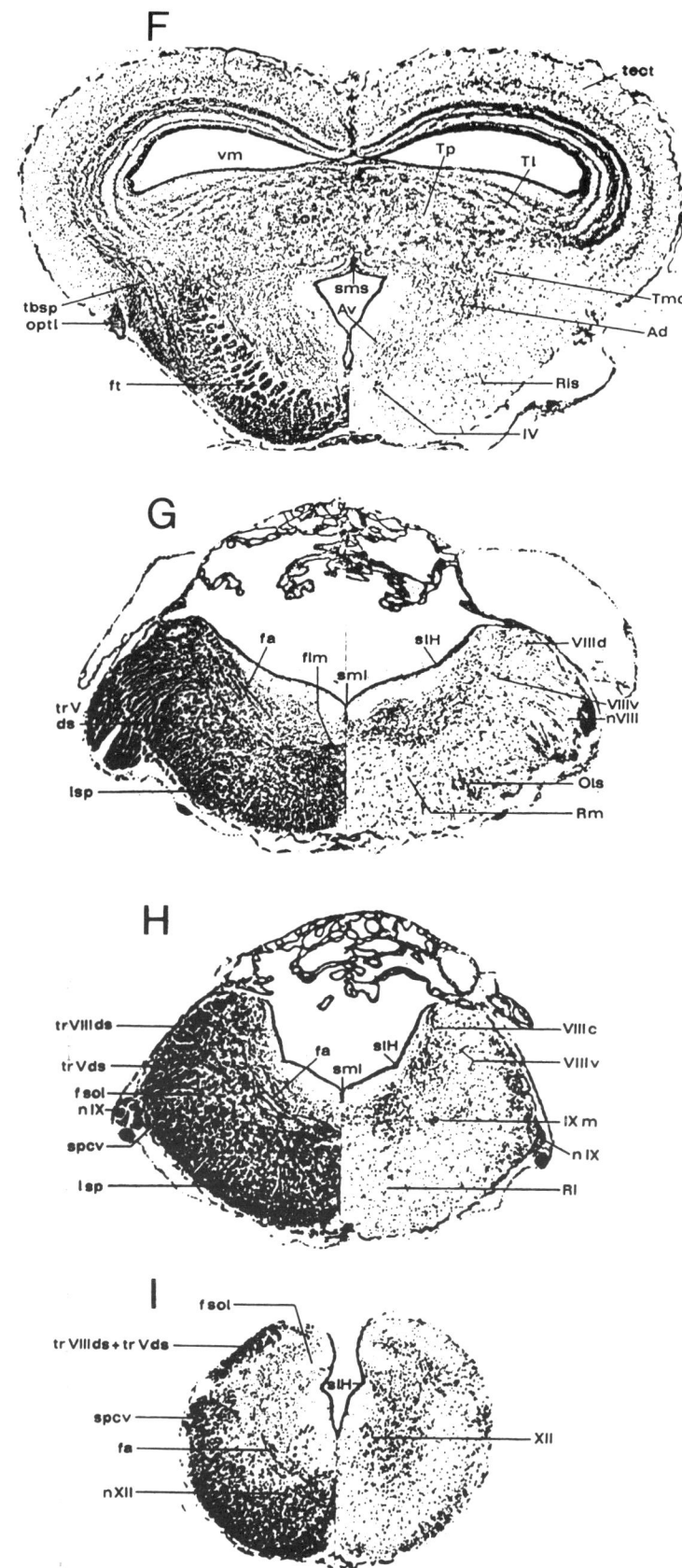

Fig. 22. F–I

The cerebellum is larger in anurans, but still small compared to that of most other vertebrates. The auricles are reduced relative to salamanders. The *corpus cerebelli* is a single transverse plate consisting of three layers: the molecular layer, the granular layer and the somewhat irregular layer of Purkinje cells, which, in contrast to those of salamanders, are oriented at right angles to the transverse plane.

5. Medulla Oblongata and Cervical Spinal Cord

The *medulla oblongata* includes the brain portion (except the cerebellum) between the isthmus and the obex as the caudal end of the fourth ventricle and, thus, of the brain. There is no pons in the amphibian brain. The rostrolateral wall and the rostral part of the underside constitute the *auricula cerebelli*.

The *medulla oblongata* is the area of termination and origin of cranial nerves V (trigeminal), VI (abducens), VII (facialis), VIII (stato-acusticus), IX (glossopharyngeus), X (vagus) and XII (hypoglossus). Its dorsal part, the alar plate of His, receives all sensory fibers from the head (except the optic and olfactory fibers), fibers from the lateral-line system, when present, and general visceral sensory and gustatory fibers. In the ventral part are situated the motor nuclei of nerves V to X, and partly the nucleus of nerve XII. The motor nuclei are surrounded by neurons of the reticular formation. This motor system is the co-ordination centre of head and neck motor function including the mouth and tongue movements involved in feeding.

B. Morphology of Visual Centres

1. Visual Afferents to the Brain

Most fibers of the optic nerve cross in the *chiasma opticum* at the ventral side of the diencephalon to the opposite side of the brain. However, a certain number of the fibers turn back to the ipsilateral side of the brain. Behind the chiasma, the main fiber pathway runs dorsally within the white matter of the diencephalon. A small number of coarser fibers separates early, forming a distinct fascicle, the basal optic tract (BOT), which extends caudally over the surface of the infundibulum of the hypothalamus. Rostral to the root of the third cranial nerve, *n. oculomotorius*, these fibers form a dense terminal field, the basal optic neuropil (BON).

In salamandrids, three visual terminal fields can be distinguished within the rostral thalamus (Fritzsch 1980). The *corpus geniculatum thalamicum* (CGT) is a large, egg-shaped neuropil situated superficially within the ventral thalamus. Immediately dorsomedial to the CGT, is situated the *neuropil Bellonci, pars lateralis* (NBl). It extends parallel to the CGT in a ventrocaudal direction from the surface into the white matter. It is smaller than the CGT. Mediodorsal to the NBl, a third thalamic field, the *neuropil Bellonci, pars medialis* (NBm), can be separated clearly. In the praetectum region, retinal afferents form the laterally oriented pretectal neuropil (P) and a small sickle-shaped neuropil, the *area uncinata* or "uncinate field" (UF) which lies directly medial to the P. These thalamic and pretectal projection sites are present both contra- and ipsilaterally.

In the contralateral tectum, the incoming fibers are arranged in several layers. In the rostral part of the tectum, which corresponds to the binocular visual field, four densely stained layers are found, separated by three layers of less intense staining. The deepest layer is constituted by the neuropil of the *area uncinata* continuing into the optic tectum. In the caudal part of the tectum, only two intensely stained layers occur.

The condition in plethodontid salamanders differs from that just described, mostly with regard to the ipsilateral retinofugal projections (Rettig and Roth 1982, 1986; Roth, unpubl. obs.) (Figs 23, 24). The ipsilateral NBm shows nearly the same intensity as the contralateral NBm. The ipsilateral NBl is less intensely stained compared to the contralateral one in the Desmognathinae, Hemidactyliini, and Plethodontini, but almost equally stained in the Bolitoglossini. The ipsilateral CGT, again, is stained massively in the Bolitoglossini. In the praetectum, the ipsilateral UF is about the same size as the contralateral one, although it is less densely stained.

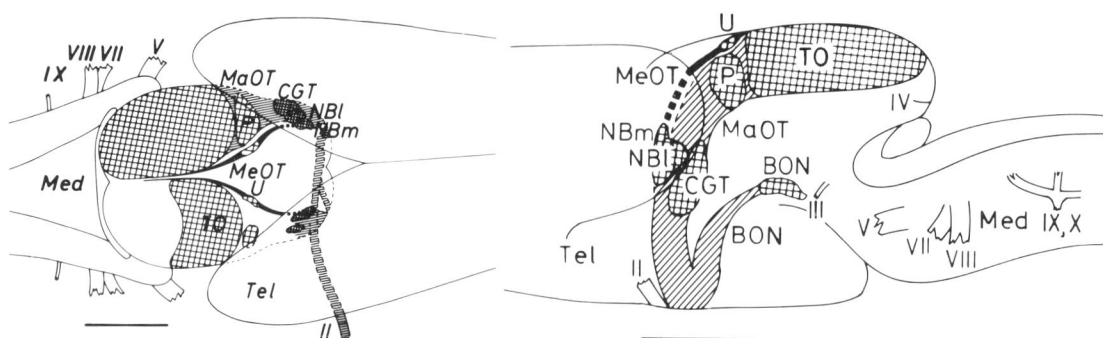

Fig. 23. Distribution of retinofugal neuropil areas and tracts in the plethodontid salamander *Plethodon cinereus.* **left:** dorsal view; **right:** lateral view. BON = basic optic neuropil; CGT = *corpus geniculatum thalamicum*; MaOT = marginal optic tract; Med = *medulla oblongata*; MeOT = medial optic tract; Nbl = *neuropil Bellonci pars lateralis*; Nbm = *neuropil Bellonci pars medialis*; P = pretectal neuropil; Tel = telencephalon; TO = *tectum opticum*; U = *area uncinata*; I–X = cranial nerves. Bar = 500 μm. From Rettig and Roth (1986).

In the Plethodontini, the ipsilateral retinofugal projections cover mostly the rostral tectum, forming two layers. The deep layer is continuous through the whole width of the tectum, whereas the superficial layer covers only the mediorostral part. In the Bolitoglossini, the deep layer extends almost to the caudal margin of the tectum. The more superficial layer extends much further caudally than in non-bolitoglossines, but is relatively indistinct in the caudal optic tectum.

In *Rana esculenta*, Lázár and Székely (1969) found the following contralateral, retinofugal projection sites: in the thalamus, the *nucleus* (= neuropil) *Bellonci* (NB) and the lateral geniculate body (LGB) (= *corpus geniculatum thalamicum*, CGT), and in the praetectum, a

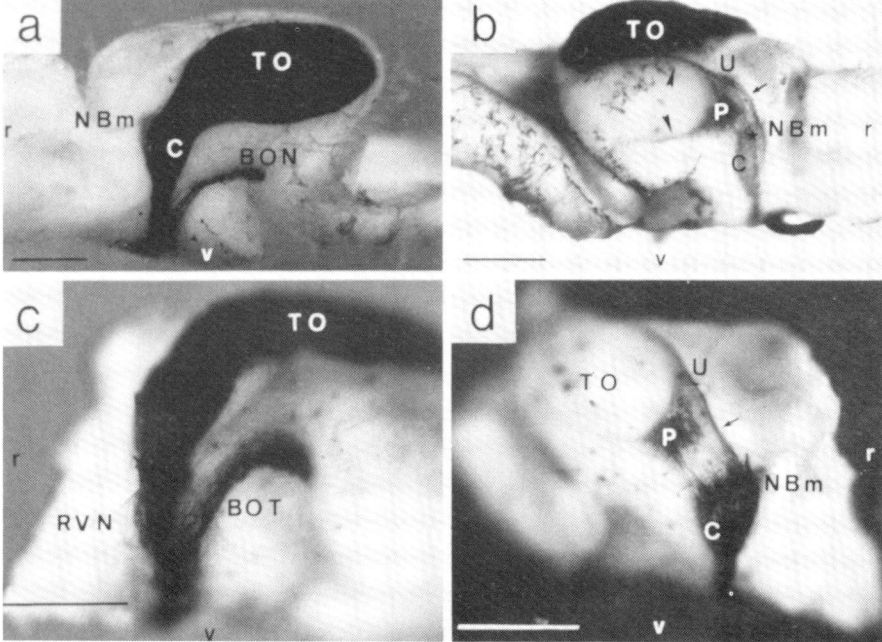

Fig. 24. Photomicrographs from retinofugal projections in plethodontid salamanders. **a** Lateral view of a brain of *Hydromantes genei*, contralateral aspect. **b** Dorsolateral view of a brain of *Eurycea bislineata*, ipsilateral aspect. Note the fasciculated projection of the *tractus opticus marginalis* in the dorsal thalamus. Asterisk indicates the *neuropil Bellonci pars lateralis*, arrow the *tractus opticus medialis*. The arrowheads show the two branches of the *tractus opticus marginalis* indicating the extension of the ipsilateral projection into the *tectum opticum*. **c** Whole mount preparation of a brain of *Bolitoglossa subpalmata*, rostrolateral view, ipsilateral aspect with the forebrain cut off. The *neuropil rostroventralis* (RVN) is weakly stained and the broad projection of the *tractus opticus basalis* (BON) is visible. **d** Whole mount preparation of a brain of *Bolitoglosa subpalmata*, rostrolateral view, ipsilateral aspect with the forebrain cut off. The *neuropil Bellonci pars medialis* (NBm) and the *neuropil posterior thalami* (P) are patchily stained. The *tractus opticus marginalis* connecting these areas is organized in fascicles (arrow). From Rettig and Roth (1986).

dorsolateral projection site. Other authors, by means of horseradish peroxidase (HRP) studies, have established evidence for the existence (as in salamanders) of two independent pretectal neuropils, the posterior thalamic neuropil and the uncinate field (Fite and Scalia 1976).

In the tectum, four laminae of retinofugal fibers are formed (Székely and Lázár 1976): lamina 1 and lamina 2 are situated immediately below the surface and consist mostly of thin, unmyelinated fibers; lamina 3 is of myelinated fibers and is situated above layer 8; lamina 4, consisting of few, thick unmyelinated fibers, is located in layer 8 and beneath it. Fite and Scalia (1976) reported ipsilateral projections in all of the diencephalic neuropils mentioned above. Lázár and Székely (1969), as well as Fite and Scalia (1976) found no evidence of direct ipsilateral fibers in the tectum of *Rana*. Singman and Scalia (1990), on the basis of retrograde HRP tracing experiments, estimated that in *Rana pipiens*, 2.3% of the overall population of ganglion cells project to the ipsilateral tectum.

2. Topic Organization of Retinal Projections to the Diencephalon and Tectum

In salamanders, fibers from the *nasal* quadrant run to the contralateral diencephalic neuropils CGT, NBl and P. In the optic tectum, nasal retinal afferents terminate in the caudal tectum. Some thick fibers run to the BON. Very few ipsilateral fibers from the nasal quadrant are found in the anterior thalamus, and the tectum is free of them.

Contralateral fibers from the *ventral* quadrant project to the CGT and NBl and terminate in their medial parts. Few fibers run to the UF. The P shows weak staining in its medial part. In the tectum, fibers from the ventral quadrant project to the medial part of the tectum. Few, but consistent fibers project to the BON. Ipsilateral projections from the ventral quadrant are detectable in the NBm and NBl and, in small numbers, in the CGT, UF, and P. In the tectum, they are located medially.

Contralateral projections from the *dorsal* quadrant are found in the lateral parts of the NBl and NBm, the CGT, and the P, in the whole UF and in the lateral tectum. Some coarse fibers consistently project to the BON.

Contralateral projections from the *temporal* quadrant are found in the caudo-axial parts of the CGT, NBl, and P, and nontopically in the UF and NBm. In the tectum, they project to the rostral and central part. The contralateral BON is well-stained.

The ipsilateral NBl, NBm, UF and CGT do not show any topic arrangement, whereas the P does. In the ipsilateral tectum, fibers are seen in the rostral part of the tectum. The ipsilateral BON is stained in most cases.

In anurans, the visual afferents to the diencephalic neuropils are retinotopically organized. In the NB, CGT, the posterior thalamic neuropil and the UF of *Rana pipiens*, the temporal and dorsal quadrants of the retina project to the posterior halves of the neuropils, and the ventral and nasal quadrants to the anterior halves. Furthermore, the ventral and temporal quadrants project to the dorsal halves and the nasal and dorsal quadrants project to the ventral halves (Fite and Scalia 1976). This means that with respect to the retino-tectal map, the nasotemporal axis of the retino-diencephalic map is reversed, but that the representation of the dorsoventral axis remains unchanged. The ipsilateral projections to the diencephalic neuropils derive almost entirely from the temporal quadrant of the retina, only a small portion comes from the dorsal and ventral quadrants. The condition in *Rana* is, therefore, very similar to that found in urodeles.

3. Organization and Cytoarchitecture of the Optic Tectum

In anurans, as well as in most other vertebrates, the *tectum* is multistratified with alternating fiber and cellular layers (Fig. 25a, 27b). Usually, nine layers are distinguished beginning from the ventricle (Székely and Lázár 1976): layer 1 of ependymal glial cells that send moderately branching, long processes toward the tectal surface, where they occur as feet-like structures forming the external limiting membrane. Cellular layers 2, 4, and 6 (*stratum griseum periventriculare* of Ariens Kappers *et al.* 1936/1960) together constitute the periventricular grey matter. Of these, layer 6 is the thickest. These cellular layers are divided by fiber layers 3 and

5, which consist of deep unmyelinated afferent and efferent fibers and basal dendrites of the periventricular neurons. Fiber layer 7 *(stratum album centrale)* contains the bulk of efferent tectal fibers and a few scattered neurons. Layer 8 *(stratum griseum centrale)* consists of loosely arranged neurons embedded in a meshwork of dendrites of tectal neurons and afferent fibers. Layer 9 *(stratum fibrosum et griseum superficiale + stratum opticum)* contains relatively few neurons dispersed in the meshwork of retinal afferents and dendrites of tectal neurons. Layer 9 is further divided into 7 laminae, A–G (Potter 1969). Of these, A (occurring only in the rostral tectum), B, D, F and G are layers consisting of myelinated and unmyelinated fibers (mostly retinal afferents), and C and E cellular layers.

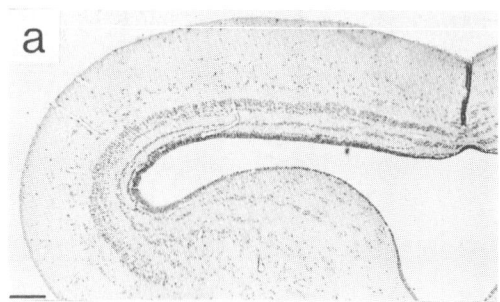 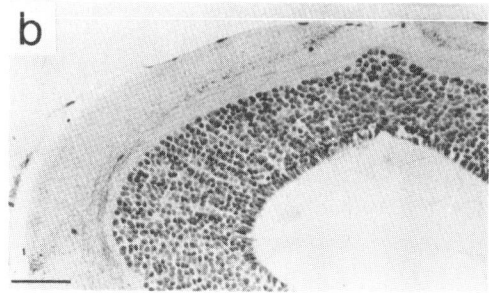

Fig. 25. Photomicrographs of the tectum and dorsal tegmentum (including *torus semicircularis*) of the frog *Limnodynastes tasmaniensis* (**a**) and the salamander *Hydromantes italicus* (**b**). While *Limnodynastes* exhibits one of the most complex tectal morphologies among anurans, with many alternating cellular and fibre layers and a high degree of migrated neurons, the tectal morphology of *Hydromantes* is secondarily simplified with a bi-layered tectum and very few, if any, migrated neurons. Bar: 100 μm.

In a recent HRP study on retinal afferents in *Rana pipiens,* Hughes (1990) found the following situation. In lamina A, medium-sized retinal axons terminate, each axon ending in a dense bush; in lamina B, bundles of medium-sized myelinated axons are found ending abruptly and carrying few terminal swellings. In lamina C, few myelinated axons and many fine unmyelinated axons beaded near their end are present, and in lamina D, bundles of medium-sized myelinated axons with collaterals and fine axons. Lamina E contains thin, unmyelinated axons, and laminae F and G large, myelinated axons (often with large terminal swellings) as well as smaller, unmyelinated axons, with many labeled boutons. In cellular layer 8, some large retinal axons are found between cell bodies, which pass between laminae F and G and give rise to many smaller branches. There are roughly four different types of axons and terminations, viz., those contained in A, in B and D, in C and E, and in F and G (embedded in layer 8).

In contrast, the salamander tectum, like that of caecilians (personal observation) and lepidosirenid lungfishes (Northcutt 1977), shows a more or less two-layered structure consisting of a periventricular cellular layer and a superficial "white matter" consisting of dendrites of tectal neurons and tectal afferent and efferent fibers, in which a few migrated neurons are dispersed (Fig. 25b, 27a). However, based on tracer experiments, the salamander tectum can be divided (from the surface to the ventricle) into 9 layers (Roth 1987). Layers 1–3 contain retinal afferent fibers and 4–5 contain efferent fibers and afferents from other senses, e.g., somatosensory, vestibular and lateral line afferents (when present). Fibers carrying non-visual (mechanoreceptive) information ascend from somata located in the dorsal part of the spinal cord (somatosensory) as well as from relais nuclei located in the medulla oblongata (somatosensory, vestibular, auditory, lateral line) and terminate predominantly in layer 4 and to a lesser degree in layer 3 of the contralateral tectal hemisphere, and in layers 4 and 5 in the ipsilateral hemisphere. Layer 6 consists of the superficial cellular layer, while layer 7 (absent in miniaturized plethodontid salamanders) contains deep unmyelinated fibers. Layer 8 is the deep cellular layer, and layer 9 contains periventricular ependymal (glial) cells. In some salamanders, e.g., *Ambystoma mexicanum,* the periventricular grey matter regionally may exhibit two to three sublayers.

The termination of retinal axons in the tectum of plethodontid salamanders were recently determined by intracellular labeling studies (Wiggers *et al.*, unpubl. data) (Fig. 26). Injection of Biocytin into retinal ganglion cells revealed three types of axonal terminals in the

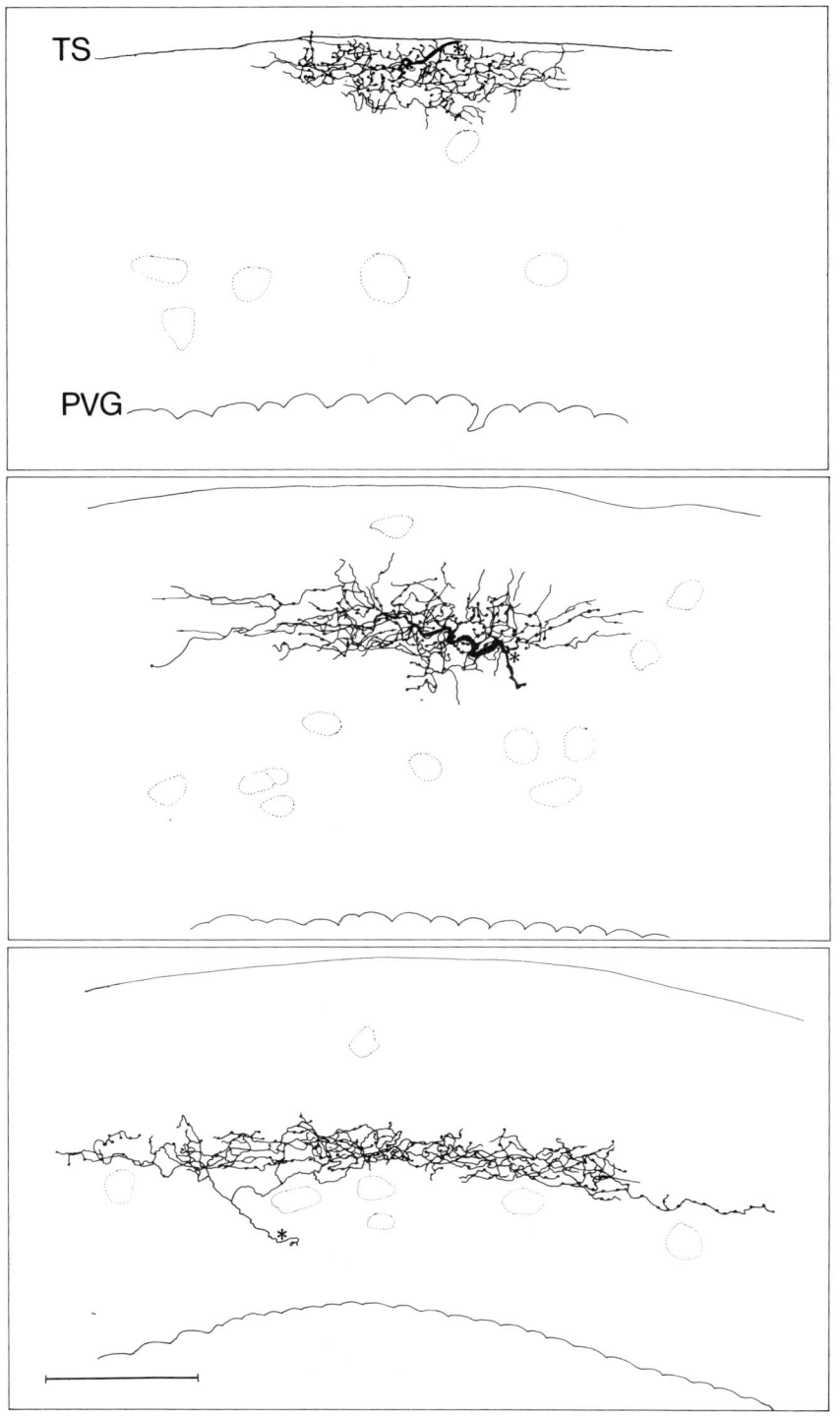

Fig. 26. Reconstruction of axonal terminals from different retinal ganglion cells in the afferent fibre layers 1, 2 and 3 of the tectum in *Plethodon jordani*, based on intracellular injection of Biocytin. The asterisk indicates the main axon of a retinal ganglion cell, the dotted lines represent the outlines of migrated tectal neurons. TS = Tectal surface; PVG = periventricular grey matter. Bar: 100 µm.

contralateral tectum. The first type contains terminals restricted to the upper region of layer 1 and covering a round patch of 70–100 µm in the horizontal plane. The second type is comprised of axons terminating in the middle of layer 2, with terminal arbors up to 200 µm in diameter. The third type of retinal terminals is restricted to layer 3 and covers an area up to 500 µm in diameter. With the exception of a few fibers, these three bands of retinal terminals are clearly separated from each other.

While type 1 axons terminate only in the tectum, type 2 and 3 axons, on their way to the tectum, form pronounced telodendra or synaptic-contact boutons in the regions of thalamic and praetectal neuropils.

Ipsilaterally projecting RGC have not yet been labeled intracellularly, but tracing of tracts with HRP and Biocytin revealed that direct ipsilateral retino-tectal afferents terminate at the border between layers 3 and 4. It is still unclear, whether direct ipsilateral and deep contralateral retinal afferents overlap.

Axonal terminals of praetectal neurons in the tectum are variable and range from non-arborizing fibers to massively arborizing terminals that cover almost one entire tectal hemisphere. No distinct patterns (patches or bands) could be observed (Luksch et al. 1995).

Termination sites of isthmic neurons within the tectum are highly specific. Ipsi- and contralateral isthmic terminals are restricted to round vertical columns of about 100 μm in diameter. While the single columns of ipsilateral telodendra pass through all three layers of retinal afferents, those of the contralateral ones are restricted to layer 1 (see below).

A. HOMOLOGY OF TECTAL LAYERS IN URODELES AND ANURANS

Layers 6–9 of the salamander tectum are homologous with layers 1–6 of the frog tectum; layers 4 and 5 of salamanders are homologous with layer 7 of frogs, and layers 1–3 of salamanders are homologous with anuran layers 8 and 9 (including laminae A–G) (Fig. 27). Thus, the salamander tectum has fewer layers than does that of frogs and has fewer neurons in a "migrated" position in the white matter (layers 7–9).

B. TECTAL CELL TYPES IN SALAMANDERS

On the basis of Biocytin studies (Dicke and Roth 1994b; Dicke 1997; Dicke and Roth 1996a; Roth et al., submitted), five types of tectal neurons could be identified in salamanders, confirming and enlarging results from Golgi and HRP studies (Roth et al. 1990). Four of them (T 1, T 2, T 2*, T 3) are projection neurons, i.e., they possess axons that leave the tectum, while the fifth type (T 4) is comprised of interneurons, i.e., of cells that either have no axon or the axon does not leave the tectum.

The somata of T 1 are always found in the first or second row of cells of the periventricular grey. Either one primary thick dendrite originates from the soma which immediately divides into several thick secondary dendrites, or several primary dendrites originate directly from the soma. The majority of these dendrites extend in a candelabra-like manner toward the tectal surface, where they arborize extensively and densely within layer 1, mostly in the dorsalmost part of it immediately below the surface. A much sparser but equally wide arborization is found in layers 3 and 4. Layer 2 is mostly devoid of dendrites. Thin dendrites extend in all directions in layer 4 and 5 parallel to the surface of the periventricular grey (Figs 28A,B, 29A).

The axon originates close to the soma from a primary or thick secondary dendrite and leaves the tectum through layer 4 or 5. It descends to the tegmentum and sends a collateral to the contralateral tegmentum via the tegmental commissure. Axon collaterals aborize extensively ipsi- and contralaterally in the entire tegmentum. The axon then descends contralaterally to the medulla oblongata and spinalis in a medial position down to the level of the second or third spinal nerves, constituting the *crossed tecto-bulbo-spinal tract*. Many collaterals are found throughout the white matter. No *ascending* axonal projections exist.

Somata of T 2 neurons are always found in the upper part of layer 6, mostly in the second to fourth row of cells. Usually, one primary dendrite extends to fiber layer 3, where it divides into secondary and tertiary dendrites which together form a wide dendritic field that extends within layers 2 and 3. Dendritic arborization is most dense in layer 2, but wider in layer 3, including axonal arborization extending along the border to the efferent fiber layers 4 and 5. Only very few fine dendrites extend into layer 1. Often, in cross sections half of one tectal hemisphere covered by the dendritic tree of this type of tectal neuron (Figs 28C, 29B, 30).

All T 2 neurons have axons ascending to the praetectum and thalamus (Fig. 30). In two-thirds of neurons, these projections are bilateral. An axon originates from secondary or tertiary dendrites in layer 3, descends to the dorsal tegmentum and then runs rostrally arborizing in the lateral praetectum, the ventral and — often less intensely — in the dorsal thalamus. An axon collateral descends to the diencephalic postoptic commissure, crosses and

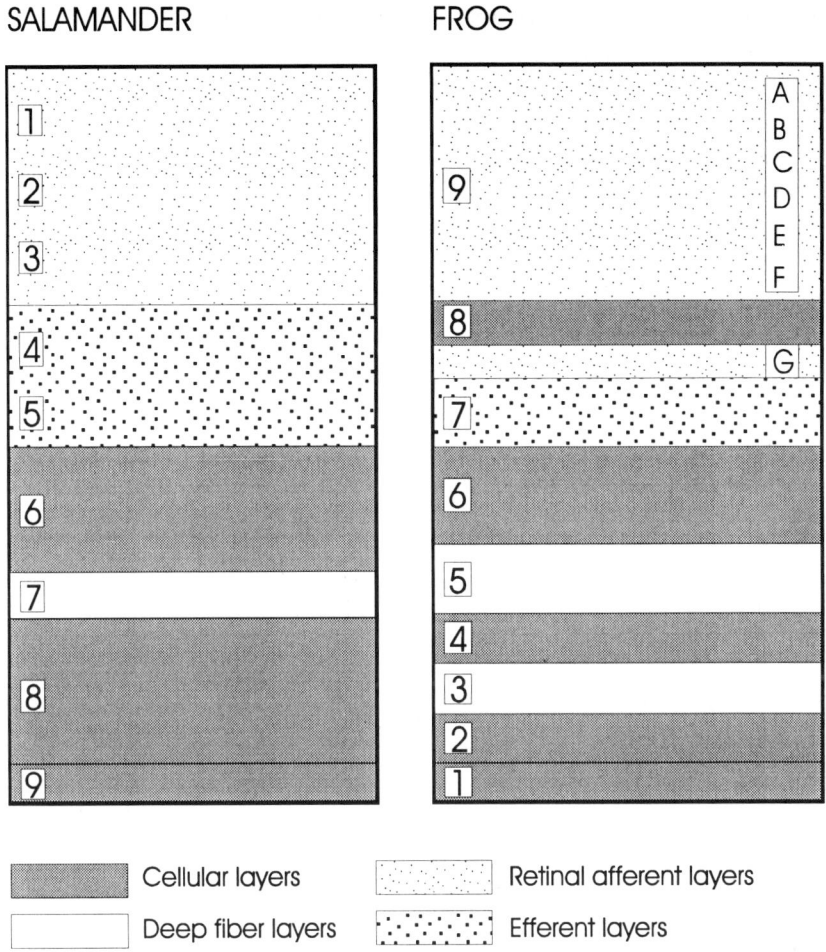

Fig. 27. Comparison of the stratification of the tectum of salamanders and frogs. For further explanation see text.

ascends to the contralateral ventral and — again less frequently — to the dorsal thalamus. In about one-fourth of these neurons, the projection is only ipsilateral. In one-third of T 2 neurons, a second ascending axon arises from medial dendrites, runs to the medial tectum and projects to the ipsilateral medial praetectum. Medial axons crossing in the postoptic commissure to the contralateral praetectum were not observed.

The descending axon descends laterally in the tegmentum without forming extensive collaterals and constitutes the *lateral uncrossed* tecto-bulbo-spinal tract, which extends at least as far as the level of the third spinal nerve (Fig. 30). In contrast to the descending T1-axons, T2-axons arborize sparsely within the *medulla oblongata*. In a few cases, the descending axon forms a dense neuropil at the level of the *nucleus isthmi*.

An infrequent subtype called T 2* exhibits the same system of ascending projections as T 2, but the laterally descending axon reaches only the level of the *nucleus isthmi*.

Somata of T 3 cells are always found in deeper positions inside the periventricular grey matter, mostly in the deep portion of layer 6. One primary dendrite extends to the border between grey and white matter, where it immediately forms a flat and wide dendritic tree mostly confined to the efferent layers 4 and 5 and the lower part of layer 3 (Figs 28A, C, D, 29C). Often, the thick primary dendrite bends laterally within these layers. Basal processes arborizing in deep fiber layer (layer 7) are often encountered. The ascending projection apparently is the same as in T 2 neurons. The axon descends ipsilaterally in a medial position, constituting the *medial uncrossed* tecto-bulbo-spinal tract.

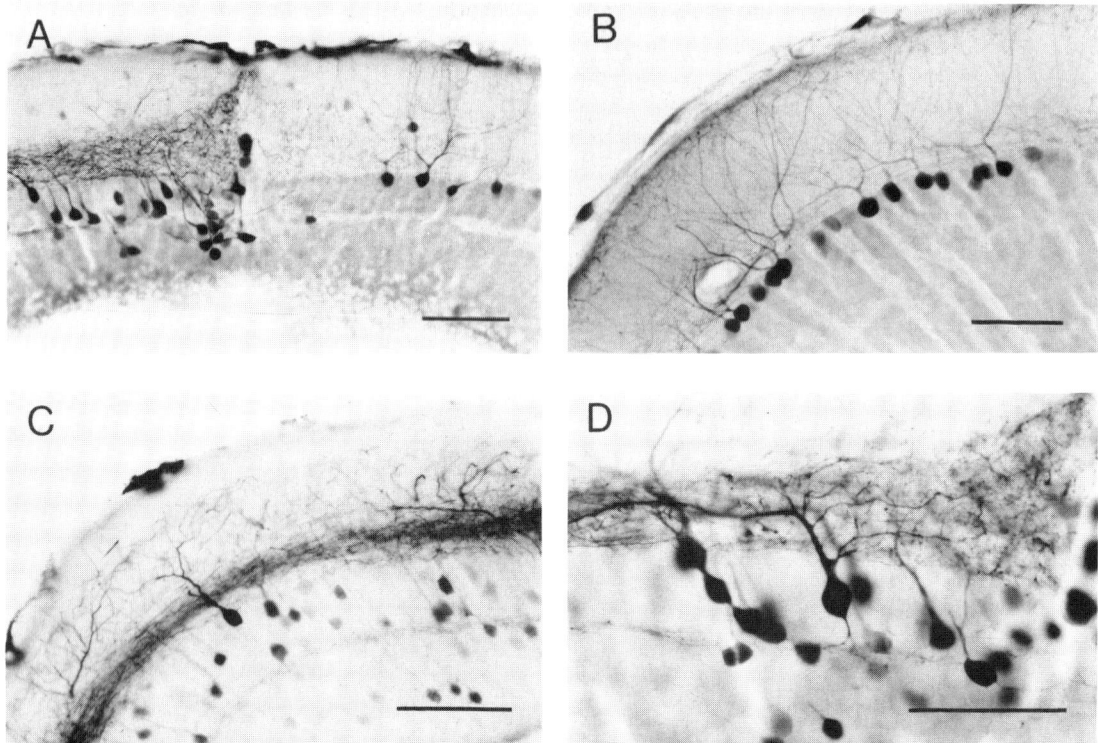

Fig. 28. Photomicrographs of transverse sections through the tectum of *Plethodon jordani* (**A, C, D**) and *Hydromantes italicus* (**B**) showing tectal projection neurons retrogradely labelled after application of Biocytin to the ventral medulla oblongata. **A:** T 1 neurons (right side, contralateral to the injection site) and T 3 neurons (left side, ipsilateral to the injection site); **B:** T 1 neurons; **C:** T 2 and T 3 neurons; **D:** T 3 neurons at higher magnification. For further explanation see text. Bars: 100 μm.

Somata of T 4 neurons are found at any depth of the periventricular grey, and dendrites arborize in various layers of the white matter. The diameter of the dendritic tree is generally smaller than that of projection neurons. In most cases, either no axon is present or it cannot be distinguished from dendrites. Therefore, these neurons are pure tectal interneurons. Various subtypes can be distinguished: (1) cells where the dendritic tree is very small and restricted to layer 1, 2 or 3; (2) cells where the dendritic tree is wide and arborizes in layer 1 and 3 or 2 and 3; and (3) cells where the dendritic tree arborizes in layers 1 and 3. In the latter two cases, dendritic arborization is always widest in layer 3.

In the intracellular staining experiments of Roth *et al.* (submitted), among 109 successful Biocytin stainings, about one-third resulted in the staining of single neurons, while in the other cases from 2 to 11 neurons were sufficiently stained to determine the cell types. In most cases, more than one T 4 neuron, partly of the same and partly of a different subtype, were stained, or a projection neuron was combined with one or more T 4 interneurons. In four cases, projection neurons of different types were combined (T 1 and T 2, T 1 and T 3, T 2 and T 3) and in two cases, two T 2 neurons were stained simultaneously.

The observed multiple staining of neurons after single injections of Biocytin is unlikely to be the consequence of uncontrolled penetration of the tracer into cell bodies, dendrites or axons. Similar results have been obtained by intracellular labeling of tectal and pretectal neurons in the same species of salamanders (Wiggers and Roth 1994) and have been reported in the anuran *torus semicircularis* and brainstem (Luksch and Walkowiak 1994) as well as in mammalian tissue, e.g., the adult rat neocortex (Schulte-Mattler and Luhmann 1995). Multiple staining of neurons is most probably due to coupling of these cells via gap junctions, which function as electrical synapses.

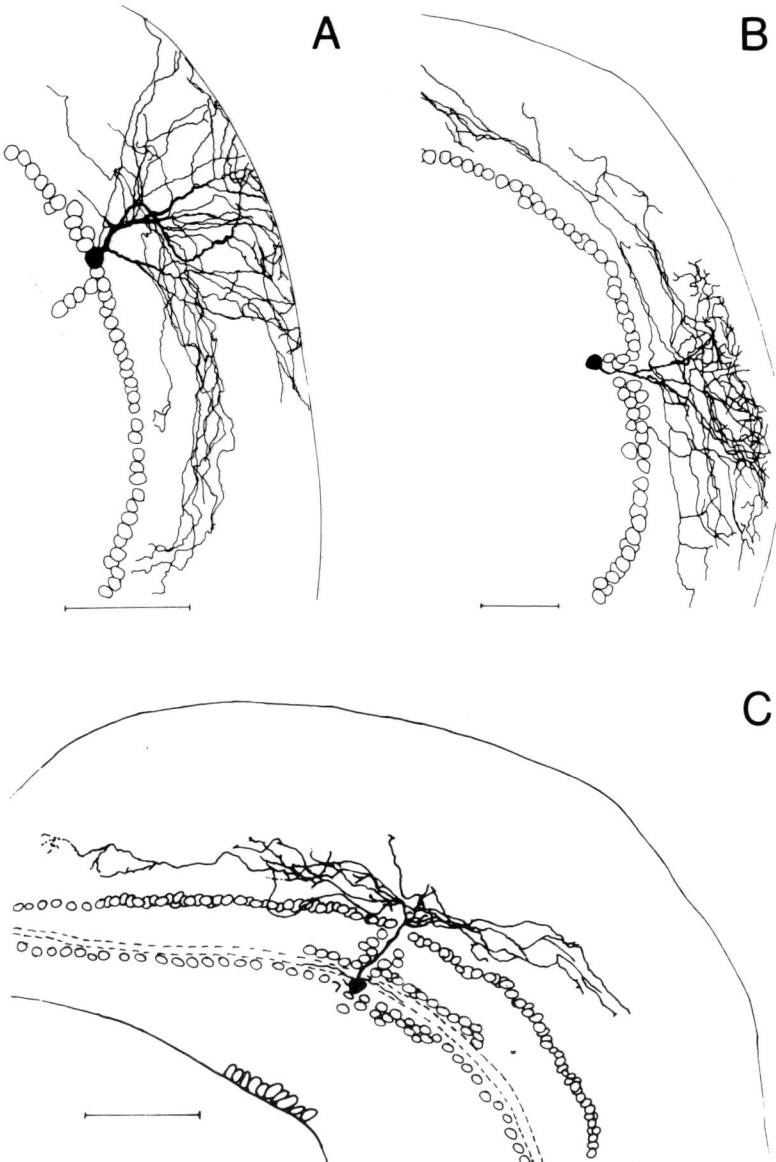

Fig. 29. Reconstruction of tectal projection neurons labelled by intracellular injection of Biocytin. **A:** T 1 neuron *(Plethodon jordani)*; **B:** T 2 neuron *(Hydromantes italicus)*; **C:** T 3 neuron *(Plethodon jordani)*. For further explanation see text. Bars: 100 μm.

The distribution of a number of transmitters was investigated in the tectum of plethodontid and salamandrid salamanders (Wallstein *et al.* 1995; Dicke *et al.* 1996). Neurons with Serotonin (5-HT)-like immunoreactivity (lir) were found in the thalamus, praetectum, midbrain and *medulla oblongata*. Neurons of the nuclei raphes, which are aligned along the midline in the ventral grey matter of the tegmentum and *medulla oblongata*, give rise to ascending fibers that in the optic tectum extend into fiber layers 2 and 3 (retinal afferents) and into layer 4 (tectal efferent and non-visual afferent fibers forming numerous boutons). Axons of retinal afferents and dendrites of tectal neurons arborizing in these layers are supposed to process information regarding object movement and changes in overall illumination. In the mammalian brain, this bulbo-tectal (bulbo-collicular) projection is involved in the control of visual attention. A similar function in the amphibian brain is likely, because movement of objects is the dominant stimulus for the release of orienting and feeding action in amphibians, and changes in overall illumination constitute the primary releaser for escape behaviour.

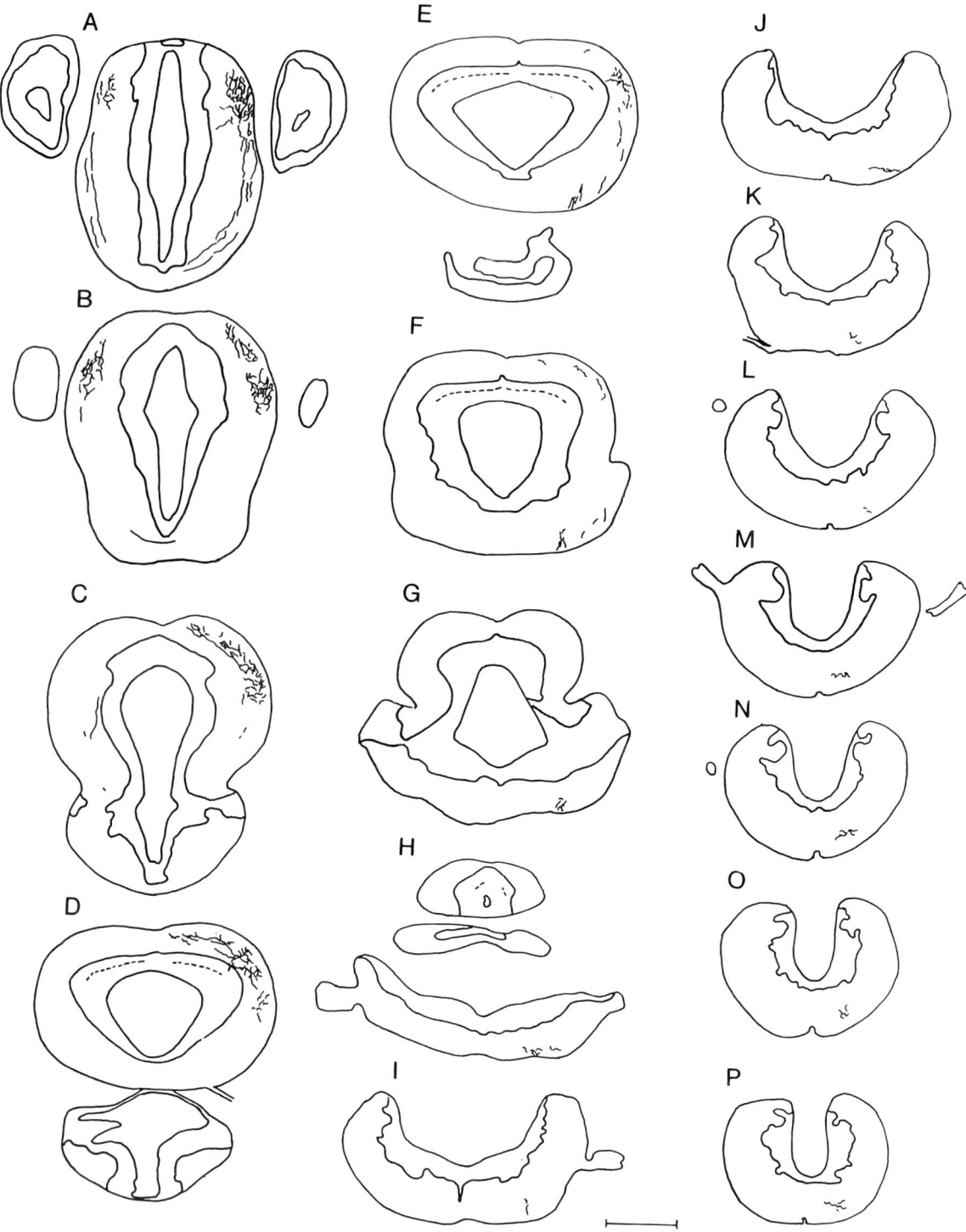

Fig. 30. Projection pattern of an intracellularly labelled T 2 neuron in *Plethodon jordani*. Transverse sections from mid diencephalon (**A**) to caudal medulla oblongata (**P**) are shown. **A–C:** Bilaterally ascending projections to the thalamus and pretectum; **D–P:** Ipsilaterally descending projection. Bar: 500 μm.

The distribution and co-localization in the optic tectum of plethodontid salamanders of the excitatory transmitter, glutamate, and the inhibitory transmitters, gamma-aminobutyric acid (GABA) and glycine, was investigated (Wallstein and Dicke 1996). GABA-lir was found in one third of tectal neurons with the majority located in cellular layer 8 and to a lesser degree in cellular layer 6. Neurons with GABA-lir were either immunoreactive for GABA only, for GABA and glutamate, or for GABA and glycine. In the first and second row of layer 6, a number of cells were clearly free of GABA.

About 80% of tectal cells revealed glutamate-lir and were situated in layers 6 and 8. Of these cells, one-fourth were also immunoreactive for GABA, but only few cells revealed glutamate- and glycine-lir. These latter groups were predominantly found in layer 8. Less than 20% of tectal cells revealed glycine-lir; most of them were also immunoreactive for GABA. Only a few revealed glycine-lir only, or glycine- and glutamate-lir. A distinct fiber bundle with glycine-lir was found in fiber layers 4 and 5. It is known from tracer studies that nuclei of the *medulla oblongata* project to the tectum with axons entering the tectum via layers 4 and 5. Somata with glycine-lir were likewise found in these nuclei.

From the specific distribution of the transmitters, it can be concluded that most tectal interneurons, in addition to glutamate, contain GABA or GABA/glycine, while the majority of projection neurons are glutamatergic.

C. TECTAL CELL TYPES IN ANURANS

On the basis of Golgi studies, Lázár and Székely (1969), Székely and Lázár (1976), and Lázár (1984) described the following major types of neurons in the *tectum opticum* of *Rana esculenta*:

Large pear-shaped cells: These are the main elements of layers 2, 4, and 6. They have round or oval somata. The most numerous of them have relatively small dendritic trees, which arborize only in the upper laminae of layer 9. Other cells of this type start arborizing above layer 8. Their axons originate from the dendritic trunk, ascend to layer 9, and terminate in the neighbourhood of the neuron. The somata show only few tiny basal dendrites. The dendrites of another group of large pear-shaped cells start arborizing widely in layer 7. Some of them arborize only in one lamina, others in more than one. The basal dendrites are well developed and extend for some distance. They are the main constituents of deep fiber layers 3 and 5.

Small pear-shaped cells: These have smaller cell bodies and narrower dendritic trees and are exclusively found in layers 8 and 9. Lázár *et al.* (1983), on the basis of cobaltic lysine studies, described two subtypes. Subtype-1 arborizes in lamina B of unmyelinated retinal afferents whereas subtype-2 arborizes in all laminae of retinal afferents except lamina D. The axon of small pear-shaped cells originates either from the soma and descend to layer 6 (subtype 2) or from a thicker dendrite and arborizes within the dendritic tree (subtype 1).

Pyramidal cells: These cells are located in deeper cellular layers, mainly in layer 6. The somata have a pyramidal or oval shape. The thick main dendrite diverges into 2–4 secondary branches in layer 7. They run horizontally for a certain distance, then extend to the tectal surface forming a wide dendritic tree. The few tertiary branches terminate mainly in laminae B, C, and F. The axon originates from the main or from a secondary dendrite and enters layer 7. There are numerous cells of intermediate morphology between pyramidal and large piriform cells.

Large ganglionic cells: These cells are the largest cells of the anuran tectum, except for mesencephalic trigeminal cells. They are exclusively (or mostly) situated in layers 6 and 7. They have extremely wide dendritic arborization. Their somata are boat-shaped or spindle-like, because two to several thick dendrites originate directly from the soma and branch into thick secondary dendrites which extend obliquely to the surface. Smaller dendrites of subtype 1, with a soma located in layer 7, preferentially terminate in lamina F, less frequently in laminae B and D, whereas those of subtype 2, with somata located in layer 6, terminate in lamina B. The axons of these large ganglionic cells leave the tectum via layer 7.

Using Biocytin for retrograde labelling and intracellular staining in *Discoglossus pictus* and *Eleutherodactylus coqui*, it was possible to further clarify the cytoarchitecture of the anuran tectum and the morphology and projection pattern of tectal neurons (Dicke and Roth 1996a), which partially confirm results from the study of Antal *et al.* (1986) using cobaltic lysine as an intracellular tracer. As in salamanders, injection of Biocytin into a single tectal neuron often resulted in clusters of labelled neurons. Five types of descending projection neurons were identified (Fig. 31).

Type 1 neurons are labelled after application of Biocytin to the ventro*medial* part of the *medulla oblongata*. The pear-shaped and pyramid-shaped somata are regularly distributed rostrocaudally in the tectal hemisphere *contralateral* to the site of Biocytin application; the majority of somata are found in layer 6. A primary dendrite quickly divides within layers 7 or 8 into two or more secondary dendrites, which extend toward the tectal surface in a candelabra-like fashion, where they form a thin, but dense neuropil in lamina A or B of layer 9. In some animals, this neuropil is found in the uppermost tectal layer, while in others it is covered either by a thin layer or by a lamina, which is thick in the lateral tectum and thins out toward the medial tectum. Fine tertiary dendrites extend horizontally from secondary dendrites in the upper part of layer 7 and in layer 8, while layer 9 is largely devoid of tertiary dendrites except for layer A. The axon originates close to the soma and leaves the tectum via layer 7. While descending, it arborizes bilaterally in the tegmentum and further descends *contralaterally* constituting the *crossed* tecto-bulbo-spinal tract. In most cases, no ascending axonal projections exist; however, two intracellularly stained neurons of this type showed ascending ipsilateral projections to the praetectum and thalamus.

After application of Biocytin to the superficial white matter of the ventro*lateral medulla oblongata*, four more types (types 2–5) of tectal neurons were labelled *ipsilateral* to the application site. Type 2 and type 3 neurons always have wide to very wide dendritic trees. Type 2 neurons have horizontally oriented spindle-shaped somata, which are exclusively found in layers 7 or 8; two to several primary dendrites originate directly from the soma and extend horizontally or obliquely toward the tectal surface. Secondary and tertiary dendrites form up to four laminae in layer 9, the uppermost lamina overlapping with lamina A of retinal afferents. Dendritic arborization is extensive in the two thick deeper laminae but sparse in the upper thin ones. This type of neuron corresponds to the large ganglionic cells of Székely and Lázár. Type 3 neurons have pear-shaped or pyramidal somata, which are situated somewhat deeper than the somata of the previous type, but still in layer 6. Their primary dendrite quickly divides into two to several secondary dendrites exhibiting the same wide dendritic patterns as type 2 neurons. Their arborization is again mostly restricted to the deeper two laminae inside layer 9, and few dendrites extend toward the tectal surface.

Type 2 and type 3 neurons have ascending and descending axons. Ascending axons extend to the ipsilateral praetectum and thalamus via the dorsomedial and/or the dorsolateral tract. In most cases, axons running within the dorsolateral tract cross in the postoptic commissure and ascend to the contralateral praetectum and thalamus. The descending axon remains ipsilateral, constituting the *uncrossed* tecto-bulbo-spinal tract. Often, small neuropils are formed at the level of the *nucleus isthmi*.

Type 4 neurons have pear-shaped somata situated in layer 6 or in deep cellular layers 2 and 4. Their long and slender primary dendrite arborizes with secondary and tertiary dendrites in the same laminae as the previous two types. They correspond to the small and large pear-shaped cells of Székely and Lázár. They appear to have the same axonal projection pattern as type 2 and type 3 neurons.

Type 5 neurons have pear-shaped somata situated in deep cellular layers 2 and 4. The primary dendrite extends to layer 7, where it divides in a T-shaped fashion into several horizontally extending secondary and tertiary dendrites, which are mostly confined to layer 7. These neurons exhibit the same pattern of ascending (mostly bilateral) and descending projections (exclusively ipsilateral) as the previous types. Neuron types 2–5 have extensive basal dendrites, which together constitute the deep fiber layers 3 and 5 of the anuran tectum.

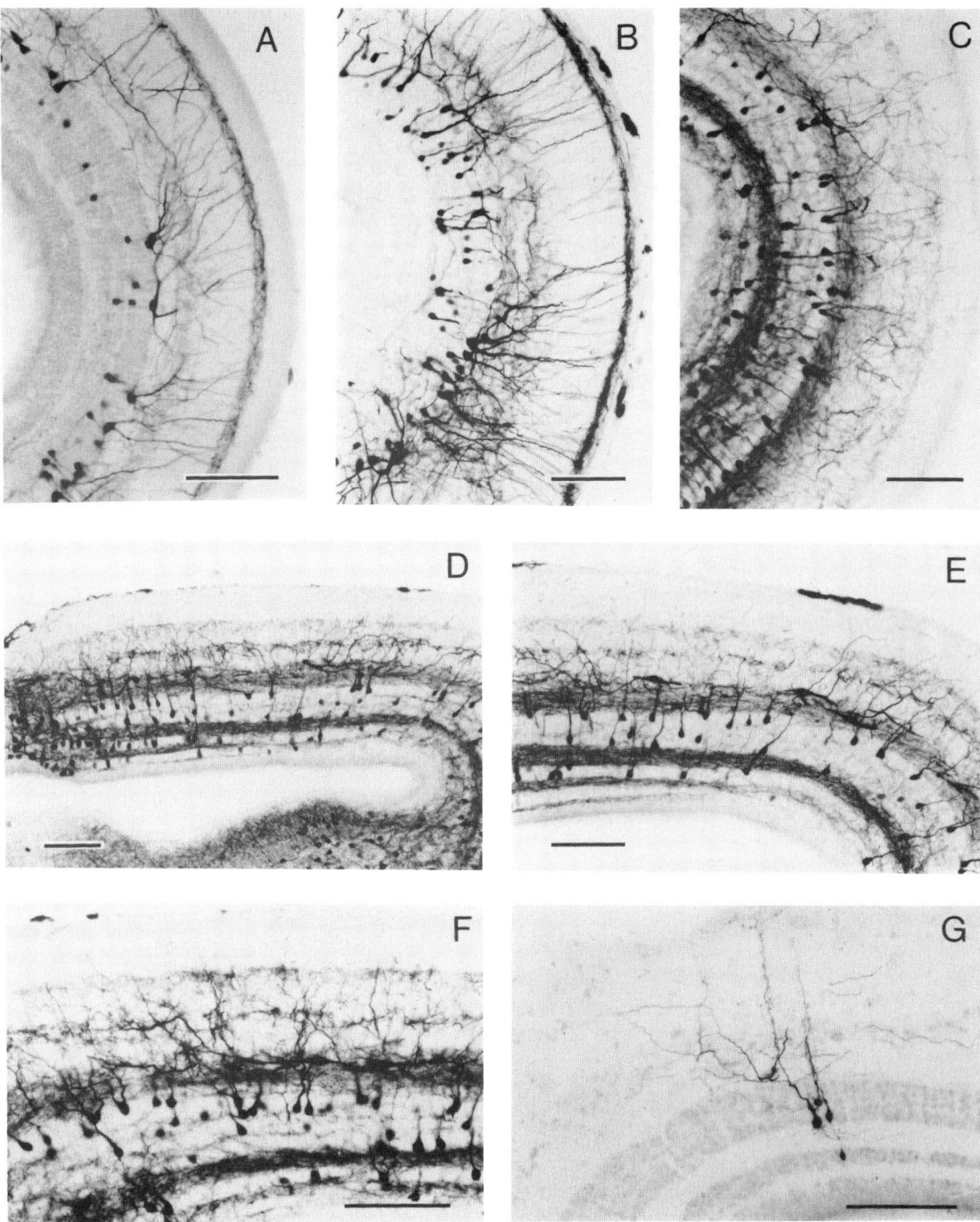

Fig. 31. Photomicrographs of transverse sections through the tectum of the frogs *Eleutherodactylus coqui* (**A**) and *Discoglossus pictus* (**B–G**) showing tectal projection neurons retrogradely labelled after application of Biocytin to the ventral medulla oblongata (**A–F**) and by intracellular injection of Biocytin (**G**). **A, B:** Type 1 neurons contralateral to the application site; **C:** Type 3, 4 and 5 neurons ipsilateral to the application site. **D–F:** Type 2, 3, 4 and 5 neurons; **G:** Cluster of a type 3 neuron and 3 interneurons labelled after intracellular injection into one single neuron. As in salamanders, the multiple staining is assumed to be due to intercellular transport of Biocytin through gap junctions. For further explanation see text. Bar: 100 μm.

Intracellular injections of Biocytin reveal a number of types of tectal neurons without axons descending to the medulla oblongata. These neurons either have only ascending projections (mostly bilateral) via the dorsolateral tract, with or without projections to the nucleus isthmi, or they project only to the *nucleus isthmi*. Their dendritic arborization is similar to that of neuron types 2–4. The majority of tectal cells are large and small pear-shaped interneurons, either with no axon or with axons that do not leave the tectum. Their somata are located at any depth of the tectum; their dendritic trees are mostly slender and arborize at various levels inside layer 9.

D. COMPARISON OF THE RESULTS FROM FROGS AND SALAMANDERS

Despite the striking differences in gross morphology of the tectum in frogs and salamanders, there is evidence that salamanders and frogs possess the same set of tectal projection neurons (Fig. 32). T 1 neurons in salamanders and type 1 neurons in frogs closely resemble each other in that (1) both types of neurons have somata situated within the superficial layer of the periventricular grey (PVG), (2) the candelabra-shaped dendritic tree arborizes predominantly in the uppermost retino-recipient part of the tectal white matter; (3) they do not project to the praetectum, thalamus (with the exception of two neurons in *Discoglossus*) or nucleus isthmi; and (4) with their descending axon they constitute the *crossed* tecto-bulbo-spinal tract.

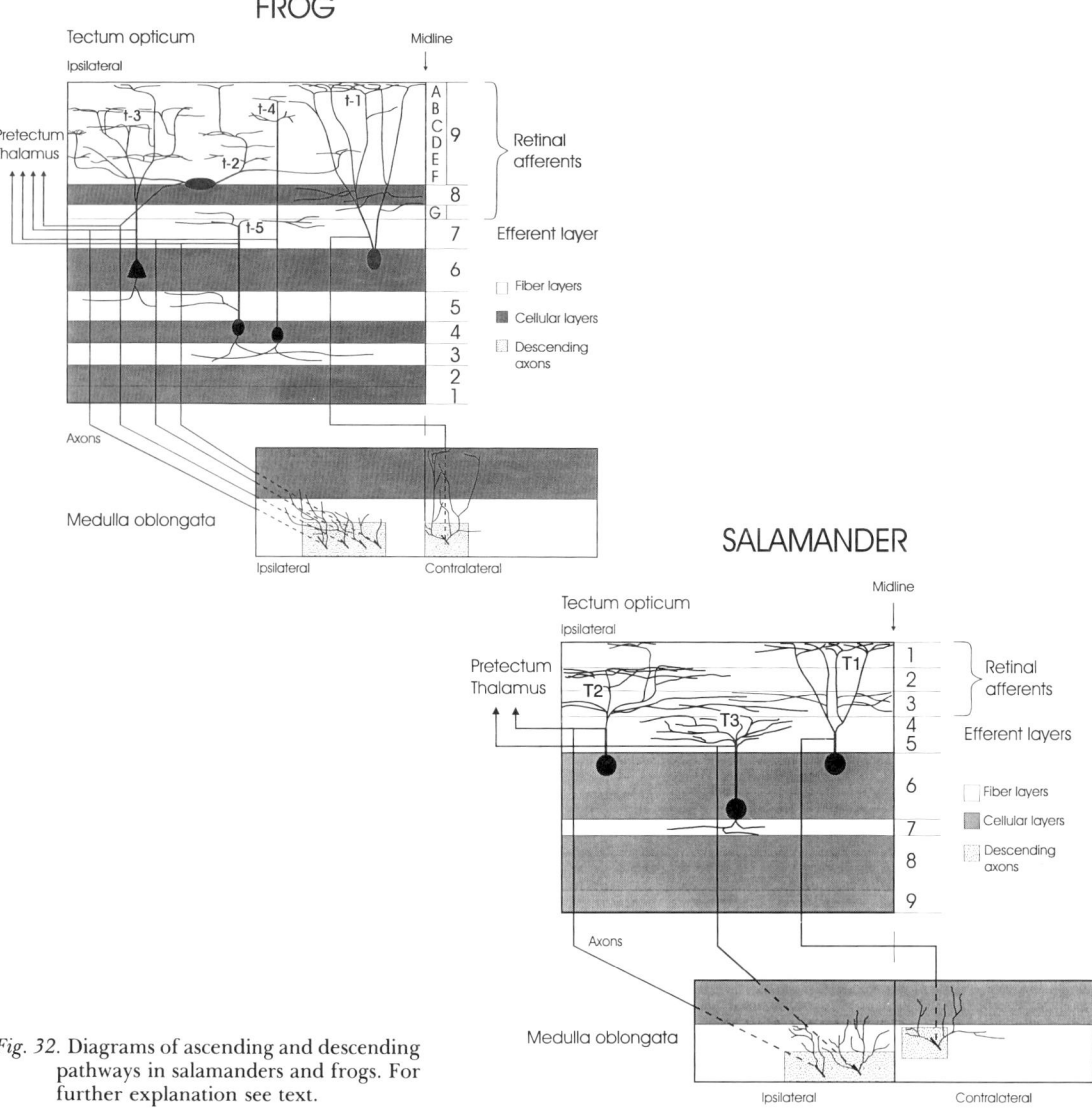

Fig. 32. Diagrams of ascending and descending pathways in salamanders and frogs. For further explanation see text.

T 2 neurons in salamanders and type 3 neurons in frogs can be regarded as homologous, because (1) their somata are situated in the superficial layer of the PVG; (2) both types have very wide dendritic trees, which arborize predominantly in the deeper retino-recipient laminae of the tectal white matter; (3) they have ascending bilateral or (less frequently only) ipsilateral projections to the praetectum and thalamus, with fibers crossing in the *commissura postoptica;* (4) descending fibers constitute part of the uncrossed tecto-bulbo-spinal pathway (in salamanders the *lateral uncrossed* tecto-bulbo-spinal tract); and (5) some of these neurons form contacts with the nucleus isthmi.

Type 2 neurons of frogs have the same pattern of descending and ascending axonal projections as type 3 cells, but have spindle-shaped somata situated in or immediately above the large efferent layer 7. This type of tectal neuron descending ipsilaterally was not found in salamanders.

Type 4 neurons in frogs characterized by a slender dendritic tree arborizing in the upper laminae of layer 9 and an ipsilaterally descending axon were less frequent. In salamanders, a comparable type of neuron was occasionally found in some animals but were not described in the results.

T 3 neurons of salamanders and type 5 neurons of frogs can be considered homologous, because (1) their somata are usually situated in the deeper part of the PVG, (2) their dendritic tree is flat and T-shaped and mostly confined to the efferent fiber layers (4 and 5 in salamanders, 7 in frogs) of the white matter; (3) they have ascending ipsilateral or bilateral projections to the praetectum and thalamus; and (4) with their descending axon they contribute to the *uncrossed* tecto-bulbo-spinal tract (in salamanders the *medial* uncrossed tract). T 2* like neurons with ipsilateral or bilateral ascending projections but no descending axons are likewise present in frogs and salamanders.

Interneurons closely resemble each other in the salamander and frog tectum because (1) they have pear-shaped somata, the majority of which are situated in the deeper part of the PVG; (2) their dendritic trees are mostly very slender; and (3) the dendritic trees arborize in different retino-recipient laminae of the tectum.

The differences in morphological complexity between frogs and salamanders essentially depend on two factors: (1) The higher rate of cell proliferation in frogs, which leads to five to ten times more neurons in the brain compared to salamanders; and (2) a greater extent of cell migration which in frogs results in multiple lamination. While early stages of tectal development are very similar in frogs and salamanders, crucial events at later developmental stages, predominantly the activity of the lateral proliferative zone and the migration of neurons from the PVG into the white matter, are largely retarded in salamanders or are completely absent (Roth *et al.* 1993). As a consequence, the salamander tectum does not develop the lateral expansion or the multiple lamination characteristic of the frog tectum.

E. NUMBER OF TECTAL CELLS

Blanke and Roth (unpubl. data) determined the number of tectal cells in a larger number of salamanders and frogs (Tables 1, 2). Among salamanders, the lowest number of tectal neurons is found in *Batrachoseps attenuatus* (35 186) and the highest in *Salamandra salamandra* (149 500). The average number of tectal neurons in the salamanders studied was 75 051.

Based on Biocytin tracing experiments, in well-stained brains about 900 descending efferent tectal cells were identified in *Plethodon jordani*, and about 600 in *Hydromantes genei* and *H. italicus*. These constitute 1.8% *(Plethodon)* and 0.7% *(Hydromantes)* of the total number of tectal cells, with one-third situated contralaterally and two-thirds ipsilaterally.

Among frogs, the lowest number of tectal neurons is found in *Arenophryne rotunda* (132 160) and the highest in *Eleutherodactylus coqui* (1 729 330). The average number of tectal neurons in the frogs is 718 959. Thus, frogs on average have about ten times more tectal neurons than do salamanders. Only the frog *Arenophryne rotunda* has fewer tectal neurons than any salamander (Tables 1, 2).

Table 1. Numbers of tectal neurons and neuron layers in anurans.

Taxon	Total No. of tectal neurons	Neurons in layers 7–9	
		Number	% of total
Bombina orientalis	258 902	45 962	17.8
Discoglossus pictus	206 285	40 975	19.9
Afrixalus fornasii	655 192	102 845	15.7
Arenophryne rotunda	132 160	28 174	21.3
Dendrobates pumilio	349 000	84 000	24.1
Eleutherodactylus coqui	1 729 330	400 967	23.2
Gastrotheca riobambae	1 312 046	361 546	27.6
Hyla septentrionalis	770 000	137 000	17.8
Hyperolius quinquevittatus	1 032 046	245 876	23.8
Limnodynastes tasmaniensis	478 297	126 856	26.5
Mantella aurantiaca	433 523	69 598	16.1
Mantella cowani	465 064	112 539	24.2
Rana temporaria	552 000	89 000	16.1
Rhacophorus leucomystax	1 500 000	265 000	17.7
Sminthillus limbatus	204 778	51 831	25.3
Xenopus laevis	1 424 725	243 004	17.1

Table 2. Numbers of tectal neurons in salamanders.

Species	No. of tectal neurons
Ambystoma opacum	48 723
Ambystoma mexicanum	124 723
Salamandra salamandra	149 500
Pleurodeles waltl	127 659
Desmognathus wrighti	80 415
Desmognathus aeneus	50 544
Desmognathus ochrophaeus	60 438
Desmognathus monticula	140 140
Desmognathus quadramaculatus	147 888
Eurycea bislineata	60 732
Plethodon cinereus	50 544
Plethodon jordani	50 733
Hydromantes italicus	91 933
Batrachoseps attenuatus	35 186
Parvimolge townsendi	37 835
Thorius narisovalis	37 906
Thorius pennatulus	56 012
Bolitoglossa subpalmata	56 012

F. PRETECTAL CELLS AND THEIR PROJECTION PATTERN

The morphology of pretectal neurons and their projection pattern was investigated in the salamanders *Plethodon jordani* and *Hydromantes italicus* by means of intracellular injection of Biocytin (Figs 33, 34). All labelled neurons (37) belonged to the deep pretectal nucleus. They all had pear-shaped somata, and their dendritic tree ramified extensively in the white matter without any apparent stratification. The main dendritic tree showed variable orientation between the pretectal neuropil and the ventral thalamus. Thus, in contrast to the situation found in the tectum, the morphology of pretectal neurons appeared to be no useful criterion for classification.

All labelled pretectal neurons turned out to be projection neurons, and most of them sent axons in parallel to more than one target. Most projected to the ipsilateral thalamus and arborized in the dorsal and ventral portions, and half of them sent axons to the contralateral diencephalon via the postoptic commissure and ramified in the contralateral praetectum, and dorsal or ventral thalamus. Again, half of pretectal neurons projected to the tectum, most of them ipsilaterally, fewer to the contralateral tectum, and only a few with bilateral projections. Inside the tectum, no regular pattern of terminal arbors (patches or bands) was observed.

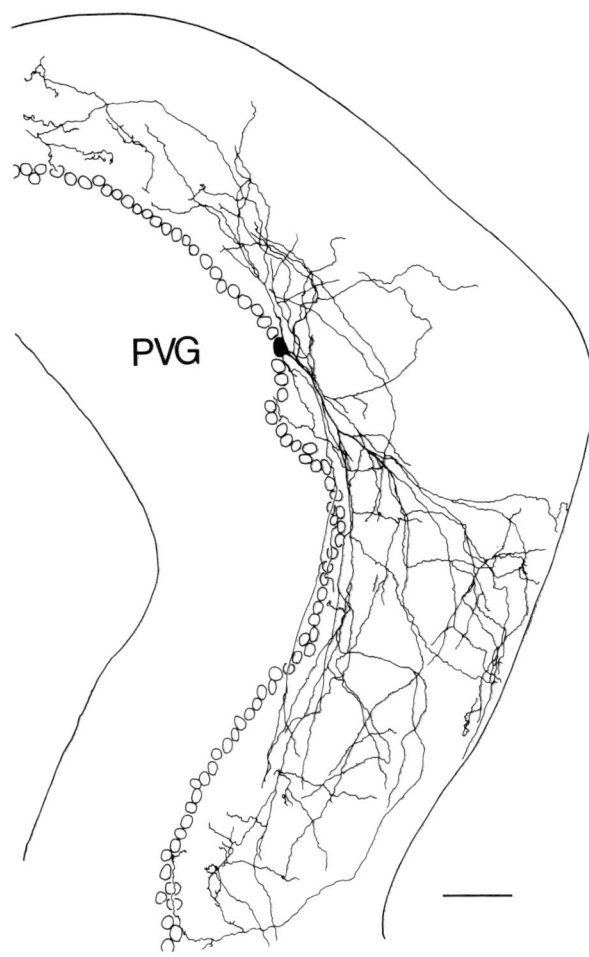

Fig. 33. Pretectal neuron *(Plethodon jordani)* labelled by intracellular injection of Biocytin. PVG = periventricular grey matter. Bar: 100 μm.

Axonal arborization was highly variable and ranged from a single fiber to the rostral tectum to massive arborizations covering almost the entire tectal hemisphere. Nearly all pretectal neurons (87%) had axons descending to the *medulla oblongata* and rostral spinal cord. Most (70%) descended to the ipsilateral, a few (17%) to the contralateral medulla and spinal cord.

In most cases, intracellular injection into single pretectal neurons resulted in the labelling of more than one neuron. As with tectal neurons, this is probably due to the presence of neuronal coupling via gap junctions.

Retrograde labelling studies of pretectal descending neurons in plethodontid salamanders revealed that the majority of pretectal neurons descend ipsilaterally. Pretectal neurons with contralaterally descending axons are situated in the pretectal grey, while those with ipsilaterally descending ones are found in the *nucleus praetectalis superficialis* as well (Fig. 35). The latter were labelled only after application of Biocytin to the ventrolateral *medulla oblongata*. The same situation after retrograde labelling of pretectal descending neurons was found in the frogs *Discoglossus pictus* and *Eleutherodactylus coqui* (Fig. 36). (Dicke, unpubl. data).

4. Central Visual and Visuomotor Pathways

A. NON-RETINAL AFFERENTS TO THE TECTUM IN SALAMANDERS

Tracer (HRP, Biocytin) studies by Finkenstädt *et al.* (1983) in *Salamandra salamandra* and by Rettig (1984), Wiggers and Roth (1991), Dicke (1997), Dicke and Roth (1996a) as well as Luksch *et al.* (unpubl. data) in plethodontid salamanders revealed the following non-retinal afferents to the tectum:

1. Telencephalic afferents from the ipsilateral amygdala *pars lateralis* and striatum via the lateral forebrain bundle.

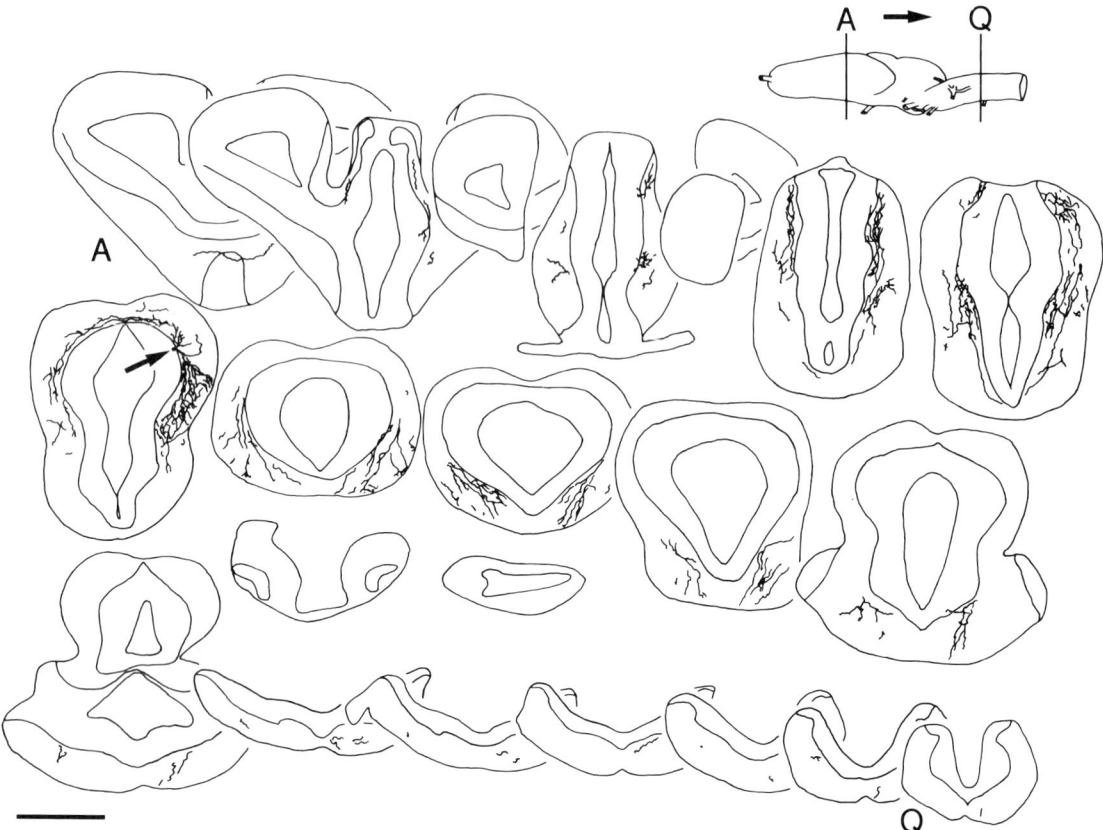

Fig. 34. Reconstruction of the projection pattern of an intracellularly labelled pretectal neuron *(Plethodon jordani)* on 17 transverse sections through the brain between the caudal telencephalon (**A**) and the caudal medulla oblongata (**Q**). Arrow indicates site of soma. Bar: 500 μm.

2. Diencephalic afferents from (a) ipsilateral and contralateral dorsal and ventral thalamus; (b) the ipsi- and contralateral praetectum.

3. Afferents from the contralateral tectum (very sparse).

4. Afferents from the dorsal tegmentum, predominantly from the ipsilateral side, fewer (via axons through the *commissura tecti mesencephali*) from the contralateral dorsal tegmentum.

5. Afferents from the ipsilateral and contralateral *nucleus isthmi*, the latter crossing in the postoptic commissure.

6. Afferents from the ipsilateral and contralateral *medulla oblongata*, originating in the *nucleus reticularis medius*, the *nucleus vestibularis*, and the *nucleus dorsalis*.

7. Afferents from the dorsal grey matter of the contra- and ipsilateral spinal cord *(nucleus lemnisci spinalis)*.

The arrangement of afferents to and efferents from the tectum in anurans (Lázár 1984; Grüsser-Cornehls 1984) is very similar to that described for salamanders. In general, there seems to be no direct tecto-telencephalic connection in amphibians. The connection is mediated in frogs by the lateral thalamic nucleus and in salamanders by the posterior thalamic cell group (Wicht and Himstedt 1988). These relay nuclei may be homologous to the mammalian pulvinar.

B. TECTAL EFFERENT PATHWAYS AND THEIR CONNECTION TO PREMOTOR AND MOTOR CENTRES

Descending efferents: In salamanders, the fibers of the crossed tract originate from type T 1 cells, those of the lateral crossed tract from T 2 and of the medial crossed tract from T 3 cells (cf. Fig. 32). Collaterals of the crossed tract arborize extensively in the ipsi- and

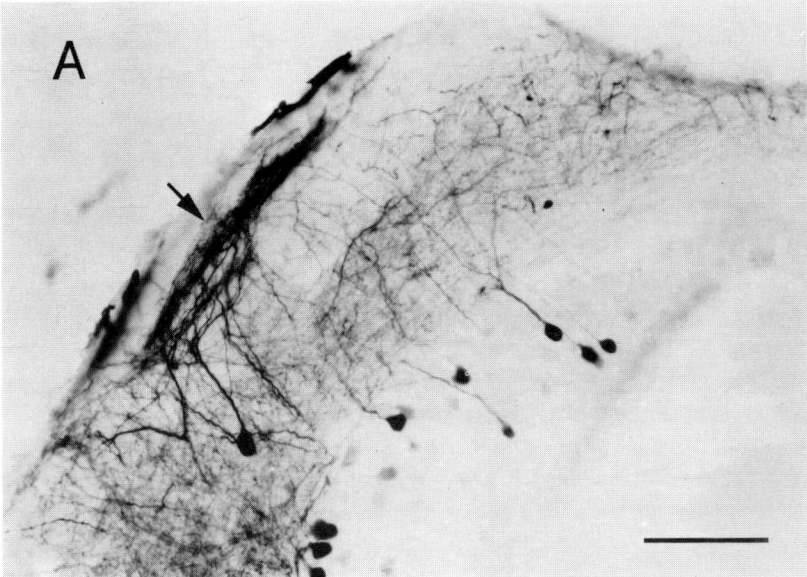

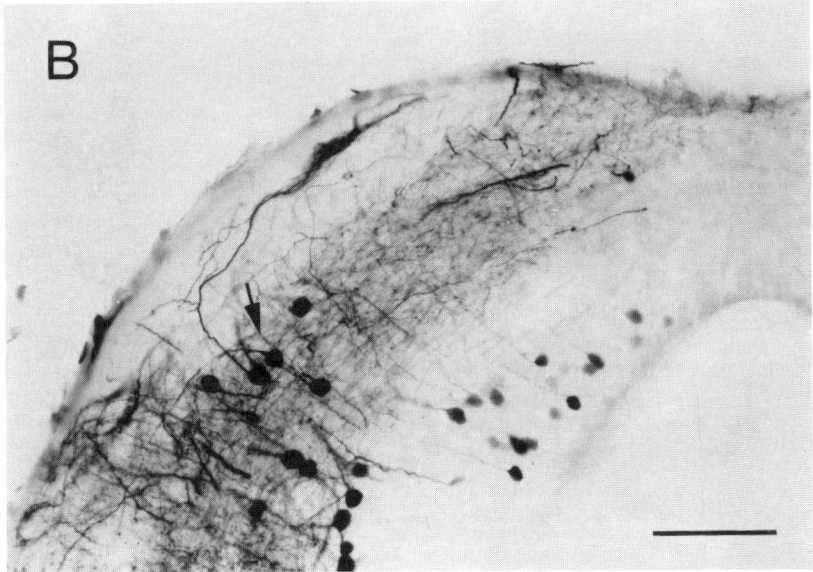

Fig. 35. Photomicrograph of transverse sections through the caudal diencephalon showing neurons in the deep and superficial pretectal nucleus of the salamander *Plethodon jordani* labelled by application of Biocytin to the ipsilateral ventrolateral *medulla oblongata*. Two consecutive transverse sections are shown. **A:** Arrow points to the dense neuropil formed by dendrites of the neurons belonging to the superficial pretectal nucleus. **B:** Arrow indicates the migrated somata of the superficial pretectal nucleus. Bar: 100 μm.

contralateral tegmentum and then descend contralaterally in the ventral part of the medulla forming a round fascicle close to the midline somewhat below the surface. Fibers of the uncrossed tracts do not arborize substantially in the tegmentum and descend superficially in the ventromedial and ventrolateral *medulla oblongata* (Fig. 37). Axons extending ventrolaterally form a dense bundle, with a sharp border at the lateral edge. In the reticular formation, axons of the uncrossed tracts form collaterals, which remain in the vicinity of the tracts and many axon collaterals of the crossed tract extend to the cellular layer and arborize among somata. The collaterals often terminate on the somata where they form beads. The crossed and uncrossed tracts extend to the level of the entrance of the third spinal nerve or beyond. The uncrossed tract thins out after the obex region more than does the crossed one.

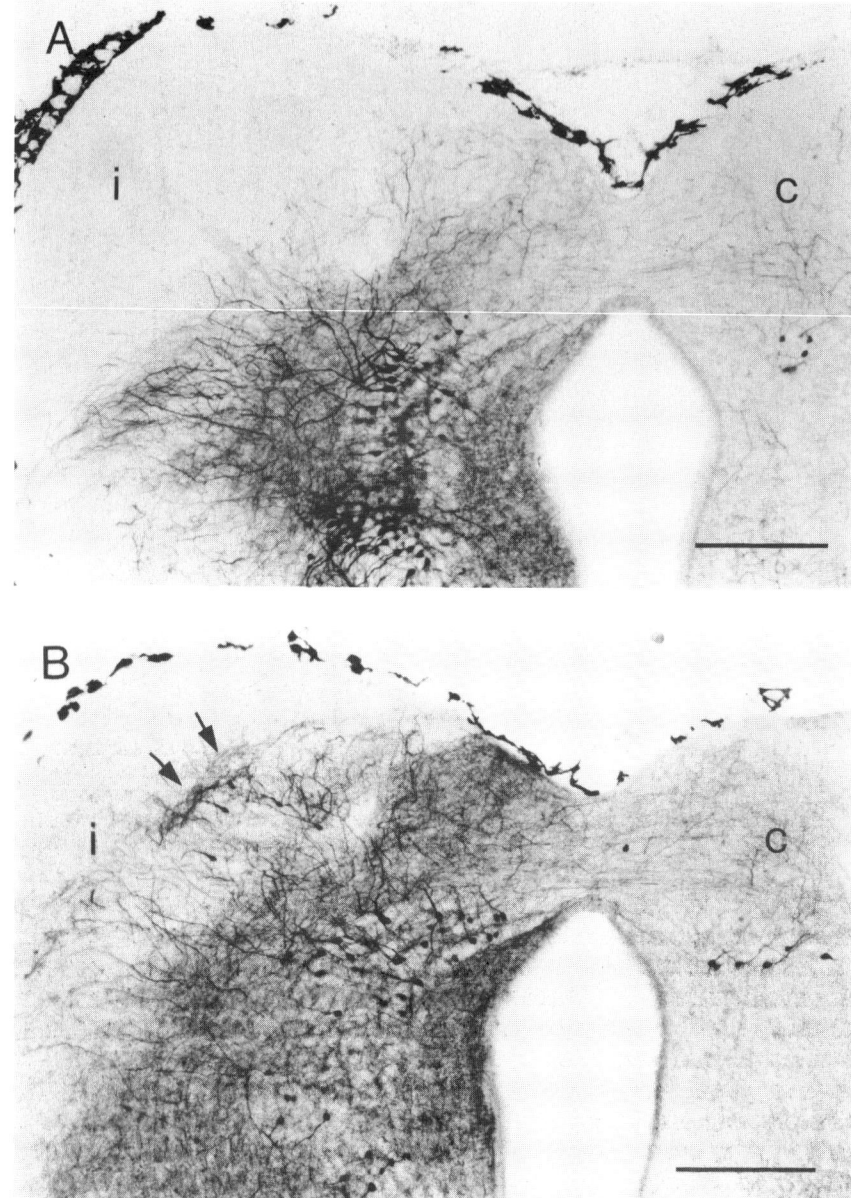

Fig. 36. Photomicrograph of transverse sections through the caudal diencephalon showing pretectal neurons in the frog *Discoglossus pictus* labelled after application of Biocytin to the ventromedial (**A**) and ventrolateral (**B**) *medulla oblongata* at the level of the entrance of the VIIth cranial nerve. Only few somata are labelled in the contralateral side. In (**A**) somata are located in the pretectal grey, while in (**B**) somata are situated in the pretectal grey in the *nucleus praetectalis superficialis*. Arrows point to the neuropil, which is formed by the migrated somata of the *nucleus praetectalis superficialis*. i = ipsilateral, c = contralateral. Bar: 200 μm.

The motor nuclei of cranial nerves V–X and the rostral parts of the motor nuclei of cranial nerves XI and XII are situated in the ventral part of the *medulla oblongata* (Figs 38, 39). The somata of the motor neurons are located at the border between the grey and white matter. In most salamander species the motor neurons are aligned in a medial and a lateral cell column (Roth *et al.* 1988; Wake *et al.* 1988; Dicke 1992). The dendrites of these motor neurons extend in a lateral direction, where they form a dense neuropil situated immediately below the surface (Fig. 39). These lateral dendritic fields slightly overlap in their medial parts with the lateralmost fibers of the uncrossed tract. Motor neurons also send dendrites to the ventrolateral and ventromedial medulla. These dendrites terminate in an area where the

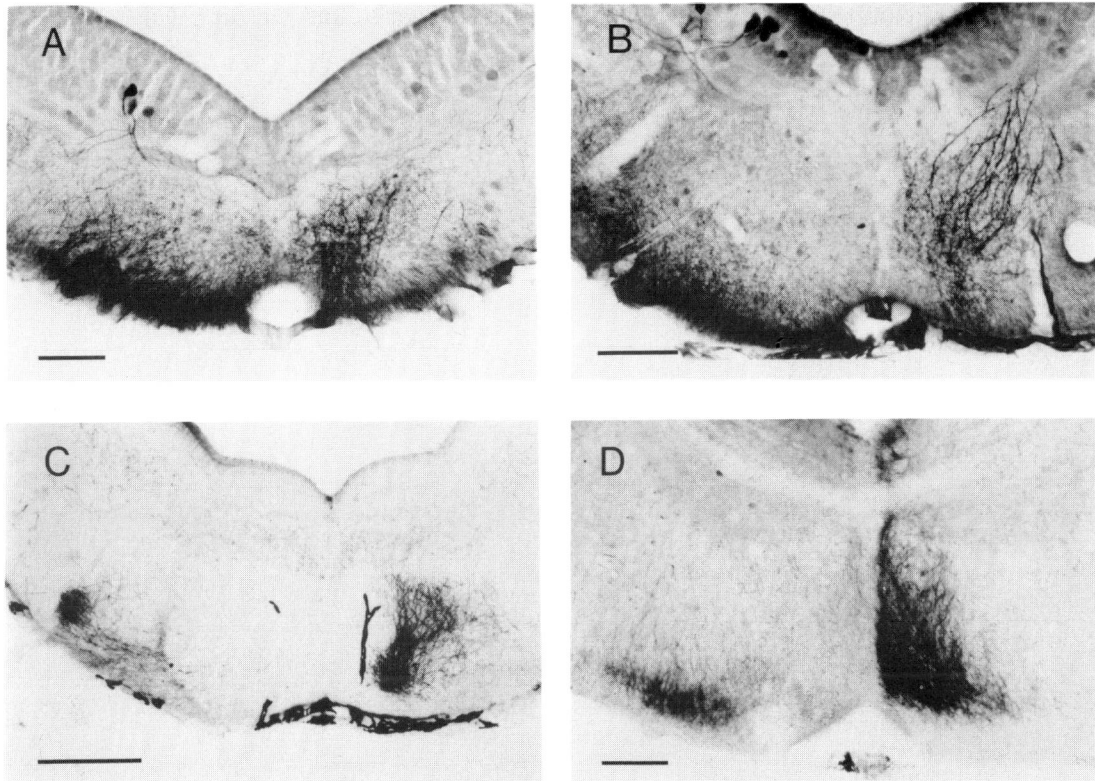

Fig. 37. Photomicrographs of transverse sections of the *medulla oblongata* showing the tecto-bulbo-spinal tracts in salamanders and frogs. **A:** *Hydromantes italicus*, at the level of the VIIth cranial nerve. **B:** *Plethodon jordani*, at the level of the IXth cranial nerve. **C:** *Discoglossus pictus*, at the level between the VIIth and IXth cranial nerves. **D:** *Eleutherodactylus coqui*, at the level of the Xth cranial nerve. The crossed tract is to the right. Note the strong branching of axons in A, B and C, whereas in D axon collaterals extend dorsally close to the midline. The fibers of the uncrossed tracts run superficially in the ventrolateral medulla and reveal few collaterals. Bar = 100 μm.

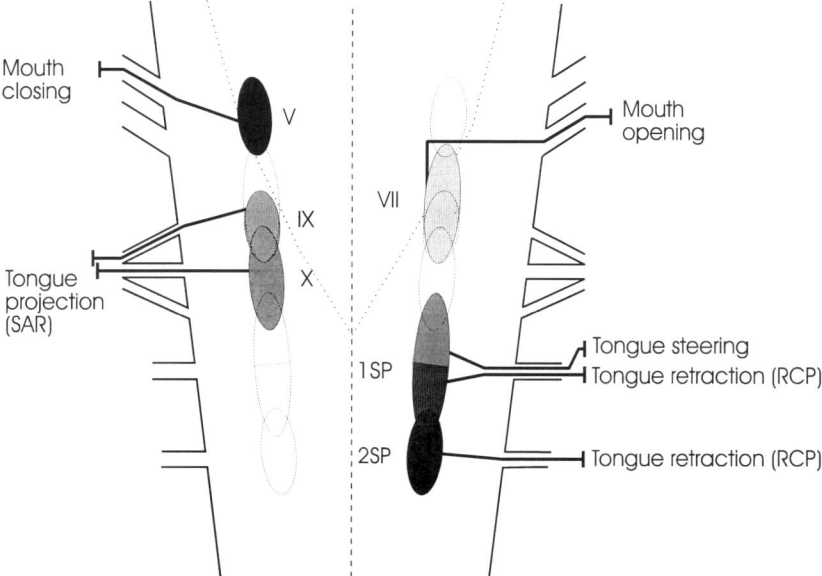

Fig. 38. Diagram of the *medulla oblongata* showing the rostrocaudal distribution of cranial and cervical spinal motor nuclei related to feeding in salamanders. Shading intensity indicates onset of activity (grey: first, black: latest). The respective contralateral nuclei are marked by dotted lines. RCP = *musculus rectus cervicis profundus;* SAR = *m. subarcualis rectus;* V–X = cranial motor nuclei; 1 sp = nucleus of first spinal nerve; 2 sp = nucleus of second spinal nerve.

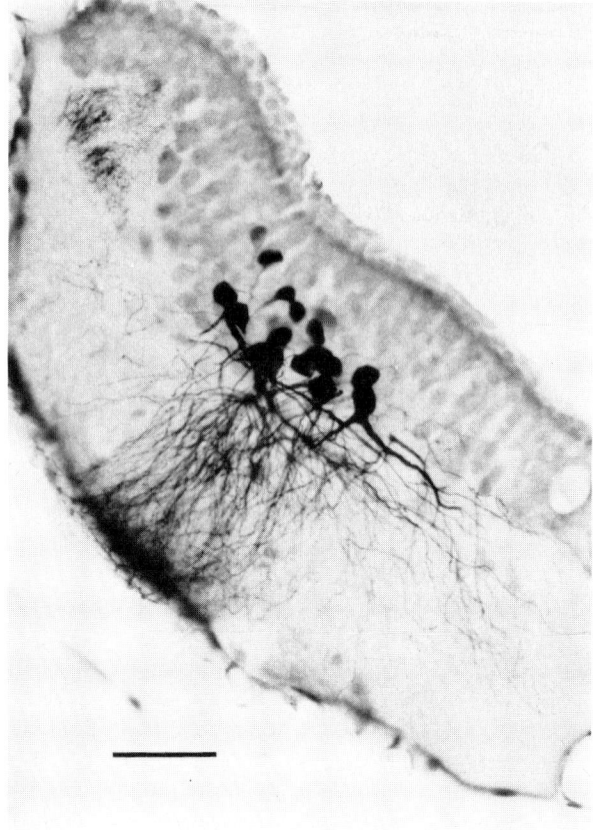

Fig. 39. Photomicrograph of a transverse section through the motor nucleus of the vagal nerve. The motor neurons are labelled by retrograde transport of Biocytin. Note the dense lateral neuropil formed by the dendrites of the motor neurons and the medial dendrites extending toward the position of the crossed tecto-bulbo-spinal tract (not labelled). Bar: 100 μm.

axons of the crossed and uncrossed tracts descend (Dicke 1992). It is likely that the dendrites of motor neurons form monosynaptic contacts with the descending fibers of the tectum and that collaterals of the descending axons directly contact at least some somata of the motor neurons.

Some dendrites of the motor neurons also extend into the dorsolateral medulla, where the sensory fibers of cranial nerves V–X enter the brain and form ascending and descending bundles. Another sensory fiber bundle, the *lemniscus spinalis*, runs in the dorsolateral part of the medulla. This bundle is constituted by axons of sensory cells located in the grey matter of the *medulla oblongata* and *spinalis*. These cells arborize in regions where the somatosensory afferent fibers extend and where the descending tracts and the dendritic fields of the motor neurons are found. Their axons ascend to the midbrain and the thalamus. In the tectum, this tract terminates in layers 3–5.

In anurans, the descending tracts take a course similar to that in salamanders, with the difference that at the level of the Vth cranial nerve root the crossed tract runs medially but not close to the midline within the white matter, while in the caudal part of the *medulla oblongata* axons descend close to the midline in a ventrodorsal band and collaterals branch heavily within the dorsal grey matter (Dicke 1997) (Fig. 37C, D). As in salamanders, the crossed tract thins out around the obex, but extends at least as far as to the second spinal nerve. Axons of the uncrossed tracts run laterally in the ventral white matter and again form a sharp lateral edge. Collaterals arise from the lateral part of the tracts and extend to the border of the grey matter. The uncrossed tracts terminate before the obex region, i.e., much earlier than in salamanders.

Projections to and from the nucleus isthmi: The tectum projects to the ipsilateral nucleus isthmi, which projects back to both hemispheres (Figs 40, 41). Intracellular labelling of tectal neurons in plethodontid salamanders (Wiggers and Roth 1994) revealed that only about 10% of efferent tectal neurons (which themselves constitute only 5–10% of the total number of tectal neurons) project to the nucleus isthmi. Their axons, while descending to the *medulla*

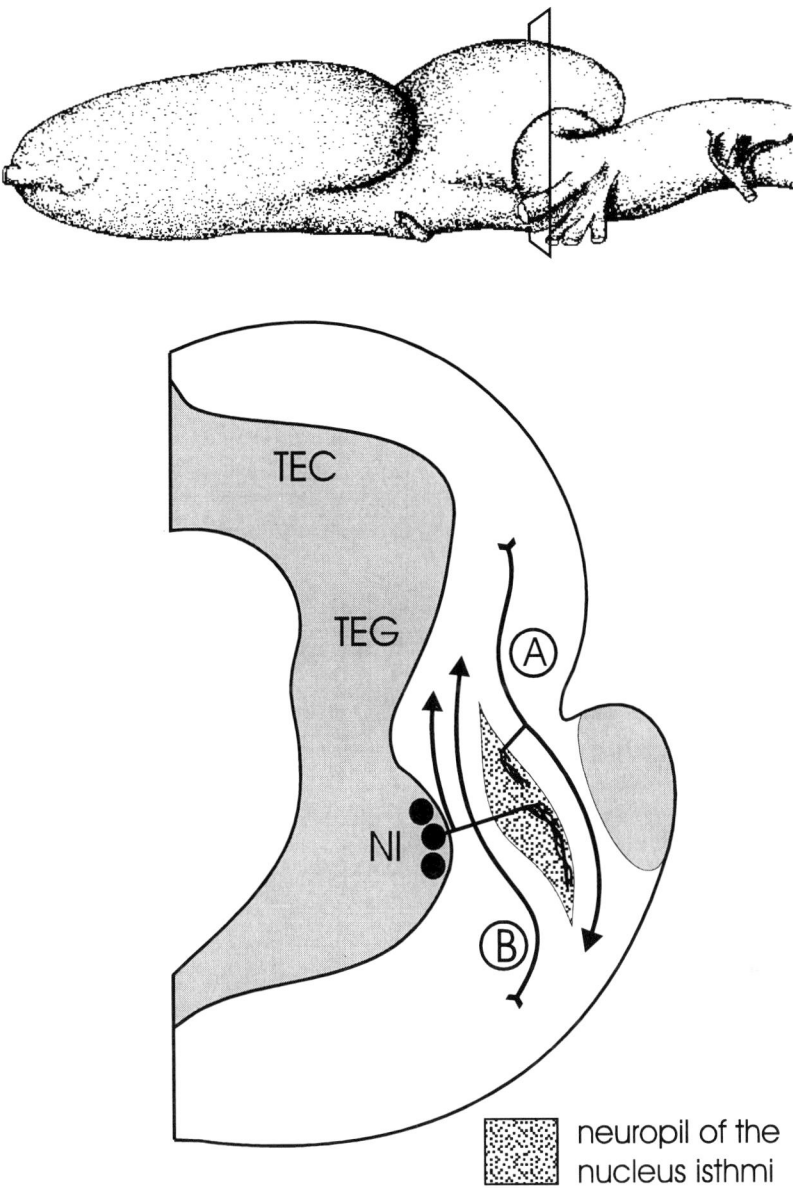

Fig. 40. Schematic diagram of the location of the *nucleus isthmi* and the isthmic neuropil in plethodontid salamanders. A = tectal efferents (tecto-bulbo-spinal tract); B = lemniscal and isthmo-tectal tract; TEC = tectum; TEG = tegmentum; NI = *nucleus isthmi*.

oblongata and cervical spinal cord, form distinct and dense telodendra within the isthmic neuropil. All other descending tectal fibers pass in the vicinity of the *nucleus isthmi* without forming collaterals or boutons. Tecto-isthmic projections probably arise from T 2 neurons.

The pattern of isthmic projections to the ipsi- and contralateral tectum was again studied by intracellular recording and Biocytin staining studies (Wiggers and Roth 1994). The isthmic axons sprout close to the soma and enter the ipsilateral tectum along the border between layers 3 and 4. Within the rostral two-thirds of the tectum, isthmic fibers give rise to conspicuous telodendric structures. Sprouting from one or two branches, a web of fine, beaded fibers ascend in a column through tectal layers 3, 2 and 1 to the surface. The diameter of this column is 50–70 μm in mediolateral, and 100–150 μm in rostrocaudal extent (Fig. 42). The main axon remains at the border between layers 3 and 4 and extends further rostrally, where it descends, crossing the pretectal and thalamic neuropils without sprouting or boutons. They cross to the contralateral side within the dorsal part of the optic chiasm. Then, the axons

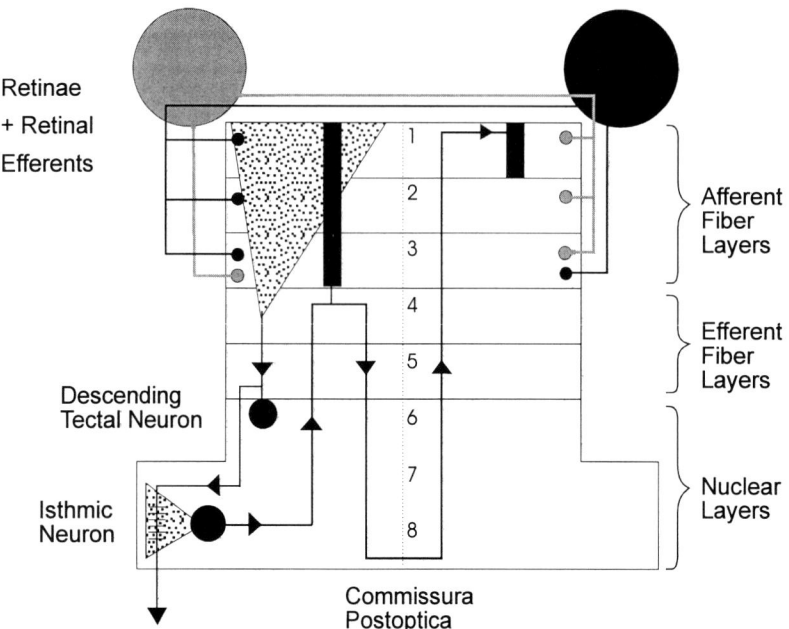

Fig. 41. Schematic diagram of tecto-isthmo-tectal connections. Dendritic trees are drawn as dotted triangles; telodendritic structures are drawn as black columns.

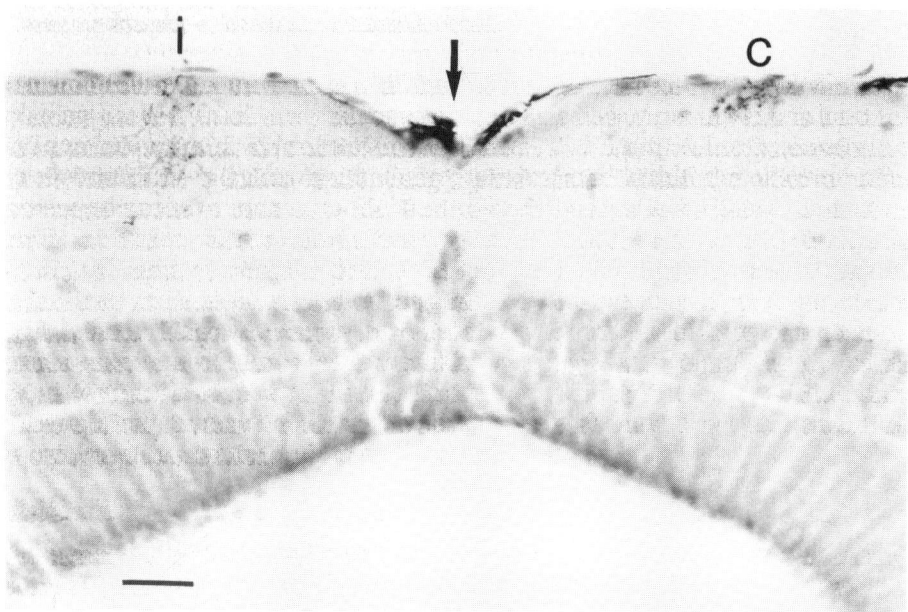

Fig. 42. Photomicrograph of the ipsi- and contralateral axonal telodendra of a single isthmic neuron in the tectum of *Plethodon jordani*. The figure shows a transverse section through the tectum. Note that the ipsilateral and contralateral telodendron are at the same distance from the midline (arrow). i = ipsilateral telodendron; c = contralateral telodendron. Bar: 100 μm.

ascend together with the retinotectal fibers, again without any visible sites of contact, through the thalamic and pretectal neuropils and enter the tectum at the very surface. They, then, terminate in the rostral two-thirds in circular patches of beaded fibers, which have the same extent as the columns in the other side, but are restricted to layer 1. The topography corresponds exactly to the retinotopy of retinal afferents (see below).

V. NEUROPHYSIOLOGY OF VISION

A. Methods

The neurophysiological data presented in this chapter were obtained by single cell extracellular or intracellular recordings from immobilized amphibians as well as from isolated brain preparations. So far, recordings from freely moving toads have yielded very few data, because of extreme technical difficulties, and these data are difficult to interpret and compare with the data from immobilized animals (but see Schürg-Pfeiffer and Ewert 1989). For the presentation of visual stimuli during the recording experiments, a so-called perimeter apparatus commonly is used. On a screen in front of the animal, visual stimuli varying in size, configuration, velocity and movement pattern are presented. For single-cell recordings, steel or tungsten microelectrodes or glass micropipettes are used. In order to determine the exact position of the recorded cells, the micropipettes are filled with alcian blue solution which, when applied iontophoretically, results in small marking points. For intracellular recordings micropipettes with a tip diameter equal to or smaller than 0.5 μm are used. For anatomical characterization of the recorded cell, the micropipette is filled with a solution of HRP, cobaltic lysine or Biocytin.

In recent times, "whole-brain *in vitro* experiments" were used for intracellular recording and labeling experiments. During the experiments, the isolated brains, often together with a retina attached, are situated in a recording chamber which is continuously perfused with oxygenated ringer solution (Fig. 43). Such preparations allow recording and labelling experiments for several days. Peripheral nerves are stimulated using suction electrodes and brain centers stimulated with bipolar stainless steel electrodes. Intracellular recording and labeling experiments are carried out as under *in vivo* conditions. A comparison between recordings and labeling under *in vivo* and *in vitro* conditions revealed no substantial differences.

B. Retinal Cells

1. Photoreceptors

Vertebrate photoreceptors do not generate action potentials. Therefore, their neurophysiology can be elucidated only by intracellular recordings. Generally, photoreceptors respond with a sustained *hyperpolarization*, when illuminated with a small spot of light. The time course of the potentials is different in rods and cones. The source of this hyperpolarization is the photoisomerization of the visual pigment (rhodopsin), which triggers a series of steps that culminate in the reduction of a steady inward current of Na^+ ions at the membrane of the outer segment (Hagins *et al.* 1970).

In amphibians, rods are strongly coupled to adjacent rods and weakly coupled to neighbouring cones by means of electrical synapses (gap junctions). Furthermore, there are feedback connections mediated by horizontal cells. This results in complex interactions within the photoreceptor layer. Passing current into a rod elicits a strong voltage response of the same sign in adjacent rods, but a lower response in cones. More distant cones respond with a transient sign-inverted signal, probably mediated by horizontal cells (Attwell *et al.* 1984). As a result, the signal may be averaged over a certain area at the first stage of visual processing. The principal advantage ascribed to this mechanism is an enhancement of the signal-to-noise ratio of the transmitted signal (Attwell *et al.* 1984).

In summary, the synaptic activity, i.e., the release of transmitter substance is coupled to the state of membrane polarization in such a way that illumination *decreases* the release of depolarizing transmitter substance (Dacheux and Miller 1976).

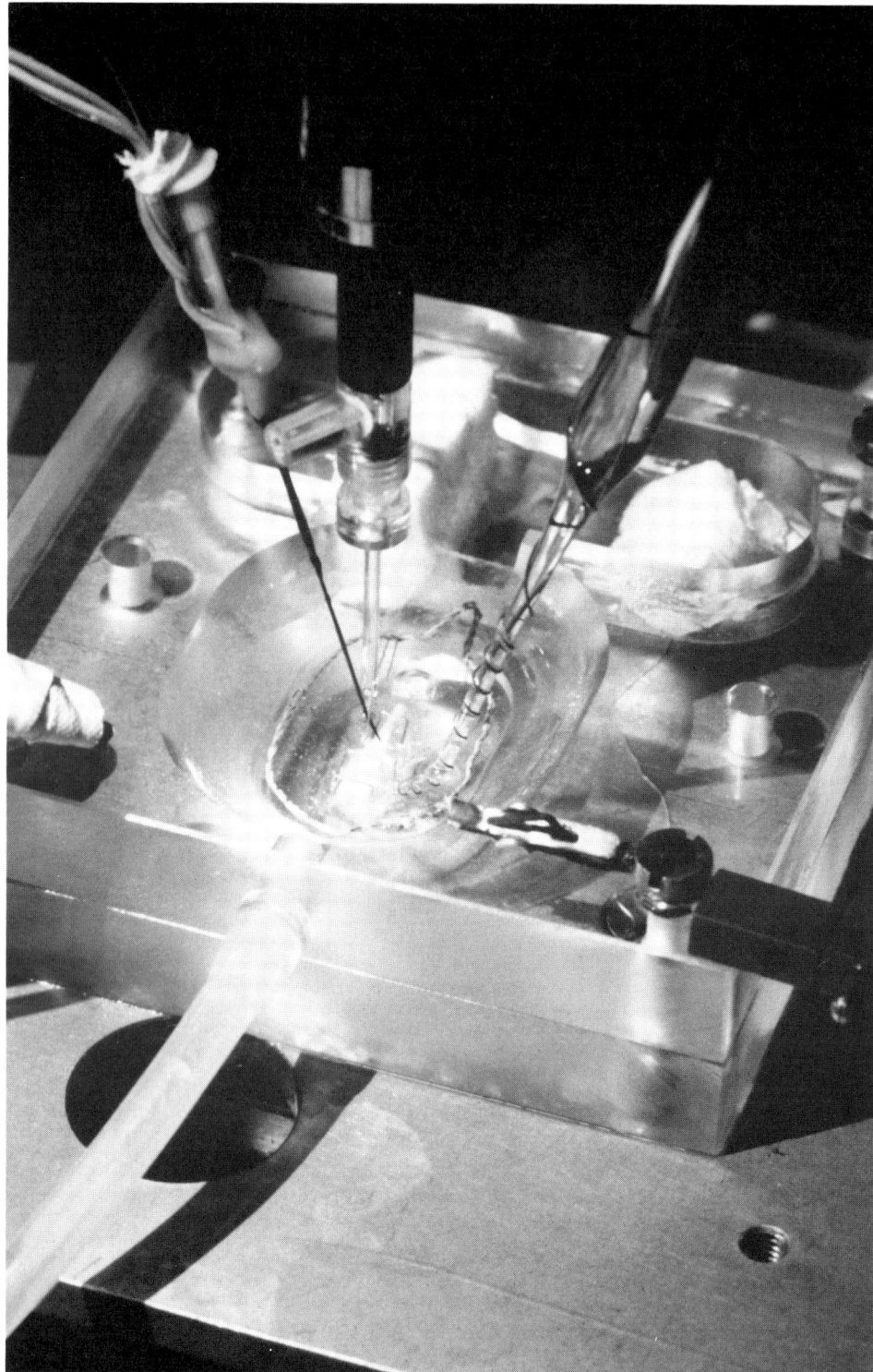

Fig. 43. Photograph illustrating the *in vitro* whole brain technique. The isolated brain is fixed with stainless insect pins in a recording chamber which is continuously perfused with oxygenated ringer solution at a temperature of 13–16°C. The contralateral optic nerve is stimulated by means of a suction electrode (right); the ipsilateral or contralateral *medulla oblongata* is stimulated using bipolar stainless steel electrodes (left) inserted into the brain at the level of the 7th nerve. Tectal neurons are recorded intracellularly by a glass micropipette.

2. Horizontal cells

Horizontal cells (HC) are the first retinal interneurons along the visual pathway. Since the response of a photoreceptor to a flash of light is a hyperpolarization and a decreased release of depolarizing transmitter to the postsynaptic horizontal cell, an illumination leads to hyperpolarization of the HC as well. Intracellular recordings in HC reveal a sustained response to illumination, which increases with light intensity. The onset and offset of hyperpolarization in HC is more rapid than in rods. The area of spatial integration is broad, and HC are thought to be responsible for the centre-surround antagonistic response in the postsynaptic neurons of the visual pathway (Werblin and Dowling 1969).

There are, however, different responses in inner and outer HC depending on the wavelength of illumination. Outer HC with a large dendritic field make synaptic contacts with blue-sensitive rods and red-sensitive cones (Witkovsky et al. 1981). This mixed rod-cone input to HC is a specialty of amphibians. The signals of both receptors to the HC interact in a nonlinear way and are balanced by the inhibitory transmitters GABA and glycine. The response of the cone system is enhanced by rod activity, but not the reverse (Witkovsky and Stone 1983, 1987). Few outer HC generate chromaticity responses, as they hyperpolarize only to illumination with short wavelength (below 500 nm). Stimulation with long wavelengths (beyond 517 nm) leads to depolarization of the outer HC (Ogden et al. 1984).

3. Bipolar Cells

Bipolar cells (BC) receive presynaptic input from receptors as well as from HC. They do not generate action potentials, but their responses to the onset and offset of light are more diverse than are those of HC. Due to at least two distinct glutamate receptors, there are two physiological types of BC: The OFF-centre bipolars, which hyperpolarize in response to light, and the ON-centre bipolars, which have sign-inverting synapses with photoreceptors and depolarize in response to light (Dowling 1987). Both types of cells show an antagonistic response to illumination of the surround, which is mediated through HC.

4. Amacrine Cells

The responses of amacrine cells (AC) are more complex than are those of other interneurons. While some of the AC respond with transient (or sustained) graded potentials to the onset and offset of illumination, others generate action potentials during the transient response. Whether an AC responds to ON, OFF or ON/OFF, depends on the shape and position of the stimulus within the receptive field (Werblin and Dowling 1969).

5. Retinal Ganglion Cells

Retinal ganglion cells (RGC) generate the output signals of the retina to the brain. Their activity can be recorded directly within the retina, in the optic nerve or in the terminal arborization of the optic axons in the diencephalon or mesencephalon (e.g., tectum). RGC exhibit the most complex response patterns of retinal neurons. The different classes of RGC can be determined by (1) the size of the excitatory receptive field (ERF) and inhibitory receptive field (IRF), (2) the response to the on- and offset of diffuse illumination, (3) the response to stationary or moving stimuli of different velocity, size, shape and background contrast, and (4) the response to stimulation with light of different wavelengths.

A. FROGS

According to various authors (Maturana et al. 1960; Grüsser and Grüsser-Cornehls 1970), recordings from the optic nerve and in the superficial layers of the optic tectum yielded five different physiological classes of RGC:

Class 0: These cells have large ERFs and respond to the onset of illumination. Their axons terminate in the diencephalon, but not in the tectum.

Class 1: RGC of this type have very small ERFs with a diameter of about 2–3° and a slightly larger IRF of about 5–6°. They do not respond to ON/OFF; they terminate in the uppermost layer of the tectum. These neurons are also called "sustained edge detectors", because they respond to stationary edges within the ERF. This response reappears after light has turned off and then on again ("non-erasability" of response; Maturana et al. 1960).

Class 2: These neurons have slightly larger ERFs of about 2.5–4° and a larger IRF of 6–20° and, therefore, show stronger inhibition. As class 1 cells, they show no ON/OFF response; they terminate in the tectum beneath class 1 terminals. These cells respond to stationary edges as do class 1 cells, but their response to such stimuli does not reappear after light ON/OFF ("erasability"). These cells are called "convex edge detectors" (Maturana et al. 1960).

Class 3: This class has an ERF of 6–8° and an IRF of 10–15°. The cells show a short ON/OFF response, and their axons terminate in the diencephalon as well as in the tectum beneath class 2 terminals. They are called "changing contrast detectors" or ON/OFF cells (Maturana et al. 1960).

Class 4: These cells have large excitatory receptive fields of about 10–15° and an inhibitory surround of the same size or no IRF. They generate a sustained ON/OFF response, and the axons terminate in the tectum at the deepest layer of retinal afferents. These RGC are called OFF-neurons or "dimmers", because dimming of light elicits a pronounced activation (Maturana et al. 1960).

Some authors found an additional class 5 of RGC. These neurons have large RFs with rather indistinct borders. They show only responses to change in light intensity and not to moving objects. These cells are the so called "dark detectors" of Maturana et al. (1960).

In the toad *Bufo bufo*, Ewert and Hock (1972) described three types of RGC: (1) "R2" with an ERF of 4°, a weak ON-response and a strong inhibitory surround, corresponding to class 2 RGC of *Rana;* (2) "R3" with an ERF of 8°, ON/OFF response and medium inhibitory surround, corresponding to class 3 RGC; and (3) "R4" with an ERF of 10–15°, OFF-response and weak inhibitory surround, corresponding to class 4 RGC. RGCs corresponding to class 1 of *Rana* only rarely were recorded in the tectum of *Bufo*.

For a description of RGC response properties in other anuran species see Grüsser and Grüsser-Cornehls (1976) and Grüsser-Cornehls and Langeveld (1985).

An increase in stimulus velocity generally elicits an increase in impulse frequency (Grüsser and Grüsser-Cornehls 1970, 1976). The slope depends on the type of RGC and on the form and configuration of the stimulus. All anuran RGC types show variation in impulse rate depending on the size and shape of the stimulus. The response to a horizontal bar ("wormlike" stimulus) does not change significantly, when edge length (EL) is varied (Grüsser and Grüsser-Cornehls 1976; Ewert and Hock 1972). In the toad *Bufo bufo*, R3 cells exhibit a slight increase followed by a decrease, and R4 cells an increase in impulse rate, when the bar is elongated in edge length from 4 to 16° (Ewert and Hock 1972). The most striking differences occur in the response to squares. Depending on the diameter of the ERF, R2 cells prefer squares of 4° EL, while R3 neurons prefers 8° EL squares, and R4 neurons exhibit a constant increase in impulse rate, when EL increases up to 16°. The response to vertical bars moved horizontally through the visual field ("antiwormlike" stimuli) is similar to that of squares (Ewert and Hock 1972).

According to these data, retinal ganglion cells do not discriminate between the different shapes of the stimuli; rather, they respond to the ratio between horizontal and vertical angular extension of the stimulus (Ewert 1984, 1989).

B. SALAMANDERS

In salamanders, the retinal afferents terminate in three distinct layers in the superficial fiber layer of the optic tectum. Electrophysiological recordings in *Salamandra salamandra* yielded the following types of response to illumination and moving stimuli (Grüsser-Cornehls and Himstedt 1973; 1976):

Layer-1 unit: This unit is similar to classes 1 and 2 in anurans, because of strong inhibitory surround and lack of response to a change in illumination, although they are characterized by a relative large ERF. These cells are best activated by small objects of 2–3° in diameter. The discharge rate increases with increasing velocity of the stimulus. Unlike anuran class 1 and class 2 cells, layer-1 units prefer rectangles oriented parallel to the direction of movement ("wormlike" stimuli).

Layer-2 units: These correspond to the anuran class 3 RGC. They have small ERFs of about 6–9° and a weak inhibitory surround. A change of diffuse light elicits a short ON/OFF response, and the preferred stimulus size is slightly larger than in layer-1 units. In contrast to layer-1 units, these neurons respond better to squares and vertical objects than to horizontal bars.

Layer-3 units: These are similar to class 4 RGC in anurans. They have large ERFs of 10–20°, respond best to larger stimuli than layer-2 units and generate a tonic ON/OFF response to change in light intensity. There is a clear dominance of the square over both the vertical and the horizontal rectangle at all velocities.

In general, RGC of salamanders have larger receptive field sizes compared to frogs and toads. In contrast to anurans, a clear directional selectivity to black and white stimuli has been found in RGC of *Salamandra salamandra* (Grüsser-Cornehls and Himstedt 1973) and other urodeles (Cronly-Dillon and Galand 1966 in *Triturus;* Norton *et al.* 1970 in *Necturus*). In some cases the directional selectivity depends upon the angular velocity of the stimulus (Grüsser-Cornehls 1985).

C. COLOUR VISION

The visual system of most vertebrates consists of a luminance-sensitive rod system and a trichromatic cone system with three different photosensitive pigments. Amphibians also have photosensitive pigments of at least three different absorption spectra. A dichromatic rod system, which is most sensitive to blue and green light (433 and 502 nm), and a cone system, which is most sensitive to yellow/red light (580 nm). The double cone contains both cone and rod pigment. The role of the amphibian "green rod" receptor remains unclear. It may enhance the sensitivity of monochromatic vision at dim light levels, or may play the role of the blue-cone trichromatic colour vision of other vertebrates (Matthews 1983). In this case, the blue-sensitive "green rods" and yellow/red-sensitive cones could act as antagonistic receptors.

Chromaticity signals are found in all types of retinal cells. The activity of some outer horizontal cells is affected by the wavelength of the stimulus. Chromatic bipolar cells in *Xenopus* are hyperpolarized by blue light and depolarized by red light (Witkovsky and Stone 1983), but most investigations concerning colour vision have been conducted on ganglion cells. In *Rana esculenta*, Grüsser-Cornehls and Langeveld (1985) found a strong correlation between the wavelength of the stimulus and the velocity function and directional selectivity in some of the ganglion cells. There is also a striking difference in the effect of achromatic (black/white) and chromatic stimuli. In some of the neurons, an achromatic stimulus exhibits no directional selectivity, whereas a coloured stimulus of the same size, velocity and form does.

D. SUMMARY

Three universal classes of RGC seem to exist in amphibians:

1. Small-field "edge-detector" cells responding either to moving or non-moving objects and requiring relatively high visual contrast. They are represented in frogs by class 1 and class 2 cells, in toads by R2 cells and by layer-1 cells in salamanders. They exhibit low conduction velocity due to unmyelinated fibers. They may include several subclasses (RGC class 1 and 2 in frogs) differing in colour sensitivity, responses to light ON/OFF (i.e., they show either no response or an "ON" response), and "erasability" or "non-erasability" of stimulation by stationary edges. Their axons terminate in the uppermost layer of the optic tectum. They are most probably involved in the detection of small, high-contrast objects such as prey. These cells are comparable to X-cells in cats and to P-cells in primates (Spillmann and Werner 1990).

2. Medium-field "movement-detector" or "ON-OFF" cells, which respond to small changes in contrast and small dislocations of edges and, thus, to movement. They exhibit high conduction velocity due to myelination. They are represented by class 3 RGC in frogs, R3 cells in toads and layer-2 cells in salamanders. They do not respond to non-moving objects. Their axons project to the thalamus, praetectum and tectum in parallel; in the tectum they terminate in the intermediate layer of retinal afferents. They are predominantly involved in movement and movement pattern detection. They correspond to Y-cells in cats and M cells in primates.

3. Large-field, "dimming-detector", "OFF"-cells. They do not respond well to small or medium-sized forms, but do respond well to large objects and to changes in illumination in larger parts of the visual field. These cells are represented by class 4 and R4 cells in frogs and toads, respectively, and by layer-3 cells in salamanders. They may be involved in predator detection, optomotor behaviour or detection of changes in overall illumination. They show high conduction velocity and have thick myelinated fibers, which project in parallel to the thalamus, praetectum and tectum. In the tectum, they terminate in the deepest layer of retinal afferents.

Thus, it seems that in amphibians, three separate retinofugal pathways exist, viz., a shape/colour pathway, a motion pathway and an ambient illumination pathway.

C. Tectum

1. Topic Organization of Retinal Afferents in the Tectum

Retino-tectal afferents form regular, two-dimensional representations or "maps", one contralateral and one ipsilateral from each retina. The situation found in the salamander *Hydromantes italicus* may serve as an example (Wiggers *et al.* 1995). Each visual hemifield is projected completely on to the *contralateral* tectal hemisphere. The nasal quadrant of the retina, corresponding to the caudal visual hemifield, is represented in the caudal tectum and the temporal quadrant, corresponding to the frontal visual hemifield, in the rostral tectum. The dorsal and ventral quadrants, corresponding to the ventral and the dorsal visual hemifield, respectively, are represented in the lateral and medial tectum (Fig. 44). From this it follows that the representation of an object moving from caudal (caudolateral) to rostral (frontal) in the left visual hemifield, is moved from caudal to rostral in the right (contralateral) tectum, and that of an object moving upwards in the left visual hemifield, is moved from the lateral to the medial part of the right tectum (adjacent to the midline). There is a relatively precise, linear correlation between the visual hemifield and the corresponding contralateral tectal retinotopic map along the rostro-caudal axis (Fig. 45), whereas the representation along the medio-lateral axis seems less precise.

The direct and indirect (via *nucleus isthmi*) *ipsilateral* projection covers roughly the rostral two-thirds of the tectal surface and includes only the temporal, dorsotemporal and ventrotemporal retinal quadrants; there is no ipsilateral projection of the nasal retina and, therefore, none from the caudal visual hemifield (including caudal parts of the dorsal and ventral visual hemifields). Furthermore, the representation of the retina to the *ipsilateral* tectum is rotated 180° compared to the situation found in the *contralateral* tectal hemisphere, in the sense that the temporal retina is projected on to the caudal parts of the region covered by ipsilateral afferents and the central part of the retina on to the rostral tectal hemisphere. Thus, the representation of an object that moves from lateral to rostral/frontal in the visual hemifield, is moved from rostral to caudal in the *ipsilateral* tectum, i.e., in a direction opposite to the *contralateral* projection from the same eye. As soon as this object reaches the binocular field, there is an additional two-fold (i.e., contralateral and ipsilateral) retinotectal projection of the other eye, in which the tectal representations move opposite to the situation in the contra- and ipsilateral representations of the eye previously considered. In the case of objects that move *latero-frontally*, this peculiar arrangement results in the ipsilateral tectal representation from one eye and the contralateral representation from the other eye being in register, whereas the left and right contralateral and the left and right ipsilateral representations move in opposite directions. The reverse occurs, when objects move *along the z-axis*, i.e., straight

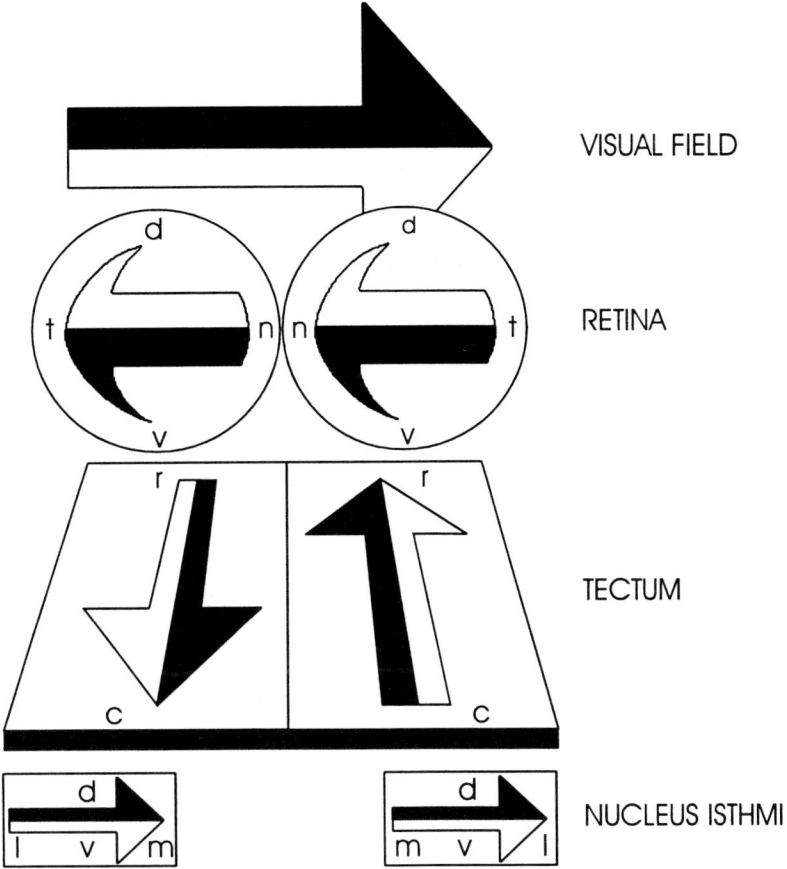

Fig. 44. Diagrammatic presentation of the topic organization of the visual field in retina, *tectum opticum* and *nucleus isthmi,* as revealed by electrophysiological recordings and neuroanatomical tracing methods. d = dorsal; r = rostral; l = lateral; m = medial; c = caudal; t = temporal; n = nasal; v = ventral. From Wiggers and Roth (1991).

toward or away from the animal. In this case, the two contralateral tectal representations move in the same direction and opposite to the two ipsilateral representations. This inverse arrangement of the contralateral and ipsilateral retinotectal projection is important for depth perception based on retinal disparity (Fig. 46).

2. Topic Organization and Receptive Field Sizes of Neurons

In contrast to the receptive fields of the afferents, those of the neurons are not distributed evenly over the tectum but concentrated in a frontal area of the visual field (Wiggers *et al.* 1995). There is an exponential correlation between the rostro-caudal position of the somata and the center of their contralateral and ipsilateral receptive fields. In *Hydromantes,* the average RF diameter of all tectal neurons was 41°, with a minimum at 10° and a maximum close to 360°. When the large-field neurons (>50°) are excluded, the average diameter of the remaining cells is 25°. There was no correlation between RF-size and position of the soma either with respect to the rostro-caudal axis of the tectum or the vertical position inside the gray matter. However, large-field neurons were situated mainly in the caudal part of the tectum and within the superficial layer of the gray matter (layer 6).

In *Hydromantes,* more than half of the recorded neurons were binocular and were located in the rostral tectum. They were found in all layers of the tectum. In most cases, there was no difference in size or shape between the monocular contralateral receptive field and the binocular field. In two-thirds of the recorded binocular neurons, the monocular ipsilateral field was smaller than the contralateral one, covering 70% of the average monocular

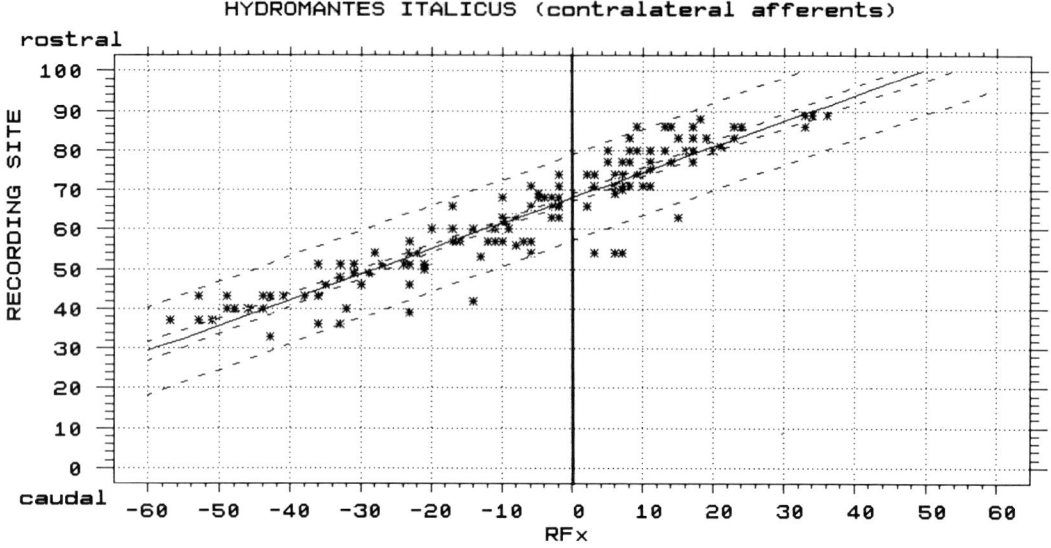
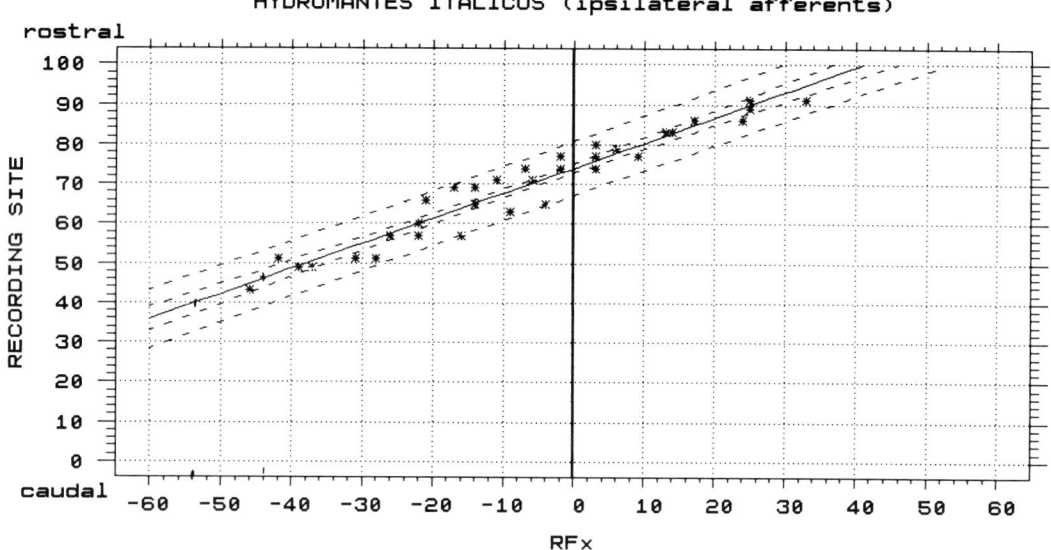

Fig. 45. Correlation between the site of a recording from an axon of a RGC in the superficial afferent tectal layers (recording site) and the centre of its receptive field along the horizontal axis in the visual field (Rfx) of *Hydromantes italicus*. The upper diagram shows recordings from *contralateral* retinal fibers in the tectum, the lower one recordings from *ipsilateral* fibers. From Wiggers *et al.* (1995).

contralateral or binocular field diameter. The monocular ipsilateral RFs had an average diameter of 21° (minimum 5°, maximum 39°). In the remaining binocular neurons, the ipsilateral and contralateral fields were either identical in size, or the ipsilateral field was larger than the contralateral one.

3. Response Properties of Tectal Cells

A. ANURANS

Grüsser and Grüsser-Cornehls (1970, 1976) were the first to conduct extensive studies on the response properties of anuran tectal cell responses using extracellular recordings (Fig. 47). On the basis of data obtained from *Rana esculenta*, *R. pipiens*, *Hyla septentrionalis* and *Bufo bufo*, they described the following main response types:

Class-T1 neurons: The ERF of these cells is 15–30° in size and has an oval shape. They are found in the binocular visual field, but some of them respond only to monocular stimulation. They are preferentially activated by small moving objects.

Class-T2 neurons: They have large ERFs beyond 90° and respond only to monocular stimulation. They exhibit directional selectivity in the sense that their discharge rate increases when the stimulus moves in a temporonasal direction within the visual field. These neurons, too, respond best to small moving objects 2–6° in size, but also to larger objects.

Class-T3 neurons: These neurons respond well to objects larger than 3° in size when moving toward the eye of the animal, while movement away from the eye slightly inhibits the spontaneous activity. Movements in a tangential plane elicits no, or only weak, responses.

Class-T4 neurons: The receptive field of these neurons covers the entire monocular, or even the entire visual field or large parts of it. They are primarily found in deeper layers of the tectum. Some of them are multisensory (tactile and auditory). They respond to small (less than 5° in diameter) as well as larger objects (up to 30°). Many of them exhibit strong adaptation to repeated presentation of stimuli and correspond to the "newness" neurons described by Lettvin *et al.* (1961).

Class-T5 neurons: These cells have ERFs of 8–30° and respond best to moving objects within a size range of 3–15°, but a 0.1° stimulus still elicits responses. T5 neurons show great variability with respect to their response properties (see below).

Class-T6 neurons: Their ERFs have sizes of at least 120° in rostrocaudal extension and 90° in horizontal extension and are located above the animal. They respond best to stimuli larger than 8°.

Class-T7 neurons: These cells have very small ERFs (2–3°) and respond best to very small objects (1–2°). The angular velocity threshold is very low (0.01°/s).

More detailed extracellular recordings from tectal cell were carried out by Ewert and von Wietersheim (1974) and Roth and Jordan (1982) in the toad *Bufo bufo* and by Schürg-Pfeiffer and Ewert (1981) in *Rana temporaria*.

In experiments of Ewert and co-workers, T5 tectal neurons were studied to a greater extent. Particular emphasis was laid on the "worm-antiworm" discrimination ability of these neurons. In these experiments, the length of the horizontal and/or vertical edges of rectangular stimuli was increased stepwise resulting in squares differing in size from 2° × 2° to 20° × 20° and "wormlike" and "antiwormlike" rectangles with a constant width of 2° and lengths of 2–20°. These stimuli were moved through the visual field of the animals at a constant velocity of 7.6°/s (Fig. 48).

T5 neurons were monocularly driven and had relatively small ERFs of about 26°. A first type of neuron, called T5-1, responded best to squares (c/S) at all edge lengths measured, with an optimum at 8°. The second-best stimulus was the "wormlike" one (a/H), again with an optimum at 8° edge length. The least effective stimulus was the "antiwormlike" one (b/V). Its effectiveness increased only slightly at edge lengths of 4° and 8°, compared to the 2° × 2° square. In a second type of tectal neuron, called T5-2, H was superior to S, while the respective response curves again showed a maximum at 8°. In contrast, the responses to V decreased continuously with increasing stimulus length. The response differences between H and V in T5-1 and T5-2 neurons proved to be statistically significant. Thus, class T5-1 neurons are in the first place sensitive to the area of moving stimuli and in the second place to extension in the direction of movement. Class T5-2 cells show the opposite rank of preferences. However, in both types, extensions perpendicular to the direction of movement are relatively ineffective, or even lower the discharge rate.

Schürg-Pfeiffer and Ewert (1981) investigated the types of responses of tectal neurons in the frog, *Rana temporaria* with the same set of stimuli. Three classes of response types were described which differed to some extent from the data obtained from *Bufo*. The T5-1 type strongly preferred the S stimulus, again with a maximum at 8°, but there was no clear difference between the responses to the H and V stimuli. In type T5-2, the responses to S were slightly better than, or equal to, those to H, and both were superior to the responses to S. In a third

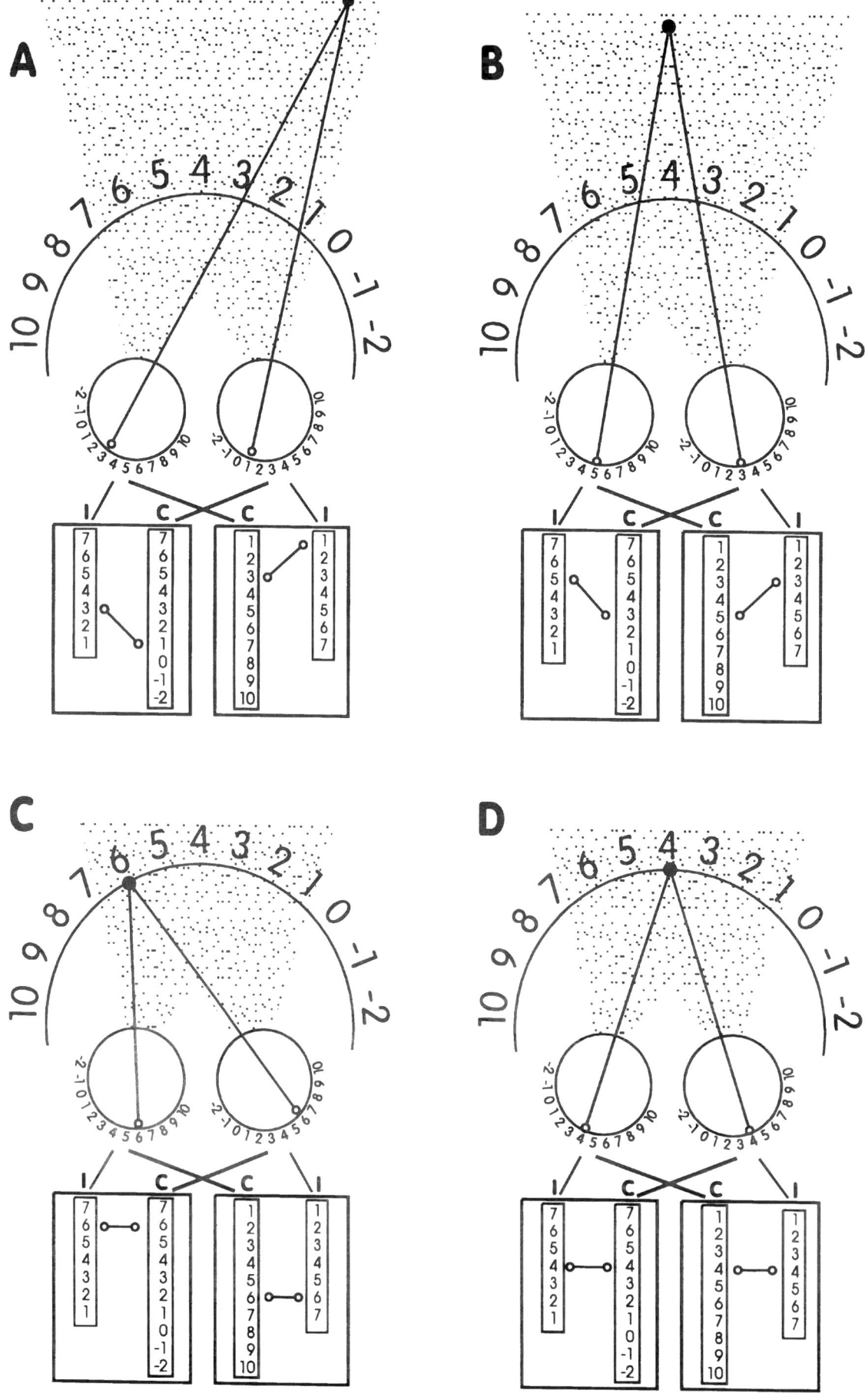

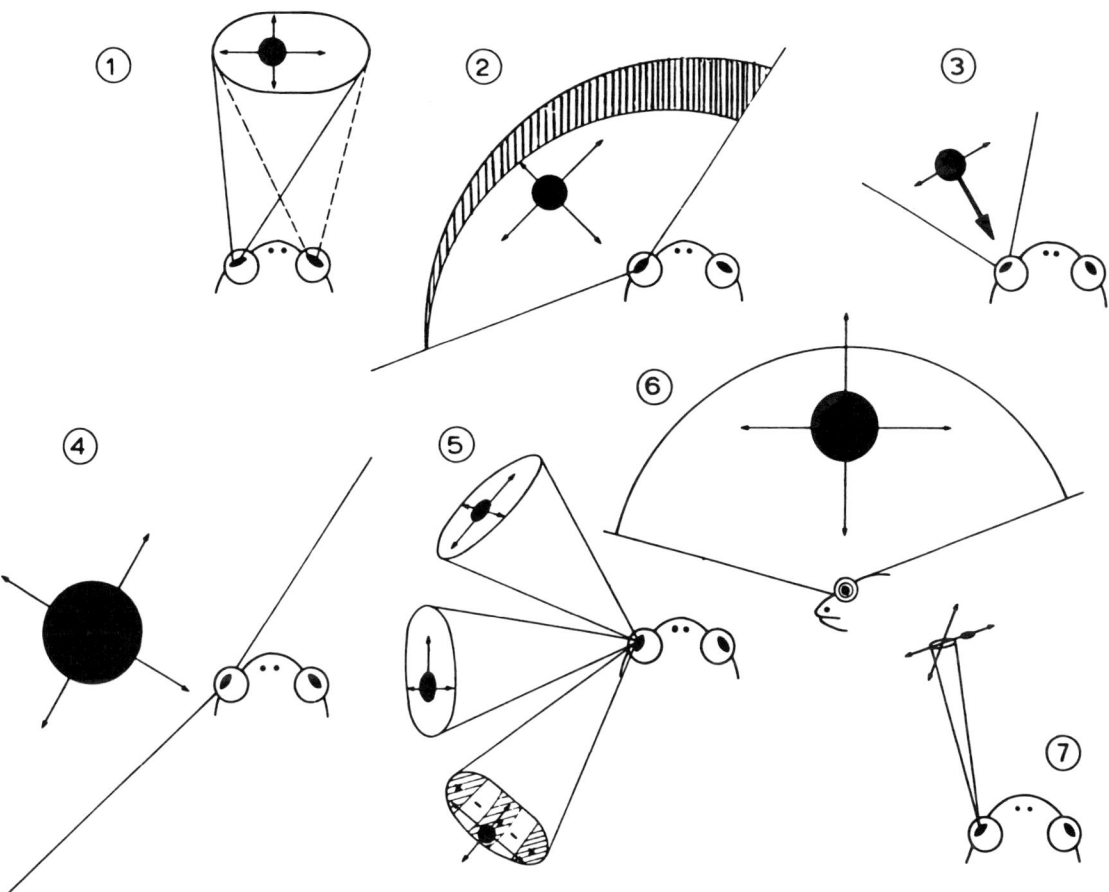

Fig. 47. Different classes of tectal neurons (T1–T7) in the frog *Rana temporaria*. For further explanation see text. From Grüsser and Grüsser-Cornehls (1976).

▶

Fig. 46. Diagram illustrating the representation of an object in the visual space in the retina and in the two tectal hemispheres via direct ipsi- and contralateral retinal efferents in tongue-projecting salamanders. The tectum is shown from dorsal view. The black dot represents an object in space; its representations in the retina and within the visual map of the tectum are drawn as open circles. The semicircle carrying numbers represents the horopter (= distance at zero disparity), numbers from −2 to 10 indicate the extent of the visual field, which correspond to the ipsilateral and contralateral representation of the visual field in each tectal hemisphere. Lines connecting open circles represent intratectal disparities, which are correlated with the distance of the object from the horopter. **A:** The object is located laterally and falls far of the horopter; accordingly, its intra- and intertectal representations are disparate. **B:** The object is in front of the animal, but falls far of the horopter; accordingly, intratectal representations are disparate, but are symmetric in the left and right hemisphere. **C:** The object is located on the horopter; accordingly, intratectal disparities are zero, but object representations in the two hemispheres are at different sites along the rostrocaudal axis. **D:** The object is on the horopter and in front of the animal; accordingly, intratectal disparities are zero and object representations are at the same site in the two hemispheres. I = ipsilateral projection; C = contralateral projection. From Wiggers *et al.* (1995).

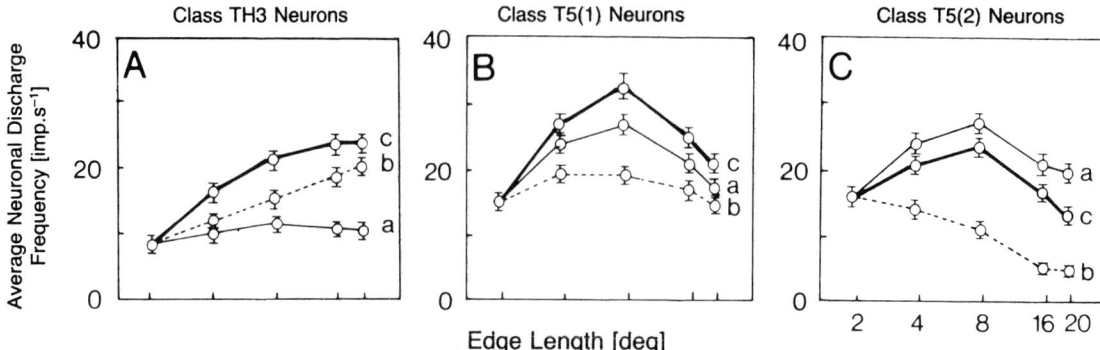

Fig. 48. Response characteristics of thalamo-pretectal (TH3, A) and tectal neurons (T-5.1, B, and T-5.2, C) in the toad *Bufo bufo*. Stimuli: Starting with a 2 × 2 mm square, edge length of the stimuli was increased parallel (a), and perpendicular (b) to the direction of movement or in both directions (c). From Roth (1987) after Ewert and von Wietersheim (1974) and Ewert (1984).

type, T5-3, the preference was S>V>H, i.e., the V stimulus always was preferred to the H stimulus. Thus, in *Rana temporaria*, the tectal cells do not demonstrate a clear "worm-antiworm" discrimination as is characteristic of *Bufo*.

In the experiments of Roth and Jordan (1982) on tectal response types in *Bufo bufo*, a different set of test stimuli was used, viz., a 8° × 8° square (S) and 2° × 8° rectangles, one oriented parallel (H) and one perpendicular (V) to the direction of movement were compared at three different velocities: 2, 6 and 20°/s. The authors found five predominant response types (Fig. 49). The largest proportion of neurons (34%) preferred S to H and H to V at all three velocities (S>H>V). This type is equivalent to the class T5-1 neurons noted by Ewert and co-workers. A second type (25%) preferred H to S and S to V (H>S>V), thus, corresponding to Ewert's T5-2 neurons. A third type (13%) showed the preference S>V>H, roughly corresponding to type T5.3 neurons as described by Schürg-Pfeiffer and Ewert (see above). While the stimulus preference of these neurons was invariant to changes in velocity, a substantial number of tectal neurons (38%) showed the phenomenon of "preference inversion". In a first type of these "inversion neurons", at low velocity S was better than V or H with the latter two of equal efficacy, while at high velocity, H and S elicited the same responses and both were better than V. In another type, H was better than S at low velocity but the reverse at high velocity, with V always being the least effective stimulus.

In intracellular recordings and staining of tectal neurons by Antal *et al.* (1986) and Matsumoto *et al.* (1986) in *Rana temporaria*, cobaltic lysine was used as a tracer. In these studies, intracellularly recorded neurons responded to (1) changes of diffuse light (light "ON/OFF"), (2) electrical stimulation of the optic nerve, and (3) visual stimuli, (a black 8° × 8° square [S], a 2° × 16° "wormlike" rectangle [H] and a 16° × 2° "antiwormlike" rectangle [V] moved at 25°/s).

Stimulation by changes in diffuse light revealed various response types: an E–E type, which showed an excitatory response at both light-ON and light-OFF; an I–I type, which showed inhibitory responses at light-ON and light-OFF. An EI–EI type and an IE–IE type responded with a combined excitation and inhibition to light ON/OFF. Stimulation with "configural" stimuli revealed (1) cells called T5-1 in which, in all of the above responses to light ON/OFF, S was preferred to both H and V, which were of equal efficiency (S>H = V), (2) cells which showed either EI–EI or I–I responses to light ON/OFF and in which S was preferred to H and H to V (S>H>V); these cells were called T5-2 cells (which, however, is not consistent with the definition of Ewert *et al.* of T5-2 cells), and (3) T5-3 like cells with a S>V>H preference and mostly E–E responses to light ON/OFF.

The cells called T5-1 were identified as large ganglionic neurons in layer 8 or 7, and pear-shaped neurons in layer 8 or the top of layer 6. No clear relationship was found between physiological and morphological properties. Only one cell, called T5-2 neuron, could be labeled and was located on top of layer 6 and resembled a pyramidal cell with an axon projecting

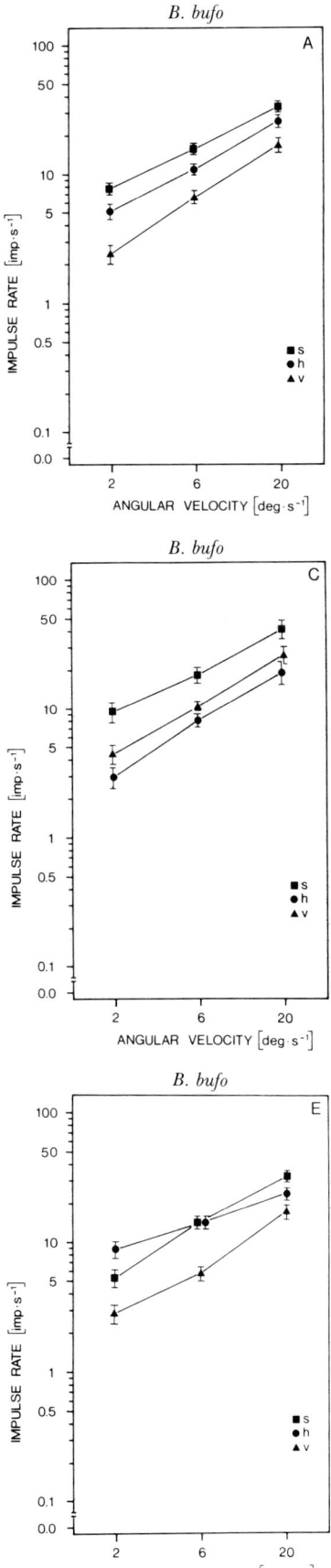
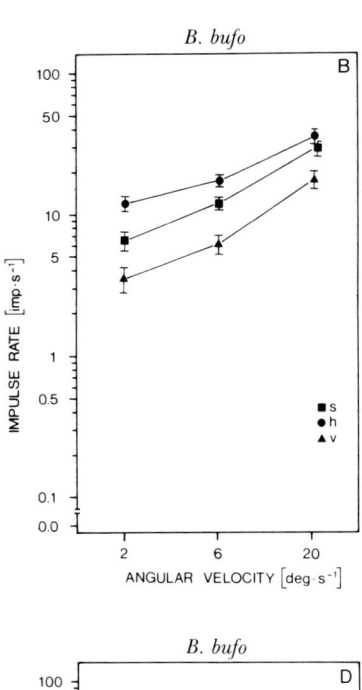
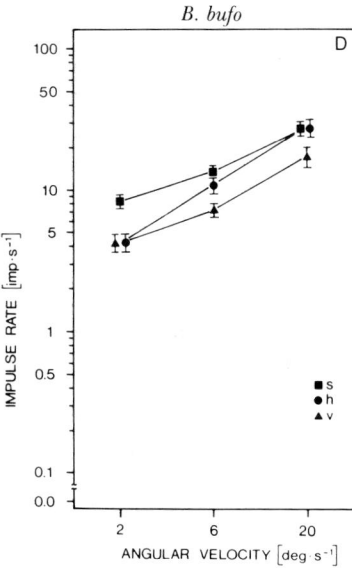

Fig. 49. Response characteristics of tectal neurons in the toad *Bufo bufo*. Neurons were stimulated with an 8 × 8° square(s) a 2 × 8° horizontal rectangle (h) and a 8 × 2° vertical rectangle (v) moved at 2°, 6° and 20°/s. A: Neurons with preference S>H>V. B: Neurons with preference H>S>V. C: Neurons with preference S>V>H. D: Neurons with change in preference regarding H. E Neurons with preference inversion HXS>V. From Roth and Jordan (1982).

down to the *medulla oblongata*. Class T5-3 neurons were identified as large ganglionic neurons in layer 8. Neurons responding predominantly by IPSPs could be identified as pear-shaped cells in layers 2, 4, or 6.

B. SALAMANDERS

A qualitative description of responses of tectal cells in salamanders was given by Grüsser-Cornehls and Himstedt (1973). They found neurons in the tectum of *Salamandra* that resembled the types T1–T5 described by Grüsser and Grüsser-Cornehls (see above). Extracellular single-cell recordings using the same experimental conditions as in the study by Roth and Jordan on *Bufo* tectal cells have been carried out on *Salamandra* (Himstedt and Roth 1980; Himstedt et al. 1987), *Hydromantes italicus* (Roth 1982) and *Hydromantes italicus* and *Bolitoglossa subpalmata* (Wiggers 1991; Wiggers et al. 1995) (Fig. 50). In all of these studies, the most common tectal response type found was S>H>V. This type, corresponding to the anuran type T5-1, had a mean RF size of 42° in *Salamandra* and 31° in *Hydromantes*. Another common type was S>V>H, corresponding to the anuran type T5-3, with a mean RF size of 48° in *Salamandra* and of 22° in *Hydromantes*. A type H>S>V, corresponding to the anuran type T5-2, was found in *Salamandra*, with a mean RF size of 32°, but not in *Hydromantes*. Likewise, a type V>S>H, i.e., a type that preferred the "antiworm" at all velocities, with a mean RF size of 34°, was found in *Salamandra* but not in any other amphibian. In addition, several types of "preference inversion neurons" were found, such as S>VXH (RF 34° in *Salamandra*, 21° in *Hydromantes*), and HXS>V (RF 27° in *Salamandra*). Finally, a number of "high velocity neurons" were found, with the characteristic S>V>H or S>H>V, and RF sizes of 29° in *Salamandra* and 19° in *Hydromantes*. In the latter species as well as in *Bolitoglossa*, a limited number of "low velocity" neurons was found. These responded better to low than to high velocity, with RF sizes between 8° and 30°.

Finkenstädt and Ewert (1983a,b) studied tectal neurons of *Salamandra* under the same conditions as in the studies on *Bufo* by Ewert and von Wietersheim (1974; see above), but varying stimulus size and configuration rather than velocity. They found tectal cells corresponding to T5-1 and T5-3 type of anurans, but — somewhat ironically — no T5-2 or H>S>V type (i.e., no "worm detectors"). The reason for this failure remains unclear, because Himstedt and Roth clearly showed the presence of T5-2 in *Salamandra*. In addition, neurons were described which corresponded to the frog/toad tectal neuron classes T2 (with RF sizes of about 100° and preferences of S>V>H), T4 neurons (with RF sizes of about 180° and responding to very large stimuli), and T3 "approach neurons".

Recently, Roth *et al.* conducted intracellular recording and labeling experiments from tectal cells under *in vivo* and *in vitro* (whole brain) conditions in the salamanders *Hydromantes genei* and *Plethodon jordani*. Latencies and response types of tectal cells were determined by means of electrical stimulation of the contralateral optic nerve. There was a clear-cut correlation between the neurophysiological data and the morphology of the recorded neurons, as revealed by Biocytin injection. Projection neurons (T1, T2, T3) generally had short latencies between 2 and 12 ms (Fig. 51), while all interneurons (T4) labelled had "long" latencies between 14 and 80 ms (average 35 ms). The first responses of projection neurons were excitatory, followed either by a second excitatory or an inhibitory response (around 36 ms). No differences in response latencies or response types could be detected among T1, T2 or T3. In the T4 neurons, the first responses were mostly inhibitory or excitatory-inhibitory, with the excitatory responses disappearing quickly, while the inhibition deepened.

In most cases of simultaneous staining of a projection neuron and one or several interneurons (due to coupling via gap junctions; see above), a combination of short excitatory and long inhibitory responses around 36 ms was observed. This indicates that the electrically coupled tectal neurons act as a functional unit, in which the short excitatory response is produced by the projection neurons, and the late inhibitory response by one or several interneurons.

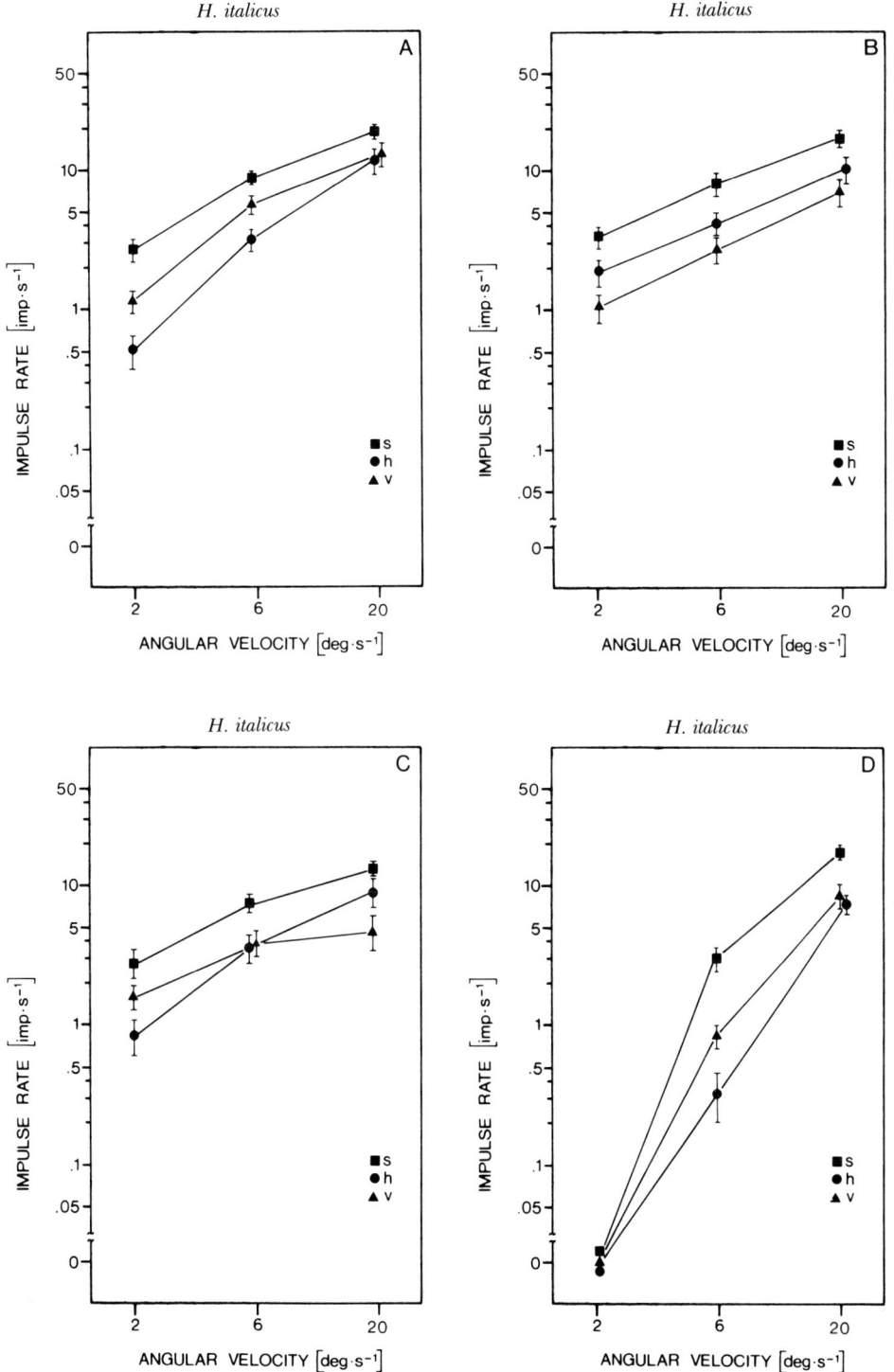

Fig. 50. Response types in the tectum of the salamander *Hydromantes italicus*. Stimulation as in Figure 49. A: Neurons with stimulus preference S>V>H. B: Neurons with preference S>H>V. C: Neurons with preference inversion S>VXH. D: "High-Velocity" neurons. From Roth (1982).

4. Disparity-Sensitive Tectal Neurons

As mentioned above, in amphibians binocular depth perception is based primarily on computation of the disparity of the two retinal images of an object as represented in the two tectal hemispheres. The horizontal disparity between the two monocular receptive fields of a binocular neuron is defined by the angle subtended by the two lines connecting the geometrical centre of each receptive field to the nodal point of the corresponding eye (Finch and Collett 1983).

In *Hydromantes italicus*, 72% of the binocular neurons exhibited a horizontal disparity of their monocular fields. In *Batrachoseps attenuatus*, the corresponding value was 65%. At a stimulus distance of 15 cm, all disparities are crossed, i.e., in the left tectal hemisphere, the centre of the ipsilateral receptive field of a given neuron is always located to the right of the centre of the contralateral receptive field, with the opposite being the case in the right tectum.

In 28 binocular neurons of *Hydromantes italicus*, the average horizontal disparity of the monocular fields was 10° (range: 4.7–21.8°). Neurons with larger fields tended to have larger disparities. In *Batrachoseps attenuatus*, 23 binocular neurons were tested. The horizontal disparity of their monocular fields was 5° on average (range: 2.8–12.4°). In three binocular neurons, the two monocular fields were displaced vertically.

On the basis of the distance between the eyes, the amount and direction of the displacement of the two monocular fields can be used to calculate the horopter, i.e., the distance at which the geometrical centres of both RFs coincide and the disparity is zero. The average distance between the two eyes is 7.5 mm in *Hydromantes italicus* and 2.5 mm in *Batrachoseps attenuatus*.

In *Hydromantes italicus*, the horopters of recorded binocular neurons are distributed over an area between 19.5 mm and 90 mm. Most of them are concentrated in an area between 20 and 50 mm, corresponding to the snapping zone as determined by the reach of the tongue. *Batrachoseps attenuatus* has a shorter tongue than does *Hydromantes* and accordingly the distances in which the geometrical centres of its receptive fields are superimposed are shorter. The average and maximal distance of tongue projection ranges from 15 mm to 45 mm in *Hydromantes*, and from 5 mm to 10 mm in *Batrachoseps*. The horopters are distributed over an area between 11.5 mm and 50 mm and concentrated in an area between 10 and 30 mm again corresponding nicely with the snapping zone. Studies on *Plethodon jordani* by Manteuffel *et al.* (1989) likewise showed a correlation between tongue length and horopter distance.

D. Neurons of the *Nucleus Isthmi*

Topic representation of the visual field, receptive field sizes and response properties of neurons of the *nucleus isthmi* were studied intensely in *Hydromantes italicus* by Wiggers and Roth (1991). The receptive fields of isthmic neurons are centred in the frontal 100° field. The visual hemifield covered by neurons of each nucleus extends horizontally from 50° contralateral to 30° ipsilateral to the nucleus, and vertically from 36° below to 50° above the horizon. Thus, the projection fields of the *nucleus isthmi* overlap by 60°. About two-thirds of recorded neurons had their RFs within the upper part of the visual field, above eye level.

The RFs of isthmic neurons had an average diameter of 17.8°, the diameter of most isthmic RFs ranged between 8 and 30° (maximum 54°). These frontally oriented RFs are about equal in size to the RFs of corresponding tectal neurons, which range between 10° and 48°. The nucleus isthmi contains a representation of the visual field, but its exact topography remains to be studied.

E. Response Properties of Isthmic Neurons

Stimulus size preferences: A majority of neurons preferred stimulus sizes between 4.6° and 9.1° edge length (EL); fewer neurons preferred the largest size tested (18°). One neuron responded best to stimulus sizes between 2.3° and 4.8° EL. This is similar to stimulus size preferences found in tectal neurons.

Velocity: In isthmic neurons, the highest impulse rates, up to 25 impulses/s, were recorded at velocities between 20 and 36°/s. At these velocities, the responses to different stimulus edge lengths were most distinct. With decreasing stimulus velocity, small stimuli tended to become more attractive, but did not reach the attractiveness of large ones. The same effect was observed in tectal neurons.

Binocularity: Half of isthmic neurons responded to stimuli from both eyes, whereas 35% responded only to contralateral stimulation. The remaining 14% responded weakly to ipsilateral stimulation, and strongly to contralateral stimulation.

In summary, in *Hydromantes italicus* (and in similar experiments on anurans) electrophysiological recordings from *nucleus-isthmi* neurons revealed response properties that are similar to those found in tectal neurons. This corroborates the view of the *nucleus isthmi* as a relay station for the tectum (see below).

In intracellular studies, isthmic neurons exhibited a "multiple-spike response" after electrical stimulation of the contralateral optic nerve (Fig. 52). Two to three spikes were generated at latencies of 10, 60 and 120 ms, which were separated by inhibition or inactivity. Tegmental neurons situated at the ventral border of the *nucleus isthmi* likewise showed single responses latencies of 10 ms on average, but no multiple spikes.

F. Thalamic and Pretectal Neurons

1. Anurans

Pretectal/caudal thalamic neurons were studied qualitatively by Ewert (1971) in the toad *Bufo americanus* and quantitatively by Ewert and von Wietersheim (1974) in *Bufo bufo*. The classes found were called TH 1–10. Of these, TH 3–TH 10 neurons could be activated visually. TH 3 neurons had RF sizes between 30° and 46° and responded best to large moving stimuli (Fig. 48). TH 4 neurons were movement-specific and had RFs that occupied either the whole contralateral or the entire visual field. Some of them showed little adaptation, whereas others quickly adapted and exhibited "newness" and "sameness" characteristics (Lettvin *et al.* 1959). TH 6 cells were mostly sensitive to object movement along the z-axis approaching the eye of the subject. TH 7 neurons are "luminance detectors" or "darkness detectors", because they responded to changes in general illumination with very long-lasting discharges. TH 10 neurons were remarkable in that they responded to large stationary stimuli within the 30–90° RFs. Some of them also can be activated by cutaneous stimuli.

Many neurons recorded from the praetectum in anurans are directionally selective and are particularly sensitive to *horizontal* optokinetic patterns moving at velocities of 5–10°/s (Katte and Hoffmann 1980). In contrast, neurons recorded in the region of the nucleus of the basal optic root (or neuropil; nBOT/nBON) respond more selectively to slowly moving *vertical* patterns, although horizontally sensitive neurons also have been reported (Katte and Hoffmann 1980; Gruberg and Grasse 1984). This provides evidence that pretectal and accessory neurons are involved in oculomotor and optokinetic behaviours (see above).

2. Salamanders

Manteuffel (1982, 1984b) and Himstedt *et al.* (1987) studied the response characteristics of thalamic and pretectal visual neurons in *Salamandra*. The latter authors used three stimuli, a 15° × 15° square, a 3° × 15° horizontal rectangle and a 15° × 3° vertical rectangle and moved them at three different velocities, 1, 10°, and 100°/s. In addition, a 60° × 60° random-dot pattern (so-called Julesz pattern) also was used and moved at the same velocities as the other stimuli.

Neurons in the rostral thalamus, at the level of the *neuropil Bellonci* and the *corpus geniculatum thalamicum* (the "*nucleus Bellonci*" of Herrick) mostly had RF size between 36° and 50°. These RF-sizes are considerably larger than those of tectal cells.

At low stimulus velocity most neurons showed either preference type H>V=S or V>S=H. There were no responses to the Julesz-pattern. At intermediate velocity, most

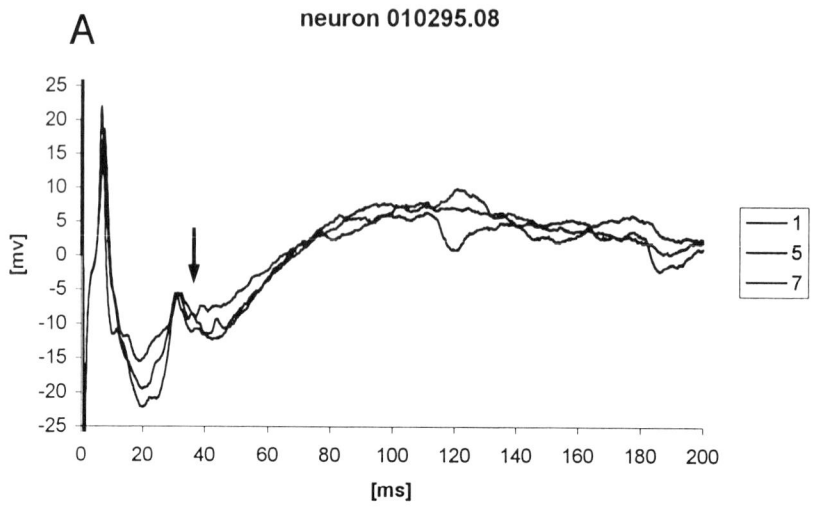
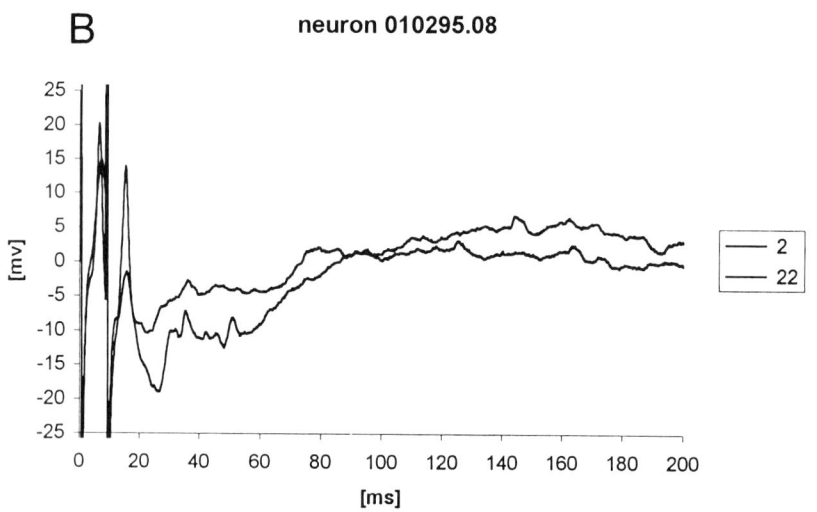
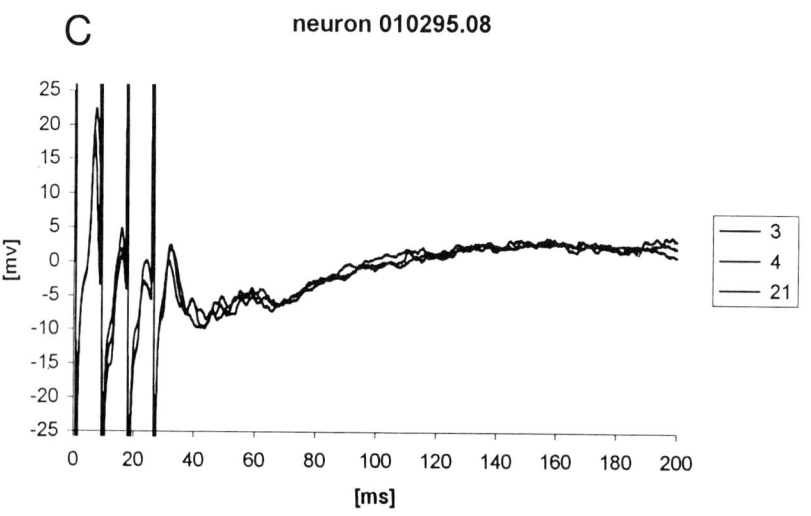

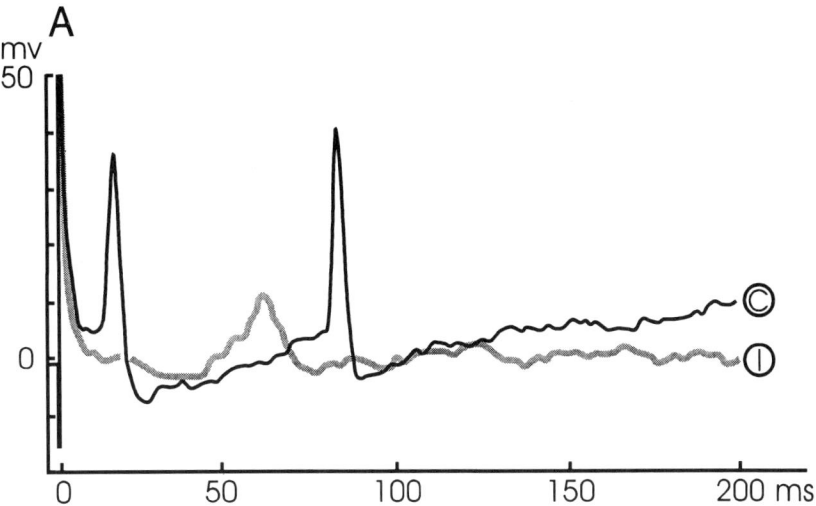

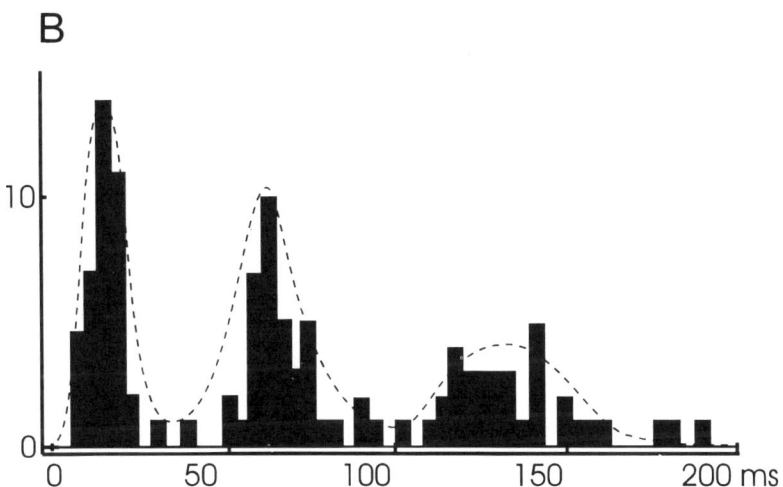

Fig. 52. **A:** Intracellular responses of an isthmic neuron after ipsi and contralateral electrical stimulation of the optic nerve. i = ipsilateral; c = contralateral. **B:** Post-stimulus time histogram of the intracellular responses of isthmic neurons after contralateral electrical stimulation of the optic nerve. The responses of 20 isthmic neurons are accumulated to visualize the three peaks of activity (dotted line).

neurons showed either type H>S=V or S>H>V, and some responded best to the Julesz-pattern. At high velocity (100°/s) most neurons responded to H only or were of type S>H>V. One neuron responded best to the Julesz-pattern.

In the caudal dorsal thalamus, rostral to the pretectal neuropil and to the *commissura posterior*, most neurons had RF sizes between 36° and 50°; in some neurons the RF size was

◄

Fig. 51. Intracellular recording from a tectal projection neuron in *Plethodon jordani* at electrical stimulation of the optic nerve. Two to three traces are superimposed. A: Excitatory response with a latency of 7 ms at single electric shock. The action potential is followed by an early inhibition starting at about 16 ms and a late inhibition (arrow) starting at about 34 ms. B: Responses at double shocks at 9 ms interval. The neuron follows exactly repetitive stimulation. Amplitudes of excitatory and inhibitory responses are diminished at later stimulation trial. C: Responses at four shocks. Again, the neurons follow exactly repetitive stimulation, but the response amplitudes are largely diminished. Stimulus artifacts are indicated by black vertical lines.

beyond 50°. Therefore, the RF sizes are comparable to those measured in the rostral thalamus. There was generally no response to the Julesz-pattern.

At a low velocity of 1°/s, the majority of neurons showed a preference H>V>S, and some one of S>V=H. At 10°/s the cells showed either a preference of H>V>S or S>H>V. At 100°/s, cells showed preference types H>S>V or S>H>V.

Recordings from pretectal cells were carried out in *S. salamandra* by Manteuffel and co-workers (Manteuffel 1979, 1982, 1987; Sperl and Manteuffel 1987; see also Manteuffel 1989). The neurons were activated by a pattern of vertical stripes, moved either linearly in one direction at different velocities or sinusoidally with an amplitude of 30° and at frequencies of 0.01–1 Hz. The recorded neurons were marked iontophoretically by means of alcian blue, some neurons were labelled by intracellular injection of HRP.

Most of the recorded and labelled neurons were situated around the commissura at a depth of 400 μm; a few cells were situated more deeply within the area of the basal optic nucleus (BON) and the area of the nucleus of the oculomotor nerve.

About two-thirds of the cells showed a sensitivity for temporonasal movement of the stimulus. They mostly possessed large RFs with a diameter of 82° and 34° in the horizontal and vertical directions, respectively. The RF centre always was situated within the contralateral visual hemifield. The majority of these cells preferred velocities lower than 10°/s, mostly between 1 and 5°/s. Nearly half of the pretectal cells were binocular and did not respond to ipsilateral stimulation.

The cells located in the BON were sensitive to a temporonasal direction of stimulus movement, as are most pretectal cells, except one cell which was sensitive for the opposite direction. All neurons recorded from the BON or the nucleus of the oculomotor nerve were strictly monocular.

In *Salamandra*, cells that are sensitive to horizontal optokinetic stimuli form a band that extends from a position rostral to the *commissura posterior* ventrocaudally to the region of the BON. This band involves the *nucleus praetectalis*, the *nucleus Darkschewitsch* and the nucleus of the *fasciculus longitudinalis medialis*. However, at deeper and more caudal sites close to the BON, cells that respond preferentially to vertical movement tend to be more numerous.

G. Lesion Experiments in Visual Centres

Removal of the tectum results in the loss of visually guided prey-catching and threat-avoidance in a number of anurans (Ewert 1970; Ingle 1973; Grobstein and Comer 1983). Experiments with lesions in the praetectum were carried out by Ingle (1980) in *Rana pipiens*. Frogs with bilateral removal of the caudal thalamus/praetectum were unable to avoid collision with a vertically striped barrier that partly surrounded the animal. At the same time, visual guidance of feeding and avoidance of threat was normal. In addition, all lesioned animals were unable to turn toward an aperture within black or white enclosures in order to escape noxious stimulation. However, the animals were able to avoid a small barrier placed before them, which indicates that some residual detection of stationary edges remains. The author concluded that pretectal visual neurons play a key role in the detection of stationary objects in contrast to the motion detection functions of the optic tectum.

Ewert (1968) using *Bufo bufo* and Finkenstädt and Ewert (1983b) using *Salamandra*, carried out brain lesion experiments using anodal DC current, radio frequency current, kainic acid microinjections or microknife cuts. Ablation of the optic tectum abolished any visually guided prey-catching and predator-avoidance behaviour. After unilateral lesions in the caudodorsal thalamus (praetectum), the animals responded with prey-capture toward moving objects of any size and configuration when the object was shown within the visual field contralateral to the lesioned brain region. However, their "worm/antiworm" discrimination was impaired. Predator-avoidance behaviour could not be elicited, i.e., the animals responded about equally to elongated stimuli oriented parallel and perpendicularly to the direction of movement. Frogs and salamanders showed a thalamo-pretectal disinhibition phenomenon ("TP-phenomenon") originally described by Ingle (1973). The size of the area in the visual field where disinhibition occurred corresponded with the size of the lesions. The authors assumed that disinhibition was due to a destruction of the somata of pretectal cells and not of fibers originating in other brain regions and passing through the praetectum. After lesions in the rostrodorsal thalamus the same disinhibition phenomenon occurred, but affected only the binocular visual field. Lesions in the ventral thalamus did not result in any disinhibition with respect to feeding. However, the animals exhibited abnormal body postures such as extreme body bending, and spontaneous motor activity increased.

In *Salamandra*, lesions in the medial pallium of the telencephalon led to strong inhibition of feeding, often combined with "aggressive" behaviour (Finkenstädt and Ewert 1983b). Escape behaviour could not be elicited. Spontaneous motor activity was observed. After lesions in the lateral pallium the animals more readily attempted to escape from small objects not normally eliciting escape reactions. The authors concluded that the pretectal and the rostrodorsal thalamic region, as well as the medial pallium, are involved in the constitution of prey-enemy distinction. After destruction of these regions all kinds of objects are regarded by the animal as prey.

In repeating this kind of experiment with frogs or salamanders using kainic acid injection or suction lesions, Luksch and Roth (in press) were unable to reproduce the behavioural disinhibition effect. Furthermore, it proved to be impossible to restrict the excitotoxic effect of kainic acid to the praetectum or even to parts of it, because of the high diffusibility of this substance. Parts of the tectum were always affected by the injection of kainic acid. Thus, the significance of these lesion experiments and of the PT-phenomenon remains unclear.

Various authors made lesions in the *nucleus isthmi* as a means of elucidating its function (Glasser and Ingle 1978; Grobstein *et al.* 1978; Caine and Gruberg 1985; Collett *et al.* 1987; Gruberg *et al.* 1991). It was demonstrated in *Rana pipiens* and in toads that lesions to the *nucleus isthmi* abolish responses of neurons in the tectum on the same side as the visual stimuli. Gruberg *et al.* (1991) demonstrated that unilateral lesions of the *nucleus isthmi* results in a scotoma to prey and threat stimuli in the contralateral monocular field; the size of the scotoma and of the amount of ablated *nucleus isthmi* are correlated. There is a regression of the scotoma in the nasal part of the visual field which then stabilizes. Whereas in some animals the remaining scotoma remains relatively stable in size for up to two years, other animals completely recover. Collett *et al.* (1987) likewise carried out bilateral lesions of the *nucleus isthmi*, but in contrast to the results of Gruberg *et al.* (1991) and those of the other authors cited above, they were unable to find any impairment of vision after lesioning the *nucleus isthmi* in *Rana pipiens*. The ability of partial or complete recovery of the *nucleus isthmi* might explain these discrepancies in results.

In salamanders, size and position of the nucleus isthmi makes precise lesions almost impossible without damaging the tegmentum, cerebellum and above all the adjacent tectobulbar tracts.

VI. EXTRAOPTIC LIGHT PERCEPTION

Extraoptic light perception in amphibians as well as in all other groups of vertebrates occurs through the activity of the pineal complex (Dodt 1966; Dodt *et al.* 1971). In anuran amphibians, this complex consists of two components, namely the intracranial pineal organ proper (the *epiphysis*) and the extracranial frontal organ *(Stirnorgan)*. Salamanders and caecilians are generally assumed to only possess a pineal organ proper and lack a frontal organ (Korf 1976; Hartwig and Korf 1978). However, recently an anterior body of the pineal organ has been described, which appears to be homologous to the frontal organ of anurans (Takahama 1993).

The pineal organ proper of anurans and urodeles extends from the region between the habenular commissure and the posterior commissure of the roof of the diencephalon. In anurans, the pineal organ proper is basically a hollow, sac-like or tube-like structure (Fig. 53), which in some species is completely subdivided by numerous septa extending from the wall into the lumen. In salamanders, it is either again a hollow and sac-like structure or a more or less solid organ with few remnants or no trace of a lumen (e.g., in *Desmognathus*). The frontal

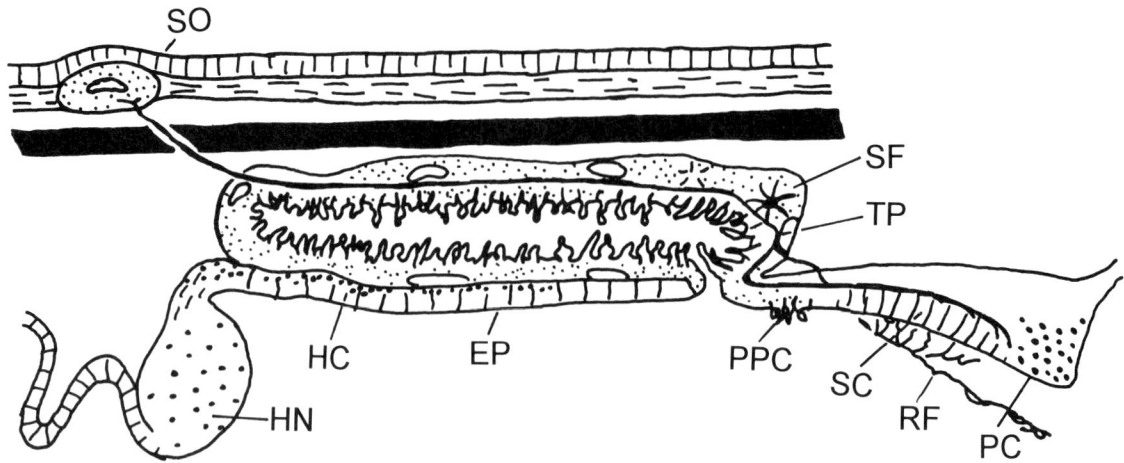

Fig. 53. Diagrammatic representation of a sagittal section through the organs of the diencephalic roof in anurans. EP = *epiphysis cerebri*; HC = habenular commissure; HN = habenular nucleus; PC = posterior commissure; PPC = posterior pineal cluster; RF = Reissner's fibre; SC = subcommissural organ; SF = secretory follicle; SO = stirnorgan; TP = pineal tract. Modified from Vollrath (1981).

organ of the anuran pineal complex occupies an extracranial position and lies between the eyes beneath the surface. In some species, its position is indicated by specializations such as a small, pale area *(Rana temporaria)* or a small skin protrusion *(Xenopus laevis, Rana esculenta)*, while in others *(Bufo bufo)* no external indication of the location of the frontal organ exists.

Both the frontal organ and the pineal organ proper exhibit the typical features of a photosensory organ. They contain photoreceptors which — like typical vertebrate photoreceptors — consist of outer and inner segments, perikarya, basal processes and terminal swellings, which are the sites of synaptic contacts. In both *Rana temporaria* and *Ambystoma tigrinum*, the majority of the pineal outer segments are arranged parallel to the roof of the skull; it is assumed that this orientation plays a role in the perception of polarized light (Hartwig and Korf 1978). The outer segments of both the frontal organ and pineal organ proper contain photopigments; those of the frontal organ a photopigment with an absorption maximum between 560 and 580 nm, and those of the pineal organ proper a rhodopsin-like pigment with an absorption maximum at 502 nm (Hartwig and Baumann 1974; Oksche and Hartwig 1979). There are photoreceptor cells whose outer segments are reduced, but it is unclear whether these represent rudimentary photoreceptors or are photoreceptors whose outer segments are in the process of renewal. Finally, there are photoreceptors, whose discs of the outer segments look atypical. Electron microscopic studies demonstrated that three types of synapses are present: ribbon synapses, conventional synapses and synapses characterized by subsurface cisternes. Ribbon synapses (which are typical of vertebrate photoreceptors; cf. Section III E 1 of this chapter) are the most numerous type of synapse and are not restricted to the basal processes of the pineal photoreceptors, but occur also on the cell body (Korf 1976).

In addition to photoreceptors (and supporting cells), both the frontal organ and the pineal organ proper contain nerve cells of various types, which comprise large and small multipolar, pseudo-unipolar and a few bipolar neurons (Wake *et al.* 1974). These neurons, however, lack a precise arrangement.

Two kinds of nervous connections exist in the pineal complex:

1. The frontal nerve is a fiber bundle originating from pseudounipolar cells lying in the wall of the frontal organ, which runs through the dorsal lymph sac, before it penetrates the skull and reaches the pineal organ proper, where it merges with the pineal tract (see below). It comprises myelinated and non-myelinated fibers and contains a number of efferent fibers originating in the medial amygdala, the preoptic area, the praetectum and the lateral mesencephalic grey matter.

2. The pineal tract originates mostly from pseudounipolar neurons inside the pineal organ proper, while a few fibers stem from the frontal nerve. It, too, consists of myelinated and non-myelinated fibers. This tract leaves the pineal organ proper near its caudal border and gives off fibers to the region of the subcommissural organ; then, it reaches the posterior commissure and the pretectum.

The pineal complex of amphibians is directly photosensitive (Dodt and Heerd 1962; Dodt and Jacobson 1963; Hamasaki 1969, 1970; Morita 1969; Morita and Dodt 1975; Vollrath 1981). In the presence of light, neurons in the frontal organ of frogs show spontaneous electrical activity (actions potentials as well as graded potentials). There are chromatic as well as achromatic responses. With respect to the former, the activity of the frontal organ is inhibited by light of short wavelengths (maximum inhibition at 355 nm) and excited by light of longer wavelength (maximum excitation at 515 nm). The achromatic response consists of inhibition independently of the wavelength applied. The pineal organ proper shows achromatic responses only.

Apparently, the pineal region is involved in phototaxis (Adler 1970, 1976) and other behavioural responses. *Xenopus* larvae initiate swimming when a shadow is cast upon them, a response that is mediated by pineal photoreceptors (Roberts 1978). When the pineal region is subjected to dimming light, an irregular discharge of biphasic impulses of 4.5–7.5 ms duration is generated in pineal sensory ganglion cells; these probably initiate swimming by neuronal pathways connecting to the hindbrain.

Compass orientation is also influenced by pineal photoreception. *Ambystoma tigrinum* trained to orient in a particular compass direction under the sun failed to orient in the trained direction, if they were blind and simultaneously had the brain covered with opaque plastic or were pinealectomized; in contrast, salamanders with either the eyes or the pineal intact continue to orient in the trained direction (Taylor and Adler 1978). Furthermore, the pineal complex, at least the pineal organ proper, seems to be involved in regulating colour change. After removal of the pineal organ proper, no lightening occurs when the animals are placed in darkness; lightening of animals in the dark does occur when only the frontal organ is destroyed.

VII. NEURONAL MECHANISMS UNDERLYING VISUALLY GUIDED BEHAVIOUR: AN OVERVIEW

Despite intensive investigation of the visual system of amphibians for more than four decades, insight into the neuronal mechanisms underlying behaviour is still limited. Some processes like orienting behaviour, depth perception or optomotor behaviour are relatively well understood, as described above, and models have been developed in analytical, mathematical and neural network terms. Computer simulations have been performed that reasonably fit the empirical neurobiological data and the observed behaviour (Manteuffel 1984a,b, 1989; Manteuffel and Roth 1993; House 1989; Wiggers *et al.* 1995; Eurich *et al.* 1995). Two major topics which still are less well understood, despite a wealth of data, are object-recognition and the transformation of central visual processes into motor action. The current state of ideas and perspectives with respect to these two topics are discussed next.

For many years, in amphibians (as well as in other animals) object-recognition was thought to be achieved by the action of "detector cells", i.e., neurons that respond selectively to particular objects or classes of objects, "filtering out" all others. Such detector cells were thought to closely resemble in their response properties the prey preferences of the behaving animal. Barlow (1953) was the first to develop a detector concept in amphibians. He started from the assumption that in frogs "there is no indication of any form discrimination" and speculated that "ON-OFF" units (corresponding to class-3 RGC, see above) seem to possess the whole of the discriminatory mechanism needed to account for this rather simple behaviour. The receptive field of an "ON-OFF" unit would be filled nicely by the image of a fly at 2 inches distance and it is difficult to avoid the conclusion that the "ON-OFF" units are matched to this stimulus and act as "fly detector" (Barlow 1953, p. 86). In addition, "OFF" units (class-4 RGC) seemed most suited for localization of prey. The conclusion was that "the retina is acting as a filter rejecting unwanted information and passing useful information" (Barlow 1953, p. 87). Lettvin, Maturana, McCulloch and Pitts (see section D 2), on the basis of their study of *Rana pipiens*, extended this concept of "early" (i.e., retinal) object-recognition. They arrived at a description of five "natural" classes of RGC that transform the pointlike information of photoreceptors about the distribution of brightness into meaningful information about size, shape, illumination, contrast, movement and position. Most importantly, RGC identify properties of objects that are *constant* in a changing environment. According to these authors, a great deal of object-recognition and prey/enemy-discrimination already takes place in the retina (e.g., prey are identified by class-1 and class-2 RGC and enemies are identified by class-3 RGC). This implies that the activity of retinal ganglion cells already resembles behaviour, and that these behavioural responses are determined primarily by one class, or maximally two classes, of RGC.

However, a few years later, Grüsser and Grüsser-Cornehls (1968) and Ewert and Hock (1972) showed that retinal ganglion cells are not sufficiently specific in their response patterns to reliably identify behaviourally relevant objects. As Grüsser and Grüsser-Cornehls stressed (1976), a complete representation of prey or predator is constituted only at later stages in the visual centres. However, Grüsser and Grüsser-Cornehls (1968, 1976) conducted only preliminary investigations on central visual neurons describing qualitatively the seven tectal response types (T1–T7) mentioned in section D3. They developed a "sequential" model for neural guidance of feeding behaviour, in which tectal neurons belonging to certain response classes elicit different kinds of behaviour, and these cells, together, guide the complete feeding sequence. For example, type T 5 and T 7 neurons which respond specifically to small moving objects in the monocular visual field signal "prey!"; then T 2 neurons which are selectively activated by objects moving in a temporo-nasal direction in the visual hemifield, will trigger an orienting turn toward the prey object, until it comes to lie inside the frontal binocular field, which is signalled by T 1 binocularity neurons. Next, T 3 neurons which respond selectively to objects moving toward the animal along the z-axis come into play by releasing approach behaviour, until the prey is within snapping distance.

The disadvantage of this model is that it repeats at the tectal level the previous concept criticized above by attributing behavioural acts to the activity of single tectal cells, and by assuming that these cells control behaviour in a more or less direct manner. Also, nothing detailed is said about object-discrimination, depth perception, the constitution of size-constancy and the manner in which sensory information is transformed into motor action.

Ewert was the first to develop more detailed ideas about prey and enemy recognition in amphibians, particularly in the toad *Bufo bufo* (Ewert 1974, 1984, 1989). According to him, in the nervous system, prey objects are identified by a three-stage system of filters that respond preferentially to a "wormlike" prey configuration; the more "wormlike" a stimulus, the higher the probability that a toad will orient and eventually snap at it. The first stage of the system is represented by two kinds of retinal ganglion cells (class 2 and class 3 cells in the latest version of this model; Ewert 1989), one type of thalamic/pretectal neuron (TH5 cells) and several types of tectal neurons (the T5.1, T5.2 and T5.3 cells in the latest version) (Fig. 54). Of these, T5.2 cells play a particular role in the model, because their response properties are said to resemble most closely the observed behavioural preferences for prey. Therefore, T5.2 cells are assumed to act as prey feature detectors and at the same time as command neurons or more precisely as a command elements (Ewert 1989). In an early version, this reads as follows: "The response of these neurons constitutes the key stimulus 'prey'. That is, they can presumably be considered the trigger system for the prey-catching response" (Ewert 1974). How is the response of this type of tectal cell constituted? There is one major problem that Ewert and his co-workers themselves had identified: as mentioned in section D5, toad retinal ganglion cells do not "prefer" wormlike objects, but respond best to edges perpendicular to the direction of movement ("antiwormlike" stimuli) that have a height identical to the diameter of the excitatory receptive field. Therefore, they are not well suited to "drive" T5-2 cells directly. Rather, they are assumed to project in parallel on to tectal T5-1 (and probably tectal T5-3) and pretectal TH5 cells, i.e., cell types which, like the RGC, "prefer" "antiwormlike" to "wormlike" objects. These two types of cell, in turn, project differently to T5-2

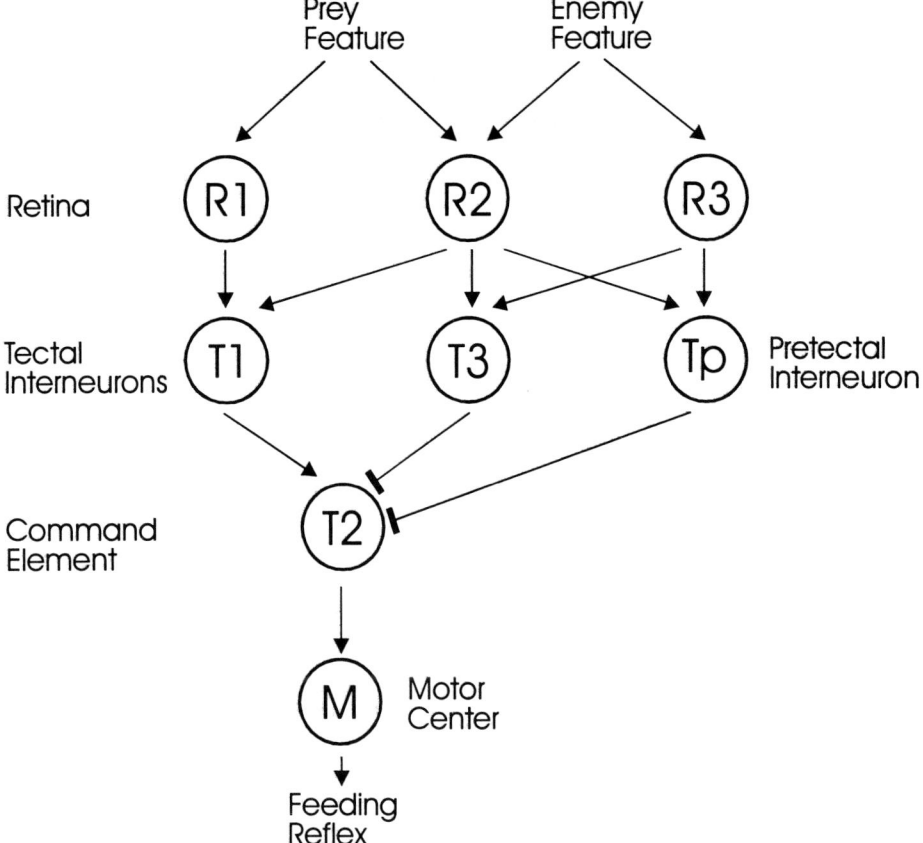

Fig. 54. Diagram of the network for visual control of feeding behaviour in the toad *Bufo bufo* as proposed by Ewert. Prey stimuli activate R1 and R2 retinal ganglion cells, while enemy stimuli activate R2 and R3. R1 in turn activates T1 (= T5.1) tectal neurons, R2 T1 (= T5.1), T3 (= T5.3) and Tp pretectal interneurons, and R3 T3 and Tp neurons. T1 neurons *activate* T2 (T5.2), while T3 and Tp *inhibit* T2 neurons. T2 neurons act as command elements influencing motor centres related to the feeding "reflex". After Ewert (1984).

cells, to T5-1 cells in an excitatory and to TH5 (and tectal T5-3) in an inhibitory way. According to Ewert's (1974) concept of a "window discriminator" when a "wormlike" stimulus appears in the monocular visual field, it will moderately activate RGC class 2 (R1) and 3 (R2), which — for reasons not explained in the model — activate relatively strongly tectal T5-1 cells, while pretectal TH5 cells will not respond, or do so only weakly. T5-1 cells will then strongly activate T5-2 cells which are not inhibited by TH5 cells (because of their inactivity). Now, if an "antiwormlike" stimulus appears, it will strongly activate RGC class 3 and class 4 (R3) as well as rectal T3 and pretectal TH5 cells and to a moderate degree tectal T5-1 cells. Thus, T5-2 cells are only moderately excited by T5-1 cells, but strongly inhibited by TH5 cells. Strangely enough, the result is not a strong hyperpolarization (= inhibition) of T5-2 cells, but very low activity. Finally, when an object appears that has both "wormlike" and "antiwormlike" properties (as is the case with a square), it again strongly excites RGC class 2, class 3 and class 4, but also T5-1 and TH5 (and T5-3) cells. At the level of T5-2 cells, a moderate activity results from the interaction of excitation and inhibition (and not zero activity, as one would expect).

In the early version of the model (Ewert 1974), T5-2 neurons were seen as a trigger system for prey recognition. In later versions (Ewert 1984, 1989), the author adopted the Kupfermann and Weiss "command neuron" concept (Kupfermann and Weiss 1978). However, for Ewert, T5-2 cells are not command neurons in the strict sense, because the simultaneous and consecutive activity of a number of other types of tectal neurons are required for the occurrence of orienting, approaching and snapping; rather, T5.2 neurons are necessary members of this ensemble and are accordingly called *command elements* which

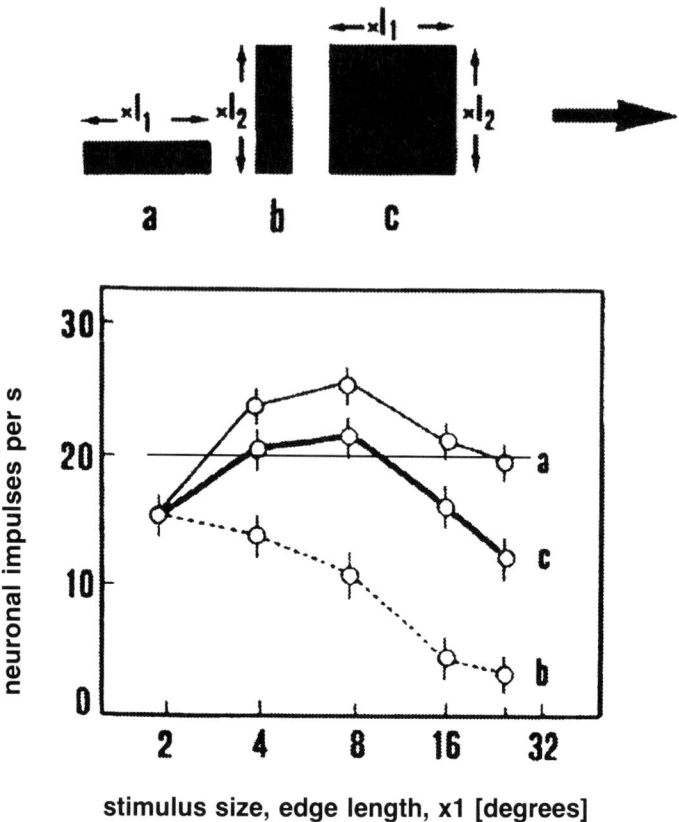

Fig. 55. Response properties of a T5.2 neuron. A horizontal line through the three response curves is drawn to illustrate the equivocality of responses at 20 impulses/s. For further explanation see Figure 48 and text. Modified from Ewert (1984).

"tell" the nervous system "this object is a prey!". Inside the tectum, T5-2 neurons integrate the information about the nature of the prey and act as the only tectal output cell in the context of prey recognition.

Although Ewert and his co-workers concentrated their work on the toad *Bufo bufo*, it was always understood that their findings could be applied to neural guidance of feeding in amphibians generally. However, there is a substantial amount of data showing that this concept, even in its most recent version, is applicable neither to the toad nor to any other frog or salamander. The major points of criticism are: (1) There is not a close correspondence between the behavioural preference of the toad and the activity of the T5.2 neuron: the toad prefers 2° × 16° or 2° × 32° dummies, while the neuron responds best to 2° × 8° dummies, and the behaviourally best stimuli occupy only the fifth and sixth "rank" in the T5.2 preference. Square stimuli are often as good as wormlike ones. (2) The supposed prey detector T5.2 neurons cannot reliably "distinguish" between wormlike and square or even "antiwormlike" stimuli, as can clearly be seen in Ewert's own data. For example, the diagram presented in Figure 55 shows that a response of 20 spikes per second is elicited by a 2° × 3° "wormlike" stimulus, a 4° × 4° square, a 10° × 10° square and a 2° × 16° "wormlike" stimulus — stimuli that in behavioural tests elicit very different responses. If the velocity were changed in addition to size and shape (let alone changes in contrast and movement pattern), the equivocality would increase further. (3) Cells of this class (like all visual cells) respond only to angular size, while the toad clearly shows size-constancy. (4) Many amphibians neither prefer wormlike prey nor do they possess T5.2-like neurons (cf. Roth 1987). (5) So far, neither the T5.2 neuron nor any other tectal neuron has been identified as a true command element. As Satou and Ewert (1985) themselves demonstrated (see below), most morphologically and physiologically identified tectal neurons send axons to the premotor regions, and not just the T5-2 cell, as initially believed.

The following is a list of suggestions and starting points for a more adequate model:

1. Earlier models did not take into account that in the amphibian visual system (as well as in those of other animals) "object" is not something directly given. Rather, the first step of object-recognition consists in the distinction between object and background and between object-movement and self-induced or passive movement of the predator. Presumably, this occurs through the interaction between praetectum and tectum. As mentioned above, the praetectum, together with the system of the basal optic nucleus/neuropil (BON), is involved in the recognition of large-structured visual patterns and in the guidance of compensatory optomotor movements. Thus, the praectectum signals what is "background", namely, everything that is large and elicits relative retinal shift only due to active or passive movement of the animal. The praetectum presumably exerts an indirect inhibitory influence on the tectum by activating inhibitory tectal interneurons (the morphological type T4 cells). During perception of an object in front of a structured background, the praetectum is activated by movement of a large-patterned area relative to the body and/or head of the salamander, and the activity of the whole tectum is suppressed by pretectal inhibition except for activation by self-moving objects like predator or prey.

Next, a "quick-and-dirty" distinction between prey, conspecific and predator takes place on the basis of the absolute size of these objects. Anything that moves independently from the background and the frog or salamander and is larger than a certain absolute size, may be an enemy, a conspecific or a neutral being; if it is smaller, then it is prey. Thus, determination of absolute size is necessary for distinction between prey and non-prey resulting from a computation of angular size and distance. As previously seen, in amphibians distance-estimation is most probably based on retinal disparities under binocular conditions and on lens accommodation under monocular conditions. At present, it is unclear how and where computation of absolute size is achieved in the amphibian visual system.

The next step of object-recognition is a more precise identification of the prey object. It is evident from *a priori* reasoning that the detector concept can be applied only in cases where the relevant stimulus has relatively simple and highly invariant properties, e.g., when everything that is relatively small or "wormlike" and moves is prey. If prey objects have properties that can change, a detector system breaks down. This happens in most cases of feeding, when the shape of a prey varies during three-dimensional movement, when prey objects are to be distinguished by finer details, or when prey is to be identified by a number of different properties rather than by just one.

As described in the first section, there are a number of properties by which an object can be identified as prey, and their relevance varies according to motivational level and environmental (e.g., light) conditions. The "coarsest" criteria for prey are size and movement: at highest motivational level, an animal will most probably feed on everything that moves and fits the range of possible prey. At lower motivation, more specific parameters become relevant (Heatwole and Heatwole 1968). Here, *movement pattern* seems to be most important. As mentioned above, most prey can be accurately recognized by the way they move. Movement pattern is a very appropriate cue for identification, because it is largely independent of variation in speed, shape and illumination level. At very low light intensities, where many frog and salamander species search for prey, movement pattern is the only reliable cue for prey identification. In contrast, shape and configuration seem to be relatively unimportant cues, particularly because they can change considerably in one and the same individual prey when it moves. However, shape and contour details are probably used under sufficient light conditions, when precise distinction between prey objects is required (e.g., in the context of learned distinction between tasteful and distasteful prey). Finally, contrast is even less important, provided the prey object can be distinguished at all from the background, because the visual system of most amphibians seems to be well suited for prey identification and localization at very low contrast because of the high light sensitivity of their photoreceptors. Colour is a cue for identification of objects only in the context of avoidance of noxious prey (warning colours) or of mating behaviour (e.g., of newts).

Thus, like the visual system of other vertebrates, that of amphibians works basically in an *opportunistic* fashion, i.e., it uses any cue that is useful to identify an object depending on the ambient conditions. This alone refutes any simple detector concept.

An important topic is the solution of the "paradox" that amphibians, on the one hand, often can be "fooled" by very simple prey dummies (giving rise to the view that amphibian prey capture is reflex-like), and on the other hand are capable of subtle visual discriminations. It is true that toads will snap many times at "wormlike" pieces of cardboard and do not seem to realize that the object is not prey. However, the solution proposed here for this paradox is simple. When confronted with unfamiliar objects, amphibians follow the simple rule: Take everything that moves and see what follows. This quickly will exclude from the diet any prey that is noxious or dangerous (e.g., by defending itself), distasteful or impossible to swallow or digest. Often, this implies one-trial learning (e.g., in the case of honeybees or wasps). A second stage of narrowing preferences for prey includes experience about the ease of catching certain prey, its abundance and nutritive value. All this requires an evaluation of the consequences of prey-catching activity. Prey dummies, as used in common behavioural tests, differ fundamentally from natural prey by the fact that they are not taken into the mouth (and swallowed) and do not become "evaluated". Therefore, amphibians are "naive" with respect to prey dummies and accordingly apply the first rule. If one lets amphibians experience distasteful or unpalatable prey dummies, they immediately stop snapping at the dummies (at least for some time).

2. Knowledge about response properties and functional classification of tectal response types is still insufficient. Almost all studies on response properties of tectal visual cells have been carried out with rectangular stimuli differing in size, configuration and velocity and believed to be "behaviourally relevant", as described in section D3. However, it is necessary to carry out complementary recordings with other types of visual stimuli, e.g., grating patterns of different waveforms and spatial and temporal frequencies which more adequately reveal the receptive field properties of the recorded cells.

3. Another important question is whether tectal cells form discrete classes as RGC apparently do. Tectal cells can be classified according to RF size (distinguishing small-field, medium-field, and large to very large field cells), optimal stimulus size and stimulus shape. But there is a bewildering number of intermediate response types. The mathematical model of an der Heiden and Roth (1987, 1989) suggests that tectal cells, rather than falling into discrete classes, form a continuum of response characteristics by combination of different retinal input to each tectal cell and the degree of inhibition between cells, with higher probabilities for some response properties than for others (Fig. 56). Further mathematical analysis of the extracellular data from Roth (1982) by Woesler, Schwegler and Roth (unpubl. data) fully confirmed this assumption.

In opposition to the detector concept, a network model for prey recognition was proposed by Roth (1987) (Fig. 57). According to this model, the tectum contains numerous recognition "modules" consisting of a minimal set of neurons, each of which responds differently (although in an overlapping fashion) to the different parameters of a prey object, such as size, shape, velocity and movement pattern. The module is composed of neurons differing in their responses to the visual properties of objects. Such a module may act as a universal "prey analyser"; any prey object will activate all (or nearly all) components of this network, due to the fact that the response properties of the components vastly overlap, but different objects will activate the module differently. For example, if one assumes a very simple recognition module consisting only of T5-1 (S>H>V), T5-2 (H>S>V) and T5-3 (S>V>H) (regardless of whether these are members of "true" classes or only points in a continuum), then a squarelike object measuring $8° \times 8°$ will activate T-5.1 and T-5.3 cells to a high, and T-5.2 cells to a medium degree, and the overall activity level will be rather high. A horizontal rectangle will activate T-5.2 to a high, T-5.1 cells to a medium, and T-5.3 cells to a low degree, and the overall activity of the module will still be high, but lower than with stimulation by the square. Even a vertical rectangular stimulus will activate all types in the network, but the level of activation will be low except for type T-5.3 where medium activation will occur. Other and more complex response types like

CONSTITUTION OF TECTAL RESPONSE TYPES

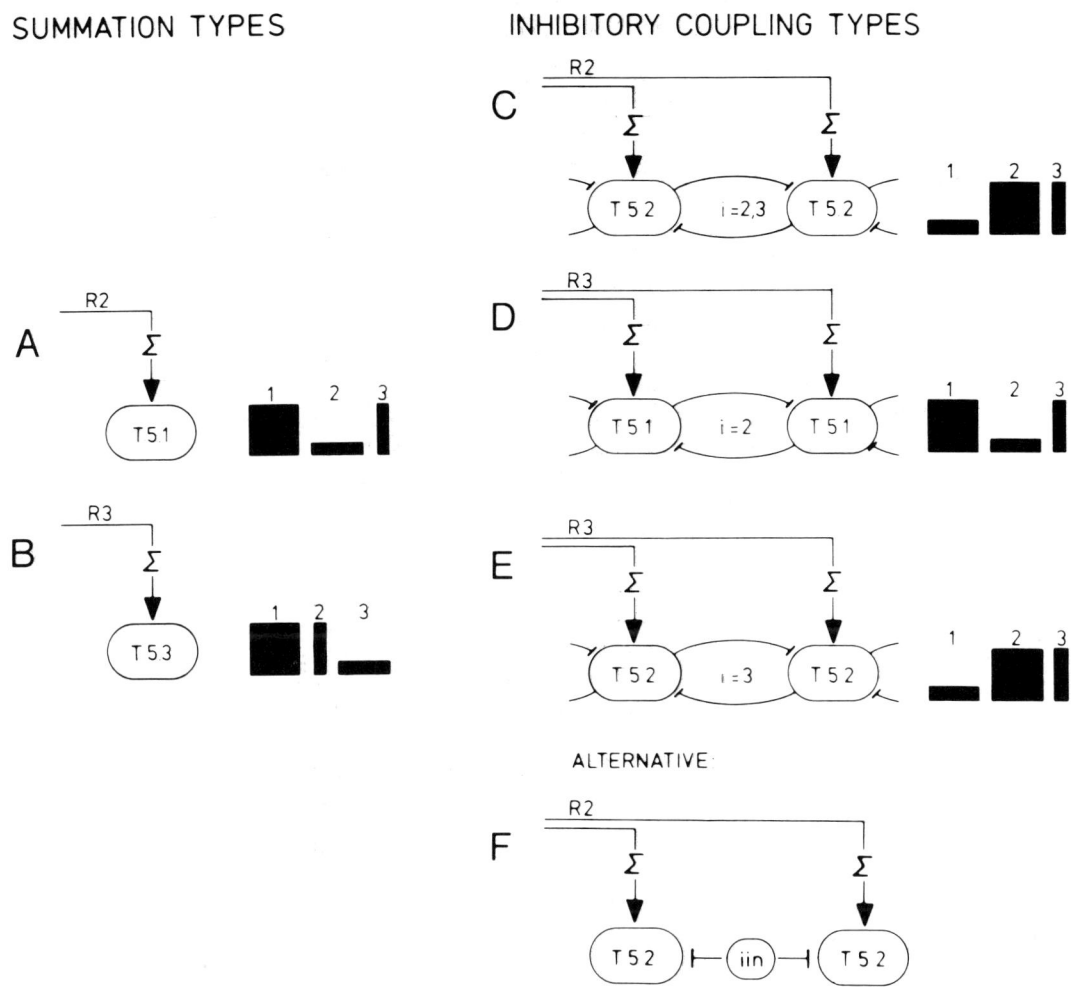

Fig. 56. Constitution of tectal response types according to the model developed by an der Heiden and Roth (1987). Left: Response types constituted by mere summation of activity of retinal ganglion cells are given. Right: Response types constituted by summation of rgc plus lateral recurrent inhibition between tectal cells are shown. Strength of inhibitory coupling is expressed in arbitrary units. Row of black stimuli indicates the respective stimulus preferences. In the bottom line an alternative for inhibitory coupling through inhibitory interneurons (iin) is presented. R2: rgc class-2; R3: rgc class-3. From Roth (1987).

the "inversion" neurons or neurons that respond differently to different movement patterns easily can be incorporated into such a network.

Thus, the components of this module act as semi-specific analyzers (because their responses are not exclusive) with respect to different parameters of the prey object such as orientation and horizontal and vertical extension of edges, size and velocity. Their *simultaneous* activity, then, encodes the prey object with regard to some relevant features. However, it cannot be considered a true "prey detector" network, because it is incapable of determining absolute size and, therefore, reliably distinguishing between a near prey and a distant predator of similar shape. Also, the concept of such a "recognition module" does not necessarily imply that its activity is summarized completely already within the tectum, i.e., that the module has just one output line (the axon of an output cell). Its cells may well send their axons separately to premotor and motor regions.

On the basis of recent tract-tracing and intracellular recording and staining experiments in salamanders and frogs (Dicke and Roth 1994a,b, 1996a; Roth *et al.*, submitted; Dicke and Roth, unpubl. data), it is assumed that in amphibians three separate retino-tectal subsystems for object-recognition exist, and that these process information about (1) size

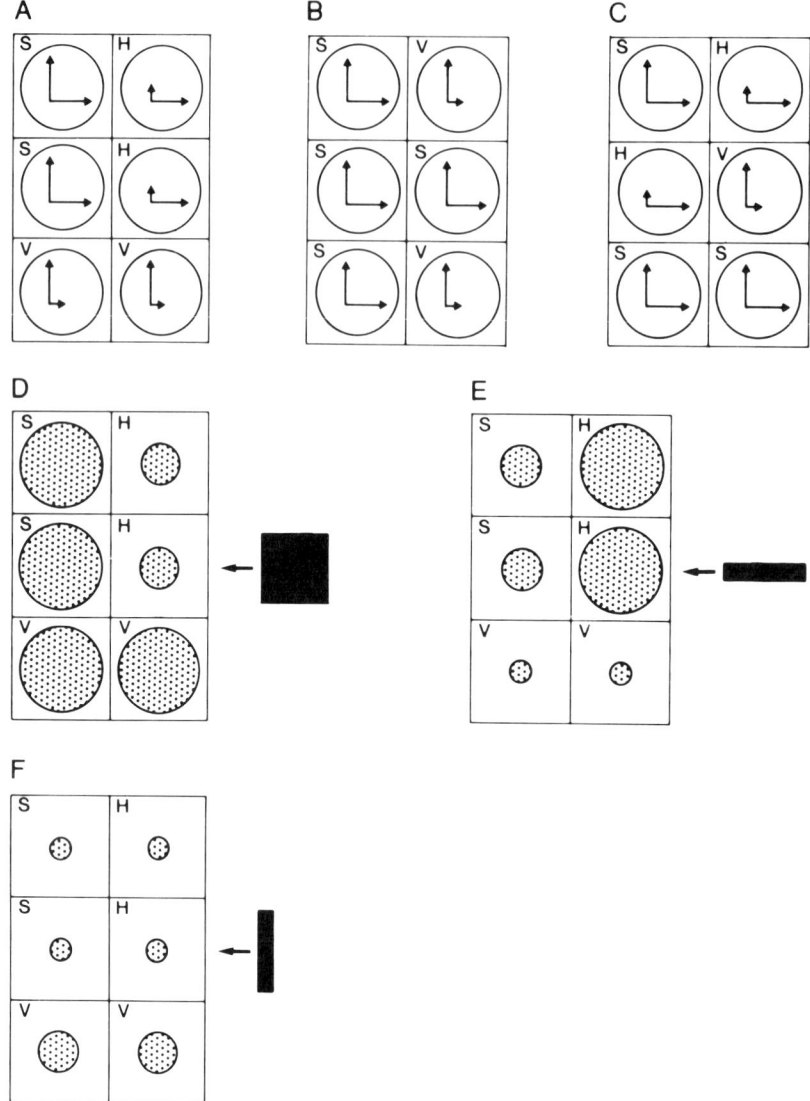

Fig. 57. Diagrammatic representation of the "recognition module" concept. **A–C** Recognition modules differing in qualitative and quantitative composition by tectal response types. Only three of tectal response types, T-5.1 (S), T-5.2 (H) and T-5.3 (V), are considered. These types differ in their responses to horizontal and vertical edges, as indicated by arrows. **A** Module with equal number of S, H, and V; **B** module with only S and V; **C** module with S is most frequent, followed by H. **D-F** The states of activity of module A during stimulation by a square (**D**), a horizontal rectangle (**E**), and vertical rectangle (**F**). The size of the circles indicates the strength of responses of the respective response types to the stimulus. From Roth (1987).

and shape, (2) velocity and movement pattern, and (3) changes in ambient illumination (such as caused by large moving objects). T1 cells receive dominant input from RGC class 1 and 2 and should, therefore, respond preferentially to small, high-contrast objects. T2 cells (type 2–4 neurons in frogs) receive dominant input from class 2 RGC and should be most sensitive to motion and movement patterns of small as well as larger objects. T3 cells (type 5 neurons in frogs) receive dominant input from class 3 RGC as well as from visual and non-visual (somatosensory, vestibular) input from other regions of the brain. They should be most sensitive to large-scale changes in ambient illumination.

These projection neurons show excitation upon stimulation of the optic nerve and have an excitatory effect on other cells (with some being inhibitory in addition). They are surrounded by a larger number of — at least partly electrically coupled — excitatory-inhibitory T4 interneurons which receive input from one or more RGC (see above). Their inhibitory and/or excitatory action might increase the response selectivity of the projection neurons in the context of shape on the one hand (T1) and recognition of movement-pattern on the other (T2). Projection neurons seem to interact with other projection neurons of the same type as well as with other types via electrical and chemical synapses (Fig. 58).

The result of this complex interaction among the three different projection neurons as well as the various subtypes of interneurons is then sent via anatomically separate pathways to different premotor and motor centers in the *medulla oblongata* and rostral spinal cord (Fig. 58). However, the information content carried via the uncrossed and the lateral and medial uncrossed descending pathways remains to be identified, and combined extra- and intracellular recordings are currently being undertaken to answer this important question.

Collaterals from the T1 pathway projects bilaterally to tegmental nuclei, whereas T2 and T3 pathways send collaterals bilaterally to thalamic and pretectal nuclei. It is important to note that these nuclei receive the same retinal input (viz., from class-3 and -4 RGC) as do T2 and T3 cells. These projection of tectal cells to the thalamus and praetectum possibly modulate the activity of these centres in the context of figure-background discrimination, prey/enemy distinction, and distinction of movement-pattern. However, this again remains to be verified experimentally.

In summary, the idea has to be abandoned that there exists in the amphibian tectum a "command neuron" or "command element" with respect to prey-recognition. At least three major streams of information meet at premotor and motor levels in order to elicit the various steps of prey-catching behaviour: (1) information about certain properties of the object perceived (size, contrast, colour, shape, velocity, movement pattern etc.); (2) information about the precise location of that object; pathways (1) and (2) need to interact in order to fully identify properties of visual objects (including absolute size); and (3) information about the level of motivation (presumably coming from the telencephalon and the hypothalamus).

One can envision the whole premotor process of visual guidance of feeding behaviour as an "AND" gate, with at least three inputs. *Motivation* sets a minimum level for the attractiveness of prey in order to activate the feeding sequence; at low motivation, the attractiveness of the prey object must be relatively high, and vice versa. The information about the spatial relationship between prey and the horopter is particularly important, because the prey-catching response is not released until the prey has reached or passed the horopter distance. However, this does not happen *automatically* as soon as prey reaches the horopter. Rather, it is again a matter of motivation, whether or not the salamander fires its tongue at maximal distance. The same is true for frontal versus lateral projection of the tongue; the higher the motivation, the more likely it is that the tongue will be projected laterally, provided the prey is within reach of the tongue (i.e., has passed the horopter).

4. The connectivity between descending visual pathways and the premotor and motor region in the *medulla oblongata* and the rostral (cervical) spinal cord is not fully clear. In the Japanese toad, *Bufo japonicus*, Matsushima *et al.* (1985) found two polysynaptic pathways from the tectum to the level of hypoglossal motor nuclei, an inhibitory one and an excitatory one. The excitatory responses of hypoglossal neurons became evident only after the tectum was stimulated repetitively with stronger stimuli, in contrast to the inhibitory responses which appeared after single or double stimulations with weaker stimuli. The existence of polysynaptic pathways and, thus, of interneurons, was shown by "spatial facilitation" after simultaneous electric stimulation of two different points within the tectum as well as by "temporal facilitation" after repetitive stimulation. The authors assumed a convergence of the excitatory and inhibitory pathways on interneurons situated in close proximity to the motor nuclei related to feeding. No monosynaptic input to the motor neurons was found.

In the bolitoglossine salamander, *Hydromantes italicus*, Matsushima and Roth (1990) found oligosynaptic and most likely monosynaptic input from tectal cells to hypoglossal motor neurons. All first responses were excitatory, in contrast to the situation in the Japanese toad. The existence of monosynaptic contacts between tectum and cervical spinal motor neurons involved in feeding in plethodontid salamanders was suggested by

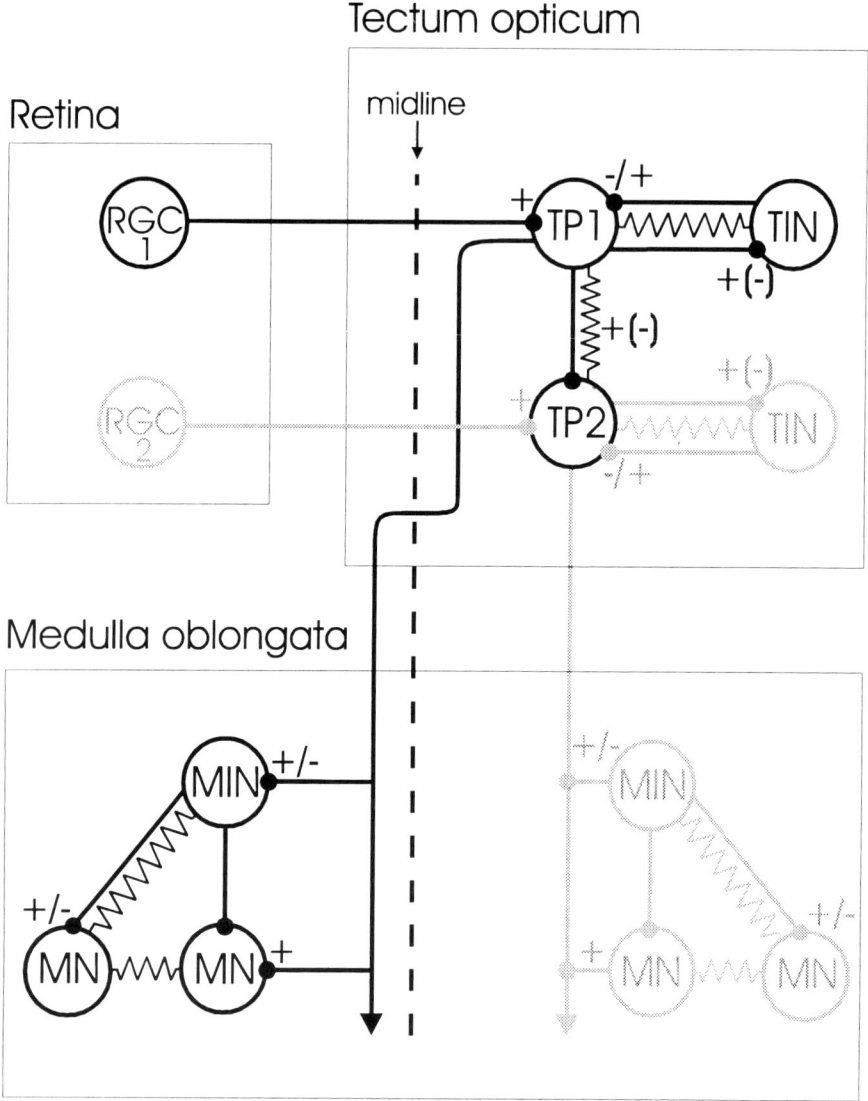

Fig. 58. Model of synaptic connectivity (electrical/chemical, excitatory/inhibitory) in the retino-tecto-bulbar system of salamanders based on retrograde labelling, intracellular recording and staining as well as transmitter immunohistochemistry. Only part of known connections are shown. The T1 pathway is drawn in black, the T2 pathway in grey. The T3 pathway is omitted for sake of clarity. From Dicke and Roth (1996b).

neuroanatomical studies involving double-labeling tectal descending pathways and motor neurons (Dicke 1992). Direct excitatory tectal pathways on to motor neurons is typical of embryonic and larval nervous systems; this may underline the paedomorphic character of the brains of salamanders in general and of bolitoglossines in particular. However, it is probable that even in salamanders the majority of descending fibers terminate on interneurons of the reticular formation which, then, influence the state of activity of motor neurons in an excitatory or inhibitory way. Little is known about the role of ascending fibers from reticular formation neurons terminating in the visual centers. However, it is safe to assume that these fibers constitute a complex feedback system between central visual and bulbar/spinal premotor regions.

It is unknown to what extent the visual centers (particularly the tectum) transform information about spatial localization and visual properties of a prey object into a motor code leading to appropriate recruitment and activation of motor neurons and eventually contraction of muscle fibers. In any case, this has to happen in a spatio-temporal fashion such that the tongue is projected out of the mouth at an appropriate strength, direction and distance. As far as is known, this process is determined by (1) the number of activated motor neurons and the related muscle fibers, and (2) the firing frequency of motor neurons.

VIII. REFERENCES

Adler, K., 1970. The role of extraoptic photoreceptors in amphibian rhythms and orientation. a review. *J. Herpetol.* **4:** 99–112.

Adler, K., 1976. Extraocular photoreception in amphibians. *Photochem. Photobiol.* **23:** 275–298.

an der Heiden, U. and Roth, G., 1987. Mathematical model and simulation of retina and tectum opticum of lower vertebrates. *Acta Biotheoretica* **36:** 179–212.

an der Heiden, U. and Roth, G., 1989. Retina and optic tectum in amphibians: a mathematical model and simulation studies. Pp. 243–267 in "Visuomotor Co-ordination, Amphibians, Comparisons, Models and Robots", ed by J.-P. Ewert and M. A. Arbib. Plenum Press, New York.

Antal, M., Matsumoto, N. and Székely, G., 1986. Tectal neurons of the frog: Intracellular recording and labeling with cobalt electrodes. *J. Comp. Neurol.* **246:** 238–253.

Arey, L. B., 1916. Changes in the rod-visual cells of the frog due to the action of light. *J. comp. Neurol.* **26:** 429–442.

Attwell, D., Wilson, M. and Wu, S. M., 1984. A quantitative analysis of interactions between photoreceptors in the salamander *(Ambystoma)* retina. *J. Physiol. (Lond.)* **352:** 703–737.

Autrum, H., 1959. Das Fehlen unwillkürlicher Augenbewegungen beim Frosch. *Naturwissenschaften* **46:** 435.

Ball, A. K. and Dickson, D. H., 1983. Displaced amacrine and ganglion cells in the newt retina. *Exp. Eye Res.* **36:** 199–214.

Barlow, H. B., 1953. Summation and inhibition in the frog's retina. *J. Physiol. (Lond.)* **119:** 69–88.

Baylor, D. A. and Fettiplace, R., 1975. Light path and photon capture in turtle photoreceptors. *J. Physiol. (Lond.)* **248:** 433–464.

Birukow, G., 1937. Untersuchungen über den optischen Drehnystagmus und über die Sehschärfe des Grasfrosches (*Rana temporaria* L.). *Z. vergl. Physiol.* **25:** 92–142.

Bousfield, J. D. and Pessoa, V. F., 1980. Changes in ganglion cell density during post-metamorphic development in a neotropical tree frog *Hyla raniceps*. *Vision Res.* **20:** 501–510.

Braekevelt, C. R., 1992. Retinal photoreceptor fine structure in the red-backed salamander *Plethodon cinereus*. *Histol. Histopath.* **7:** 463–470.

Burghagen, H. and Ewert, J.-P., 1983. Influence of the background for discriminating object motion from self-induced motion in toads *Bufo bufo* (L.). *J. Comp. Physiol.* **152:** 241–249.

Caine, H. and Gruberg, E. R., 1985. Ablation of nucleus isthmi leads to loss of specific visually elicited behaviors in the frog *Rana pipiens*. *Neuroscience Letters* **54:** 307–312.

Collett, T. S., 1977. Stereopsis in toads. *Nature* **267:** 349–351.

Collett, T. S. and Harkness, L. I. K., 1982. Depth vision in animals. Pp. 111–176 in "Advances in the Analysis of Visual Behavior", ed by D. J. Ingle, M. A. Goodale and R. J. W. Mansfield. Masachusetts Institute of Technology Press, Cambridge.

Collett, T. S., Udin, S. B. and Finch, D. J., 1987. A possible mechanism for binocular depth judgements in anurans. *Exp. Brain Res.* **66:** 35–40.

Collewijn, H., 1981. "Studies of Brain Functions", Volume 5, "The Oculomotor System of the Rabbit and its Plasticity", ed by V. Braitenberg. Springer Verlag, Berlin.

Cronly-Dillon, J. R. and Galand, G., 1966. Analyse des réponses visuelles unitaires dans le nerf optique et le tectum du Triton, *Triturus vulgaris*. *J. Physiol. (Lond.)* **58:** 502–503.

Dacheux, R. F. and Miller, R. F., 1976. Photoreceptor-bipolar cell transmission in the perfused retina eyecup of the mudpuppy. *Science* **191:** 963–964.

Deban, S. M. and Nishikawa, K., 1992. The kinematics of prey capture and the mechanism of tongue protraction in the green tree frog *Hyla cinerea*. *J. Exp. Biol.* **170:** 235–256.

Dicke, U., 1992. Neuroanatomische Untersuchungen zum Aufbau und zur Entwicklung der cervicospinalen Motornuclei und ihrer Verknüpfung mit den visuellen Zentren bei Salamandern. Doctoral thesis, University of Bremen.

Dicke, U., 1997. Tectobulbospinal and spinobulbotectal pathways in amphibians and their cells of origin. *J. Comp. Neurol.*, in press.

Dicke, U. and Roth, G., 1994a. Parallel processing in the visuomotor system of amphibians. Proceedings of the World Congress on Neural Networks, San Diego IV, 340–345.

Dicke, U. and Roth, G., 1994b. Tectal activation of premotor and motor networks during feeding in salamanders. *Europ. J. Morphol.* **32**: 106–116.

Dicke, U. and Roth, G., 1996a. Similarities and differences in the cytoarchitecture of the tectum of frogs and salamanders. *Acta Biol. Hung.*, in press.

Dicke, U. and Roth, G., 1966b. Visually guided behavior in amphibians: anatomical, physiological and functional aspects. Extended abstract, *Workshop Sensorimotor Co-ordination*, Sedona.

Dicke, U., Wallstein, M. and Roth, G., 1996. 5-HT-like immunoreactivity in the brain of plethodontid and salamandrid salamanders *(Hydromantes italicus, Hydromantes genei, Plethodon Jordani, Desmognathus ochrophaeus, Pleurodeles waltl)*: An immunohistochemical and Biocytin double-labelling study. *Cell Tissue Res.*, in press.

Dieringer, N., 1986. Image fading — a problem for frogs? *Naturwissenschaften* **73**: 330.

Dieringer, N., 1987. The role of compensatory eye and head movements for gaze stabilization in the unrestrained frog. *Brain Res.* **404**: 33–38.

Dodt, E., 1966. Vergleichende Physiologie de lichtempfindlichen Wirbeltier-Epiphyse. *Nova Acta Leopoldina* **31**: 21–235.

Dodt, E. and Heerd, E., 1962. Mode of action of pineal nerve fibers in frogs. *J. Neurophysiol.* **25**: 405–429.

Dodt, E. and Jacobson, M., 1963. Photosensitivity of a localized region of the frog diencephalon. *J. Neurophysiol.* **26**: 752–758.

Dodt, E., Ueck, M. and Oksche, A., 1971. Relations of structure and function: The pineal organ of lower vertebrates. Proceedings of the Centenary Symposium, Prague, Pp. 253–278.

Donner, K. O. and Reuter, T., 1976. Visual pigments and photoreceptor function. Pp. 251–277 *in* "Frog Neurobiology", ed by R. Llinas and W. Precht. Springer, Berlin.

Douglas, R. H., Collett, T. S. and Wagner, H. J., 1986. Accomodation in anuran amphibian and its role in depth vision. *J. Comp. Physiol.* **158**: 133–143.

Dowling, J. E., 1977. Receptoral and network mechanisms of visual adaptation. *Neurosci. Res. Prog. Bull.* **15**: 397–407.

Dowling, J. E., 1987. "The Retina: an Approachable Part of the Brain". Harvard University Press, Cambridge, Mass.

Dowling, J. E. and Werblin, R. S., 1969. Organization of the retina of the mudpuppy. I. Synaptic structure. *J. Neurophysiol.* **32**: 315–338.

Duellman, W. E. and Trueb, L., 1986. "Biology of Amphibians". McGraw-Hill, New York.

Dunlop, S. A. and Beazley, L. D., 1981. Changing retinal ganglion cell distribution in the frog, *Heleioporus eyrei*. *J. Comp. Neurol.* **202**: 221–237.

Dunlop, S. A. and Beazley, L. D., 1984. A morphometric study of the retinal ganglion cell layer and optic nerve from metamorphosis in *Xenopus laevis*. *Vision Res.* **24**: 417–427.

Eurich, C., Roth, G., Schwegler, H. and Wiggers, W., 1995. Simulander: A neural network model for the orientation movement of salamanders. *J. Comp. Physiol. A* **176**: 379–389.

Ewert, J.-P., 1968. Der Einfluss von Zwischenhirndefekten auf die Visuomotorik im Beute- und Fluchtverhalten der Erdkröte (*Bufo bufo* L.). *Z. vergl. Physiol.* **61**: 41–70.

Ewert, J.-P., 1970. Aufnahme und Verarbeitung visueller Informationen im Beutefang- und Fluchtverhalten der Erdkröte *Bufo bufo* (L.). Pp. 218–226 *in* "Proc. 64th Conf. Dt. Zol. Ges.". Gustav Fischer, Stuttgart.

Ewert, J.-P., 1971. Single unit response of the toad's *(Bufo americanus)* caudal thalamus to visual objects. *Z. Vergl. Physiol.* **74**: 81–102.

Ewert, J.-P., 1974. The neural basis of visually guided behavior. *Sci. Amer.* **230**: 34–42.

Ewert, J.-P., 1984. Tectal mechanisms that underlie prey catching and avoidance behaviors in toads. Pp. 247–416 *in* "Neurology of the Optic Tectum", ed by H. Vanegas. Plenum Press, New York.

Ewert, J.-P., 1989. The release of visual behavior in toads: Stages of parallel/hierarchical information processing. Pp. 39–120 *in* "Visuomotor Co-ordination. Amphibians, Comparisons, Models, and Robots", ed by J.-P. Ewert and M. Arbib. Plenum, New York.

Ewert, J.-P. and Gebauer, I., 1973. Grössenkonstanzphänomen im Beutefangverhalten der Erdkröte (*Bufo bufo* L.). *J. Comp. Physiol.* **85**: 303–315.

Ewert, J.-P. and Hock, F. J., 1972. Movement sensitive neurons in the toad's retina. *Exp. Brain Res.* **16**: 41–59.

Ewert, J.-P. and von Wietersheim, A., 1974. Der Einfluß von Thalamus/Prätectum — Defekten auf die Antwort von Tectum Neuronen gegenüber bewegten visuellen Mustern bei der Kröte (*Bufo bufo* L.). *J. Comp. Physiol.* **92**: 149–160.

Finch, D. J. and Collett, T. S., 1983. Small-field, binocular neurons in the superficial layers of the frog optic tectum. *Proc. R. Soc. Lond.* **217B**: 491–497.

Finkenstädt, T. and Ewert. J.-P., 1983a. Processing of area dimensions of visual key stimuli by tectal neurons in *Salamandra salamandra*. *J. Comp. Physiol.* **153**: 85–98.

Finkenstädt, T. and Ewert, J.-P., 1983b. Visual pattern discrimination through interaction of neural networks: A combined electrical brain stimulation, brain lesion, and extracellular recording study in *Salamandra salamandra*. *J. Comp. Physiol.* **153**: 99–110.

Finkenstädt, T., Ebbesson, S. O. E. and Ewert, J.-P., 1983. Projections to the midbrain tectum in *Salamandra salamandra* L. *Cell Tiss. Res.* **234**: 39–55.

Fite, K. V., 1973. The visual fields of the frog and toad: A comparative study. *Behav. Biol.* **9**: 707–718.

Fite, K. V. (ed), 1976. "The Amphibian Visual System. A Multidisciplinary Approach". Academic Press, New York.

Fite, K. V. and Scalia, F., 1976. Central visual pathways in the frog. Pp. 87–118 in "The Amphibian Visual System. A Multidisciplinary Approach", ed by K. V. Fite. Academic Press, New York.

Frank, B. D. and Hollyfield, J. G., 1987. Retina of the tadpole and frog: delayed dendritic development in a subpopulation of ganglion cells coincident with metamorphosis. *J. Comp. Neurol.* **266**: 435–444.

Freed, A. N., 1982. A treefrog's menu: selection for an evening's meal. *Oecologia (Berl.)* **53**: 20–26.

Fritzsch, B., 1980. Retinal projections in european salamanders. *Cell Tiss. Res.* **213**: 325–341.

Glasser, S. and Ingle, D., 1978. The nucleus isthmi as a relay station in the ipsilateral visual projection to the frog's optic tectum. *Brain Res.* **159**: 214–218.

Gordon, J. and Hood, D. C., 1976. Anatomy and physiology of the frog retina. Pp. 29–86 in "The Amphibian Visual System: A Multidisciplinary Approach", ed by K. Fite. Academic Press, New York.

Grobstein, P., Comer, C. and Kostyk, S., 1978. A crossed isthmo-tectal projection in *Rana pipiens* and its involvement in the ipsilateral visuo-tectal projection. *Brain Res.* **156**: 117–123.

Grobstein, P. and Comer, C., 1983. The nucleus isthmi as an intertectal relay for the ipsilateral oculotectal projection in the frog, *Rana pipiens*. *J. Comp. Neurol.* **217**: 54.

Grobstein, P., Reyes, A., Zwanziger, L. and Kostyk, S. K., 1985. Prey orientation in the frog: accounting for variations in output with stimulus distance. *J. Comp. Physiol.* **156**: 775–785.

Gruberg, E. R., 1972. Optic fiber projections in the tiger salamander *Ambystoma mexicanum*. *J. Hirnforsch.* **14**: 399–411.

Gruberg, E. R. and Grasse, K. L., 1984. Basal optic complex in the frog *(Rana pipiens)*: a physiological and HRP study. *J. Neurophysiol.* **51**: 98–110.

Gruberg, E. R., Wallace, M. T., Caine, H. S. and Mote, M. I., 1991. Behavioral and physiological consequences of unilateral ablation of the nucleus in the leopard frog. *Brain Behav. Evol.* **37**: 92–103.

Grüsser, O.-J. and Grüsser-Cornehls, U., 1968. Neurophysiologische Grundlagen visuell angeborener Auslösemechanismen beim Frosch. *Z. vergl. Physiol.* **59**: 1–24.

Grüsser, O.-J. and Grüsser-Cornehls, U., 1970. Die Neurophysiologie visuell gesteuerter Verhaltensweizen bei Anuren. *Verh. Dtsch. Zool. Ges.* **64**: 201–218.

Grüsser, O.-J. and Grüsser-Cornehls, U., 1976. Neurophysiology of the anuran visual system. Pp. 298–385 in "Frog Neurobiology", ed by R. Llinas and W. Precht. Springer, Berlin.

Grüsser-Cornehls, U., 1984. The neurophysiology of the amphibian optic tectum. Pp. 211–245 in "Comparative Neurology of the Optic Tectum", ed by H. Vanegas. Plenum Press, New York.

Grüsser-Cornehls, U. and Himstedt, W., 1973. Responses of retinal and tectal neurons of the salamander *(Salamandra salamandra* L.) to moving visual stimuli. *Brain Behav. Evol.* **7**: 145–168.

Grüsser-Cornehls, U. and Himstedt, W., 1976. The urodele visual system. Pp. 203–266 in "The Amphibian Visual System. A Multidisciplinary Approach", ed by K. V. Fite. Academic Press, New York.

Grüsser-Cornehls, U. and Langeveld, S., 1985. Velocity sensitivity and directional selectivity of frog retinal ganglion cells depend on chromaticity of moving stimuli. *Brain Behav. Evol.* **27**: 165–185.

Hagins, W. A., Penn, R. D. and Yoshikami, S., 1970. Dark current and photocurrent in retinal rods. *Biophys. J.* **10**: 380–412.

Hamasaki, D. I., 1969. Pre-excitatory inhibition in the stirnorgan of the bullfrog. *Vision Res.* **9**: 1305–1307.

Hamasaki, D. I., 1970. Interaction of excitation and inhibition in the stirnorgan of the frog. *Vision Res.* **10**: 307–316.

Hartwig, H. G. and Baumann, C., 1974. Evidence for photosensitive pigments in the pineal complex of the frog. *Vision Res.* **14**: 597–598.

Hartwig, H. G. and Korf, H. W., 1978. The epiphysis cerebri of poikilothermic vertebrates: a photosensitive neuroendocrine circumventricular orga. *Scannin Electr. Microsc.* **2**: 163–168.

Heatwole, H. and Heatwole, A., 1968. Motivational aspects of feeding behavior in toads. *Copeia* **1968**: 692–698.

Hendrickson, A., 1966. Landolt's club in the amphibian retina: A Golgi and electron microscope study. *Invest. Ophthalmol.* **5**: 484–496.

Herrick, C. J., 1948. "The Brain of the Tiger Salamander *Ambystoma tigrinum*". University of Chicago Press, Chicago.

Himstedt, W., 1967. Experimentelle Analyse der optischen Sinnesleistungen im Beutefangverhalten der einheimischen Urodelen. *Zool. Jb. Physiol.* **73**: 281–320.

Himstedt, W., 1973a. Die spektrale Empfindlichkeit von Urodelen in Abhängigkeit von Metamorphose, Jahreszeit und Lebensraum. *Zool. Jb. Physiol.* **77**: 246–274.

Himstedt, W., 1973b. Die spektrale Empfindlichkeit von *Triturus alpestris* (Amphibia, Urodela) während des Wasser- und Landlebens. *Pflüger's Arch.* **341**: 7–14.

Himstedt, W., 1982. Prey selection in salamanders. Pp. 47–66 in "Analysis of Visual Behavior", ed by D. Ingle, M. A. Goodale and R. J. W. Mansfield. Massachusetts Institute of Technology Press, Cambridge.

Himstedt, W. and Roth, G., 1980. Neuronal responses in the tectum opticum of *Salamandra salamandra* to visual prey stimuli. *J. Comp. Physiol.* **135**: 251–257.

Himstedt, W., Tempel, P. and Weiler, J., 1980. Responses of salamanders to stationary visual patterns. *J. Comp. Physiol.* **124**: 49–52.

Himstedt, W., Heller, K. and Manteuffel, G., 1987. Neuronal responses to moving visual stimuli in different thalamic and midbrain centers of *Salamandra salamandra* (L.) *Zool. Jb. Physiol.* **91**: 243–256.

Hinsche, G., 1935. Ein Schnappreflex nach "Nichts" bei Anuren. *Zool. Anz.* **111**: 113.

House, D., 1989. Depth perception in frogs and toads (Lecture notes in biomathematics 80). Springer, New York.

Hughes, T. E., 1990a. Light and electron microscopical investigations of the optic tectum of the frog, *Rana pipiens*. I. The retinal axons. *Visual Neurosci.* **4:** 499–518.

Hughes, T. E., 1990b. Light and electron microscopical investigations of the optic tectum of the frog, *Rana pipiens*. II. The neurons that give rise to the crossed tecto-bulbar pathway. *Visual Neurosci.* **4:** 519–531.

Ingle, D., 1968. Visual releasers of prey catching behavior in frog and toads. *Brain Behav. Evol.* **1:** 500–518.

Ingle, D., 1972. Depth vision in monocular frogs. *Psychon. Science* **29:** 37–38.

Ingle, D., 1973. Two visual systems in the frog. *Science* **181:** 1053–1055.

Ingle, D., 1976. Spatial vision in anurans. Pp. 119–140 *in* "The Amphibian Visual System. A Multidisciplinary Approach", ed by K. V. Fite. Academic Press, New York.

Ingle, D., 1980. Some effects of pretectum lesions in the frog's detection of stationary objects. *Behav. Brain Res.* **1:** 139–163.

Jaeger, R. G., 1972. Food as a limited resource in competition between two species of terrestrial salamanders. *Ecology* **53:** 535–546.

Jaeger, R. G. and Barnard, D. E., 1981. Foraging tactics of a terrestrial salamander: Choice of diet in structurally simple environments. *Am. Nat.* **117:** 639–664.

Jaeger, R. G. and Rubin, A. M., 1982. Foraging tactics of a terrestrial salamander: Judging prey profitability. *J. Anim. Evol.* **51:** 167–176.

Jaeger, R. G., Barnard, D. E. and Joseph, R. G., 1982. Foraging tactics of a terrestrial salamander: Assessing prey density. *Am. Nat.* **119:** 885–890.

Jordan, M., Luthardt, G., Meyer-Naujoks, C. and Roth, G., 1980. The role of eye accommodation in depth perception of common toads. *Z. Naturforsch.* **35:** 851–852.

Källén, B., 1951. Some remarks on the ontogeny of the telencephalon in some lower vertebrates. *Acta Anat.* **11:** 537–548.

Katte, O. and Hoffmann, K.-P., 1980. Direction specific neurons in the pretectum of the frog *(Rana esculenta)*. *J. Comp. Physiol.* **140:** 53–57.

Korf, H. W., 1976. Histological, histochemical and electron microscopical studies on the nervous apparatus of the pineal organ in the tiger salamander, *Ambystoma tigrinum*. *Cell Tissue Res.* **174:** 475–497.

Kupfermann, I. and Weiss, K. R., 1978. The command neuron concept. *Brain Behav. Sci.* **1:** 3–39.

Larsell, O., 1967. "The Comparative Anatomy and Histology of the Cerebellum from Myxinoids through Birds", ed by J. Jansen. University of Minnesota Press, Minneapolis.

Larsen, L. O. and Pedersen, J. N., 1982. The snapping response of the toad, *Bufo bufo*, towards prey dummies at very low light intensities. *Amphibia-Reptilia* **2:** 321–327.

Lázár, G., 1984. Structure and connections of the frog optic tectum. Pp. 185–210 *in* "Comparative Neurology of the Optic Tectum", ed by H. Vanegas. Plenum Press, New York.

Lázár, G. and Szekely, G., 1969. Distribution of optic terminals in the different optic centers in the frog. *Brain Res.* **16:** 1–14.

Lázár, G., Toth, P., Csank, E. and Kicliter, E., 1983. Morphology and location of tectal projection neurons in frogs: A study with HRP and cobalt filling. *J. Comp. Neurol.* **215:** 108–120.

Lettvin, J. Y., Maturana, H. R., McCulloch, W. S. and Pitts, W. H., 1959. What the frog's eye tells the frog's brain. *Proc. Inst. Radio Engrs. NY* **47:** 1940–1951.

Lettvin, J. Y., Maturana, H. R., McCulloch, W. S. and Pitts, W. H., 1961. Two remarks on the visual systems of the frog. Pp. 757–776 *in* "Sensory Communication", ed by W. Rosenblith. Massachusetts Institute of Technology Press, Cambridge, Mass.

Linke, R. and Roth, G., 1989. Morphology of retinal ganglion cells in lungless salamanders (Fam. Plethodontidae): An HRP and Golgi study. *J. Comp. Neurol.* **289:** 361–374.

Linke, R. and Roth, G., 1990. Optic nerves in plethodontid salamanders (Amphibia, Urodela): neuroglia, fiber spectrum and myelination. *Anat. Embryol.* **181:** 37–48.

Linke, R., Roth, G. and Rottluff, B., 1986. Comparative studies on the eye morphology in lungless salamanders, family Plethodontidae, and the effect of miniaturization. *J. Morphol.* **189:** 131–143.

Llinas, R. and Precht, W. (eds), 1976. "Frog Neurobiology". Springer, Berlin.

Lock, A. and Collett, T., 1979. A toad's devious approach to its prey: A study of some complex uses of depth vision. *J. Comp. Physiol.* **131:** 179–189.

Luksch, H. and Walkowiak, W., 1994. The nucleus laminaris of the anuran auditory midbrain: Part of a parallel organized audiomotor interface? P. 356 *in* Proceedings of the 22nd Goettingen Neurobiology Conference, ed by N. Elsner and H. Breer. Thieme, Stuttgart.

Luksch, H., Kahl, H., Wiggers, W. and Roth, G., 1995. The pretectum of salamanders: An intracellular study on the morphology and physiology of pretectal neurons and there influence on tectal evoked potentials. P. 253 *in* Proc. IVth Int. Conf. Neuroethol., ed by M. Burrows, T. Matheson, P. L. Newland and H. Schuppe. Thieme, Stuttgart.

Luthardt, G., 1981. "Verhaltensbiologische Untersuchungen zum visuell gesteuerten Beutefangverhalten von *Salamandra salamandra* (L.)". Minerva, München.

Luthardt, G. and Roth, G., 1979. The relationship between stimulus orientation and stimulus movement pattern in the prey catching behavior of *Salamandra salamandra*. *Copeia* **1979:** 442–447.

Luthardt-Laimer, G., 1983. Distance estimation in binocular and monocular salamanders. *Z. Tierpsychol.* **63:** 233–240.

Lynch, J. F., 1985. The feeding ecology of *Aneides flavipunctatus* and sympatric plethodontid salamanders in Northwestern California. *J. Herpetol.* **19:** 328–352.

Maglia, A. M. and Pyles, R. A., 1995. Modulation of prey-capture behavior in *Plethodon cinereus* (Green) (Amphibia: Caudata). *J. Exp. Zool.* **272**: 167–183.

Maiorana, V. C., 1978a. Difference in diet as an epiphenomenon: Space regulates salamanders. *Can. J. Zool.* **57**: 1017–1025.

Maiorana, V. C., 1978b. Behavior of unobservable species. Diet selection by a salamander. *Copeia* **1978**: 664–672.

Manteuffel, G., 1979. Transition of optomotor behavior to fixation in the crested newt. *Zool. Jb. Physiol.* **83**: 526–539.

Manteuffel, G., 1982. The accessory optic system in the newt, *Triturus cristatus*: Unitary response properties from basal optic neuropil. *Brain Behav. Evol.* **21**: 175–184.

Manteuffel, G., 1984a. Electrophysiology and anatomy of direction specific pretectal units in *Salamandra salamandra*. *Exp. Brain Res.* **54**: 415–425.

Manteuffel, G., 1984b. A physiological model for the salamander horizontal optokinetic reflex. *Brain Behav. Evol.* **25**: 197–205.

Manteuffel, G., 1987. Binocular afferents to the salamander pretectum mediate ritation sensitivity of cells selective for visual background motions. *Brain Res.* **422**: 381–383.

Manteuffel, G., 1989. Compensation of visual background motion in salamanders. Pp. 311–340 *in* "Visuomotor Coordination", ed by J.-P. Ewert and M. A. Arbib. Plenum, New York.

Manteuffel, G., Wess, O. and Himstedt, W., 1977. Messungen am dioptrischen Apparat von Amphibienaugen und Berechnung der Sehschärfe in Wasser und Luft. *Zool. Jahrb. Physiol.* **81**: 395–406.

Manteuffel, G., Kopp, J. and Himstedt, W., 1986. The amphibian optokinetic afternystagmus: properties and comparative analysis in various species. *Behav. Brain Res.* **28**: 186–197.

Manteuffel, G. and Naujoks-Manteuffel, C., 1987. Synergistic visual and vestibular self motion related inputs to the optic tectum of salamanders. *Neurosci. Suppl.* **22**: 737.

Manteuffel, G., Fox, B. and Roth, G., 1989. Topographic relationships of ipsi- and contralateral visual inputs to the rostral tectum opticum indicate the presence of a horopter. *Neuroscience Lett.* **107**: 105–109.

Manteuffel, G. and Roth, G., 1993. A model of the saccadic sensorimotor system in salamanders. *Biol. Cybernetics* **68**: 431–440.

Matsumoto, N., Schwippert, W. W. and Ewert, J.-P., 1986. Intracellular activity of morphologically identified neurons of the grass frog's optic tectum in response to moving configurational visual stimuli. *J. Comp. Physiol.* **159**: 721–739.

Matsushima, T. and Roth, G., 1990. Fast excitatory action of optic tectum on tongue motoneurons in the tongue-projecting salamander *Hydromantes italicus*. P. 87 *in* "Brain-Cognition-Perception", ed by N. Elsner and G. Roth. Thieme, Stuttgart.

Matsushima, T., Satou, M. and Ueda, K., 1985. An electromyographic analysis of electrically-evoked prey-catching behavior by means of stimuli applied to the optic tectum in the Japanese toad. *Neurosci. Res.* **3**: 154–161.

Matthews, G., 1983. Physiological characteristics of single green rod photoreceptors from toad retina. *J. Physiol. (Lond.)* **342**: 347–359.

Maturana, H. R., 1959. Number of fibers in the optic nerve and the number of ganglion cells in the retina of anurans. *Nature (Lond.)* **183**: 1406–1407.

Maturana, H. R., Lettvin, J. Y., McCulloch, W. S. and Pitts, W. H., 1960. Anatomy and physiology of vision in the frog *(Rana pipiens)*. *J. Gen. Physiol.* **43**: Suppl. 2: 129–175.

Morita, Y., 1969. Wellenlängen-Diskriminatoren im intrakranialen Pinealorgan von *Rana catesbyana* [sic.]. *Experientia* **25**: 1277.

Morita, Y. and Dodt, E., 1975. Early receptor potential from the pineal photoreceptor. *Pflügers Arch.* **354**: 273–280.

Naujoks-Manteuffel, C. and Manteuffel, G., 1988. Origins of descending projections to the medulla oblongata and rostral medulla spinalis in the urodele *Salamandra salamandra* (Amphibia). *J. Comp. Neurol.* **273**: 187–206.

Neary, T. J., 1975. Architectonics of the thalamus of the bullfrog *(Rana catesbeiana)*. A histochemical analysis. *Anat. Rec.* **181**: 434–435.

Neary, T. J. and Northcutt, R. G., 1983. Nuclear organization of the bullfrog diencephalon. *J. Comp. Neurol.* **213**: 262–278.

Nieuwenhuys, R. and Opdam, P., 1976. Structure of the brain stem. Pp. 811–855 *in* "Frog Neurobiology", ed by R. Llinás and W. Precht. Springer-Verlag, Berlin.

Nguyen, V. S. and Straznicky, C., 1989. The development and the topographic organization of the retinal ganglion cell layer in *Bufo marinus*. *Exp. Brain Res.* **75**: 345–353.

Northcutt, R. G., 1977. Retinofugal projections in the lepidosirenid lungfishes. *J. Comp. Neurol.* **174**: 553–574.

Northcutt, R. G. and Kicliter, E., 1980. Organization of the amphibian telencephalon. Pp. 203–255 *in* "Comparative Neurology of the Telencephalon", ed by S. O. E. Ebbesson. Plenum Press, New York.

Norton, A. L., Spekreijse, H., Wagner, H. and Wolbarsht, M., 1970. Responses to directional stimuli in retinal preganglionic units. *J. Physiol.* **206**: 93–107.

Ogden, T. E., Mascetti, G. G. and Pierantoni, R., 1984. The internal horizontal cell of the frog retina: Analysis of receptor input. *Invest. Ophthalmol. Vis. Sci.* **25**: 1382–1394.

Oksche, A. and Hartwig, H. G., 1979. Pineal sense organs — components of photoneuroendocrine systems. *Prog. Brain Res.* **52**: 113–130.

Podufal, G., 1971. Zur Entfernungsmessung und Grössenbeurteilung durch die Erdkröte *Bufo bufo* L. Thesis, Göttingen.

Potter, H. D., 1969. Structural characteristics of cell and fiber populations in the optic tectum of the frog *Rana catesbeiana*. *J. Comp. Neurol.* **136**: 202–232.

Ramon y Cajal, S., 1892. "The Structure of the Retina". Translated 1972 by M. Glickstein and S. A. Thorpe. Thomas, Springfield.

Rettig, G., 1984. Neuroanatomische Untersuchungen der visuellen Projektionen bei Salamandern (Ordnung Caudata). Dissertation. Universität Bremen.

Rettig, G. and Roth, G., 1982. Afferent visual projections in three species of lungless salamanders (Family Plethodontidae). *Neurosci. Lett.* **31:** 221–224.

Rettig, G. and Roth, G., 1986. Retinofugal projections in salamanders of the family Plethodontidae. *Cell Tiss. Res.* **243:** 385–396.

Roberts, A., 1978. Pineal eye and behaviour in *Xenopus* tadpoles. *Nature* **273:** 774–775.

Roth, G., 1976. Experimental analysis of the prey catching behavior of *Hydromantes italicus* Dunn (Amphibia, Plethodontidae). *J. Comp. Physiol.* **109:** 47–58.

Roth, G., 1978. The role of stimulus movement patterns in the prey capture behavior of *Hydromantes italicusenei* (Amphibia, Plethodontidae). *J. Comp. Physiol.* **123:** 261–264.

Roth, G., 1982. Responses of the optic tectum of the salamander *Hydromantes italicus* to moving prey stimuli. *Exp. Brain Res.* **45:** 386–392.

Roth, G., 1987. "Visual Behavior in Salamanders. Studies of Brain Function". Springer, Berlin.

Roth, G. and Dicke, U., 1994. Is fixed action pattern a useful concept? Pp. 2–14 *in* "Dahlem Workshop on "Flexibility and Constraints in Behavioral Systems", ed by R. J. Greenspan and C. P. Cyriacou. Wiley and Sons, Chichester.

Roth, G. and Jordan, M., 1982. Response characteristics and stratification of tectal neurons in the toad *Bufo bufo* L. *Exp. Brain Res.* **45:** 393–398.

Roth, G. and Wiggers, W., 1983. Response of the toad *Bufo bufo* to stationary prey stimuli. *Z. Tierpsychol.* **61:** 225–234.

Roth, G., Grunwald, W. and Naujoks-Manteuffel, C., 1990. Cytoarchitecture of the tectum mesencephali in salamanders. A Golgi and HRP study. *J. Comp. Neurol.* **291:** 27–42.

Roth, G., Naujoks-Manteuffel, C., Nishikawa, K., Schmidt, A. and Wake, D. B., 1993. The salamander nervous system as a secondarily simplified, paedomorphic system. *Brain Behav. Evol.* **42:** 137–170.

Roth, G., Nishikawa, K., Dicke, U. and Wake, D. B., 1988. Topography and cytoarchitecture of the motor nuclei in the brainstem of salamanders. *J. Comp. Neurol.* **278:** 181–194.

Satou, M. and Ewert, J.-P., 1985. The antidromic activation of tectal neurons by electrical stimuli applied to the caudal medulla oblongata in the toad *Bufo bufo* L. *J. Comp. Physiol. A* **157:** 739–748.

Schipperheyn, J. J., 1965. Contrast detection in frog's retina. *Acta Physiol. Pharmacol. Neerlandica* **13:** 231–277.

Schmidt, A. and Roth, G., 1990. Central olfactory and vomeronasal pathways in salamanders. *J. Hirnforsch.* **31:** 543–553.

Schneider, D., 1954. Beitrag zu einer Analyse des Beute- und Fluchtverhaltens einheimischer Anuren. *Biol. Zbl.* **73:** 225–282.

Schulte-Mattler, M. and Luhmann, H. J., 1995. Dye coupling reveals intrinsic microcircuitry in rat neocortex. P. 559 *in* "Proceedings of the 23rd Goettingen Neurobiology Conference", Vol. 2, ed by N. Elsner and R. Menzel. Thieme, Stuttgart.

Schürg-Pfeiffer, E., 1989. Behavior-correlated properties of tectal neurons in freely moving toads. Pp. 451–480 *in* "Visuomotor coordination. Amphibians, Comparisons, Models, and Robots", ed by J.-P. Ewert and M. Arbib. Plenum, New York, London.

Schürg-Pfeiffer, E. and Ewert, J.-P., 1981. Investigations of neurons involved in the analysis of gestalt prey features in the frog *Rana temporaria*. *J. Comp. Physiol.* **141:** 139–152.

Singman, E. L. and Scalia, F., 1990. Quantitative study of the tectally projecting retinal ganglion cells in the adult frog: I. The size of the contralateral and ipsilateral projections. *J. Comp. Neurol.* **302:** 792–809.

Sites, J. W., 1978. The foraging strategy of the dusky salamander, *Desmognathus fuscus* (Amphibia, Urodela, Plethodontidae): An empirical approach to predation theory. *J. Herpetol.* **12:** 373–383.

Skrzypek, J., 1984. Electrical coupling between horizontal cell bodies in the tiger salamander retina. *Vis. Res.* **24:** 701–711.

Sperl, M. and Manteuffel, G., 1987. Directional selectivities of visual afferents to the pretectal neuropil in the fire salamander. *Brain Res.* **404:** 332–334.

Spillmann, L. and Werner, J. S. (eds), 1990. "Visual Perception. The Neurophysiological Foundations". Academic Press, San Diego.

Stenner, G., 1976. "Untersuchung über die Funktion von Reizfiltermechanismen im Beutefangverhalten des Feuersalamanders nach der Metamorphose". Staatsexamensarbeit, Technische Hochschule Darmstadt.

Stephan, P. and Weiler, R., 1981. Morphology of horizontal cells in the frog retina. *Cell Tiss. Res.* **221:** 443–449.

Straznicky, C. and Straznicky, I. T., 1988. Morphological classification of retinal ganglion cells in adult *Xenopus laevis*. *Anat. Embryol.* **178:** 143–153.

Szekely, G. and Lázár, G., 1976. Cellular and synaptic architecture of the optic tectum. Pp. 407–434 *in* "Frog Neurobiology", ed by R. Llinas and W. Precht. Springer, Berlin.

Takahama, H., 1993. Evidence for a frontal-organ homolgue in the pineal complex of the salamander, *Hynobius dunni*. *Cell Tiss. Res.* **272:** 575–578.

Taylor, D. G. and Adler, K., 1978. The pineal body: Site of extraocular perception of celestial cues for orientation in the tiger salamander *(Ambystoma tigrinum)*. *J. Comp. Physiol.* **124:** 357–361.

Tinbergen, N., 1951. "The Study of Instinct". Oxford Universitiy Press, New York.

Trachtenberg, M. C. and Ingle, D., 1974. Thalamo-tectal projections in the frog. *Brain Res.* **79:** 419–430.

Vollrath, L., 1981. The pineal organ. Pp. 1–665 *in* "Handbuch der mikroskopischen Anatomie des Menschen" VI/7. Springer, Berlin.

Wake, K., Ueck, M. and Oksche, A., 1974. Acetylcholinesterase containing nerve cells in the pineal complex and subcommissural area in the frogs, *Rana ridibunda* and *Rana esculenta*. *Cell Tiss. Res.* **154:** 423–442.

Wake, M. H., 1986. The comparative morphology and evolution of the eyes of caecilians (Amphibia, Gymnophiona). *Zoomorphology* **105:** 277–295.

Wake, D. B., Nishikawa, K., Dicke, U. and Roth, G., 1988. Organization of the motor nuclei in the cervical spinal cord of salamanders. *J. Comp. Neurol.* **278:** 195–208.

Wallstein, M., Dicke, U. and Roth, G., 1995. Distribution of 5-HT-like immunoreactivity in visual and visuomotor brain centres of salamanders. P. 472 *in* "Proc. 23rd Göttingen Neurobiol. Conf. (I)", ed by N. Elsner and R. Menzel. Georg Thieme, Stuttgart.

Wallstein, M. and Dicke, U., 1996. Distribution of glutamate-, GABA- and glycine-like immunoreactivity in the optic tectum of plethodontid salamanders. P. 87 *in* "Proc. 24th Göttingen Neurobiol. Conf. (I)", ed by N. Elsner and H. U. Schnitzler. Georg Thieme, Stuttgart.

Werblin, F. S. and Dowling, J. E., 1969. Organization of the retina of the mudpuppy *Necturus maculosus*. II. Intracellular recording. *J. Neurophysiol.* **32:** 339–355.

Werner, C., 1983. Zielorientierung im Beutefangverhalten des Feuersalamanders *Salamandra salamandra* L. Dissertation. Technische Hochschule Darmstadt.

Wicht, H. and Himstedt, W., 1988. Topologic and connectional analysis of the dorsal thalamus of *Triturus alpestris* (Amphibia, Urodela, Salamandridae). *J. Comp. Neurol.* **267:** 545–561.

Wiggers, W., 1991. Elektrophysiologische, neuroanatomische und verhaltensphysiologische Untersuchungen zur visuellen Verhaltenssteuerung bei lungenlosen Salamandern. Thesis, Universität Bremen.

Wiggers, W. and Roth, G., 1991. Anatomy, neurophysiology and functional aspects of the nucleus isthmi in salamanders of the family Plethodontidae. *J. Comp. Physiol.* **169:** 165–176.

Wiggers, W. and Roth, G., 1994. Depth perception insalamanders: The wiring of visual maps. *Europ. J. Morphol.* **32:** 311–314.

Wiggers, W., Roth, G., Eurich, C. and Straub, A., 1995. Binocular depth perception mechanisms in tongue-projecting salamanders. *J. Comp. Physiol.* **176:** 365–377.

Wilczynski, W. and Northcutt, R. G., 1977. Afferents to the optic tectum in the leopard frog: An HRP study. *J. Comp. Neurol.* **173:** 219–229.

Wilson, M. A., 1971. Optic nerve fibre counts and retinal ganglion cell counts during development of *Xenopus laevis* (Daudin). *Q. J. exp. Physiol.* **56:** 83–91.

Witkovsky, P., Yang, C. Y. and Ripps, H., 1981. Properties of blue-sensitive rod in the *Xenopus* retina. *Vision Res.* **21:** 875–883.

Witkovsky, P. and Stone, S., 1983. Rod and cone inputs to bipolar and horizontal cells of the *Xenopus* retina. *Vision Res.* **23:** 1251–1258.

Witkovsky, P. and Stone, S., 1987. GABA and glycin modify the balance of rod and cone inputs to horizontal cells in the *Xenopus* retina. *Exp. Biol.* **47:** 13–22.

Wong, R. O. L., 1989. Morphology and distribution of neurons in the retina of the american garter snake *Thamnophis sirtalis*. *J. Comp. Neurol.* **283:** 587–601.

Wong-Riley, M. T. T., 1974. Synaptic organization of the inner plexiform layer in the retina of the tiger salamander. *J. Neurocytol.* **3:** 1–33.

Zhang, Y. and Straznicky, C., 1991. The morphology and distribution of photoreceptors in the retina of *Bufo marinus*. *Anat. Embryol.* **183:** 97–104.

Zhu, B. S., Hiscock, J. and Straznicky, C., 1990. The changing distribution of neurons in the inner nuclear layer from metamorphosis to adult: a morphometric analysis of the anuran retina. *Anat. Embryol.* **181:** 585–594.

APPENDIX 1

Glossary of abbreviations used in the text

AC — amacrine cells
BC — bipolar cells
BON — basal optic neuropil
BOT — basal optic tract
CGT — *corpus geniculatum thalamicus*
E — excitatory
EI — excitatory-inhibitory
EL — edge length
ERF — excitatory receptive field
5-HT — serotonin
GABA — gamma-aminobutyric acid
H — wormlike configuration
HC — horizontal cells
HRP — horseradish peroxidase
I — inhibitory
INL — inner nuclear layer
IPL — inner plexiform layer
IRF — inhibitory receptive field
IS — inner segment

LGB — lateral geniculate body
LGT — *corpus geniculatum thalamicus*
NBl — *neuropil Bellonci, pars lateralis*
NBm — *neuropil Bellonci, pars medialis*
NB — *neuropil Bellonci (nucleus Bellonci)*
nBON — neurons of the basal optic neuropil
nBOT — neurons of the basal optic tract
ONL — outer nuclear layer
OPL — outer plexiform layer
OS — outer segment
P — pretectal neuropil
PVG — periventricular grey
RF — receptive field
RGC — retinal ganglion cells
S — squares
TP — thalamo-pretectal
UF — *area uncinata* (uncinate field)
V — Antiwormlike configuration

CHAPTER 4

The Octavolateralis System of Mechanosensory and Electrosensory Organs

Bernd Fritzsch and Timothy J. Neary

I. Introduction
   A. Definition of the Octavolateralis System
   B. Developmental, Structural and Evolutionary Relationships of the Octavolateralis Organs
      1. Of Dorsolateral and Epibranchial Placodes, their Differentiation, Migration and Invagination: The Possible Role of Selector Genes for the Formation of Octavolateralis Organs
      2. Of Mechanosensory, Electrosensory and Gustatory Secondary Sensory Cells: Sructural, Functional and Developmental Similarities and Differences
      3. Commonalities and Differences in the Central Projections of Octavolateralis Organs: What Makes them Segregate in the Alar Plate?
      4. Evolution of the Amphibian Octavolateralis System: Losses and Gains and their Functional Implications
   C. The Octavolateralis Efferent System: Structural Diversity of a Functionally Enigmatic Mechanosensory Control System

II. The Amphibian Ear
   A. The Inner Ear, its Sensory Epithelia and their Innervation Patterns
      1. Overview of Sensory Organs in the Ear of Various Amphibians: Two Functional Systems in One Organ
      2. Generalized Organization of a Mechanosensory Transducer System: The Hair-cell, its Apical Specializations and their Functions
   B. Gravistatic and Acceleration-sensitive Parts of the Vestibular System: Different Functions are Determined by Accessory Structures
      1. The Semicircular Canals, their Structure and Transducer Functions
      2. The Three Organs of the Gravistatic System: Utricle, Saccule and Lagena
   C. The Auditory End-organs of Amphibians: A Unique Experiment in Terrestrial Hearing
      1. The Periotic System Provides a Pathway for Sound from the Oval to the Round Window
      2. The Basilar Papilla: Structure and Function
      3. The Amphibian Papilla: Structure and Function
      4. Development of the Ear: Segregation of Epithelia, Formation of Recesses and Addition of Hair-cells
      5. The Middle Ear: Structure and Function
      6. Central Projections and their Function
      7. The Amphibian Auditory System: Variation within a Network of Constraints

III. The Lateral Line: Mechanosensory and Electrosensory Organs
   A. Mechanosensory Receptor Organs: Structure and Function
      1. Distribution of Neuromasts in Amphibians
   B. The Electroreceptive Ampullary Organ: Structure and Function
   C. Evolution of Lateral Lines among Amphibians
   D. Central Projections of the Lateral Line: Generalizations and Differences
      1. Projection to the Midbrain: Generation of a Spatial Map and its Orientation with the Visual Map
   E. Predation and Predatory Avoidance: Is *Xenopus* a Good Model for Lateral Line Function in Amphibians?
   F. Development of the Lateral Line
      1. Induction of Placodes
      2. Formation and Migration of the Placodes and Formation of Neuromasts and Electroreceptors
      3. Regeneration of Neuromasts and their Dependence on Afferent Innervation
   G. The Octavolateralis System as a Model for Development and Neurobiology

IV. References

V. Abbreviations Used in the Text

I. INTRODUCTION

A. Definition of the Octavolateralis System

AMPHIBIANS, like other vertebrates, have (1) primary sensory cells, which send their own axon to the brain, and (2) secondary sensory cells, which have no axon of their own. The latter cells are connected to the brain by ganglion cells situated in cranial nerves VII–X. Whereas primary sensory cells are restricted in craniate vertebrates to the olfactory organ, secondary sensory cells are found in four different organs: ampullary electroreceptors, mechanosensory lateral line organs, the inner ear (with its vestibular and its auditory sensory parts) and the taste buds. Based on developmental and connectional data, the first three of these collectively are distinguished from taste buds as the octavolateralis system (Popper *et al.* 1992). However, this distinction has limitations as described in the older literature. Therefore, the validity of the term "octavolateralis system" is explored first.

Throughout this chapter when recent, excellent reviews are available, they are cited rather than the entire body of literature.

B. Developmental, Structural and Evolutionary Relationships of the Octavolateralis Organs

Several epidermal thickenings called placodes (von Kupffer 1895; Northcutt 1992a) develop along the head of most vertebrates, including amphibians (Northcutt *et al.* 1994). These placodes are (from rostral to caudal): (1) the olfactory placode, which produces the primary sensory cells of the main and the accessory olfactory organ, the terminal nerve ganglion and Schwann-like glia cells, (2) the lens placode which produces only non-neuronal cells, (3) the profundus placode, which gives rise to the ophthalmic subdivision of the trigeminal nerve, (4) the dorsolateral placodes (which form all lateral line ganglia, the inner ear ganglia and the secondary sensory cells of the ear and the lateral line, including ampullary electroreceptors), and (5) the epibranchial placodes, which give rise to the gustatory ganglia (Farbman and Mbiene 1991; Northcutt 1992a; Webb and Noden 1993; Barlow and Northcutt 1995). Whereas the first two placodes are associated with the forebrain, all of the remaining ones produce ganglia which provide afferent input to the hindbrain (Fig. 1). The latter placodes obviously evolved in the earliest craniates (Fritzsch 1993) and are essential parts of the "new vertebrate head" (Northcutt and Gans 1983; Fritzsch and Northcutt 1993). However, it is unclear whether they are serial homologues or represent independent evolutionary events. Only the dorsolateral and epibranchial placodes and their derivatives are treated here.

1. Of Dorsolateral Placodes, their Differentiation, Migration and Invagination: the Possible Role of Selector Genes for the Formation of Octavolateralis Organs

The dorsolateral placodes form early ganglion cells which innervate the organs derived from the same placode through migration or elongation (Northcutt *et al.* 1994). Preliminary data indicate that this neurogenesis may occur as early as stage 21 in the octaval placode of *Xenopus laevis* (Winklbauer 1989). Data from the axolotl indicate that formation of the ganglion cells occurs only a few stages after the placodes can be recognized unambiguously (Northcutt *et al.* 1994).

The lateral line placodes undergo either elongation (in the head) or migration (in the trunk), depositing groups of cells which eventually differentiate into mechanosensory lateral line organs (i.e., neuromasts) and ampullary organs (Winkelbauer 1989; Northcutt 1992a; Northcutt *et al.* 1994). The ampullary organs form late in development (Fritzsch and Bolz 1986; Northcutt *et al.* 1994). The octaval placode, which always is the first to appear, undergoes a process of invagination which leads first to the formation of the otic cup and subsequently to the otocyst (primordium of the inner ear). This process is largely completed before most other placodes can be recognized. Subsequent development of the otocyst has been studied in a number of amphibians and it first forms a single sensory epithelial anlage which undergoes repeated subdivisions until all sensory epithelia of the inner ear are formed (Yntema 1955; Fritzsch and Wake 1988; Fritzsch 1996).

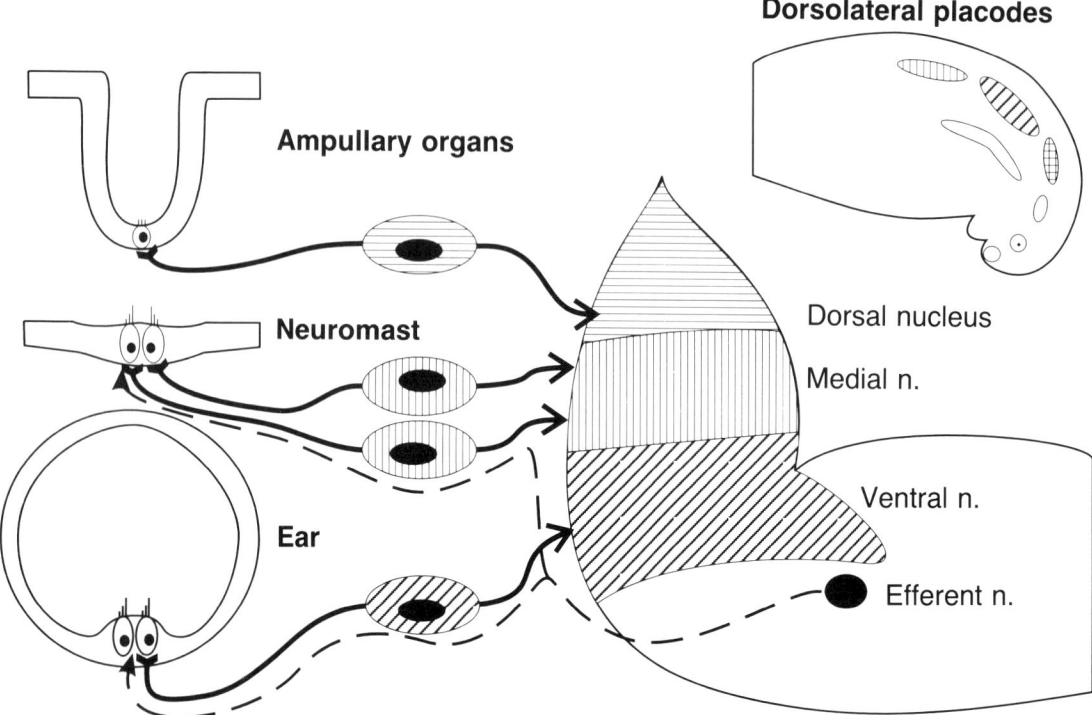

Fig. 1. The three octavolateral organs (ampullary electroreceptors, mechanosensory lateral line neuromasts, inner ear) and their primary connections in the hindbrain of a salamander. In addition, the position of efferent cells is shown as well as their axonal trajectories to the two sets of mechanosensory organs. The insert on the right shows the approximate position of the dorsolateral placodes that give rise to these organs and ganglia. Hatching is identical for ganglia and their central nuclei in the hindbrain and the placodes of origin.

While most of the molecular mechanisms leading to the formation of an ear or a lateral line are not yet understood, there is hope for a radical change in this matter. Within the last few years, a striking correlation in the spatial expression of some selector genes and the position of placodes has been reported for mammals (Noden and van de Water 1992) and bony fish (Effer et al. 1992), leading to speculation that there is a causal relationship between the topology of expression of certain genes and the induction and formation of the ear. This assumption is supported by the finding that changes in the expression of some genes have strong effects on the formation and differentiation of the ear. In chickens, blockade of the gene *Int-2* suppressed invagination of the octaval placode (Repressa et al. 1991). In mice, however, a genetically engineered destruction of *Int-2* did not suppress invagination but did alter ear differentiation (Mansour et al. 1993). In mice, blockade of product formation of the homeobox gene *Hox*A1 led to abnormal development of the inner ear, middle ear and the adjacent parts of the brainstem (Carpenter et al. 1993). This relation of gene expression to the developing ear has been tested in *Xenopus laevis* by changing the expression pattern of certain selector genes using the teratogen retinoic acid. This substance is known to change the expression of some genes in *Xenopus*; for example, *Xhox 1.6* is upregulated almost 10-fold (Sive and Cheng 1991) and application of retinoic acid can block various aspects of ear formation in a dose- and stage-dependent fashion (Neary and Fritzsch 1992). From these data one may speculate that the pattern of selector gene expression along the hindbrain provides not only a distinct code for the formation of the rostrocaudal differences of the hindbrain (Hunt and Krumlauf 1992) but also for the different placodes. It would be interesting to know whether and how this pattern of expression differs in various amphibians, given that only salamanders and many caecilians, but no frogs, form both ampullary organs and neuromasts in their anterior lateral line system (Fritzsch and Münz 1986). In summary, a start has been made towards an understanding of crucial molecular steps in the development of the octavolateralis system, but much remains to be learned.

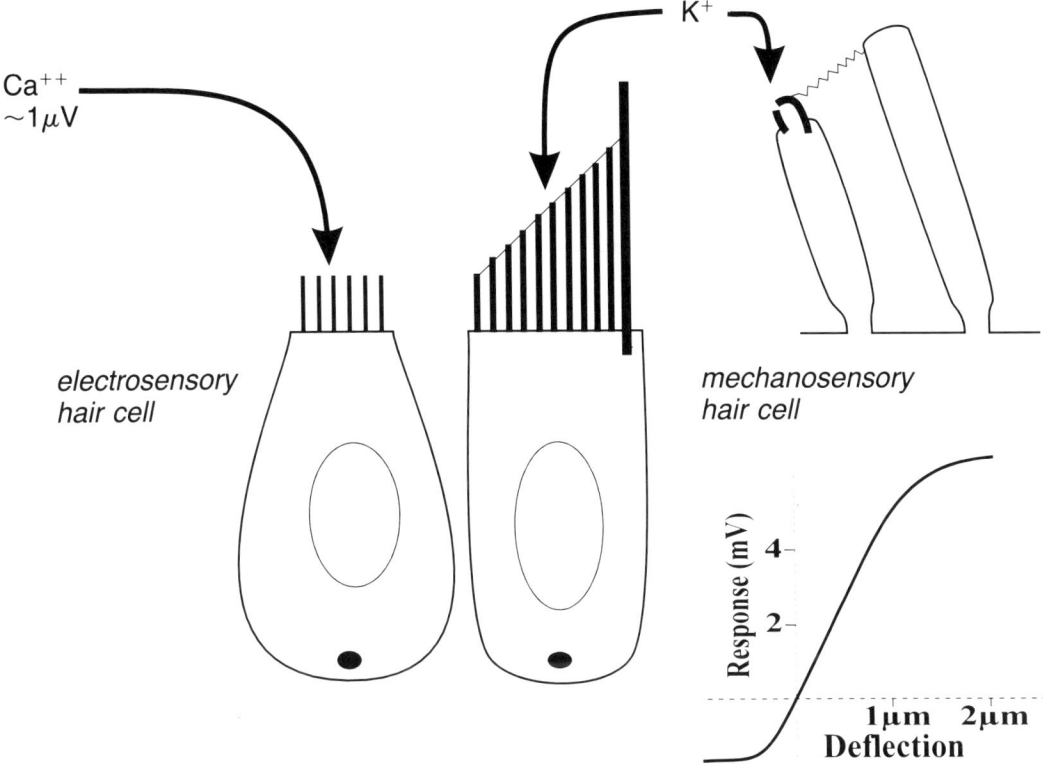

Fig. 2. Major differences between an electrosensory hair-cell and a mechanosensory hair-cell. The structural differences are most pronounced in the apex. Electrosensory hair-cells may bear microvilli or a kinocilium which show no sign of an asymmetric organization. In contrast, the mechanosensory hair-cell displays a marked asymmetry of the stereovilli which are always smaller away from the eccentric kinocilium. Mechanoelectric transduction apparently requires the tip links between microvilli (insert upper right). If deflection causes the tip links to relax the mechanically gated cation channels close. If the tip link tension is increased through appropriate deflection, mechanically gated channels open and the influx of cations cause a proportional change in the resting potential of about 3 mV per 1 μm deflection (inset lower right). In contrast, electrosensory hair-cells have likely divalent cation channels which change the stochastic nature of their gating properties under the influence of electric fields with a strength of as little as 1–10 μV/cm.

2. Of Mechanosensory, Electrosensory and Gustatory Secondary Sensory Cells: Structural, Functional and Developmental Similarities and Differences

The evolutionary origin of secondary sensory cells is as unresolved as the evolutionary origin of placodes (Northcutt 1992a). They may have evolved either from pre-existing primary sensory cells (Jorgensen 1989) or *de novo* from placodes (Fritzsch 1993). Consequently, the ancestry of the three kinds of secondary sensory cell is also unknown. It is uncertain which kind evolved first and whether there were transformations of one kind into another (Fritzsch 1993).

The mechanosensory hair-cell of the inner ear and lateral line is characterized by an "organ-pipe" arrangement of stereovilli (or stereocilia) and a single kinocilium (Fig. 2). The longest stereovilli are adjacent to the kinocilium, the shortest are furthest away. The tip of each stereovillus is connected to the next higher stereovillus by so-called tip links (Fig. 2) which insert into a mechanically gated cation channel at the shorter stereovillus. These asymmetric tip connections, together with the asymmetric arrangement of stereovilli with respect to the kinocilium provide a functional asymmetry which determines the characteristics of mechanosensory hair-cells (Corwin and Warchol 1991). If the kinocilium is deflected for only a few nm towards the bundle of stereovilli, all tip links relax and all mechanically gated channels close, thus hyperpolarizing the resting potential of the hair-cell. In contrast, deflection of the kinocilium away from the bundle of stereovilli tightens the tip links, thereby opening the mechanically gated cation channels on the stereovilli. This causes potassium and calcium, the prominent cations in the endolymph of the ear (Lewis *et al.* 1985), to flow into the hair-cell, thereby causing a depolarization of the hair-cell. Transmitter release at the base of the hair-cell reflects changes in the resting potential and, concomitantly with a deflection of the kinocilium away from the bundle of stereovilli, excites the distal process of a ganglion cell. This principle has been investigated extensively in the saccular hair-cells of frogs (Corwin and Warchol 1991; Lewis 1992) and apparently holds for the mechanosensory lateral line organs as well (Kroese and van Netten 1989).

In contrast to the functional asymmetry of the apical part of the mechanosensory hair-cell, the electrosensory hair-cell has no asymmetric specialization (Fig. 2). In caecilians these cells have a central kinocilium surrounded by a few microvilli (Wahnschaffe *et al.* 1985) and in some salamanders there are only microvilli (Fritzsch and Wahnschaffe 1983). Other salamanders may lose the kinocilium during development (Northcutt *et al.* 1994). Ampullary hair-cells, like all non-teleostean ampullary organs (Bodznick 1989), are exited by minute electric gradients (as little as 5 μV/cm) (Fritzsch and Münz 1986) which likely open voltage-gated Ca^{2+} channels in the apical membrane (Bennett and Obara 1986). The influx of cations causes a depolarization of the hair-cell that in turn leads to a concomitant release of transmitter at its base and thus to an excitation of the afferent fibre. Unfortunately, the molecular relationship of these channels to other calcium channels, such as the voltage-gated ones (with very different sensitivities) located at the base of the hair-cells (Fig. 2), is unknown.

Secondary sensory cells of taste buds also lack a functional asymmetry in their apical part (Roper 1989). Like some electroreceptive sensory cells, these hair-cells have microvilli which protrude into the mouth cavity in frogs (Sato 1978). These microvilli apparently have calcium channels which are gated by their interaction with certain ligands (Roper 1992). The influx of calcium (and perhaps other cations) leads to the depolarization of the hair-cell which causes a proportional transmitter release and excitation of the gustatory afferents impinging on them.

Clearly, similarities between secondary sensory cells exist past their origin from placodes which is now doubtful for taste buds (Barlow and Northcutt 1995). All of these sensory cells have microvilli although they are modified in mechanosensory hair-cells by the accumulation of actin filaments. All seem to have calcium or cation channels in the apical specialization. These channels are, however, very differently gated: chemically in taste buds, electrically in electroreceptors and mechanically in mechanoreceptors. Knowledge about the molecular biology of these channels may be very important to an understanding of the evolutionary relationships (if any) among themselves as well as with other cation channels.

3. Commonalities and Differences in the Central Projections of Octavolateralis Organs: What Makes them Segregate in the Alar Plate?

Beyond the ontogenetic and structural arguments of the similarities of sensory cells, there is one feature that seems to unite all secondary sensory cell organs; their placodally derived ganglion cells are restricted to cranial nerves VII to X and their central projection largely is restricted to the alar plate of the hindbrain and the most rostral cervical spinal cord (Figs 1, 3). In contrast even to the placodally derived ophthalmic ganglia of the trigeminal nerve, many (if not all) ganglia of secondary sensory cells have axons that bifurcate after entering the brain. The ascending part of the mechanosensory and the gustatory projections reach anteriorly up to the cerebellum (Fig. 3). However, both electroreceptive and auditory afferents tend to be shorter in their rostro-caudal extent.

Within the alar plate, the afferent projections show a discrete segregation (Figs 1, 3). Double-labelling studies show little overlap between afferents of different modalities. This segregation holds whether afferents are compared from different sources (i.e., lateral line and gustatory afferents) or from a single source (auditory and vestibular afferents of the inner ear). Similarities in rostro-caudal extent and position are particularly obvious when vestibular and lateral line, on one hand, or vestibular and gustatory projections, on the other, are compared. The reason(s) for this segregation could be biochemical specifications, timing during development, or patterns of activity. Moreover, ganglion cells could be autonomous

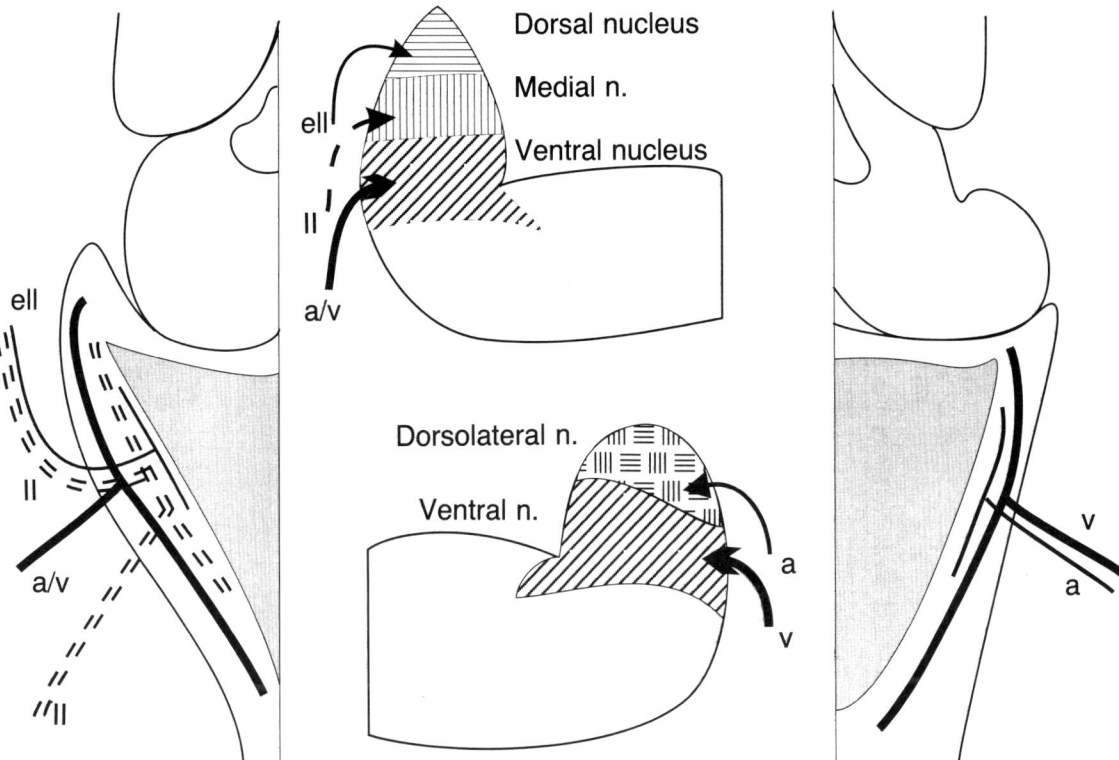

Fig. 3. The longitudinal and transverse distribution of octavolateral organ projection in adult salamanders (left and top centre) and adult frogs (right and bottom centre). Many adult salamanders have three octavolateral organ systems, the electroreceptive ampullary organs (ell), the mechanosensory lateral line organs (ll) and the ear with the auditory and vestibular component (a/v). The longitudinal extent of the projection falls off from the most dorsal to the most ventral projection. Most adult frogs have only an inner ear projection. Note that the dorsolateral (auditory) nucleus occupies a position that is topologically comparable to both dorsal and intermediate nuclei. Whether or not some of the cells of the ancestral nuclei become incorporated into the dorsolateral (auditory) nucleus once the organs are lost is a matter of conjecture.

in their projection choice or could be specified by the sensory cells they connect. Experimental analysis of these possibilities is needed to understand how segregated projections of these ganglia are achieved. It has been suggested recently that this may occur in a central to peripheral progression, e.g., those fibres that reach the electroreceptive dorsal nucleus will project to ampullary electroreceptors (Northcutt *et al.* 1994). Alternatively, the timing of ganglion cell differentiation could lead to stratified projection. The timing of the development of afferent projections suggests that the latter may indeed be the case (Fritzsch, in prep.).

In conclusion, based on the placodal origin of most secondary sensory cells and the ganglia, four sensory systems may be grouped together: (1) gustatory, (2) electrosensory, (3) mechanosensory lateral line and (4) inner ear. Only the last two share, in addition, the apical specialization for mechanoelectric transduction. The central projections of all these sensory systems are segregated about equally well and provide no clues either to ancestry or to the mechanisms for this segregation. Nevertheless, the term octavolateralis system may be useful to distinguish all extero- and proprioreceptive sensory systems (inner ear, mechanosensory lateral line, electrosensory system) from the enteroreceptive gustatory system. Used in its purely descriptive anatomical sense, the term does not imply or exclude any evolutionary affinity between all these sensory systems. It is in this sense that the term octavolateralis system is used here.

4. Evolution of the Amphibian Octavolateralis System: Losses and Gains and their Functional Implications

Comparative data imply that the plesiomorphic feature of the ancestral amphibians likely was the presence of electroreceptive ampullary organs on the head, mechanosensory lateral line organs on the head and trunk and an inner ear having both a vestibular and an auditory

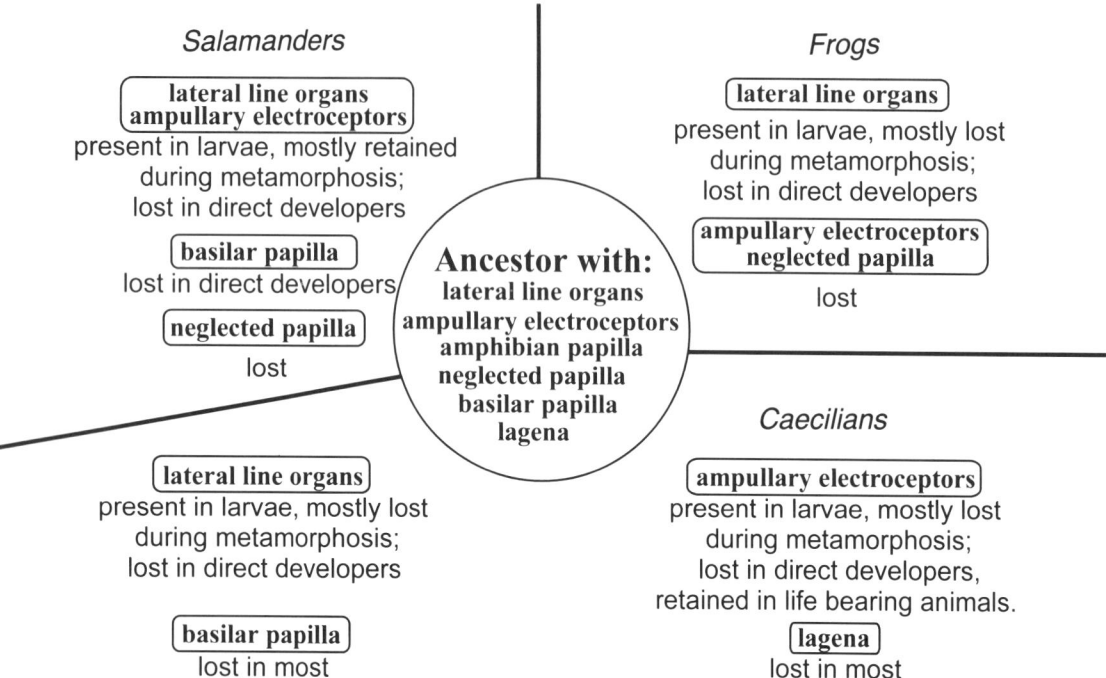

Fig. 4. Octavolateral sensory organs in a hypothetical amphibian ancestor and in the three Orders of living amphibians. Note that most losses of organs occur within a lineage and can be studied in their developmental and molecular causality. However, some losses happened in the extinct ancestors of a taxon (e.g., ampullary electroreceptors in frogs) and can not be studied directly.

part (Fritzsch 1989, 1992). In contrast, many amphibian lineages have lost one or the other part of the octavolateral system (Figs 3, 4). These losses, if interpreted in the context of functional constraints, may offer insight into the largely neglected area of regressive, adaptive evolution (Fritzsch and Wake 1988; Fritzsch 1996).

The electroreceptive ampullary system is present in salamanders and caecilians (Fritzsch and Münz 1986; Northcutt et al. 1994). This likely primitive pattern was modified among amphibians several times through losses. Electroreceptive ampullary organs are absent in all modern anurans, a loss that is likely related to a change in feeding behaviour (Fritzsch 1989). Electroreceptive organs also were lost independently several times in salamanders and caecilians. This loss appears to occur in direct-developing species. However, intraoviducal larvae can retain the ampullary electroreceptors and one aquatic caecilian family is the only vertebrate known that develops only electroceptive ampullary organs in its lateral line (Fritzsch and Wake 1986; McCormick et al. 1994). In addition, the adult, permanently aquatic salamander *Andrias* appears to lose its electroreceptive organs (Cheng et al. 1995; however, von Düring and Andres [1993] described ampullary organs) and some salamanders with direct development may at least retain the lateral line projection (Wake et al. 1988). It appears that absence of an aquatic stage in the life cycle is reasonably well correlated with loss of electroreception but does not guarantee it.

The mechanosensory lateral line is lost in many, but not all frogs during metamorphosis, but seems not to develop at all in species with direct development (Fritzsch 1989). To the contrary, metamorphosing salamanders usually retain their lateral line organs except in a few species. However, no lateral line organs seem to develop in salamanders with direct development, but are known in at least one ovoviviparous salamander. This area has been insufficiently examined and some surprises may be found on closer scrutiny. Among caecilians, there appears to be loss of both organs during metamorphosis and no formation of organs during direct development or in intraoviducal animals (Fritzsch 1990). Absence of formation of lateral line organs is reasonably correlated with the lack of an aquatic stage, whereas retention of lateral line organs during metamorphosis is strongly correlated with an aquatic habit in adult animals at least during some portion of their life cycle.

Within the amphibian ear three sensory epithelia, the basilar papilla (BP), the lagena and the neglected papilla, are known to be lost in various lineages (Lewis and Lombard 1988; Fritzsch and Wake 1988). The pattern that emerges is a multiple, likely independent loss among many salamanders and caecilians of the basilar papilla, a multiple loss of the lagena in caecilians and a likely single loss or transformation of the neglected papilla in salamanders and frogs (Fritzsch and Wake 1988). The losses of both the lagena and the basilar papilla likely are correlated with the differences in properties of sound on land, as compared to water, and may indicate a loss of receptors evolved in an aquatic environment but unsuitable under terrestrial conditions. These losses will be discussed again after the functions of these organs have been presented.

C. The Octavolateralis Efferent System: Structural Diversity of a Functionally Enigmatic Mechanosensory Control System

Most mechanosensory hair-cells of the ear have an efferent innervation (Roberts and Meredith 1992). The location of the perikarya of these cells has been studied in all three orders of amphibians and has been found to be a single nucleus, called the octavolateral efferent nucleus, which partly (salamanders, caecilians) or completely (frogs) overlaps with the facial nucleus. In addition, the mechanosensory hair-cells of the lateral line, but not the electrosensory hair-cells, have an efferent innervation (Fig. 1). The perikarya of these lateral line efferent cells may be coextensive with those to the ear or they may be clustered in a population around the glossopharyngeal motoneurons (Will and Fritzsch 1988). Some of the lateral line efferent cells also can send branches to one or both ears (Claas et al. 1981; Fritzsch and Wahnschaffe 1987). The numbers of efferent perikarya ranges between approximately 10 in salamanders and caecilians to around 20–30 in frogs (Will and Fritzsch 1988). Their location also varies; all efferent perikarya in salamanders have a likely primitive, bilateral distribution but have a (perhaps independently derived) unilateral distribution in frogs and caecilians. Axonal trajectories also differ considerably (Will and Fritzsch 1988). In salamanders and caecilians all axons form a loop to the midline of the hindbrain before turning laterally but in frogs they run directly toward the nerve root. These differences are comparable to those described in lampreys, birds and mammals (Fritzsch 1996) and may come about through similar developmental pathways.

Efferent terminals occur in all inner ear epithelia of salamanders and caecilians but are absent in the basilar papilla of ranid frogs (Will and Fritzsch 1988). This apparent loss of efferent innervation to this particular sensory epithelium may have evolved within anuran amphibians, because efferents to the basilar papilla are present in *Xenopus* (Hellmann and Fritzsch 1996) and efferent terminals in the basilar papilla are found in salamanders (Fritzsch and Wahnschaffe 1987) and caecilians (White and Baird 1982). Within the sensory epithelia, the efferent terminals generally contact hair-cells and typically have subsurface cisternae (Hillman 1976). Contacts with afferent fibres do occur but are less frequent (Lewis et al. 1985).

The function of efferents still is incompletely understood. From the available data it appears that efferents to the ear or the lateral line can function in two radically different ways: a feedback mode and a feedforward mode (Roberts and Meredith 1992). In feedback mode, efferents are activated by the various sensory systems impinging upon their vast dendritic trees (Münz and Claas 1991; Roberts and Meredith 1992). In feedforward mode they are activated by the initiation of movements. Whatever activates the efferent cells, their small calibre, slowly conducting fibres apparently either inhibit or excite at the level of the frog's ear. Recent molecular evidence suggests that one kind of activity may involve unique properties of a novel acetylcholine receptor found on hair-cells that is formed from possibly 5–9 subunits (Elgoyhen et al. 1994) while the other kind of activity may involve other, not yet characterized acetylcholine receptors on afferents.

II. THE AMPHIBIAN EAR

The amphibian ear can be subdivided into three compartments, (1) the otic capsule, (2) the periotic tissue and spaces and (3) the otic labyrinth. The otic capsule consists of hyaline and fibrous cartilage and/or bone and surrounds the membranous labyrinth or inner ear.

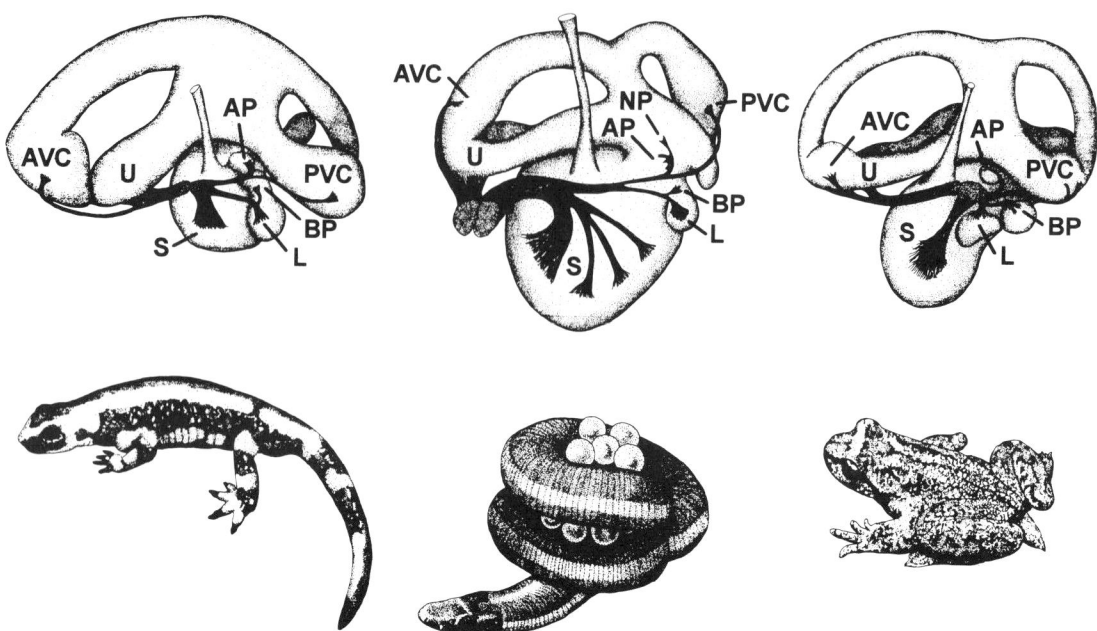

Fig. 5. Right inner ears (viewed medially) of a salamander (*Salamandra salamandra*, left), a caecilian (*Ichthyophis glutinosus*, centre) and a frog (*Ascaphus truei*, right). All species have the three canals (AVC, anterior vertical canal; PVC, posterior vertical canal; horizontal canal not shown), a utricle (U), saccule (S), lagena (L), basilar papilla (BP) and amphibian papilla (AP). Only caecilians have a neglected papilla (NP) which appears to be the primitive condition in the out-groups. Despite major differences in the form of the labyrinth there is a marked constancy in the pattern of innervation of these sensory epithelia.

Between the otic labyrinth and the otic capsule is the periotic (perilymphatic or limbic) tissue that in places forms tubes and sacks filled with perilymph (Hillman 1976; Capranica 1976). The otic capsule usually has four or more foramina: (1) a medial foramen (which may be extensively subdivided) (de Burlet 1934) for cranial nerve VIII, i.e., the nerve innervating the sensory epithelia of the inner ear, (2) a lateral foramen, the "oval window" with the stapes foot plate and (if present) the operculum, (3) a caudomedial foramen, the "round window" towards either the brain or the middle ear cavity, and (4) a dorsomedial foramen through which the endolymphatic duct exits into the braincase. Adjacent to the inner ear in most frogs is the tympanic cavity with a stapes that transmits vibration from the tympanic membrane to the ear. The tympanic cavity and the tympanum are reduced by a variable degree in many amphibian lineages (Jaslow *et al.* 1988).

A. The Inner Ear, its Sensory Epithelia and their Innervation Patterns

The amphibian inner ear (or otic labyrinth) is a system of interconnected tubes filled with endolymph (Lewis *et al.* 1985). Traditionally, the otic labyrinth is subdivided into an upper part (pars superior) and a lower part (pars inferior) which are connected by a more or less constricted junction, the utriculo-saccular foramen. The upper part consists of three semicircular ducts (anterior vertical canal [AVC], posterior vertical canal [PVC] and horizontal canal [HC]), as well as the utricule to which these ducts are joined (Fig. 5). The lower part contains the saccular recess with the otic (or perilymphatic) duct and three other, variably present recesses: a lagenar recess (absent in some caecilians), a recess for the amphibian papilla (AP) (absent in many caecilians) and a basilar recess (absent in many salamanders and caecilians). Within this system of communicating tubes there are nine sensory epithelia which fall into three categories: (1) Three sensory epithelia with hair-cells polarized into one direction only (the ampullary cristae) are covered by gelatinous cupulae and are within the ampulary enlargements of the semicircular canals. (2) Three sensory epithelia with hair-cells polarized toward or away from a dividing line (the equivalent of the mammalian striola). These are the lagena, the saccule and the utricle. They have a gelatinous matrix with calcium carbonate crystals, the otoliths, and are called maculae. (3) Three sensory epithelia with hair-cells in either single or multiple directions and which are covered by a gelatinous tectorium: these are the amphibian papilla, the basilar papilla and the neglected papilla. The latter two are reduced or absent in many amphibians.

1. Overview of Sensory Organs in the Ear of Various Amphibians: Two Functional Systems in one Organ

The nine sensory epithelia within the amphibian ear subserve the perception of two different kinds of stimuli: (1) gravistatic and dynamic vestibular stimuli that monitor the position and movement of the head (and body) in space and (2) auditory/ vibratory stimuli of the air or the substrate functioning in species-specific communication as well as in predator/ prey interactions. While in general these functions are associated with certain sensory epithelia alone (e.g., the semicircular canals perceive the dynamic component of movement in space), some epithelia can function both in vestibular sensing of position in space and in detection of low frequency substrate vibration (e.g., the saccule and lagena) (Lewis *et al.* 1985). Vibration sensitivity is known for the saccule in some fish and some mammals (Lewis *et al.* 1985) and may in fact reflect a primitive feature that is derived in frogs only in so far as they have gained an exquisite sensitivity to substrate vibration (Lewis *et al.* 1985).

2. Generalized Organization of a Mechanosensory Transducer System: The Hair-cell, its Apical Specializations and their Functions

All hair-cells of the amphibian inner ear display the generalized vertebrate apical specialization: a single kinocilium flanked on one side by a bundle of increasingly shorter stereocilia (or stereovilli). As discussed above, this functional asymmetry is the basis for the mechanoelectric transduction processes in any mechanosensitive hair-cell. Any stimulus that will result in a shearing force acting in the appropriate direction on the apical ciliary bundle can stimulate hair-cells. From this it follows that it is not the hair-cell *per se* but rather its associated structures that determine the kind of stimulus which is effective, e.g., a gravistatic, an accelerative or an auditory stimulus. Understanding the multiple functional subcomponents of the ear requires, therefore, a thorough understanding of how associated structures make certain epithelia specifically sensitive to a unique set of stimuli.

This is not to say that there are no functional differences in hair-cells themselves. In fact there is ample evidence that the length of the apical specializations of hair-cells varies and may play a role in frequency representation. Another feature unique to some hair-cells involves ringing or electric resonation at a specific frequency, again perhaps proving the discreteness of frequency responsiveness (Lewis 1992). Thus, hair-cells show some specialization of their own that may provide additional filter systems in the sensory context in which these hair-cells are involved, in particular auditory perception.

B. Gravistatic and Acceleration-sensitive Parts of the Vestibular System: Different Functions are Determined by Accessory Structures

The gravistatic and acceleration-sensitive part of the ear was described last century by Retzius (1881; reviewed by Hillman 1976). Within the pars superior are three semicircular canals, each with an ampullary enlargement harbouring a sensory epithelium (ampullary crista). In addition, the pars superior also contains the utricle, one of the two major otolithic receptor systems. The pars inferior harbours the second otolithic receptor system, the saccule and also (if present) the lagena (third otolithic receptor system) (Lewis *et al.* 1985; Fritzsch and Wake 1988). In addition, the pars inferior contains the other sensory epithelia found in the amphibian ear, (1) the amphibian papilla, (2) the basilar papilla and (3) in caecilians, the neglected papilla located near the utriculo-saccular foramen (Lewis and Lombard 1988; Fritzsch and Wake 1988).

1. The Semicircular Canals, their Structure and Transducer Functions

All semicircular canals emerge and end as tubes of variable diameter from the elongated utricle. The ACV, PVC and HC are in approximately orthogonal planes to each other. In addition, the left and right horizontal canals are in about the same plane (approximately 18°–20° of the plane of the upper jaw) (Hillman 1976). Similarly, the AVC of the right side and the PVC of the left side are in about the same plane. The anterior and posterior vertical canals are joined to the utricle medially by a common crus from which they project at about 45° towards anterior and posterior, respectively.

There is one ampullary enlargement in each of the three individual canals (except for the common crus), each with a ridge of sensory epithelia, the crista. The crista and the attached gelatinous cupula obstruct the ampullary enlargement apparently completely in frogs (Hillman 1976) and probably in other amphibians as well. The horizontal crista is different from the two vertical canal cristae. The latter are composed of two dumb-bell-shaped planar ends with a constricted region between them, the isthmus. In contrast, the horizontal crista appears to be only a half-epithelium with one dumb-bell-shaped planar end and an isthmal region attached to it (Hillman 1976). This feature appears to be a derived character of tetrapods, shared with most amniotic vertebrates except mammals, which may have secondarily modified this feature (Lewis *et al.* 1985). The sensory cells of a given crista have a single polarity, i.e., the organ-pipe specializations are pointing all in one direction: away from the utricle in the vertical canals, towards the utricle in the horizontal canal.

The hair-cells come in three different kinds with respect to their ciliary tufts, i.e., kinocilium and stereocilia. The cells in the centre of the crista have thick and long stereovilli; those in the periphery have short and slender stereovilli and cells in between are intermediate in size and length of stereovilli (Myers and Lewis 1990). The innervation pattern consists of thick fibres innervating single or few central hair-cells with long and thick stereocilia, whereas thinner fibres innervate many peripheral cells. Afferent and efferent synapses have been described (Myers and Lewis 1991). Electrophysiological evidence suggests that these structural variations provide the basis for the differences of single afferents with respect to resting activity, stimulus-response relationships and adaptation.

Semicircular canals monitor angular accelerations in space of 0.3 to 2.5°/sec^2 (in ranid frogs) using the physical principle of the relative inertia of the endolymphatic fluid behind the fast-moving canal walls. In effect, such a movement will cause the endolymph to push against the cupula, largely blocking the ampullary enlargements. With increasing acceleration, cupular movement apparently extends beyond the central region to the planar ends. It is likely that the cellular specializations, as well as the size and dimensional variations of the cupula and hair-cells, all add up to form an organ that has profound capacities to monitor onset and continuous acceleration over a rather wide range (Lewis *et al.* 1985). Acceleration in the direction of pitch and/or roll preferentially stimulates the vertical canals whereas acceleration in the yawing (horizontal) plane stimulates the horizontal canals.

A. VARIATION OF THE SEMICIRCULAR CANAL SYSTEM OF AMPHIBIANS: CURVATURE AND DIAMETER OF SEMICIRCULAR CANALS ARE RELATED TO LIFE STYLE

Semicircular canals can be modeled as a fluid-filled pipe, a viscoelastic diaphragm (the cupula) and a large reservoir of fluid (the utricle). These model systems allow for at least two distinct interpretations of the outcome of various acceleration experiments. In one model the coupling between the cilia of the receptor cells and the cupula is considered to be weak and provides for a passively responding system. Alternatively, the cilia-cupula coupling can be considered to be strong and the micromechanics of the ciliary bundle require an active element (Lewis *et al.* 1985). Exact knowledge of acceleration responses is available only from a limited sample of ranid frogs (Lewis *et al.* 1985). However, it has been shown that the length and thickness of canals vary independently among vertebrates; there are long, slender canals in some whereas others have comparatively shorter and wider ones. The case of birds is particularly interesting in that the most airborne and arboreal birds have long, thin canals whereas the land and water dwellers have shorter, wider canals (Lewis *et al.* 1985).

Among extant amphibians, the canals appear to be proportionally longer and thinner in frogs and appear to be shortest and widest in caecilians (Retzius 1881; Fritzsch and Wake 1988). It seems that the more active amphibians, such as arboreal frogs, have the longest and thinnest canals whereas the fossorial caecilians should have the shortest and widest canals. Following the current mechanical models of fluid movement, it appears that the more slender canals perhaps would allow more precise monitoring of the very small accelerations (below $0.5°/sec^2$) important for fast-moving animals with high agility. It would be interesting to see whether arboreal frogs approach the organization of birds in this respect. If so, this would be a case of convergent evolution in response to similar physical principles.

B. THE PERIPHERAL FIBRE SUPPLY AND CENTRAL PROJECTIONS OF SEMICIRCULAR CANALS

The cristae of the semicircular canals are innervated in the bullfrog by approximately 1 300–1 500 afferent fibres (Dunn 1978) with the bipolar perikarya in the anterior (AVC, HC) or the posterior ramus (PVC) of the VIIIth nerve. At the basement membrane of the cristae, the 1–3 μm thick myelinated afferents lose their sheath to innervate a few or many hair-cells, typically of one particular type (Lewis *et al.* 1985). Centrally, the afferents of the semicircular canals travel in a topological order through the octaval nerve to enter the medulla oblongata of the brain dorsal to the facial root. Within the brain, probably all fibres bifurcate to extend along the alar plate of the hind brain from the cerebellum to the obex region where they terminate in a ventral zone (Will and Fritzsch 1988). Selective projections of the semicircular canals have been mapped only in two frogs (Matesz 1979; Will *et al.* 1985) and partly in one salamander (Will and Fritzsch 1988). Nothing is known about projections from the semicircular canals of caecilians.

The data compiled thus far show a large degree of overlap of all semicircular canal afferents among themselves and with the utricular afferents. However, in certain areas one or the other of the semicircular canal projections appears to dominate (Will *et al.* 1985). A more detailed mapping employing double labelling is necessary to evaluate critically the degree of overlap in different areas. In summary, the central projections of the semicircular canals largely confirm what is known in other vertebrates (McCormick 1992).

C. THE SEMICIRCULAR CANALS AND THE VESTIBULAR-OCULAR REFLEX

The function of the angular acceleration detection system of amphibians has been studied extensively in frogs (Dieringer and Precht 1986) and in salamanders (Manteuffel *et al.* 1986). The data accumulated thus far imply some commonalties with other vertebrates, but there are also some differences (Dieringer *et al.* 1992). Amphibians react to angular acceleration by moving the eyes and the head. However, amphibians may have, presumably primitively, a much longer delay of the onset of responses than is true of mammals. In addition, frogs apparently lag in velocity-storing and consequently they show no after-nystagmus, i.e., after termination of movement their heads do not snap back. Interestingly, it appears that some salamanders may have such a system, however rudimentary (Manteuffel *et al.* 1986). In that respect, more systematic, comparative work is necessary to determine the abilities of key groups (e.g., leiopelmatid frogs). Depending on the outcome of such a cladistic analysis one may either be able to provide evidence for a parallel evolution of such a network in some salamanders or, alternatively, evidence that it is a retention of a network already present in their common ancestors with frogs. There appears to be a parallel case in bottom-dwelling as compared to open-water fish (Dieringer *et al.* 1992).

2. *The Three Organs of the Gravistatic System: Utricle, Saccule and Lagena*

Otoliths can serve two functions. They may provide the capability of measuring position in space with respect to gravity, or they may measure linear accelerations, including movements of the otolith relative to the sensory epithelium, such as those caused by vibrations of low frequency (Lewis *et al.* 1985). The latter aspect seems to be restricted in frogs to the saccule and the lagena. However, no data exist on aquatic amphibians, in which the ear should function much like a fish ear. Herrings, at least, use the utricle for sound perception whereas other bony fish use the saccule (Schellart and Popper 1992).

A. THE UTRICLE: STRUCTURE AND FUNCTION

In amphibians, as in most vertebrates, the utricle is an elongated sac with a single sensory epithelium in its anterior aspect, the macula of the utricle. This sensory epithelium is set in a horizontal plane and slightly curved. A curved striola, the boundary separating two sets of hair-cells with opposing polaritiy, is present. The kinocilia are oriented towards the striola. Thus, the entire set of sensory cells allows for sensitivity in about 360° of arc. Hair-cells display two types of tufts. A set of long-tufted hair-cells extends along the striola. An extrastriolar region is characterized by hair-cells with short stereovilli and long kinocilia (Hillman 1976; Lewis et al. 1985).

The otoconial mass consists of barrel-shaped crystals of calcite, each from 20 to 50 μm long and 12–20 μm wide (Wiederhold et al. 1992). The individual otoconia are embedded in a gelatinous membrane that completely covers the macula. In bullfrogs the striolar cell density is lower than in the extrastriolar region (Lewis et al. 1985).

Other than the general shape of the macula, there are few comparative data available. It would appear that the utricle, given its function as the major horizontally oriented gravistatic receptor, should be conservative among amphibians. Differences may be found with respect to sound or vibration sensitivity, especially between aquatic and terrestrial amphibians.

Pattern of peripheral innervation and central projection: The utricular macula is innervated by fibres from the anterior branch of the octavus nerve. This nerve carries both afferent and efferent fibres to the sensory macula. It appears that afferents are associated with specific subsets of hair-cells (Baird and Lewis 1986) and have physiological properties which seem to be determined by the apical specializations of the cells they contact. These morphological variations may reflect variations in the viscoelastic properties of the coupling of the hair-cell tuft to the otoconial membrane.

There are distinct differences between the striolar and extrastriolar innervations. In particular, it has been shown that the utricle of a caecilian has no efferent innervation in the striola but does in the extrastriolar region, a relationship that should be examined in other amphibians. Fibres have been characterized by intracellular recordings and it appears that there is a functional distinction between the responses of the striolar and extrastriolar fibres to sustained tilt of the head in gravity. The fibres innervating peripheral hair-cells in extrastriolar regions are phasic whereas fibres innervating the striola are tonic.

The central projections have been described in two species of frogs. It appears that the utricle projects most prominently at the root entrance to the medial and lateral vestibular nuclei as well as sparsely to the lateral line nucleus. In addition, there appears to be some overlap with the saccular projection (Will et al. 1985). Other areas of projection are found both anterior and posterior to the octaval root. Only a sparse projection to the cerebellum has been reported. Topologically there appears to be considerable overlap with the afferents of the semicircular canal cristae (Matesz 1979; Will et al. 1985).

B. THE SACCULE: STRUCTURE AND FUNCTION

Among vertebrates in general and amphibians in particular the saccule is a more variable structure than the utricle. The saccular chamber is usually the largest of all otic structures. A few addenda emanate from the saccule, such as the amphibian papilla near the utriculo-saccular foramen (Fritzsch and Wake 1988), the lagenar recess and, in anuran amphibians, the basilar recess near the orifice of the lagenar recess. In addition, the saccule gives rise to the endolymphatic duct which greatly enlarges around the spinal cord and brain in salamanders and anurans. The saccule may be applied medially to the otic capsule (salamanders, caecilians, some frogs) or may be separated by a portion of the periotic system. In addition, the saccule is always in contact with the periotic system, especially laterally.

The saccule contains a kidney-shaped sensory macula which may be very elongated in caecilians (White and Baird 1982). As in most other vertebrates, the macula has a striola. However, in contrast to the utricle, the polaritiy of the hair-cells points away from the dividing line (Lewis et al. 1985). In some salamanders a marginal population may display an erratic orientation. The macula stands nearly vertically in bullfrogs and has a 45° rotation away from the midline. The overall directionality ranges through a full circle (Hillman 1976). The ciliary specializations display variation in length but usually bear a bulbous kinocilium (Lewis and Lombard 1988). The bulb is attached to the pore of the gelatinous otoconial membrane in which the apical tufts are embedded.

The saccular otolith is an ovoid mass of aragonite crystals, 1.5–20 μm long and 0.5–6 μm wide (Wiederhold *et al.* 1992), supported by an otolithic membrane 10 μm thick (Hillman 1976).

Peripheral innervation and central projection: modern frogs deviate from the primitive pattern. The pattern of innervation of the saccule is much more variable than in any other end-organ of the ear. The data can be grouped into (1) a primitive pattern shared by salamanders, caecilians and some frogs and (2) a derived pattern found in other frogs. Out-group comparisons with lungfish and *Latimeria* suggests that an innervation of the saccule with the posterior octaval branch is primitive (Fritzsch and Wake 1988). This pattern is found in salamanders, caecilians and some anurans. Caecilians have additional rami (de Burlet 1934; Fritzsch and Wake 1988) which emerge from the posterior root. Among anuran amphibians there is a morphological clade of changes; saccular fibres either come from the posterior ramus *(Ascaphus, Bombina)* (Will and Fritzsch 1988), both the anterior and the posterior ramus *(Xenopus,* Will and Fritzsch 1988; *Rana,* Capranica 1976) or exclusively from the anterior ramus *(Bufo,* Will and Fritzsch 1988). This shift in the pattern of innervation appears to be related to a more anterior position of the saccule in anuran species, and it maybe related to perception of sound and vibration.

The approximately 1 500 hair-cells of the adult saccule (toad; Corwin 1985) receive about 1 000 afferents, of which approximately 80% are myelinated (Dunn 1978). The ratio of hair-cells to afferent fibres is approximately 1:0.67. A given afferent typically innervates about 30 hair-cells but may rarely contact up to 200 hair-cells. Thus, there is a high degree of divergence of afferents to many hair-cells. Recordings from afferent fibres indicate sensitivity to low-frequency substrate vibrations, low frequency (below 200 Hz) airborne sound, rotational acceleration and gravity (Lewis *et al.* 1985). Gravistatic sensitivity has not been reported in some amphibian species. As with the utricle, there appears to be a discrete innervation of specific classes of hair-cells by a specifically responding afferent. Particularly interesting is the fact that only afferents which contact hair-cells with a bulbous kinocilium respond to vibration (Lewis *et al.* 1985), suggesting a specialization of certain hair-cells to perceive such signals.

As with other end-organs, the saccule afferents course in a discrete fascicle within the more proximal part of the octaval nerve (Lewis *et al.* 1985). The afferents enter the brainstem wedged between those of the anterior vestibular ramus which supply the AVC, HC and utricle, as well as the lagena (Will *et al.* 1985). The fibres end predominately in two areas. One of these, the nucleus saccularis (Matesz 1979) is medial to the dorsolateral auditory nucleus (Will and Fritzsch 1988). In addition, there is a projection to other areas of the vestibular nuclei, largely in parallel to those from the lagena but less extensive. While these data are derived from anurans, preliminary evidence indicates a comparable projection pattern in one salamander and in one caecilian (Will and Fritzsch 1988).

The multiple functions of the saccule: retention of a primitive pattern or a novel adaptation? Clearly, the saccule is a multifunctional organ in anurans. Obviously, gravistatic and acceleration sensitivity must be regarded as a shared primitive character; it is also found in non-amphibian vertebrates. It is more difficult to assess the sensitivity to airborne sound and the exquisite sensitivity to substrate vibration (Lewis and Lombard 1988). The saccule of any aquatic vertebrate will function as a near-field detector of particle motion using the relatively high density of the otolith (Lewis *et al.* 1985; Schellart and Popper 1992). It appears likely that amphibians have retained this "acoustic function" from their aquatic ancestors but with an adaptive change in function, the perception of airborne and substrate-borne sound. If this is true, similar capabilities may be found in many non-anuran amphibians as well. However, sensitivity to a given stimulus should vary dramatically among species owing to different degrees of adaptation of an aquatic sensor to terrestrial conditions.

C. THE LAGENA: STRUCTURE AND FUNCTION

The lagenar sensory epithelium is a posterior evagination of the saccule which also contains the orifice of the basilar papillar recess (in salamanders and caecilians, when present) (White and Baird 1982; Lombard and Bolt 1979; Fritzsch and Wake 1988) or (in anurans) is located nearby (Wever 1985). The lagenar recess is in contact with the periotic system, the saccule and the otic capsule. A gelatinous membrane covers the entire lagena macula and is associated with a curved layer of otoconia. The sensory epithelium forms an elongated strip at the largest diameter of the lagenar recess. The orientation of the hair-cell is away from the striola, as it is in the saccule. Some cells near the striola have bulbed kinocilia and respond uniquely to vibration (Baird and Lewis 1986). Other areas with a different morphology of the hair-cells respond to gravity. The lagenar recess may be reduced greatly and its sensory epithelium is absent in some caecilians (White and Baird 1982; Fritzsch and Wake 1988).

Pattern of innervation and central projection: The innervation of the lagena runs through the posterior branch of the octaval nerve. These fibres typically are accompanied for some distance by other fibres that course to the PVC, the amphibian, basilar and neglected papillae. As in the saccule, there is a high degree of convergence and divergence. A given afferent fibre may supply between 6 and 20 hair-cells. Responses of afferents to various stimuli suggest similarity to the saccule, with the exception that the lagena does not respond to airborne sound. The central projection of the lagena is closely adjacent to the saccule but nevertheless distinct. For example, the lagena does not project to the nucleus saccularis but contributes more extensively to the lateral aspect of the dorsolateral auditory nucleus. The lagena projects more extensively than does the sacccule to vestibular areas, to the reticular formation and to the cerebellum (Will *et al.* 1985).

Evolution of the tetrapod lagena and multiple regression among amphibians: Apparently the lagena evolved at least three times independently among aquatic vertebrates: once in the elasmobranch lineage, once within the actinopterygian lineage and once within the sarcopterygian lineage (Fritzsch 1992). While it is unclear how the lagena, including its recess, evolved, it is likely that the lagena may play a role in directional analysis of underwater hearing as do other otolithic sensory organs (Schellart and Popper 1992). If that is the case, it is clear that this function is possible in terrestrial vertebrates only if associated structures provide the relevant stimulus to the lagena. Whereas this may have been achieved in some anurans in which the lagena functions as a low-frequency vibration detector in addition to its role as a gravistatic receptor (Lewis *et al.* 1985) it may not have occurred in other amphibian lineages. Accordingly, the only function that stabilized the developmental programme in the formation of a lagena was the gravistatic part. With respect to that stimulus, the lagena offers little advantage compared to the sacculus and its loss apparently is not detrimental. Consistent with such a scenario is the regression of a lagena in some caecilians (Fritzsch and Wake 1988). It should be noted that the lagena also was lost in mammals (Lewis *et al.* 1985). This regression both in amphibians and in mammals leads to the prediction that in land vertebrates which have retained it, the lagena will play a dual role, one in gravistatic perception and one in detecion of low-frequency vibration (Lewis *et al.* 1985).

C. The Auditory End-organs of Amphibians: A Unique Experiment in Terrestrial Hearing

In contrast to other tetrapods, amphibians have two sensory organs apparently exclusively dedicated to the perception of airborne sound; these are the amphibian papilla and the basilar papilla. In addition, amphibians have both the saccule and the lagena involved in the perception of low-frequency sound and substrate vibration. The latter aspect is not unique to amphibians because it appears also in mammals and may represent retention of a primitive function by aquatic ancestors. The accessory structures of the amphibian saccule suggest that the exquisite seismic sensitivity of some anuran ears may be a uniquely derived feature.

1. The Periotic System Provides a Pathway for Sound from the Oval to the Round Window

Perception of airborne sound requires at least a low resistance of sound channelling through the ear in compensation for the impedance mismatch created by attempting to drive the incompressible perilymph of the ear by airborne sound (Jaslow *et al.* 1988). Reduction of impedance mismatch is achieved in most amphibians by a low-resistance flow from one or two middle ear ossicles inserted into the ear capsule toward one or more pressure release windows. Both openings into the ear capsule are connected by a system of perilymphatic spaces, the periotic labyrinth. In amphibians, the periotic duct connecting the periotic cistern near the oval window (stapes) to the periotic sac inside the brain case, passes posteriorly around the ear

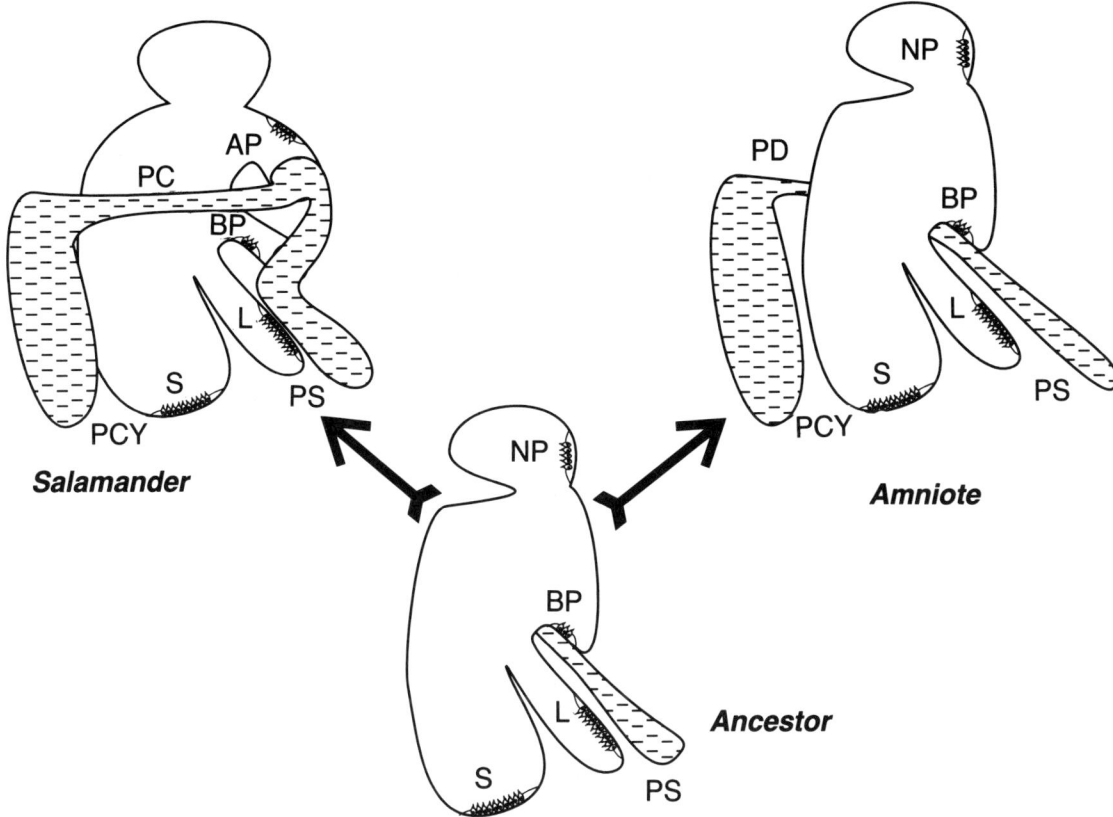

Fig. 6. Diagram of a posterior view of the left periotic labyrinth (hatched) of a hypothetical ancestor and of independently derived amphibians and amniotes. It is assumed that the ancestor had an ear much like *Latimeria* with a basilar papilla (BP) and a perilymphatic sac passing through a foramen of the otic capsule. Amphibians and amniotes are independently derived by completing a periotic canal (PC) posteriorly around the ear (amphibians) or a periotic duct (PD) anteriorly around the ear (amniotes). Note that the different pathway taken by the amphibian periotic labyrinth brings it close to the amphibian papilla (AP) and its recess, but changes the access to the basilar papilla.

(Lombard and Bolt 1979; Lewis and Lombard 1988). At least three organs are directly or indirectly related to the perilymphatic canals: the basilar papilla, the amphibian papilla and the saccule. Through this system sound may reach the sound-detecting perilymphatic end-organs with comparatively little attenuation (Jaslow *et al.* 1988). Viewed from this perspective, the periotic system is paramount in the evolution of the detection of sound pressure in air, the major way hearing is achieved in land vertebrates.

A. EVOLUTION OF THE PERIOTIC SYSTEM AND ITS VARIATION AMONG AMPHIBIANS

While not analyzed in full detail in all amphibian taxa, the data thus far indicate that in salamanders the periotic duct and its interaction with the sensory epithelia varies considerably, but that there is more uniformity among anurans and caecilians (Lombard and Bolt 1979; Lewis and Lombard 1988). Clearly, the pathway of the periotic duct posteriorly around the ear of all amphibians, as compared to the anterior pathway of the periotic canal in amniotes (Fig. 6) strongly suggests that the periotic system evolved independently in these two groups. It has been suggested that the entire system evolved in paleozoic amphibians (Bolt and Lombard 1992). Alternatively, the periotic sac and cistern may have evolved in amphibian ancestors and the completion of the periotic pathway around the ear by a periotic duct (amphibians) or a periotic canal (amniotes) may be the only part that evolved independently in these lineages (Fritzsch 1992). In any case, it appears that the differences in pathways and diameters of the periotic duct of amphibians may be related to the channeling of specific sound attributes toward epithelia. However, neither the detailed morphology nor its functional implications are known sufficiently to explain the remarkable variation in the periotic labyrinth of amphibians.

2. The Basilar Papilla: Structure and Function

The basilar papilla was first described in amphibians by Retzius (1881) and he believed this organ to have originated in salamanders. Wever (1974) pointed out that the basilar papilla of anurans is positioned differently relative to the periotic system than it is in amniotes. He concluded that this structural difference had functional significance and suggested a parallel evolution of these end-organs. Wever later (1985) described the basilar papilla in salamanders but renamed it the "lagenar papilla" to set it apart from the "basilar papilla" of anurans. He did not recognize a basilar papilla or lagenar papilla in caecilians but renamed the neglected papilla of caecilians as the "utricular papilla". Together, these structures were used to propose a multiple, parallel evolution of the basilar papilla, probably in conjunction with the parallel evolution of the periotic labyrinth (Lewis and Lombard 1988). Most recently, however, the alternative view that all basilar papillae of tetrapods are homologous and represent transformations of one end-organ that evolved prior to tetrapods, has been revived (Fritzsch and Wake 1988; Fritzsch 1992). Given these conflicting interpretations, it appears worthwhile to examine this end-organ in more detail. For the sake of simplicity this end-organ will be denoted as the basilar papilla.

The basilar papilla is found in all anurans, many salamanders and some caecilians. It is located in the vicinity of the orifice of the lagenar recess, and is a sensory epithelium occurring either in its own recess or in one emanating from the lagenar recess. This epithelium is small with a known maximum of 210 *(Pipa)* and a minimum of 10 cells *(Triturus)*. Typically, there are between 30 and 100 hair-cells. The basilar papilla is in contact with the periotic labyrinth by way of a membrane (all anurans, most salamanders). In caecilians and some salamanders (Fritzsch and Wake 1988), the basilar papilla partially overlies the periotic duct. Similar relationships were described in earlier studies (Lombard and Bolt 1979; Wever 1985) but without comment. The sensory epithelium is covered by a gelatinous tectorial membrane (Lewis *et al.* 1985). Two kinds of hair-cells with differing apical specializations have been described (Lewis *et al.* 1985). The polarity of hair-cells can be unipolar (all in the same direction), bipolar (with opposing directions pointing away or toward a dividing line) or chaotic (Lewis and Lombard 1988).

It appears that the basilar papilla of anurans functions as a receiver for sound of 1 000 Hz or higher (Fig. 7). While details are not yet clear in salamanders, recordings in one species showed a frequency response clearly above 1 000 Hz (Manteuffel and Naujoks-Manteuffel 1990), and it seems reasonable that this may be perceived by the basilar papilla. Nothing is known about the frequency response characteristics among caecilians.

Details on how sound is perceived are not yet understood, but conceivably there are two mechanisms at work: in caecilians and some salamanders the basilar papilla may move against the resting tectorial membrane whereas in anurans and other salamanders the tectorial membrane may move against the basilar papilla, the latter resting on thick limbic tissue.

A. PATTERN OF INNERVATION AND CENTRAL PROJECTION

The pattern of innervation may be the strongest argument for the presumed homology of the basilar papilla of tetrapods and *Latimeria* (Fritzsch and Wake 1988; Fritzsch 1992). In salamanders, caecilians and the primitive anuran *Ascaphus* the basilar papilla is innervated by fibres that run together with the octaval twig to the lagena. Out-group comparison with *Latimeria* and amniotes show a similar feature and suggest that this pattern of innervation is likely the plesiomorphic condition (Fig. 8). Most anurans are derived from this pattern by shifting the twig to the ramus supplying the PVC in neobatrachians (Will and Fritzsch 1988).

An additional argument may be found in the ratio of hair-cells to afferent fibres. In salamanders, caecilians and archeobatrachians this ratio is similar in the basilar papilla and the saccule (1:0.6) (Will and Fritzsch 1988). This ratio changes concomitantly with the change in fibre pathways in neobatrachians to reach 1:4 or higher. In the bullfrog, each afferent fibre contacts a single hair-cell and at the most 2 to 4 (Lewis *et al.* 1985). It would be interesting to know how many hair-cells are contacted in species with fewer afferents. Recordings and dye fillings in anurans show that the afferents of a given basilar papilla all are tuned to a narrow

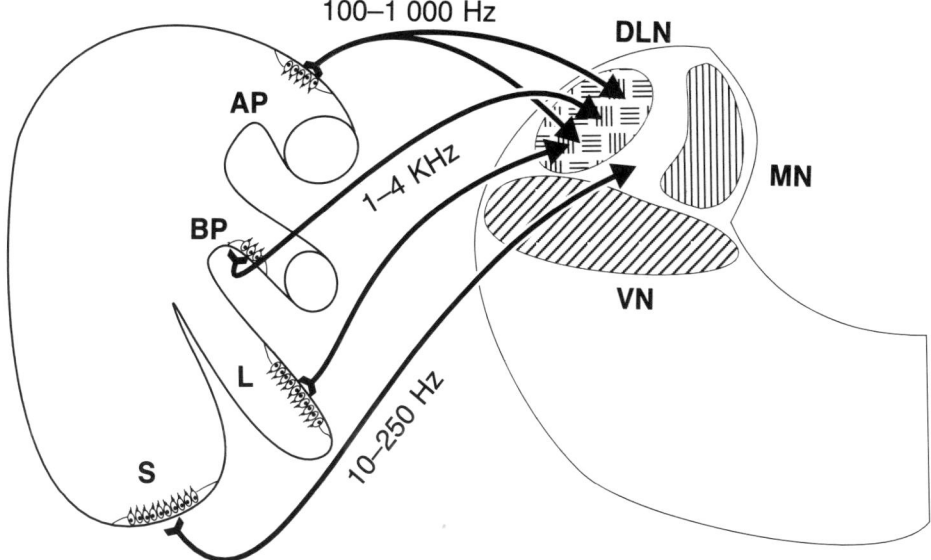

Fig. 7. Sound frequencies mediated by projections of the auditory system on to the brain. AP = amphibian papilla; BP = basilar papilla; L = lagena; S = saccule; DLN = dorsolateral nucleus; MN = medial nucleus; VN = ventral (vestibular) nucleus. Note that anatomic data suggest that the basilar papilla has a projection between two patches formed by the amphibian papilla, and that the medial nucleus, which receives the mechanosensory lateral line afferents, may be present in frogs like *Xenopus*. The saccule appears to project to an area adjacent to the medial nucleus. Thus, low-frequency vibration as perceived by either the lateral line or the saccule could converge on to the same neurons in that area.

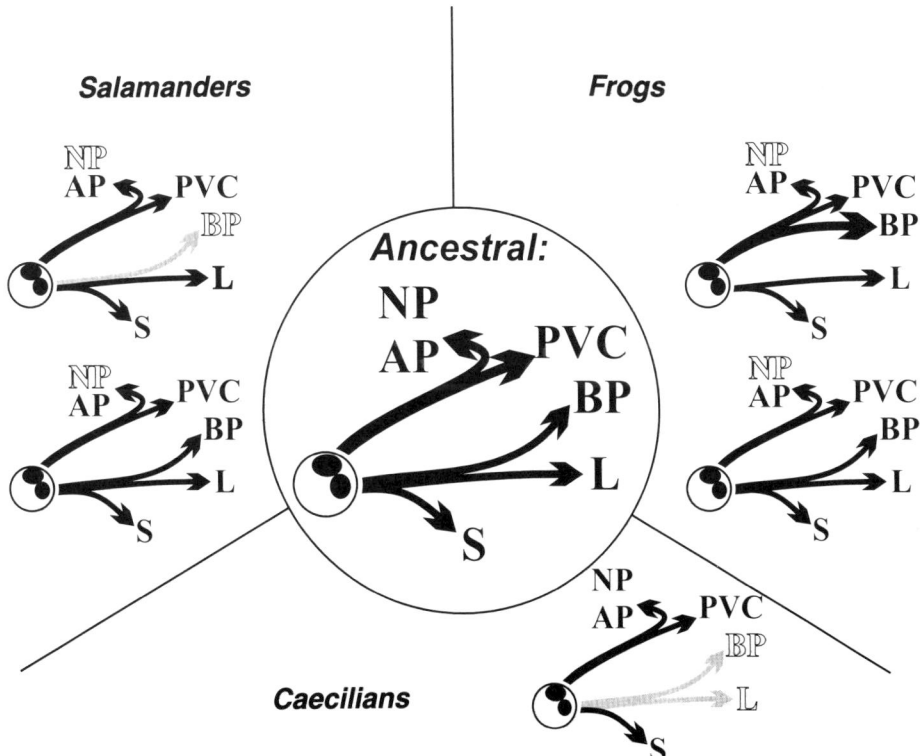

Fig. 8. Ancestral and derived patterns of posterior ear innervation. Solid symbols indicate presence, open symbols indicate loss of the sensory organ and its innervation. There is only one change in the pattern of innervation: within frogs the innervation of the basilar papilla changes from the lagenar branchlet to the PVC/AP branchlet.

frequency band which is typically between 1 and 4 kHz (Zakon and Wilczynski 1988). There is a variation in frequency selectivity which suggests that larger animals with a larger tectorial body covering the basilar papilla respond to lower frequencies. This also implies that frequency selection is a function of a mechanical resonator (Zakon and Wilczynski 1988). Such simple mechanical properties may underlie the apparent sexual dimorphism encountered in many species of anurans.

The central projections of the basilar papilla have been studied both anatomically and electrophysiologically in anurans. The afferents travel in the posterior branch of the octaval nerve adjacent to those from the amphibian papilla and the PVC. The fibres predominantly end at the dorsolateral auditory nucleus. The afferents end between two terminal fields provided by the afferents from the amphibian papilla (Will *et al.* 1985) (Fig. 7). In addition there are minor projections to restricted areas of the vestibular nuclei. With respect to its projection, the basilar papilla differs clearly from the vestibular end-organs and shows some similarities with the saccular and lagenar projections, receptors known to react to low-frequency vibration or sound. Unfortunately, comparable data are not available for other amphibian orders.

B. EVOLUTION OF THE AMPHIBIAN BASILAR PAPILLA: ARGUMENTS FOR AND AGAINST ITS HOMOLOGY WITH THE BASILAR PAPILLA OF *LATIMERIA* AND OTHER TETRAPODS

Clearly, the morphocline of pattern changes among anurans and the in-group comparison among amphibians, as well as the out-group comparison with amniotes and *Latimeria*, is explained most parsimoniously by assuming a single origin for the basilar papilla. All the differences in fibre pathway, fibre to hair-cell proportions and positioning either on limbic tissue or on a basilar membrane are morphological changes in this homologous organ. The strongest arguments against this concept were presented by Wever (1974, 1985); he contended that it is important whether hair-cells move against a stationary tectorial membrane (as in the mammalian cochlea) or whether a tectorial membrane moves against stationary hair-cells (as in the anuran basilar papilla). However, it is now known that mechanoelectric transduction is accomplished by shearing of the apical stereovilli. It appears to be irrelevant whether this is done by moving hair-cells against the tectorial membrane, a tectorial membrane against hair-cells, or by a mixture of both movements. In fact, in birds the basilar papilla largely rests on limbic tissue and is stimulated by movements against the tectorial membrane or of the tectorial membrane against the hair-cells (Manley and Gleich 1992). Moreover, Wever (1985) depicted the presence of a basilar papilla (his lagenar papilla) in salamanders correctly as sitting on a periotic duct (his figures 14–5 and 14–6). This feature also is found in caecilians (Fritzsch and Wake 1988) and consequently may be the primitive organization of the amphibian basilar papilla; this arrangement also is found in both out-group taxa, *Latimeria* and amniotes.

Another argument for a parallel evolution of the basilar papilla of amphibians and tetrapods was based on the different course taken by the periotic duct system in amphibians and amniotes (Lewis and Lombard 1988; Bolt and Lombard 1992). This is a valid argument but as such does not prove the point. It is equally likely that the completion of the periotic labyrinth occurred after a basilar papilla already had evolved (Fritzsch 1992) and that the differences in the periotic duct of amphibians and amniotes are more related to the morphological changes observed in the amphibian papilla.

These remarkable changes in topology of the basilar papilla and fibre pathways, previously so misleading for the interpretation of the basilar papilla, can now be described. However, the developmental and genetic background of these changes still are not fully understood. In this context, it might be interesting to perform transplantation experiments between ears of *Bombina* and *Bufo* to ascertain the extent to which these characteristics are endogenous to the ear as opposed to being instructed by the surrounding tissue. This may help to specify the genetic basis for these changes.

C. DIVERSIFICATION OF THE AMPHIBIAN BASILAR PAPILLA: MULTIPLE LOSSES

An argument that has been neglected in the debate on the evolution of the ear is the apparent multiple loss of the basilar papilla in salamanders and caecilians (Lombard and Bolt 1979; White and Baird 1982; Lewis and Lombard 1988; Fritzsch and Wake 1988). While there is agreement that absence of the basilar papilla in certain salamanders and caecilians should be considered as a derived rather than a primitive feature, it appears worthwhile to speculate about the reasons for these losses. One may ask: Why should a system that evolved *de novo* in comparable species be lost without major changes in life history? A simpler and more consistent answer is possible if the proposition of a single evolutionary origin of the basilar papilla in the aquatic ancestor of tetrapods is accepted (Fritzsch 1992). It appears likely that in some amphibians the adaptation to the perception of airborne sound was less successful than in others. As a consequence, these species may have lost an organ that had evolved in different functional contexts and was playing a minimal functional role in the new life style.

3. The Amphibian Papilla: Structure and Function

In contrast to the checkered distribution of the basilar papilla, all amphibians studied thus far have an amphibian papilla. However, in some caecilians this end-organ may be rather small and is not in a recess of its own (Fritzsch and Wake 1988). In all other amphibians the amphibian papilla is in its own recess which opens to the saccule or into the utriculo-saccular foramen in some salamanders (Lombard and Bolt 1979). The amphibian papilla is always in contact with the periotic labyrinth through a contact membrane (Wever 1985). The single or (in most frogs) double sensory epithelium is on the dorsal aspect of this recess. A gelatinous body underlies the papilla and fills most of the recess in salamanders, caecilians and at least one frog, *Ascaphus* (Lewis *et al.* 1985). The second, elongated sensory patch of most frogs is covered by a tectorial curtain. The polarity of hair-cells in the single patch of salamanders, caecilians and *Ascaphus* and the anterior patch of all other anurans is of opposing directions; the posterior, elongated patch has hair-cells predominantly parallel to its curved tail. It is the length of this tail that varies considerably among anuran amphibians (Lewis *et al.* 1985). Variable types of hair-cells are found in the amphibian papilla. Most anurans have hair-cells with a bulb but *Ascaphus*, salamanders and caecilians have unbulbed kinocilia. Most remarkable is the variation in size of the amphibian papilla. It ranges from six hair-cells in some caecilians to 1 500 hair-cells in the bullfrog (Fritzsch and Wake 1988).

The amphibian papilla of anurans is known to respond to sound predominately below 1 kHz (Lewis *et al.* 1985; Fuzessery 1988) (Fig. 7). It appears that the amphibian papilla is tonotopically organized with the small anterior patch representing the lower frequencies (100 to 300 Hz) and the curved posterior patch the higher frequencies (200 to 1 000 Hz in bullfrogs). It is not completely clear how this frequency segregation along the posterior patch is accomplished but both local variation in the tectorial curtain, as well as mechanoelectric properties of the hair-cells themselves, are thought to play a role (Lewis and Lombard 1988; Lewis 1992).

A. PERIPHERAL INNERVATION AND CENTRAL PROJECTION: PARALLELS TO THE BASILAR PAPILLA

Peripherally the afferents to the amphibian papilla travel with the afferents to the PVC. If present, the afferents to the neglected papilla (caecilians; White and Baird 1982; Fritzsch and Wake 1988) travel with those of the amphibian papilla. Only in neobatrachians are the afferents to the basilar papilla adjacent to those of the amphibian papilla in the twig to the PVC. Thus, the relationship of afferents changes through loss of one fibre bundle (present only in caecilians) and gain of another fibre bundle (present only in neobatrachians). As with the basilar papilla, the ratio of hair-cells to afferents changes from 1:0.6 in caecilians and salamanders to a ratio of 1:1 or better in neobatrachians (Will and Fritzsch 1988). Each afferent appears to innervate between one and 15 hair-cells with those in the anterior patch contacting more hair-cells than those in the posterior patch (Lewis and Lombard 1988).

The central projection is known in several anurans and in one salamander. In anurans the fibres enter the medulla dorsal to those of the basilar papilla. They end predominantly in the dorsolateral auditory nucleus in which they form a dorsal and medial terminal surrounding the fibres of the basilar papilla (Will *et al.* 1985) (Fig. 7). This subdivision may represent the two hair-cell patches of the amphibian papilla. The absence of such a division in salamanders is consistent with the suggestion of a dual terminal field being derived from the two sensory patches. It would be desirable to know how the projection of the basilar papilla is arranged with respect to the amphibian papilla in any amphibian which has a single patch of sensory epithelium in the amphibian papilla (e.g., salamanders, caecilians or *Ascaphus*). Like the basilar papilla the amphibian papilla has an additional projection to the vestibular nuclei. The projection of the amphibian papilla of the axolotl shows a comparable topological relationship to the basilar projection as do those of anurans. It would be important to ascertain the position of the projection of the neglected papilla of caecilians.

B. EVOLUTION OF THE AMPHIBIAN PAPILLA: ARGUMENTS FOR ITS ORIGIN FROM THE NEGLECTED PAPILLA OF OTHER VERTEBRATES

All orders of amphibians have a sensory organ that is unique to amphibians, the amphibian papilla. The name was created by de Burlet (1928) who believed that this sensory epithelium must have been present in all vertebrates but was lost in all groups but amphibians. He conjectured correctly on the auditory function of the amphibian papilla, noting its proximity to the periotic labyrinth. A simpler scheme of an affinity between the neglected and amphibian papillae originally was proposed by Sarasin and Sarasin (1888, 1892); they conjectured that in various amphibians either parts (caecilians) or the entire neglected papilla (salamanders, anurans) was shifted into its new position to become the amphibian papilla. The arguments in favour of this view were summarized recently (Fritzsch and Wake 1988) and are as follows: (1) Both end-organs receive fibres from the same, isolated branch in all vertebrates. (2) Both end-organs are close to or within the utriculo-saccular foramen. (3) Both end-organs are separated late in development through the formation of the utriculo-saccular foramen. (4) Both end-organs may come as a double or as a single sensory epithelium; except for the horizontal canal, they are the only end-organs to do so.

All these data can be explained more parsimoniously by assuming the scheme originally proposed by the Sarasins, i.e., a partial or complete shift of the neglected papilla into the saccule. This shift may be associated with the unique course of the periotic labyrinth which brings it close to the amphibian papilla only in amphibians (Lombard and Bolt 1979; Lewis and Lombard 1988) (Fig. 6). Thus, the functional differences between the neglected and the amphibian papilla may represent the transformation of an organ evolved in the context of other sensations into an organ that in amphibians became the primary auditory receptor once it shifted its position and achieved proximity to the periotic sound-conducting pathways. Evolution of the amphibian periotic labyrinth would have happened before this transformation took place.

4. Development of the Ear: Segregation of Epithelia, Formation of Recesses and Addition of Hair-cells

The ear, like other sensory systems of the octavolateralis system, develops from a placode. The induction of this placode requires the mesoderm and, to a lesser extent, the neuro-ectoderm (Yntema 1955; Noden and van de Water 1986; Fritzsch 1996). After induction, the ear undergoes fixation of polarity axes at the time it invaginates. Following invagination, the ear largely is fixed in its development but requires some interaction with the mesenchyme around it for full differentiation. Thus, the sensory epithelia segregate from the common anlage but the formation of recesses may be more or less suppressed in the absence of specific otic mesenchyme (Swanson *et al.* 1990). The sequence of segregation of sensory epithelia in a salamander suggests a surprisingly conservative evolutionary sequence. For example, the first epithelia to segregate are the vertical canal cristae, thus forming an ear with a common macula and two canal epithelia, much like the ear of hagfish, probably the most primitive ear of craniate vertebrates (Baird 1974). It is important to note that in salamanders segregation of the lagena and the basilar papilla occurs last (Norris 1892) whereas the amphibian and neglected papilla are the last to segregate from each other in caecilians (Fritzsch and Wake 1988). In addition, comparative data show that the basilar papilla and lagena, on the one hand, and the amphibian and neglected papillae on the other hand receive fibres from the same twig of octaval fibres unless secondarily modified (Fritzsch and Wake 1988; Will and Fritzsch 1988). These data, together with topological arguments (Fritzsch and Wake 1988), strongly suggest that these double epithelia represent subdivisions of a single ancestral epithelium.

More complicated is the formation of recesses. Apparently these form from different aspects of the ear (e.g., basilar recess from the lagena or the saccule close to the lagena, amphibian recess from the saccule or the utriculo-saccular foramen). In addition, a recess may have disappeared (e.g., amphibian recess in some caecilians) or the sensory epithelia may have disappeared but not the recess (e.g., lagenar recess in some caecilians). Thus, both features are somewhat independent in their development and may vary individually. An additional variation is provided by the periotic labyrinth. These perilymph channels are particularly variable in their proximity to the amphibian and basilar papillae (Lombard and Bolt 1979). It is particularly important to know what makes these three independent structures develop in such a way that they form a well organized functional entity. Given the degree of their independence, one should not be surprised that they have evolved rather rapidly into different patterns. Whether each of these patterns can be explained in terms of its adaptive value, or merely represent just another permutation of ontogenetic randomization, remains to be shown.

As in most anamniotic vertebrates, the ear of amphibians undergoes a long-lasting proliferation of new hair-cells (Corwin 1985). Moreover, it has been suggested that this is actually a turnover of hair-cells (Jorgensen 1989). There appears to be little increase in the number of ganglion cells from larval to adult stages. Thus, the ratio of hair-cells to afferents changes (Fritzsch 1988). Presumably, extensive branching of the afferents compensates for the lack of new formation of afferents. As a consequence, there would be a convergence of more hair-cells onto a given afferent fibre. It is important to know whether the different ratios of hair-cells to afferents are a function of the differential regulation of hair-cell and ganglion cell precursors for a given epithelium, and in particular, whether or not any clonal relationship exists between the sensory epithelium and the set of ganglia innervating it.

5. The Middle Ear: Structure and Function

Among amphibians, only postmetamorphic anurans possess a complete tympanic ear (tympanum, middle ear cavity with Eustachian tube, stapes and oval window). However, parts of it (stapes, oval window) are found in many amphibians including many salamanders and caecilians (Jaslow et al. 1988). The middle ear of amphibians is unique among tetrapods in having a second ossicle or cartilage, the operculum, which is connected to the scapula by the opercularis muscle (Hetherington 1992). The operculum is present in postmetamorphic frogs and salamanders but is absent in completely aquatic (neotenic) salamanders. Its absence in caecilians may be derived, probably in conjunction with the recent loss of legs in these amphibians (Jenkins and Walsh 1993). Alternatively, the opercularis muscle connecting the operculum to the shoulder girdle may not be homologous, thus forcing the view that it is functionally important and hence evolved convergently, perhaps serving in the detection of ground vibrations (Hetherington 1992) by the saccule or the amphibian papilla (Christensen-Dalsgaard and Narins 1993).

While a general principle is emerging that higher frequencies are transmitted through the tympanic ear whereas lower frequencies are transmitted through the opercular system, there are alternative pathways that need to be taken into account (review by Hetherington 1992). One such pathway is the lung/lateral body wall (Narins et al. 1988). It appears that this system may be particularly important in small frogs and for low frequencies (Jorgensen 1991). In addition, this pathway may play a role in directional hearing (Schmitz et al. 1992). Thus, access of sound (both aerial and substrate) to the ear may make use of different parts of the middle ear in different species. As a consequence, it is impossible at the moment to match the variation of the middle ear (tympanic ear present or absent, opercular system present or absent, sound perception by the lung/lateral body wall present or absent) consistently to the variation of the inner ear with its three receptors (amphibian and basilar papillae, saccule and lagena) and the variations in the periotic labyrinth. Moreover, the body size as well as the habitat (aquatic versus terrestrial or fossorial) may additionally, or even predominantly, influence the formation of one or the other pattern (Hetherington 1992). For example, smaller amphibians within a given taxon are more likely to lack a tympanic ear. In addition, loss of a tympanic ear and lack of access to a lung (lungless salamanders, caecilians) correlates with reduction of the basilar papilla, as does a fossorial life style. Physiological tests are needed to show the frequency sensitivity of these species to verify the predictable reduction of high-frequency perception.

A. EVOLUTION OF THE MIDDLE EAR

The otic and periotic characters of the inner ear likely evolved prior to a terrestrial middle ear in an inside-out fashion (Bolt and Lombard 1992; Fritzsch 1992). The inner ear, being encapsulated by bone and filled with incompressible fluid, would reflect 99.9% of all airborne sound were it not for low-resistance pathways into and through the inner ear. Thus, a middle ear is only one part of a functionally interconnected system that allows sound to enter the ear in a way minimizing the impedance mismatch between the ear and the air (Jaslow *et al.* 1988; Hetherington 1992). Obviously, having an amplifier system that further overcomes the impedance mismatch would be even more beneficial. The middle ear, in particular the tympanic ear of anurans (Hetherington 1992), may represent such an amplifier.

Further, being amphibians, it would be beneficial to have a receiver that functions both in terrestrial and aquatic environments. The latter is accomplished in bony fish by associating the lung (or swimbladder) with the ear (Fritzsch 1992). Such an association also is known in tadpoles and adult anurans, including terrestrial ones (Narins *et al.* 1988; Hetherington 1992). Thus, what began primarily as an aquatic adaptation may have continued to be exploited in a terrestrial context. The extent to which the middle ear of amphibians is an evolutionary exploitation of aquatic adaptations is a matter of conjecture and has been extensively debated (Bolt and Lombard 1992; Clack 1992; Fritzsch 1992). Most of the discussion revolves around the presumed function(s) of the hyomandibular/stapes ossicle, the oval window in the otic wall and the absence or presence of a tympanum in fossils. Differing interpretations largely are driven either by implementing the theoretical considerations of van Bergijk (1967) (Fritzsch 1992) or by focusing predominantly on available fossil data (Bolt and Lombard 1992; Clack 1992; Hetherington 1992). It seems fair to conclude that at present a number of critical issues remain unclear. Important questions are: (1) When did the autostylic skull of terrestrial vertebrates evolve thus freeing the hyomandible from its functional constraints? (2) When and how many times did the stapes adapt to new functions, including its insertion into the oval window? At least two independent insertions currently are favoured (Bolt and Lombard 1992; Fritzsch 1992). (3) Did the tympanum evolve in ancestral, aquatic tetrapods, one or several times in terrestrial tetrapods, or was it lost and re-evolved several times? Because of the paucity of features, it will be particularly difficult to distinguish between homologous and homoplastic tympana.

Given the checkered distribution of a middle ear in recent amphibians (Jaslow *et al.* 1988; Hetherington 1992) one has to be aware that known fossils may represent secondary modifications of an ear that was only partly adapted to the task of terrestrial hearing in their respective ancestors. More details about the fossil history of the middle ear and its functional interpretation are available in papers by Bolt and Lombard (1992) and Clack (1992).

B. DEVELOPMENT OF THE MIDDLE EAR

Given that most interpretations tend to agree that the structures of the anuran middle ear, including the tympanum, are novel evolutionary achievements paralleling those of amniotes (Bolt and Lombard 1992; Hetherington 1992), one may expect some unique patterns of development to be found in amphibians. In fact, a number of differences from amniotic development have been noticed. For example, the relative contribution of the otic capsule and neural crest to the formation of the stapes differs among vertebrates (Fritzsch 1992). Moreover, the timing of development of the stapes differs in caecilians and salamanders as compared to anurans; in the first two groups, it forms prior to metamorphosis when it fuses with the otic capsule (for salamanders, see Hetherington 1988), whereas in the latter group, the stapes (if it develops at all) forms at or after metamorphosis. Interestingly, the development of the operculum in salamanders and anurans is reversed in order; it develops before the stapes in anurans but after the stapes in salamanders. Together with the different muscles inserting in the operculum this suggests that the operculum in the two groups may not necessarily be homologous (Hetherington 1988). Information on the innervation pattern and the position of motoneurons in a wider range of species is needed in order to reach a more conclusive interpretation. In any case, the otic development in anuran tadpoles is so remarkably different from the formation of the tympanic ear of amniotes that it strengthens the interpretation of an independent evolutionary origin.

6. Central Projections and their Function

Among amphibians, the anatomy and physiology of the central auditory system have been most thoroughly investigated in anurans. Far fewer studies have been conducted on salamanders and virtually nothing is known about these aspects in caecilians. In anurans, the major components of the central auditory system are the dorsolateral nucleus (DLN), superior olivary nucleus (SO) and secondary isthmal nucleus of the medulla, the torus semicircularis of the midbrain, the central thalamic and ventral hypothalamic nuclei of the diencephalon, and the striatum of the telencephalon. Figure 9 shows a highly simplified version of their connections. A more detailed summary of their relations and physiology follows, along with an accounting of other structures which also contribute to the central auditory system.

A. THE DORSOLATERAL (AUDITORY) NUCLEUS

In anurans, afferent fibres from the BP and AP end in segregated terminal fields within the DLN of the medulla (Fig. 7). Basilar papillar fibres occupy a small circumscribed region in the medial or dorsomedial part of the DLN (Will and Fritzsch 1988) and recordings in this region indicate that units there respond to high frequency stimuli (Fuzessery 1988). The BP terminal area is surrounded by the AP terminal field (Will and Fritzsch 1988) and recordings through this field suggest that it is organized tonotopically with mid-range units located dorsally and low-range units ventrally (Fuzessery 1988). In addition to these VIIIth nerve afferents, the DLN also receives input either laterally or ventrolaterally from the lagena (Will and Fritzsch 1988) and these fibres may convey vibratory information to this part of the nucleus. Saccular fibres terminate either ventrally or ventromedially to the DLN (Will and Fritzsch 1988). It is uncertain whether they contact DLN dendrites.

Neurons in the DLN also receive input from the contralateral DLN (this input appears to be topographically and tonotopically organized), the SO of both sides, and from the opposite superficial reticular nucleus, a population located partly in the lateral lemniscus and spanning the ventrolateral isthmus and midbrain (Wilczynski 1988).

Efferent projections from the DLN pass to the opposite DLN, bilaterally to the SO, to the contralateral superficial reticular nucleus, bilaterally to the torus semicircularis, and, possibly, to the contralateral anterodorsal tegmental nucleus, a group that may represent a ventrocaudal extension of the laminar nucleus (Feng 1986a; Wilczynski 1988). The projection to the SO appears to be topographically and tonotopically organized and this projection is reciprocated. The efferents to the torus end primarily in the contralateral principal nucleus, although some fibres also terminate in the ipsilateral principal nucleus, contralateral laminar nucleus, and bilaterally in the anterodorsal tegmental nucleus (Feng 1986a, b). Recently, Feng and Lin (1991) labelled cells bilaterally, with an ipsilateral predominance, in the DLN following HRP applications in the magnocellular toral nucleus.

The spectral response properties of DLN neurons in ranid frogs are very similar to those of primary VIIIth nerve fibres. These neurons generally have narrow "V-shaped" tuning curves and units tuned to low frequencies display two-tone inhibition when higher frequency stimuli are presented (Fuzessery 1988).

Many DLN units in ranid frogs also have temporal response properties similar to those of primary VIIIth nerve fibres and are able to follow sinusoidal AM stimuli up to 250 Hz (Walkowiak 1988). However, some units in the DLN show response properties that suggest tuning to temporal aspects of the mating call. Notably, units selective for certain AM rates are also present in the DLN and the majority of these band-pass units are tuned to AM rates below 40 Hz (Walkowiak 1988; Feng *et al.* 1990; Hall and Feng 1991). The large majority of DLN units, like VIIIth nerve fibres, display tonic responses, but units displaying phasic

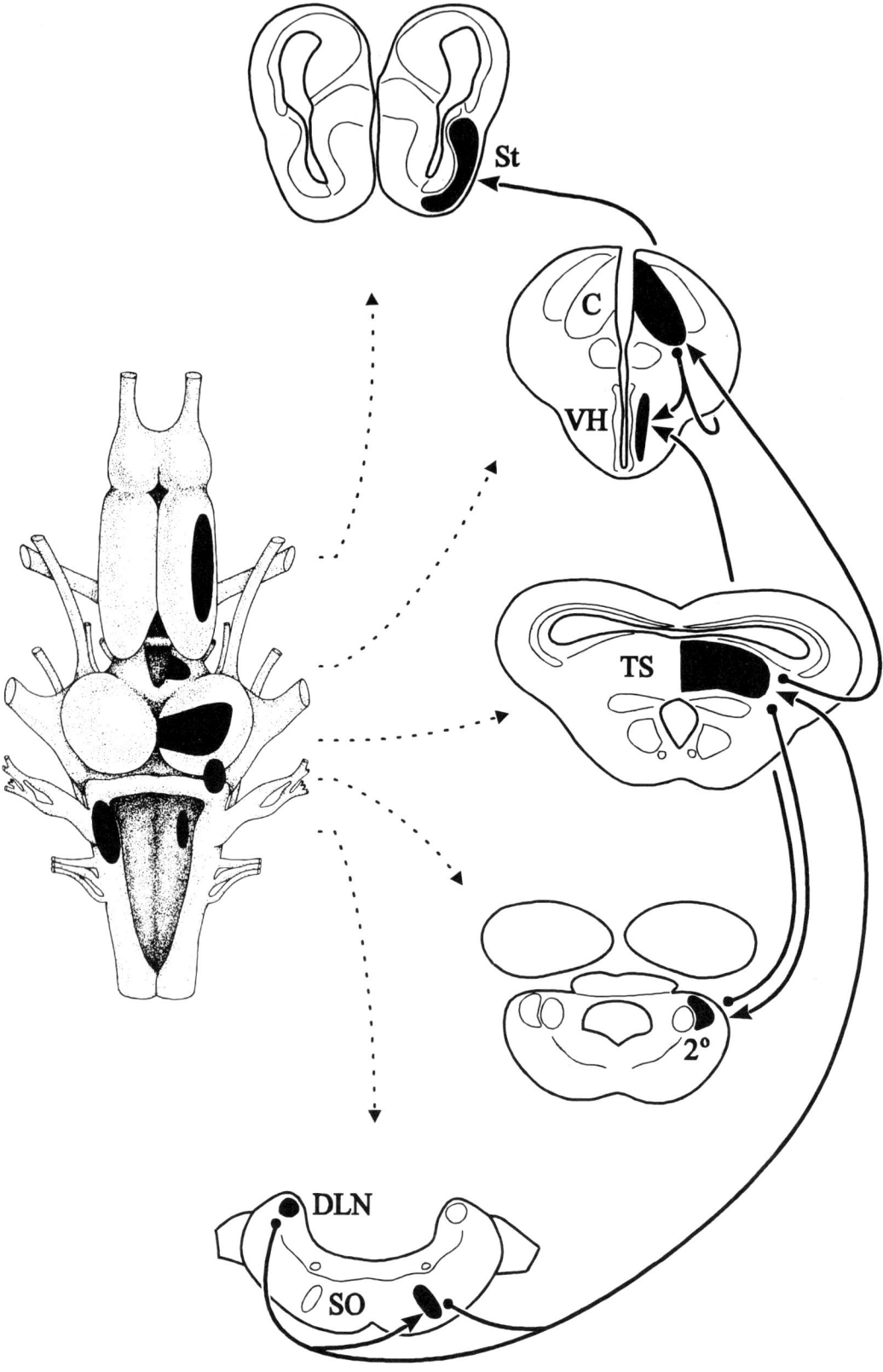

Fig. 9. Highly simplified diagram showing the major components of the central auditory system and their ascending connections. Left: Dorsal view of a bullfrog brain with the major components indicated by black fills. Dashed arrows indicate the levels of the schematic cross-sections in which the components are illustrated. 2° = secondary isthmal nucleus; C = central thalamic nucleus; DLN = dorsolateral nucleus; SO = superior olivary nucleus; St = striatum; TS = torus semicircularis; VH = ventral hypothalmic nucleus.

responses are first encountered at this level in the ascending auditory system (van Stokkum 1987; Hall and Feng 1988, 1990, 1991; Feng et al. 1990). Phasic units preferentially respond to stimuli with rapid rise times and generally display band-pass characteristics when presented with AM stimuli (Hall and Feng 1988; Feng et al. 1990).

Given the bilateral interconnections between the DLN and the SO, it is not surprising that binaural interactions have been recorded in the ranid DLN. Nearly half of the units in the DLN have been found to display binaural responses and the vast majority of these are excited by ipsilateral stimuli and inhibited by contralateral stimuli (Eggermont 1988).

Salamanders and caecilians do not possess a distinct DLN and the auditory functions of the octavolateral area are likely to be performed by the ventral part of the intermediate and the ventral octavolateral nuclei (Will and Fritzsch 1988; Manteuffel and Naujoks-Manteuffel 1990). The anuran DLN is generally regarded as a homologue of the mammalian cochlear nuclei. However, Will (1988) has argued that the anuran DLN may be an independently evolved structure (see above).

B. SUPERIOR OLIVARY NUCLEUS

The SO in anurans receives bilateral input from the DLN and input from the opposite SO (Feng 1986b). The mapping of bilateral input from the DLN appears to form the basis for the tonotopy exhibited in the SO where a high-frequency region is located ventromedially and a low-frequency region dorsolaterally (Fuzessery 1988). In ranids, the SO also receives direct input from lagenar and saccular fibres (Will and Fritzsch 1988) and from the nucleus caudalis (Feng 1986b; Will 1988). This last projection may also convey information from the saccule and lagena (Will and Fritzsch 1988).

Several rostral auditory structures provide input to the SO: the ipsilateral superficial reticular nucleus, the torus semicircularis bilaterally, possibly the ipsilateral anterodorsal tegmental nucleus, and the ipsilateral posterior and central thalamic nuclei (Feng 1986b; Hall and Feng 1987; Wilczynski 1988). Toral input is derived from the three major toral nuclei with ipsilateral input coming from the laminar and magnocellular nuclei and bilateral input from the principal nucleus (Feng 1986b).

Ascending projections from the SO course forward to the ipsilateral superficial reticular nucleus, predominantly ipsilaterally to the torus semicircularis, and to the ipsilateral posterior and central thalamic nuclei (Feng 1986b; Hall and Feng 1987; Wilczynski 1988; Feng and Lin 1991). The ipsilateral principal toral nucleus is the primary target of these ascending fibres.

The spectral response properties of units in the SO of ranid frogs are similar to those of units in the DLN. However, some SO units are inhibited by tones on either side of their best excitatory frequency and, as a result, have highly restricted excitatory tuning curves. Other SO units may have broad unimodal or bimodal excitatory tuning curves, suggesting that they receive convergent input from DLN neurons (Fuzessery 1988).

As in the DLN, some neurons in the ranid SO display selective responses to temporal aspects of acoustic stimuli that suggest a role in the detection of species-specific calls. Band-pass responses to stimuli of varying AM rate can be recorded in the SO, but band-suppression responses have also been found (Walkowiak 1988; Feng et al. 1990; Condon et al. 1991). Band-pass units tend to display phasic responses and show preferences to stimuli with fast rise times (Feng et al. 1990; Condon et al. 1991).

Binaural interactions have been recorded in the SO of *H. cinerea*, where approximately two-fifths of the units were binaural and the majority of these were excited by contralateral stimuli and inhibited by ipsilateral stimuli (Eggermont 1988).

An SO has been identified in the salamander, *Salamandra salamandra*, and it had ipsilateral projections to the midbrain (Manteuffel and Naujoks-Manteuffel 1990). At present, there is no experimental evidence for an SO in caecilians. The anuran SO has been compared to the medial superior olivary nucleus of mammals (Wilczynski 1988), but Will (1988) has suggested that it may be an independently derived population.

C. HIGHER ORDER CONNECTIONS TO THE HINDBRAIN AND MIDBRAIN

Superficial Reticular Nucleus: This anuran nucleus (apparently equivalent to the nucleus of the lateral lemniscus of Feng (1986a,b) and Feng and Lin (1991) receives input from the DLN, SO, and the principal, laminar, and magnocellular nuclei of the torus (Feng 1986a,b; Wilczynski 1988; Feng and Lin 1991). Hall and Feng (1987) reported a projection to a "nucleus of the lateral lemniscus" from the central thalamic nucleus, but in that study the group outlined was directly lateral to nucleus isthmi, a position occupied by the secondary isthmal nucleus (nomenclature of Neary and Wilczynski 1986).

Efferent projections of the superficial reticular nucleus go to the contralateral DLN, ipsilateral laminar and magnocellular toral nuclei, and bilaterally, with an ipsilateral predominance to the principal toral nucleus (Feng 1986a; Wilczynski 1988; Feng and Lin 1991).

Although responses to acoustic stimuli have been recorded from the superficial reticular nucleus (Wilczynski 1988), no systematic exploration of their spectral and temporal response properties has been performed. An equivalent to the anuran superficial reticular nucleus has been identified in a salamander (Manteuffel and Naujoks-Manteuffel 1990) but not, as yet, in any caecilian. The superficial reticular nucleus has been compared to mammalian nuclei of the lateral lemniscus (Wilczynski 1988).

Secondary Isthmal Nucleus: This anuran nucleus receives a strong input from the torus (Neary 1988) and may also receive input from the SO (Rubinson and Skiles 1975). It provides a major input to the ventral hypothalamic neuropil, but also appears to supply the torus, central thalamic nucleus, anterior preoptic area, striatum, and septum (Hall and Feng 1987; Neary 1988, 1990; Neary and Northcutt 1990) These connections are largely ipsilateral.

No examination of the acoustic response properties of secondary isthmal neurons has been performed in anurans and an equivalent population in salamanders and caecilians has yet to be identified. No equivalent to the secondary isthmal nucleus has been identified in reptiles or mammals, and it may represent an independently derived population in anurans.

Torus Semicircularis: The torus semicircularis is a target, often the major one, for ascending projections from all of the brainstem nuclei discussed previously. It also receives a predominantly contralateral input from nucleus caudalis that may convey low frequency auditory and seismic information transmitted by afferents from the saccule and lagena (Feng and Lin 1991). In addition to these afferents related to the VIIIth nerve, the lateral torus also receives ascending input from the obex region (Wilczynski and Neary 1986) that may account for the demonstrated responsiveness of this part of the torus to somatosensory stimulation (Comer and Grobstein 1981). The torus also receives commissural projections from the opposite laminar, principal, and magnocellular toral nuclei and descending projections from a wide variety of structures including populations associated with the medial forebrain bundle (the lateral septal nucleus, anterior preoptic area, dorsal part of the magnocellular preoptic nucleus, suprachiasmatic nucleus, and ventral hypothalamic nucleus), populations associated with the lateral forebrain bundle (the anterior and posterior entopeduncular nuclei), all of the ventral thalamic groups (ventromedial and ventrolateral thalamic nuclei and the nucleus of Bellonci), and most of the dorsal thalamic and pretectal nuclei (central, lateral and posterior thalamic nuclei) (Neary 1988; Feng and Lin 1991). The descending projections to the torus are largely ipsilateral and most appear to terminate in either the laminar or magnocellular toral nuclei (Wilczynski and Neary 1986; Feng and Lin 1991).

Targets for the ascending projections (largely ipsilateral) of the torus are also extraordinarily widespread and include the optic tectum, nearly all thalamic and pretectal nuclei, and most of the structures associated with the medial and lateral forebrain bundles, with the exception of the ventral hypothalamic nucleus (Neary 1988; Zittlau *et al.* 1988; Masino and Grobstein 1990; Feng and Lin 1991). Of these targets, only the following areas receive attention in the present chapter: the posterior, central, lateral, and anterior thalamic nuclei in the dorsal thalamus, the nucleus of Bellonci and ventromedial nucleus in the ventral thalamus, the anterior preoptic area, and the striatum. Feng and Lin (1991) found that most of these areas received input from either the magnocellular or laminar toral nuclei and that the principal toral nucleus had only a very light and restricted projection to the posterior and central thalamic nuclei. However, Neary and Wilczynski (1994) found that the anterior thalamic nucleus and middle thalamic zone (central thalamic nucleus and the anterior and posteroventral parts of the lateral thalamic nucleus) received their toral input largely from the principal nucleus and that laminar and magnocellular toral nuclear projections to this region were sparse.

Evoked potential studies in ranids and bufonids have suggested that a rough tonotopy is present in the torus. In general, higher frequencies are represented rostrally and dorsally and lower frequencies are represented caudally and ventrally. However, more than one peak of spectral sensitivity may be present at a single recording site (Fuzessery 1988). Diekamp and Gerhardt (1992) were unable to demonstrate tonotopy in the torus of *Pseudacris crucifer*. In contrast, others claim to have demonstrated a tonotopic organization of frequencies in the torus (Walkowiak and Luksch 1994).

Spectral response properties of toral neurons are generally more complex and diverse than those of medullary auditory neurons. Units that respond to single tones may be either narrowly or broadly tuned. Narrowly tuned units generally exhibit two-tone inhibition, while broadly tuned units may have bimodal or even discontinuous tuning curves (Fuzessery 1988). Units that respond best to double tone combinations in *Rana pipiens* have been grouped into two classes: those that respond best when the tones are narrowly separated (Class I) and those that respond best when the tones are widely separated (Class II). There is often a close match between the frequencies exciting both Class I and II neurons and the spectral peaks in the mating call (Fuzessery 1988; Feng *et al.* 1990).

Studies on the response properties of toral neurons to temporally modulated stimuli have demonstrated that many toral units (25%–60%) show band-pass responses. The optimal AM rates for eliciting these responses often match the AM rates of various calls in the species studied (Walkowiak 1988; Feng *et al.* 1990; van Stokkum and Melssen 1991). Again, as with spectral responses properties, the temporal response properties of toral neurons are considerably more complex than those of neurons in medullary acoustic centers (Feng *et al.* 1990).

The great majority of toral units (76%–88%) are binaurally responsive (Eggermont 1988; Melssen and Epping 1990; Melssen *et al.* 1990) and the majority of these cells are excited by contralateral stimuli and inhibited by ipsilateral stimuli (Eggermont 1988; Melssen and Epping 1990). The responses of binaural units to stimuli with varying conditions of interaural time differences (ITDs), interaural intensity differences (IIDs), AM modulation rates, and frequencies are complex. In *Rana temporaria*, asymmetric preferences for ITDs were most pronounced using stimuli with AM rates of 36 Hz, a frequency approximating the pulse repetition rate of the mating call (van Stokkum and Melssen 1991). In this same species, Melssen *et al.* (1990) found that the majority of binaural units were sensitive to both ITD and IID, and that changes in one parameter could be compensated by changes in the other. Again in this species, changes in the IID may often either narrow or broaden a toral unit's bandwidth and, in many cases, shift its best frequency (Melssen and Epping 1990). Changes in the spectral properties of toral neurons in response to shifts in sound direction have been reported for *Rana pipiens*. Shifting speaker location from the contralateral to the ipsilateral side increased thresholds and decreased bandwidth, but had little effect on best frequency (Gooler *et al.* 1993).

In salamanders, a narrow zone of periventricular cells lying just ventral to the optic tectum has been identified as the torus semicircularis. This zone receives ascending input from the octavolateralis area and units responsive to auditory and vibratory stimuli have been recorded in this zone (Gonzalez and Muñoz 1987; Manteuffel and Naujoks-Manteuffel 1990).

D. THE DIENCEPHALON

In addition to receiving ascending auditory information from the torus and SO, the anuran posterior thalamic nucleus, part of the pretectum, may also receive auditory information from the pretectal grey and central thalamic nucleus (Hall and Feng 1987). Additionally, it receives significant inputs, largely ipsilateral, from several populations that are likely to convey visual, somatosensory, and polymodal information. Visual information probably reaches the posterior nucleus from cells in the nucleus of Bellonci, the dorsal part of the ventrolateral thalamic nucleus, and the posterodorsal part of the lateral thalamic nucleus (Scalia and Gregory 1970; Scalia 1976; Hall and Feng 1987; Montgomery and Fite 1991). It may also receive visual information from the optic tectum, since part of it lies immediately adjacent to a tectal terminal field (Masino and Grobstein 1990; Montgomery and Fite 1991). Two more indirect pathways for tectal visual information may involve relays through the anterior and posterodorsal divisions of the lateral thalamic nucleus (Scalia 1976; Hall and Feng 1987; Masino and Grobstein 1990; Montgomery and Fite 1991). Somatosensory information is likely to reach the posterior thalamic nucleus from the spinal cord, obex region, ventromedial thalamic nucleus and ventral part of ventrolateral thalamic nucleus (Neary and Wilczynski 1977a; Hall and Feng 1987; Muñoz *et al.* 1994). Polymodal sensory information is probably conveyed to the posterior thalamic nucleus via connections from the anterior thalamic nucleus (Hall and Feng 1987).

Posterior thalamic nucleus efferents are widely distributed and predominantly ipsilateral. They project back to the anterior and central thalamic nuclei and to all the ventral thalamic populations from which they receive input. There is a particularly strong projection to the nucleus of Bellonci. There is also a projection to the optic tectum and to several tegmental nuclei (Hall and Feng 1987; Zittlau *et al.* 1988; Masino and Grobstein 1990).

Posterior thalamic neurons appear to be particularly responsive to auditory stimuli consisting of tone combinations resembling those frequencies found in the mating call (Hall and Feng 1987; Mudry and Capranica 1987a,b; Fuzessery 1988; Feng *et al.* 1990). The posterior thalamic nucleus represents the most rostral locality where selectivity for the spectral components of acoustic stimuli has been consistently demonstrated.

The central thalamic nucleus receives a strong ipsilateral toral projection and lighter projections from the ipsilateral SO, posterior thalamic nucleus, and probably from the ipsilateral secondary isthmal nucleus (Hall and Feng 1987; Neary 1988). Although its efferents are widespread, it has particularly strong, nearly completely ipsilateral projections to the ventral hypothalamus and striatum (Hall and Feng 1987; Neary 1988; Wilczynski and Allison 1989; Allison and Wilczynski 1991; Scalia *et al.* 1991; Neary 1995). It also has a weaker projection to the anterior preoptic area (Neary 1988; Allison and Wilczynski 1991) and projections to all divisions of the lateral thalamic nucleus (Hall and Feng 1987).

Neurons in the central thalamic nucleus in *Rana pipiens* appear particularly sensitive to two temporal properties of auditory stimuli (AM rate and stimulus duration) and they exhibit a wide range of responses when these parameters are varied (Feng *et al.* 1990). Sensitivity to the spectral properties of stimuli has also been demonstrated in the central nucleus of *Rana catesbeiana* and *Hyla cinerea* in evoked potential studies using combinations of high- and low-frequency tones (Megela and Capranica 1983; Mudry and Capranica 1987a,b). Responses to tone combinations in bullfrogs show a strong degree of habituation that may be related to the habituation shown in the evoked vocal response (Megela and Capranica 1983).

Salamanders possess a tectorecipient region in the posterodorsal part of the dorsal thalamus (Wicht and Himstedt 1988) that is likely to represent a homologue of the anuran middle thalamic zone, which includes the central nucleus. It is not known if this or any other dorsal thalamic area receives a toral input or responds to auditory stimuli. The anuran central thalamic nucleus may be homologous to the reptilian nucleus medialis and posteroventral thalamic nucleus and to portions of the posterior intralaminar complex of mammals (Bruce and Neary 1995a).

The anuran lateral thalamic nucleus consists of three divisions. The anterior and posterodorsal divisions are part of the middle thalamic zone and receive some toral input as well as input from the central thalamic nucleus (Hall and Feng 1987; Neary 1988; Feng and Lin 1991). The major target of these divisions is the striatum (Wilczynski and Northcutt 1983a), although some efferents from the posteroventral division also pass to the optic tectum and torus (Wilczynski 1981; Neary 1988; Masino and Grobstein 1990). Evoked potential responses to tone combinations have been recorded from the lateral nucleus in *Rana catesbeiana* and habituate poorly (Megela and Capranica 1983).

The anterior thalamic nucleus of anurans is likely to be a polymodal area, receiving visual inputs from the retina, somatosensory inputs from the obex region, and auditory inputs from the torus and pretectal grey (Neary 1988, 1990; Montgomery and Fite 1991). The toral input may carry somatosensory, as well as auditory information since it appears to be derived primarily from the lateral part of the torus (Neary and Wilczynski 1994).

The anterior nucleus has largely ipsilateral projections to nearly all telencephalic pallial and subpallial fields, but its connections with the medial half of the hemisphere, particularly the lateral septal nucleus and medial pallium are especially well developed, as is its projection to the dorsal pallium (Neary 1990). It also has projections to the anterior preoptic area and ventral hypothalamus (Neary 1988, 1995; Wilczynski and Allison 1989; Allison and Wilczynski 1991).

Evoked potential studies in *Rana catesbeiana* indicate that the anterior nucleus responds best to combinations of high- and low-frequency tones and that these responses habituate, although not to the degree shown by responses in the central nucleus (Megela and Capranica 1983).

Salamanders possess an anteroventral part of the dorsal thalamus that is located medial to retinal terminations and projects to the medial pallium (Wicht and Himstedt 1988). This area is almost certainly homologous to the anuran anterior thalamic nucleus. The anterior thalamic nucleus has been compared to several reptilian dorsal thalamic populations, including nucleus dorsolateralis anterior and nucleus dorsomedialis anterior (Neary 1990). A recent analysis by Butler (1994) suggests that it may be related to several dorsal thalamic populations in mammals, including the anterior and ventrobasal nuclear complexes.

The nucleus of Bellonci, a migrated population of the ventral thalamus (Neary and Northcutt 1983), is closely associated with a circumscribed retinal terminal field, the neuropil of Bellonci (Scalia and Gregory 1970), and its projections are largely ipsilateral and descending to the pretectum and tectum (Neary and Wilczynski 1980; Hall and Feng 1987).

Evoked potentials in response to high- and low-frequency tone combinations and to mating calls have been recorded from the nucleus of Bellonci in *Rana pipiens* (Hall and Feng 1988). With the artificial stimuli, the magnitude of response was dependent on the rise time of the stimulus envelope and was optimal for rise times of 25 ms. In this same study, little, if any, response to the mating calls of *Rana catesbeiana* (rise time >100 ms) was found in this nucleus.

The ventromedial thalamic nucleus receives a presumed somatosensory input from the obex region and spinal cord (Neary and Wilczynski 1977a; Muñoz et al. 1994) in addition to its toral input. It has projections back to the torus and to the optic tectum (Neary 1988; Masino and Grobstein 1990). Evoked potentials showing poor habituation have been recorded in this nucleus (Megela and Capranica 1983).

The anuran ventral hypothalamic nucleus receives most of its input, with the exception of the secondary isthmal nucleus, from ipsilateral telencephalic and diencephalic structures. Aside from the secondary isthmal and central thalamic nuclear inputs that are probably largely auditory, the ventral hypothalamus receives strong projections from structures associated with the medial forebrain bundle (lateral septal nucleus, anterior preoptic area, magnocellular preoptic nucleus, and suprachiasmatic nucleus), a polymodal sensory pallial field (medial pallium), and from structures associated with the main and accessory olfactory systems (ventral lateral pallium and lateral amygdala) (Wilczynski and Allison 1989; Allison and Wilczynski 1991; Neary 1995). It projects back to the same structures associated with the medial forebrain bundle, to the same structures associated with the main and accessory olfactory systems, the anterior thalamic nucleus, and to the deep layers of the optic tectum and laminar nucleus of the torus (Neary and Wilczynski 1977b; Allison and Wilczynski 1991).

Wilczynski and Allison (1989) and Allison (1992) recorded responses to conspecific mating calls in the ventral hypothalmus. These responses generally consisted of tonic or phasic increases in the firing rate of spontaneously active units. The ventral hypothalmus is considered an important centre for the regulation of gondadotropins and auditory input to this area may contribute to the regulation of gonadal functions (Wilczynski 1992; Wilczynski *et al.* 1993).

The anterior preoptic area in anurans is likely to receive auditory input from the secondary isthmal nucleus, torus semicircularis, and the central thalamic nucleus (Hall and Feng 1987; Neary 1988; Allison and Wilczynski 1991). Polymodal, including auditory, sensory inputs likely arise from the anterior thalamic nucleus. Other sources of input to this area include the medial and lateral septal nuclei, medial pallium, and the dorsal, lateral, and ventral hypothalamic nuclei (Neary 1988; Allison and Wilczynski 1991). All of these connections are primarily ipsilateral.

Efferent connections of the anterior preoptic area have not been studied in detail, but projections to the ventral hypothalamus, laminar toral nucleus, and the dorsal tegmental nucleus of the medulla (DTAM) have been reported (Wetzel *et al.* 1985; Wilczynski and Neary 1986; Wilczynski and Allison 1989; Neary 1995). The projection from the anterior preoptic area to the DTAM in *Xenopus* appears to be stronger in males than in females (Wetzel *et al.* 1985).

Responses to mating calls have been recorded in the anterior preoptic area (Urano and Gorbman 1981; Allison 1992). As with the ventral hypothalamus, these are largely modulations of ongoing activity. The anterior preoptic area contains steroid hormone-concentrating neurons (Kelley *et al.* 1975, 1978; Morrell *et al.* 1975) and pituitary hormone treatments enhance the responsiveness of neurons in this area to acoustic stimuli (Urano and Gorbman 1981). The anterior preoptic area has been strongly implicated in the regulation of gonadal steroid levels and the control of vocal behaviour. Auditory pathways to this area are thought to provide avenues whereby acoustic communication signals may influence these two important aspects of reproduction in anurans (Wilczynski 1992; Wilczynski *et al.* 1993).

E. THE TELENCEPHALON

The lateral septal nucleus in anurans may receive an auditory input via projections from the central thalamic nucleus (Neary 1990). It also receives afferents from the medial pallium, anterior thalamic nucleus, and ventral hypothalamus (Neary and Wilczynski 1977b; Neary 1990; Northcutt and Ronan 1992). Acoustic responses have not been recorded as yet from this nucleus.

The anuran medial pallium is a major target of the anterior thalamic nucleus and this nucleus is likely to convey not only auditory, but somatosensory and visual information to this pallial field. It also receives significant input from two secondary olfactory structures, the dorsal and lateral pallia (Neary 1990; Northcutt and Ronan 1992). The medial pallium has widespread connections to all remaining pallial fields, both septal nuclei, the striatum, structures associated with the medial forebrain bundle (including the anterior preoptic area and ventral hypothalamic nucleus), and to dorsal thalamic and pretectal structures, including the anterior and posterior thalamic nuclei and, possibly, the central thalamic nucleus (Neary 1988, 1990; Northcutt and Ronan 1992).

Auditory, visual, and somatosensory stimuli evoke responses in the anuran medial pallium (Neary 1990). As noted earlier, the medial pallium of salamanders receives a projection from the anterior part of the dorsal thalamus (Wicht and Himstedt 1986), but it is uncertain whether the medial pallium of salamanders is responsive to acoustic stimuli.

The medial pallium of amphibians has traditionally been considered homologous to the medial cortex of reptiles and the hippocampus of mammals. Recent studies on its connections and on pallial connections in reptiles have expanded the number of possible homologies. Northcutt and Ronan (1992) suggested that the medial pallium is homologous to both hippocampal and subicular cortices of mammals. Bruce and Neary (1995b) suggested that it is homologous to both the medial and dorsal cortex of reptiles.

Although responses to acoustic stimuli have not been reported in the dorsal pallium, it is likely that this area does respond to such stimuli since, like the medial pallium, it receives a largely ipsilateral input from the anterior nucleus (Neary 1990). Somatosensory responses have been demonstrated in this pallial field (Northcutt 1970) and it also receives a main olfactory bulb input (Northcutt and Royce 1975; Neary 1990; Scalia et al. 1991). Activity in the dorsal pallium would appear to only directly affect other telencephalic populations — it does not appear to project into the diencephalon. Its primary targets are the ipsilateral medial pallium and diagonal band nucleus (Neary 1990).

Acoustic responses have been reported in the striatum (Mudry and Capranica 1980). These are likely to be largely the result of input from the ipsilateral central thalamic nucleus (Vesselkin et al. 1980; Wilczynski and Northcutt 1983a; Hall and Feng 1987; Neary 1988), but the striatum also appears to receive a light ipsilateral input from the torus, secondary isthmal nucleus, and superficial reticular nucleus (Wilczynski and Northcutt 1983a; Neary 1988) and these projections may also contribute to its responsiveness. The striatum additionally receives ipsilateral input from the medial pallium, anterior entopeduncular nucleus, anterior preoptic area, suprachiasmatic nucleus, ventral hypothalamic nucleus, posterior tuberculum, and the anterior and posteroventral parts of the lateral nucleus (Vesselkin et al. 1980; Wilczynski and Northcutt 1983a; Neary 1988). These parts of the lateral nucleus almost certainly convey tectal visual information to the striatum (Gruberg and Ambros 1974; Scalia 1976; Masino and Grobstein 1990; Montgomery and Fite 1991).

The striatum possesses long descending connections to several tegmental nuclei that, in turn, project to the spinal cord — the anterodorsal, anteroventral, and posteroventral tegmental nuclei — and its caudal part also projects to the isthmal region (DTAM), caudal medulla, and spinal cord (ten Donkelaar et al. 1981; Wilczynski and Northcutt 1983b; Toth et al. 1985; Wetzel et al. 1985). Wetzel et al. (1985) suggested that the striatum, via its connections with the DTAM and caudal medulla, may subserve acoustically-evoked or modified calling behaviours in *Xenopus*.

The striatum in salamanders receives an input from the tectorecipient posterodorsal division of the dorsal thalamus (Wicht and Himstedt 1988). It is not known whether auditory information is conveyed by this input.

The amphibian striatum traditionally has been considered homologous to the striatum of reptiles and mammals. Its location within the hemisphere, strong projections to the midbrain tegmentum, and histochemistry support this homology (Northcutt 1974; Vesselkin et al. 1980; Wilczynski and Northcutt 1983b). However, the long projections to the isthmus, rhombencephalon, and spinal cord originating from the caudal part of the striatum in anurans suggest that this part of the striatum may be homologous with the striatoamygdalar area of reptiles and with portions of the central amygdalar nucleus and bed nucleus of the stria terminalis in mammals (Bruce and Neary 1995b). A portion of the striatum in salamanders also possesses these long descending connections (Kokoros and Northcutt 1977; Naujoks-Manteuffel and Manteuffel 1988).

7. The amphibian auditory system: variation within a network of constraints

From the above examples it is clear that the ear of amphibians is extremely variable. The inner ear probably evolved in the context of perception of sound pressure in water (van Bergijk 1967; Fritzsch 1992). Its adaptation to the rather different impedance of airborne sound required reorganization of the inner and middle ear in ways that would achieve a reasonable signal-to-noise ratio and would permit signal detection and directional hearing. Apparently, this was not accomplished in a coherent way. Instead, each combination of sensory epithelia, periotic labyrinth, middle ear ossicles and tympanum (if present) may be viewed as a unique solution to the problem through modification of developmental pathways. Alternatively, these may be regressions of an organ incompletely adapted to the challenges of terrestrial hearing.

Apparently, the anuran ear is better adapted to the perception of substrate and airborne sound then is either the salamander or caecilian ear and it shows much less interspecific variation with respect to sensory epithelia, periotic labyrinth and middle ear ossicles. The multiple reductions of sensory epithelia, in particular of the basilar papilla in caecilians and salamanders, is in stark contrast to the progressive development of the amphibian papilla in anurans. Moreover, in anurans there may be a close match of the evolution of the ear and the complexity of vocalization (Ryan 1988). However, before suggestions like this can be accepted one needs to resolve two questions: What developmental events underlie these changes (terminal deletion, reorganization, terminal additions) and do these changes have any significance for the life history of the animal? One needs to know the kinds of ontogenetic reorganizations responsible for these changes and the advantage (if any) accruing from them. Given that one cannot predict in detail the pathway taken by sound through the ear and how the distribution of frequencies to different end-organs is achieved, it is difficult to assess the individual consequences of each of these changes. Nevertheless, available data imply that the neobatrachian ear is a unique adaptation to the constraints of perception of airborne sound. Comparing the function of the amphibian papilla with that of the amniotic basilar papilla (both have a tonotopic representation) may help in understanding the physical constraints leading to functional similarities in these homoplastic organs. Some of these constraints are becoming apparent (Lewis 1992). For example, interaural contrast of intensity or time differences permitting directional hearing requires comparison of the same signal. Thus, directional hearing may be performed by comparing the output of an organ that (1) perceives only a limited frequency (e.g., basilar papilla), or (2) has a tonotopic organization (e.g., the amphibian papilla of most anurans or the basilar papilla of amniotes) (Manley and Gleich 1992; Carr 1992).

That frequency also is a factor in directional hearing in anurans is now well established (review by Eggermont 1988). However, it is performed rather differently than in amniotes, even if there is a tonotopic representation within one organ (Lewis 1992). In addition, anurans have to localize sound that offers only small differences in interaural time and intensity, cues that are used in large birds and mammals (Carr 1992), thus restricting sound localization to a pressure gradient system otherwise known from insects and small birds. This was achieved in anurans through modification of the middle ear such that sound reaches the eardrum from both sides and probably from the lung as well. In addition, there is a binaural interaction already at the level of the dorsolateral auditory nucleus in the hindbrain (Will 1989) and a segregation of differently responding binaural units in papillae with high- (basilar papilla) and low- (amphibian papilla) frequency ranges (review by Eggermont 1988).

Many, but not all, anurans use the presence of these two receivers' distinct, non-overlapping frequencies in conjunction with the functioning of the stapes and operculum, also with distinct frequency transmission, for discrimination of interspecific calls. However, other species may use only one or the other channel for species-specific communication. Thus, while this peripheral selection model is valid for some frogs, it is not applicable for all, again pointing out the large degree of variation in a system that still has evolutionary flexibility. Clearly, a lot has been learned about a few "model" frog species (Ryan 1988; Narins 1992) and the interplay of anuran communication and their sensory capacities (Rand 1988; Gerhardt 1988; Wells 1988). However, the level of understanding is not yet sufficient to predict the interplay of many frogs in a chorus in nature. Factoring of all variables into a consistent and predictive model of all frog's calling strategies may be an unrealistic goal owing to the great degree of variation.

III. THE LATERAL LINE: MECHANOSENSORY AND ELECTROSENSORY ORGANS

The lateral line system of mechanosensory organs has been investigated extensively in adult and developing amphibians over the last century (reviews by Russel 1976; Northcutt 1992b). There are a number of deviations from the general type exemplified by the lateral line system of salamanders (Fritzsch 1989). Probably the most important finding in the last 100 years of lateral line research was the (re)discovery of a second set of organs (Sarasin and Sarasin 1888), the ampullary electroreceptors (Hetherington and Wake 1979; Fritzsch and Münz 1986; McCormick *et al.* 1994).

A. The Mechanosensory Receptor Organ: Structure and Function

Two types of neuromast organs have been described in the axolotl, neuromasts and pit organs (Northcutt 1992b) (Fig. 10). Both are mechanosensory organs with the mechanosensory hair-cells organized in two groups of opposing polarity (Russell 1976; Wahnschaffe

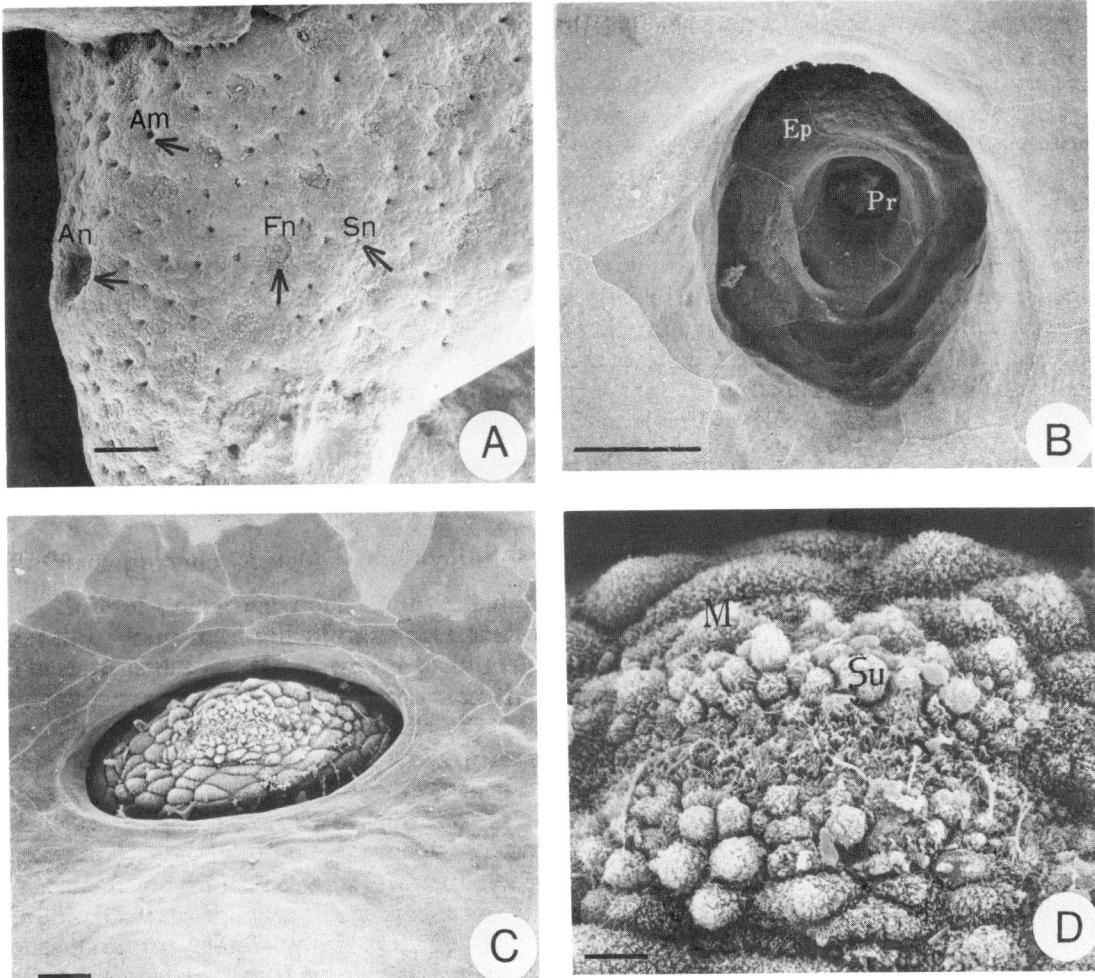

Fig. 10. Mechanosensory and electrosensory organs of the Chinese giant salamander, *Andrias davidianus*. From Cheng et al. (1995). **A:** Distribution of lateral line sense organs on the rostrum of an early larva. Am = ampullary organ; An = external naris; F = neuromast; Sn = pit organ. Bar = 0.2 mm. **B:** Pit organ of a juvenile showing the epidermal cells (Ep) delineating the wall of the canal and the apex (Pr) of the organ at the bottom of the canal. Bar = 20 μm. **C:** Neuromast from the head of a juvenile. Bar = 20 μm. **D:** Enlargement of part of a neuromast from the head of a juvenile showing an oval zone of supporting cells (Su) and elliptical mantle cells (M) with apical microvilli surrounding the central zone of sensory cells. Bar = 5 μm.

et al. 1985). Pit organs are smaller and less regularly organized then are neuromasts and thus far have been described only from the heads of axolotls (Northcutt 1992b).

The sensory cells of neuromasts are pear-shaped, have opposite polarity and are surrounded and mostly separated by supporting cells (Wahnschaffe *et al.* 1985). The assembly of up to 25 hair-cells and their associated supporting cells that make up a neuromast is separated from the surrounding epidermis by mantle cells. In many salamanders and in some anurans, but not in caecilians, as the animal grows its neuromasts divide to form multiple organs of parallel neuromasts, called stitches. A gelatinous cupula of variable length extends from the neuromast into the surrounding water. Neuromasts are innervated by at least two afferents, each of which likely ends at a subset of hair-cells sharing the same polarity. However, many neuromasts receive more than two afferent fibres and a fibre may branch to reach more than a single neuromast (Fritzsch 1989). This pattern of innervation raises the issue of funtional units within the lateral line system, i.e., how many hair-cells of how many neuromasts and of what polarity are innervated by a single afferent fibre? It is likely that in most cases there is a single afferent for hair-cells of a particular polarity in each neuromast, but the functional significance of variations needs to be explored. In addition, neuromasts of most, but not all, amphibians receive a thinner, cholinergic efferent fibre (Fritzsch 1989). The cells of these fibres receive various inputs and appear to downregulate the activity in the neuromast afferents (Münz and Claas 1991).

Neuromasts function as a "distant touch" system allowing the animal to perceive very small movements of water with respect to the body surface at distances of about 10 times or less of the body length. As a whole, the lateral line system behaves like a velocity transducer (Coombs *et al.* 1992). It appears that the spatially distributed system of neuromasts recognizes spatial irregularities in the surrouding water movements and computes the direction of a stimulus. This has been demonstrated elegantly in behavioural experiments on *Xenopus* (Elepfandt 1989; Görner and Mohr 1989). In contrast to the ear, the lateral line organs respond to a narrow range of frequencies of about 10–60 Hz (Münz 1989). Behaviourally, the lateral line system plays a role in detection of prey and perhaps also in avoidance of predators. Moreover, the system is able to perform this task with many fewer neuromasts than it actually contains (Elepfandt 1989; Görner and Mohr 1989). The role of the efferent system in behaviour is not yet settled (Russel 1976; Münz and Claas 1991). However, given that there is a wide variety of inputs to efferent cells, it appears likely that they play a role other then merely being part of a simple reflex arc.

1. Distribution of Neuromasts in Amphibians

Neuromasts, singly or in stitches, are distributed in discrete lines mostly laterally over the head and body. On the body there are typically three lines, dorsal, middle and ventral (Russel 1976; Northcutt 1992b; Cheng *et al.* 1995). However, there is some variation. In a few salamanders and frogs there is a fourth line extending from the occipital region on to the body. In some frogs the lines are reduced to two and caecilians have only a single line on the body (Fritzsch 1989). The lines on the body are all innervated by nerve cells within the combined cranial ganglia of the glossopharyngeal and vagal nerves. In addition, two somewhat separate nerves, the middle and the supratemporal nerve, innervate neuromasts and pit organs in the occipital group of organs in salamanders (Northcutt *et al.* 1994). *Xenopus* (Russel 1976) also has several subgoups in the occipital region as do other members of the family Pididae. Like many other lateral line rows of the head, the homology with lines in other amphibians is not yet fully established (Escher 1926; Fritzsch 1989).

On the head four major lines are always distinguished in all amphibians (Fritzsch 1989): a supraorbital, an infraorbital, an oral and a mandibular line. Although nomenclature varies substantially among authors, these differences may be resolved by examining the pattern of innervation. Escher (1925) and later Fritzsch (1981) and Northcutt (1992b) defined four nerves and two ganglia associated with the trigeminal and the facial nerve, respectively. The anterio-dorsal lateral line ganglion, associated with the trigeminal ganglion, innervates the supra- and infraorbital line with the superficial ophthalmic and the buccal nerve, respectively. The anterio-ventral ganglion, associated with the facial ganglion, innervates the oral and mandibular lines with the dorsal and ventral ramus of the external mandibular nerve. In addition, there are a number of other shorter or longer lines on the head containing neuromasts and pit organs (Northcutt 1992b). The lines of caecilians appear to be homologous with those of salamanders and likely with those of anurans as well (Lannoo and Smith 1989). However, among anurans there is a reorganization in the distribution of lines, perhaps associated with the reorganization of the mouth (Escher 1925).

B. The Electroreceptive Ampullary Organs: Structure and Function

Electroreceptive ampullary organs (Fig. 10) have been misinterpreted for almost 100 years (Sarasin and Sarasin 1888; review by Fritzsch and Münz 1986). These small organs are distributed alongside the lateral lines of the head including the occipital group (Fritzsch 1988; Northcutt 1992b; Cheng *et al.* 1995) (Fig. 11). Ampullary organs may be single or in groups but always are circularly organized rather then having an obvious polarity. The apical specializations are microvilli, with or without a kinocilium, but they never have the typical organization of mechanosensory organs (Wahnschaffe *et al.* 1985; Fritzsch and Münz 1986). The approximately eight hair-cells of each organ are more ovoidal than in mechanosensory organs; like those of neuromasts they are surrounded by mantle and supporting cells.

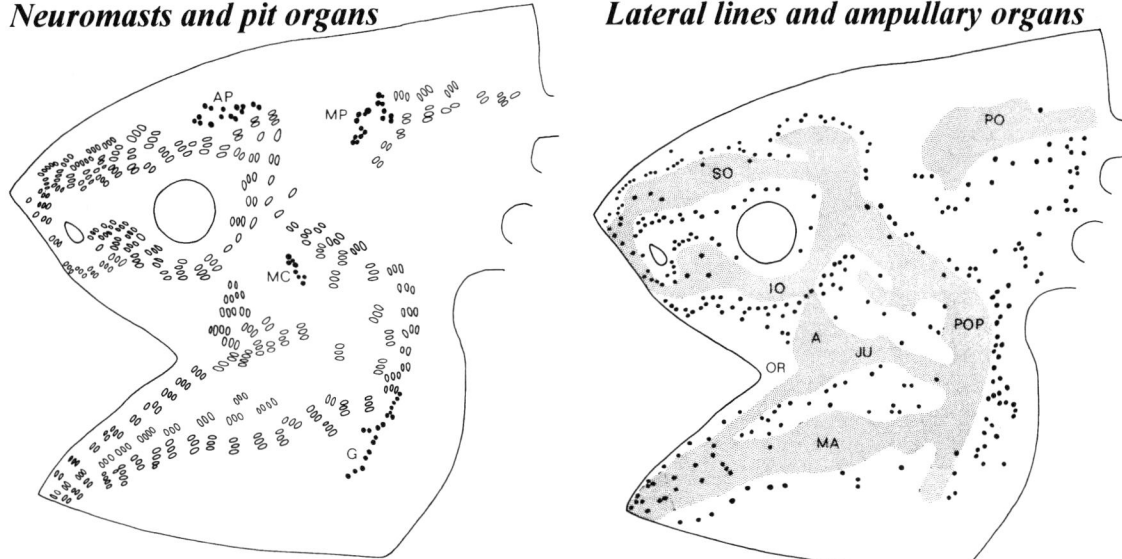

Fig. 11. Distribution of lateral line neuromasts (open ovals; A), pit organs (solid circles; A) and ampullary electroreceptors (B) in the skin of the head of a juvenile axolotl. The stippled areas in B indicate the extent of the neuromast and pit lines illustrated in A. A = angular line; AP = anterior pit line; G = gular pit line; IO = infraorbital line; JU = jugal line; MA = mandibular line; MC = middle cheek pit line; MP = middle pit line; OR = oral line; PO = postotic line; POP = preopercular line; SO = supraorbital line. Modified from Northcutt (1992).

Each ampullary organ or group of organs is innervated by a single afferent fibre. These fibres all run in the antero-dorsal and antero-ventral nerves, including those in the postotic group (Northcutt 1992b). The ampullary organs of caecillians are distributed similarly, except that they form more discrete rows in parallel to those of neuromasts (Hetherington and Wake 1979). Some caecilians are the only vertebrates known which have only ampullary organs (McCormick et al. 1997). All anurans lack ampullary organs.

Ampullary organs have been shown to be sensitive to changes in the electric fields of as little as 5 μV/cm (Fritzsch and Münz 1986). Several lines of evidence support the idea that in salamanders and caecillians they function in detection of live prey (Himstedt and Fritzsch 1990). Since almost all anuran tadpoles are herbivores or scavengers, their lack of ampullary organs may be linked to their non-predatory life style.

C. Evolution of Lateral Lines among Amphibians

The lateral line systems of appropriate out-groups, lungfish and *Latimeria*, indicate the primitive condition to be (1) ampullary organs on the head and occipital region, (2) neuromasts and pit organs in at least three lines on the body and (3) a minimum of four lateral lines on the head and the cheek (Fritzsch 1989; Northcutt 1989). Clearly, loss of canals is correlated with the loss of body armour and may represent paedomorphic truncation of lateral line development (Fritzsch 1989). Absence of neuromasts, pit organs and ampullary organs in anurans and of pit organs in caecilians appears to be a derived feature. In addition, lateral lines may be shortened or absent or the formation of stitches may be suppressed. The presence of ampullary organs alone in some aquatic caecilians and of neuromasts alone in frogs (Fritzsch and Wake 1986; McCormick et al. 1996) clearly indicates that the development of ampullary organs and neuromasts which appear tightly coupled (Northcutt et al. 1994), is essentially independent. This may be the basis for the variation in occurrences of these organs in amphibians. It would be important to demonstrate the adaptive nature of these variations in relation to different larval ecological niches. However, to do so one would need to explore in much more detail the behavioural significance of these organs in functions other then predation or predator avoidance.

Some salamanders, caecilians and anurans may have independently suppressed formation of all lateral line organs in conjunction with direct development or with intraoviducal development (Duellman and Trueb 1985). The formation even of rudimentary placodes in these animals needs to be explored in order to understand the changes in development that suppress formation of lateral line organs, while nevertheless retaining some lateral line ganglia (Wake *et al.* 1987).

A. EVOLUTION OF THE METAMORPHIC LOSS OF THE LATERAL LINE SYSTEM

It is not only the distribution of the three kinds of lateral line organs that is highly variable but also the behaviour of those organs during metamorphosis. Recently, a scheme was proposed that suggested a stepwise evolution of metamorphic loss. In this scheme aestivating salamanders, such as *Siren*, display the primitive retention of all organs during aestivation whereas in anurans a rapid loss of organs, ganglion cells and second-order neurons in the medulla represents a derived conditon (Wahnschaffe *et al.* 1987; Fritzsch 1990). This grade of changes appears much like the progressive loss of organs within lineages of salamanders, and between salamanders and anurans. Retention of part or all of the lateral line system correlates with at least some aquatic activity in adult amphibians; the function of the retained organs may be related to mediation of information during aquatic courtship.

D. Central Projection of the Lateral Line: Generalizations and Differences

The two anterior lateral line roots of larval and many adult salamanders and of larval caecilians project with a dorsal and a ventral root to the dorsal and medial nucleus of the alar plate (Gonzalez and Muñoz 1988; Fritzsch 1988). In contrast, the posterior lateral line nerves of all amphibians and all lateral line nerves of anurans project only to the medial, mechanoreceptive nucleus (Fritzsch *et al.* 1984; Gonzalez and Muñoz 1988) (Figs 1, 3). The dorsal root consists of afferents from the ampullary organs, the ventral root of afferents from the mechanosensory neuromasts and pit organs. In salamanders and caecilians the lateral line afferents project into discrete fascicles, two for a given nerve (Fig. 3). It is possible that each fascicle contains afferents contacting hair-cells of a given polarity (Fritzsch 1981). Such fasciculation is rare among anurans (Fritzsch 1989).

1. Projection to the Midbrain: Generation of a Spatial Map and its Orientation with the Visual Map

Projections form the medullary alar plate have been studied almost exclusively in *Xenopus*, the axolotl and one caecilian (Will 1989; Bartels *et al.* 1990). Neurons with dendrites to the mechanosensory projection area have been labelled from the midbrain and the spinal cord. No connections with the dorsal nucleus were reported for neurons projecting from the spinal cord in salamanders and caecilians (Will 1989). Tracer injections into the tegmental midbrain or the torus filled neurons with dendrites in the lateral line projection zone. In addition, tracer injections into the alar plate revealed fibres to the contralateral midbrain with very few continuing into the tectum and the thalamus (Will 1989). The connections to the torus and the midbrain have been considered extensively by Claas *et al.* (1989). They found that the projection of the torus onto the tectum provides directional information from stimuli registering with the retinal map on the tectum. Moreover, the authors claimed that there was an absence of orientational behaviour in a given direction after small tectal ablations, a finding that was not reported by others (Elepfandt 1989). More recently, this finding was extended to electroreception in the axolotl as well (Bartels *et al.* 1990). Like the auditory system, the mechanosensory and electrosensory lateral line system requires centrally generated information about the location of a stimulus. It is even less clear than in the auditory system (Eggermont 1988) how the electrosensory and mechanosensory spatial map is generated. It is equally unknown how similar or unrelated the generation of a spatial map is in these three modalities of the octavolateral system. Electrophysiological evidence in the axolotl has also demonstrated that in this species both electroreceptive and lateral line information reaches the forebrain (Northcutt and Plassmann 1989). It is not known whether auditory information also reaches the forebrain in these animals.

E. Predation and Predatory Avoidance: Is *Xenopus* a Good Model for Lateral Line Function in Amphibians?

Almost all of the behavioural evidence for the function of the lateral line has been generated from studies of *Xenopus* (Görner and Mohr 1989). While this animal provides a reasonable model and reliably reacts to stimuli, it is an oddity by being one of the few permanently aquatic frogs. The larvae and adults probably use the lateral line system for different purposes, predator evasion and prey detection, respectively. Unfortunately, little is known about the function of the larval lateral line system even in this model animal and more information is necessary to assess the relative importance of the system in predation and predator avoidance; the latter is believed to be its major function in larvae. Given the rich variety of ways the lateral line is organized in different amphibians, it would be worthwhile to explore the behavioural context of the lateral line in a wider range of species.

F. Development of the Lateral Line

The development of the lateral line system of salamanders and anurans has been studied extensively, whereas almost nothing is known about it in caecilians. The most recent data and reviews present a more complicated picture then previously drawn (Winklbauer 1989; Northcutt et al. 1994). It should be noted that the lateral line placodes of amphibians were the first experimental proof for cellular migration (Harrison 1904).

1. Induction of Placodes

Experimental evidence suggests that lateral line placodes are induced by the same structures that induce the ear placode, although somewhat later in development (Yntema 1939, 1955). Neither the agent(s) inducing lateral line placodes nor the precise time that induction takes place are known. The ear is either already formed before other dorsolateral placodes appear (Northcutt et al. 1994) or the ear placode is at least induced earlier then the lateral line placodes (Winklbauer 1989). The discovery of ampullary organs and their apparent origin from the same placodes as neuromasts (Fritzsch and Bolz 1986; Northcutt et al. 1995) renders the older literature insufficient for resolving the problem of ampullary organ induction. Heterochronic transplantations like the ones performed by Yntema (1939) may be helpful to clarify this point despite the apparent lack of success in more recent attempts (Smith et al. 1990).

2. Formation and Migration of the Placodes and Formation of Neuromasts and Electroreceptors

Because the time span between induction and first morphological appearance of placodes is unknown, the entities identified first as placodes vary somewhat between studies (Smith et al. 1990; Northcutt et al. 1994). This is important because it is not known whether placodes are actually induced where they first appear as morphological entities or whether they have already migrated some distance before cells accumulate as recognizable placodes. The interpretations of Northcutt et al. (1994) are followed here. These authors identified two preotic and at least three postotic placodes in a stage 35 axolotl. The preotic system consists of the anterodorsal and anteroventral placode. All placodes first form ganglia and, with some delay, most then elongate (head) or migrate (body) to form individual organs. This process of multiplication of the anlage and elongation has been studied extensively in Xenopus (Winklbauer 1989) and in the axolotl (Northcutt et al. 1994). During this process, the placodes retain contact with their ganglion cells and, in addition, deposit other groups of cells to form the first set of neuromasts, called primary organs (Winklbauer 1989). These may undergo further growth while the developing animal still depends on yolk. When feeding starts, the growth pattern changes from stem cell growth to the production of secondary organs to form stitches (Winklbauer 1989). Regeneration experiments indicate that supporting cells can produce new sensory cells (Corwin and Warchol 1991) and this may underlie the switch from primary to secondary organ formation. Thus, there are four phases in lateral line development from a placode: (1) formation of ganglion cells, (2) elongation/migration with deposition of primary neuromasts, (3) growth of primary neuromasts from stem cells and (4) formation of secondary neuromasts through proliferation, probably by newly recruited supporting cells.

Ampullary organs arise from the more lateral parts of placodes (Northcutt et al. 1994). When these organs can be identified next to the neuromasts, they consist of a single sensory cell and one supporting cell (Fritzsch and Bolz 1986). Experimental evidence suggests they are indeed derived from the same placodal material (Northcutt et al. 1995).

Ganglion cells derived from the placodes send their central process into the alar plate of the medulla before the organs erupt through the skin. Thus, the central connection is established before the periphery has achieved its function. The earliest connections have not been studied by experimental tracing techniques. It is important to find out whether the earliest projections are different from those of adults, i.e., is there a phase of overgrowth followed by correction or are the fibres initially restricted to their ultimate targets, with little or no reorganization? Such information would be important for an understanding of the nature and mechanisms of the specificity of these projections (Fritzsch 1993; Northcutt *et al.* 1994).

3. Regeneration of Neuromasts and their Dependence on Afferent Innervation

Because the neural tube plays a role in induction of the otic placode to form the ear (Yntema 1955; Jacobson and Sater 1989), it seems reasonable to assume that this is also true for the lateral line system. Logically, then, one would like to understand how long the placodes require the influence of neurons for their formation and maintenance. Experimental data concerning the ear seems to indicate a large degree of autonomy of hair-cell differentiation after the initial induction (Swanson *et al.* 1990) and similar data exist for the lateral line system. It has been shown that the organs can form autonomously in the absence of nerves after the placode has been induced. This would suggest that all mechanosensory organs may be able to differentiate their hair-cells, including the appropriate polarity (Russel 1976), without any further interaction with nerves. It is possible that once they are formed, lateral line placodes are as compartmented and autonomous in their further differentiation as is the otocyst. Transplantations of ampullary anlage to the lateral body wall indicates a similar scenario for these organs (Northcutt *et al.* 1995).

Similarly, lateral line neuromasts can regenerate and do so even in the absence of innervation (Corwin *et al.* 1989). However, in contrast to the ear, it appears that neuromasts need innervation for long-term maintenance. An even faster reaction to denervation has been reported for bony fish electroreceptors which otherwise also appear to undergo regeneration when deprived of innervation (Fritzsch *et al.* 1990). It would be important to know the behaviour of ampullary organs of salamanders in this respect. It is possible that they react faster to denervation then do neuromasts.

G. The Octavolateralis System as a Model for Development and Neurobiology

Clearly, the octavolateralis system of amphibians has had major impacts on general issues of neurobiology such as migration of cells (the first experimental demonstration), physiology (the saccular hair-cell as a model for vertebrate hair-cells), anatomy (the formation of a second middle ear ossicle, a distinct periotic labyrinth and a unique sound-pressure receiver) and evolution (the progressive and regressive evolutionary diversity of the ear and lateral line system). The demonstration that the mechanosensory and electrosensory lateral line system computes a spatial map and projects it onto the tectum in register with the visual map is confirmation of convergent evolution driven by comparable needs. Explaining the puzzling diversity of the octavolateralis system in relation to trophic habits still remains to be achieved. To do so, animals other than the standard laboratory amphibians (*Xenopus*, the axolotl and a limited number of terrestrial frogs) need to be studied. In particular, electrophysiological experiments on the octavolateralis system of caecilians would be important. Such data could offer a more refined perspective of what was attempted in this chapter, an understanding of the morphological and physiological basis of adaptation of this system in various ecological contexts, for which amphibians are the outstanding vertebrate example.

IV. REFERENCES

Allison, J. D., 1992. Acoustic modulation of neural activity in the preoptic area and ventral hypothalamus of the green treefrog (Hyla cinerea). *J. Comp. Physiol.* **171**: 387–395.

Allison, J. D. and Wilczynski, W., 1991. Thalamic and midbrain auditory projections to the preoptic area and ventral hypothalamus in the green treefrog (Hyla cinerea). *Brain Behav. Evol.* **38**: 322–331.

Baird, I. L., 1974. Anatomical features of the inner ear in submammalian vertebrates. Pp. 159–212 in "Handbook of Sensory Physiology, Vol. 1", ed by W. D. Keidel and W. D. Neff. Springer, Berlin.

Baird, R. A. and Lewis, E. R., 1986. Correspondences between afferent innervation patterns and response dynamics in the bullfrog utricle and lagena. *Brain Res.* **369**: 48–64.

Barlow, L. A. and Northcutt, R. G., 1995. Embryonic origin of amphibian taste buds. *Dev. Biol.* **169**: 273–285.

Bartels, M. Münz, H. and Claas, B., 1990. Representation of lateral line and electrosensory systems in the midbrain of the axolotl, Ambystoma mexicanum. *J. Comp. Physiol.* **167**: 347–356.

Bennett, M. V. L. and Obara, S., 1986. Ionic mechanisms and pharmacology of electroreceptors. Pp. 483–496 in "Electroreception", ed by T. H. Bullock and W. Heiligenberg. Wiley and Sons, New York.

Bodznick, D., 1989. Comparison between electrosensory and mechanosensory lateral line. Pp. 99–115 in "The Mechanosensory Lateral Line: Neurobiology and Evolution", ed by S. Coombs, P. Görner, H. Münz. Springer, New York.

Bolt, J. R. and Lombard, E. R., 1992. Nature and quality of the fossil evidence of otic evolution in early tetrapods. Pp. 377–403 in "The Evolutionary Biology of Hearing", ed by D. Webster, A. Popper and R. R. Fay. Springer, New York.

Bruce, L. L. and Neary, T. J., 1995a. Afferent projections to the ventromedial hypothalamic nucleus in a lizard, Gekko gecko. *Brain Behav. Evol.* **46**: 14–29.

Bruce, L. L. and Neary, T. J., 1995b. Afferent projections to the lateral and dorsomedial hypothalamus in the lizard, Gekko gecko. *Brain Behav. Evol.* **46**: 30–42.

Butler, A. B., 1994. The evolution of the dorsal thalamus of jawed vertebrates, including mammals: cladistic analysis and a new hypothesis. *Brain Res. Rev.* **19**: 29–65.

Capranica, R. R., 1976. Morphology and physiology of the auditory system. Pp. 551–575 in "Frog Neurobiology", ed by R. Llinás and W. Precht. Springer, Berlin.

Carpenter, E. M., Goddard, J. M., Chisaka, O., Manley, N. R. and Capecchi, M. R., 1993. Loss of Ho-A1 (Hox-1.6) functions results in the reorganization of the murine hindbrain. *Development* **118**: 1063–1075.

Carr, C. E., 1992. Evolution of the central auditory system in reptiles and birds. Pp. 511–544 in "The Evolutionary Biology of Hearing", ed by D. Webster, A. Popper and R. R. Fay. Springer, New York.

Cheng, H., Huang, S.-Q. and Heatwole, H., 1995. Ampullary organs, pit organs, and neuromasts of the Chinese giant salamander, Andrias davidianus. *J. Morphol.* **226**: 149–157.

Christensen-Dalsgaard, J. and Narins, P. M., 1993. Sound and vibration sensitivity of VIIIth nerve fibers in the frogs Leptodactylus albilabris and Rana pipiens pipiens. *J. Comp. Physiol.* **172**: 653–662.

Claas, B., Fritzsch, B. and Münz, H., 1981. Common efferents to the lateral-line and labyrinthine systems in aquatic vertebrates. *Neurosci. Lett.* **27**: 231–235.

Claas, B., Münz, H. and Zittlau, K. E., 1989. Direction coding in central parts of the lateral line system. Pp. 99–115 in "The Mechanosensory Lateral Line: Neurobiology and Evolution" ed by S. Coombs, P. Görner and H. Münz. Springer, New York.

Clack, J. A., 1992. The stapes of Acanthostega gunnari and the role of the stapes in early tetrapods. Pp. 405–419 in "The Evolutionary Biology of Hearing", ed by D. Webster, A. Popper and R. R. Fay. Springer, New York.

Comer, C. and Grobstein, P., 1981. Involvement of midbrain structures in tactually and visually elicited prey acquisition behavior in the frog, Rana pipiens. *J. Comp. Physiol.* **142**: 151–160.

Condon, C. J., Chang, S. H. and Feng, A. S., 1991. Processing of behaviorally relevant temporal parameters of acoustic stimuli by single neurons in the superior olivary nucleus of the leopard frog. *J. Comp. Physiol.* A **168**: 709–725.

Coombs, S., Jansen, J. and Montgomery, J., 1992. Functional and evolutionary implications of peripheral diversity in lateral line systems. Pp. 267–293 in "The Evolutionary Biology of Hearing", ed by D. Webster, A. Popper and R. R. Fay. Springer, New York.

Corwin, J. T., 1985. Perpetual production of hair-cells and maturational changes in hair-cell ultrastructure accompany postembryonic growth in an amphibian ear. *Proc. Natl. Acad. Sci. USA* **82**: 3911–3915.

Corwin, J. T. and Warchol, M. E., 1991. Auditory hair-cells: structure, function, development and regeneration. *Annu. Rev. Neurosci.* **14**: 301–333.

Corwin, J. T., Balak, K. J. and Borden, P. C., 1989. Cellular events underlying the regenerative replacement of lateral line sensory epithelia in amphibians. Pp. 161–183 in "The Mechanosensory Lateral Line: Neurobiology and Evolution", ed by S. Coombs, P. Görner and H. Münz. Springer, New York.

de Burlet, H. M., 1928. Über die Papilla neglecta. *Anat Anz.* **66**: 199–209.

de Burlet, H. M., 1934. Vergleichende Anatomie des statoakustischen Organs. a) Die innere Ohrsphäre; b) Die mittlere Ohrsphäre. Pp. 1293–1432 in "Handbuch der Vergleichenden Anatomie der Wirbeltiere, Vol. V/2", ed by L. Bolk, E. Göppert, E. Kallius and W. Lubosch. Urban and Schwarzenberg, Berlin.

Diekamp, B. M. and Gerhardt, H. C., 1992. Midbrain auditory sensitivity in the spring peeper (Pseudacris crucifer): correlations with behavioral studies. *J. Comp. Physiol.* A **171**: 245–250.

Dieringer, N. and Precht, W., 1986. Differences in the central organization of gaze stabilizing reflexes between frog and turtle. *J. Comp. Physiol.* **153**: 495–508.

Dieringer, N., Reichenberger, I. and Graf, W., 1992. Differences in optokinetic and vestibular ocular reflex performance in teleosts and their relationship to different life styles. *Brain Behav. Evol.* **39**: 289–304.

Duellman, W. E. and Trueb, L., 1985. P. 670 *in* "Biology of Amphibians". McGraw-Hill, New York.

Dunn, R. F., 1978. Nerve fibers of the eighth nerve and their distribution to the sensory nerve of the inner ear in the bullfrog. *J. Comp. Neurol.* **182**: 621–636.

Eggermont, J. J., 1988. Mechanisms of sound localization in anurans. Pp. 561–586 *in* "The Evolution of the Amphibian Auditory System", ed by B. Fritzsch, M. Ryan, W. Wilczynski, T. Hetherington and W. Walkowiak. Wiley and Sons, New York.

Ekker, M., Akimenko, M. A., Bremiller, R. and Westerfield, M., 1992. Regional expression of three homeobox transcripts in the inner ear of zebrafish embryos. *Neuron* **9**: 27–35

Elepfandt, A., 1989. Wave analysis by amphibians. Pp. 527–542 *in* "The Mechanosensory Lateral Line: Neurobiology and Evolution", ed by S. Coombs, P. Görner and H. Münz. Springer, New York.

Elgoyhen, A. B., Johnson, D. S., Boulter, J., Vetter, D. E. and Heinemann, S., 1994. a 9: An acetylcholine receptor with novel pharmacological properties expressed in rat cochlear hair-cells. *Cell* **79**: 705–716.

Escher, K., 1925. Das Verhalten der Seitenorgane der Wirbeltiere und ihrer Nerven beim Übergang zum Landleben. *Acta Zool. (Stockh.)* **6**: 307–414.

Farbman, A. I. and Mbiene, J. P., 1991. Early development and innervation of taste bud-bearing papillae on the rat tongue. *J. Comp. Neurol.* **304**: 172–186.

Feng, A. S., 1986a. Afferent and efferent innervation patterns of the cochlear nucleus (dorsal medullary nucleus) of the leopard frog. *Brain Res.* **367**: 183–191.

Feng, A. S., 1986b. Afferent and efferent innervation patterns of the superior olivary nucleus of the leopard frog. *Brain Res.* **364**: 167–171.

Feng, A. S. and Lin, W., 1991. Differential innervation patterns of three divisions of frog auditory midbrain (torus semicircularis). *J. Comp. Neurol.* **306**: 613–630.

Feng, A. S., Hall, J. C. and Gooler, D. M., 1990. Neural basis of sound pattern recognition in anurans. *Prog. Neurobiol.* **34**: 313–329.

Fritzsch, B., 1981. The pattern of lateral-line afferents in urodeles. A horseradish peroxidase study. *Cell Tissue Res.* **218**: 581–594.

Fritzsch, B., 1988. The lateral-line and inner ear afferents in larval and adult urodeles. *Brain Behav. Evol.* **31**: 325–348.

Fritzsch, B., 1989. Diversity and regression in the amphibian lateral line system. Pp. 99–115 *in* "The Mechanosensory Lateral Line: Neurobiology and Evolution", ed by S. Coombs, P. Görner and H. Münz. Springer, New York.

Fritzsch, B., 1990. The evolution of metamorphosis in amphibians. *J. Neurobiol.* **21**: 1011–1021.

Fritzsch, B., 1992. The water-to-land transition: Evolution of the tetrapod basilar papilla, middle ear and auditory nuclei. Pp. 351–375 *in* "The Evolutionary Biology of Hearing", ed by D. Webster, A. Popper and R. R. Fay. Springer, New York.

Fritzsch, B., 1993. Evolutionary gain of non-teleostean electroreceptors. *J. Comp. Physiol.* **173**: 710–712.

Fritzsch, B., 1996. How does the urodele ear develop? *Int. J. Dev. Biol.* **40**: 763–771.

Fritzsch, B. and Bolz, D., 1986. On the development of the ampullary organs in the mountain newt, *Triturus alpestris*. *Amphibia–Reptilia* **7**: 1–9.

Fritzsch, B. and Münz, H., 1986. Electroreception in amphibians. Pp. 483–496 *in* "Electroreception", ed by T. H. Bullock and W. Heiligenberg. Wiley and Sons, Chichester.

Fritzsch, B. and Northcutt, R. G., 1993. Cranial and spinal nerve organization in amphioxus and lampreys. *Acta Anat.* **148**: 96–110.

Fritzsch, B. and Wahnschaffe, U., 1983. The electroreceptive ampullary organs of urodeles. *Cell Tissue Res.* **229**: 483–503.

Fritzsch, B. and Wahnschaffe, U., 1987. Electron microscopical evidence for bilateral and common lateral-line and inner ear efferents in urodeles. *Neurosci. Lett.* **81**: 48–52.

Fritzsch, B. and Wake, M. H., 1986. A note on the distribution of ampullary organs in Gymnophions. *J. Herpetol.* **20**: 90–93.

Fritzsch, B. and Wake, M. H., 1988. The inner ear of gymnophione amphibians and its nerve supply: a comparative study of regressive events in a complex sensory system. *Zoomorphology* **108**: 210–217.

Fritzsch, B., Nikundiwe, A. M. and Will, U., 1984. Projection patterns of lateral-line afferents in anurans. A comparative study using transganglionic transport of HRP. *J. Comp. Neurol.* **229**: 451–469.

Fritzsch, B., Zakon, H. H. and Sanchez, D. Y., 1990. The time course of structural changes in regenerating electroreceptors of a weakly electric fish. *J. Comp. Neurol.* **300**: 386–404.

Fritzsch, B., Christensen, M. A. and Nichols, D. H., 1993. Fiber pathways and positional changes in efferent perikarya of 2.5 to 7 day chick embryos as revealed with DiI and dextran amines. *J. Neurobiol.* **24**: 1481–1499.

Fuzessery, Z. M., 1988. Frequency tuning in the anuran central auditory system. Pp. 253–274 *in* "The Evolution of the Amphibian Auditory System", ed by B. Fritzsch, M. Ryan, W. Wilczynski, T. Hetherington and W. Walkowiak. Wiley and Sons, New York.

Gerhardt, H. C., 1994. Selective responsiveness to long-range acoustic signals in insects and anurans. *Amer. Zool.* **34**: 706–714.

Gonzalez, A. and Muñoz, M., 1987. Some connections of the area octavolateralis of *Pleurodeles waltlii*. A study with horseradish peroxidase under *in vitro* conditions. *Brain Res.* **423**: 338–342.

Gonzalez, A. and Muñoz, M., 1988. The area acustico-vestibularis of *Discoglossus pictus*. II. The primary afferent system. *J. Hirnforsch.* **29**: 421–434.

Gooler, D. M., Condon, C. J., Xu, J. H. and Feng, A. S., 1993. Sound direction influences the frequency-tuning characteristics of neurons in the frog inferior colliculus. *J. Neurophysiol.* **69**: 1018–1030.

Görner, P. and Mohr, C., 1989. Stimulus localization in *Xenopus*: Role of directional sensitivity of lateral line stitches. Pp. 543–560 *in* "The Mechanosensory Lateral Line: Neurobiology and Evolution", ed by S. Coombs, P. Görner and H. Münz. Springer, New York.

Gruberg, E. R. and Ambros, V. R., 1974. A forebrain visual projection in the frog *(Rana pipiens)*. *Exp. Neurol.* **44**: 187–197.

Hall, J. C. and Feng, A. S., 1987. Evidence for parallel processing in the frog's auditory thalamus. *J. Comp. Neurol.* **258**: 407–419.

Hall, J. C. and Feng, A. S., 1988. Influence of envelope rise time on neural responses in the auditory system of anurans. *Hearing Res.* **36**: 261–276.

Hall, J. C. and Feng, A. S., 1990. Classification of the temporal discharge patterns of single auditory neurons in the dorsal medullary nucleus of the northern leopard frog. *J. Neurophysiol.* **64**: 1460–1473.

Hall, J. C. and Feng, A. S., 1991. Temporal processing in the dorsal medullary nucleus of the northern leopard frog *(Rana pipiens pipiens)*. *J. Neurophysiol.* **66**: 955–973.

Harrison, R. G., 1904. Experimentelle Untersuchung über die Entwicklung der Sinnesorgane der Seitenlinie bei den Amphibien. *Arch. Mikrosk. Anat.* **63**: 35–149.

Hellmann, B. and Fritzsch, B., 1996. Neuroanatomical and histochemical evidence for the presence of common lateral line and inner ear efferents and of efferents to the basilar papilla in a frog, *Xenopus laevis*. *Brain Behav. Evol.* **47**: 185–194.

Hetherington, T. E., 1988. Metamorphic changes in the middle ear. Pp. 339–358 *in* "The Evolution of the Amphibian Auditory System", ed by B. Fritzsch, M. Ryan, W. Wilczynski, T. Hetherington and W. Walkowiak. Wiley and Sons, New York.

Hetherington, T. E., 1992. The effects of body size on the evolution of the amphibian middle ear. Pp. 421–438 *in* "The Evolutionary Biology of Hearing", ed by D. Webster, A. Popper and R. R. Fay. Springer, New York.

Hetherington, T. E. and Wake, M. H., 1979. The lateral line system in larval *Ichthyophis* (Amphibia: Gymnophiona). *Zoomorphol.* **93**: 209–225.

Hillman, D. E., 1976. Morphology of peripheral and central vestibular systems. Pp. 452–480 *in* "Frog Neurobiology", ed by R. Llinás and W. Precht. Springer, Berlin.

Himstedt, W. and Fritzsch, B., 1990. Behavioural evidence for electroreception in larval gymnophionans. *Zool. Jahrb. Physiol.* **94**: 486–492.

Hunt, P. and Krumlauf, R., 1991. Deciphering the Hox code: clues to patterning branchial regions of the head. *Cell* **66**: 1975–1078.

Jacobson, A. G. and Sater, A. K., 1988. Features of embryonic induction. *Development* **104**: 341–359.

Jaslow, A. P., Hetheringon, T. E. and Lombard, R. E., 1988. Structure and function of the amphibian middle ear. Pp. 69–92 *in* "The Evolution of the Amphibian Auditory System", ed by B. Fritzsch, M. Ryan, W. Wilczynski, T. Hetherington and W. Walkowiak. Wiley and Sons, New York.

Jenkins, F. A. and Walsh, D. M., 1993. An early Jurassic caecilian with limbs. *Nature* **365**: 246–250.

Jorgensen, J. M., 1989. Evolution of octavolateralis sensory cells. Pp. 17–78 *in* "The mechanosensory Lateral line: Neurobiology and Evolution", ed by S. Coombs, P. Görner and H. Münz. Springer Verlag, New York.

Jorgensen, M. B., 1991. Comparative studies of the biophysics of directional hearing in anurans. *J. Comp. Physiol.* **169**: 591–598.

Kelley, D. B., Morrell, J. I. and Pfaff, D. W., 1975. Autoradiographic localization of hormone-concentrating cells in the brain of an amphibian, *Xenopus Laevis*. I. Testosterone. *J. Comp. Neurol.* **164**: 47–62.

Kelley, D. B., Lieberburg, I., McEwen, B. S. and Pfaff, D. W., 1978. Autoradiographic and biochemical studies of steroid hormone-concentrating cells in the brain of *Rana pipiens*. *Brain Res.* **140**: 287–305.

Kokoros, J. J. and Northcutt, R. G., 1977. Telencephalic efferents of the tiger salamander *Ambystoma tigrinum tigrinum* (Green). *J. Comp. Neurol.* **173**: 613–628.

Kroese, A. B. A. and van Netten, S. M., 1989. Sensory transduction in lateral line hair-cells. Pp. 17–78 *in* "The Mechanosensory Lateral Line: Neurobiology and Evolution", ed by S. Coombs, P. Görner and H. Münz. Springer, New York.

Lannoo, M. J. and Smith, S. C., 1989. The lateral line system. Pp. 176–186 *in* "Developmental biology of the axolotl", ed by J. B. Armstrong and G. M. Malacinski. Oxford Univeristy Press, Oxford.

Lewis, E. R., 1992. Convergence of design in vertebrate acoustic sensors. Pp. 163–184 *in* "The Evolutionary Biology of Hearing", ed by D. Webster, A. Popper and R. R. Fay. Springer, New York.

Lewis, E. R. and Lombard, E. R., 1988. The amphibian inner ear. Pp. 93–124 *in* "The Evolution of the Amphibian Auditory System", ed by B. Fritzsch, M. Ryan, W. Wilczynski, T. Hetherington and W. Walkowiak. Wiley and Sons, New York.

Lewis, E. R., Leverenz, E. L. and Bialek, W., 1985. Pp. 256 *in* "The vertebrate inner ear". CRC Press, Boca Raton.

Lombard, R. E. and Bolt, J. R., 1979. Evolution of the tetrapod ear: an analysis and reinterpretation. *Biol. J. Lin Soc.* **11**: 19–76.

Manley, G. A. and Gleich, O., 1992. Evolution and specialization of function in the avian auditory periphery. Pp. 561–580 *in* "The Evolutionary Biology of Hearing", ed by D. Webster, A. Popper and R. R. Fay. Springer, New York.

Mansour, S. L., Goddard, J. M. and Capecchi, M. R., 1993. Mice homozygous for a targeted disruption of the proto-oncogene int-2 have developmental defects in the tail and inner ear. *Development* **117**: 13–28.

Manteuffel, G., Kopp, J. and Himstedt, W., 1986. Amphibian optokinetic afternystagmus: properties and comparative analysis in various species. *Brain Behav. Evol.* **28**: 186–197.

Manteuffel, G. and Naujoks-Manteuffel, C., 1990. Anatomical connections and electrophysiological properties of toral and dorsal tegmental neurons in the terrestrial urodele *Salamandra salamandra*. *J. Hirnforsch.* **31**: 65–76.

Masino, T. and Grobstein, P., 1990. Tectal connectivity in the frog *Rana pipiens*: Tectotegmental projections and a general analysis of topographic organization. *J. Comp. Neurol.* **291**: 103–127.

Matesz, C., 1979. Central projection of the VIIIth cranial nerve in the frog. *Neurosci.* **4**: 2061–2071.

McCormick, C. A., 1992. Evolution of central auditory pathways in anamniotes. Pp. 323–349 *in* "The Evolutionary Biology of Hearing", ed by D. Webster, A. Popper and R. R. Fay. Springer, New York.

McCormick, C. A., Braford, M. and Seidel, W., 1997. The lateral line of *Thyphlonectes natans*. *Brain Behav. Evol.*, in press.

Megela, A. L. and Capranica, R. R., 1983. A neural and behavioral study of auditory habituation in the Bullfrog, *Rana catesbeiana*. *J. Comp. Physiol.* **151**: 423–434.

Melssen, W. J. and Epping, W. J. M., 1990. A combined sensitivity for frequency and interaural intensity difference in neurons in the auditory midbrain of the grassfrog. *Hearing Res.* **44**: 35–50.

Melssen, W. J., Epping, W. J. M. and van Stokkum, I. H. M., 1990. Sensitivity for interaural time and intensity difference of auditory midbrain neurons in the grassfrog. *Hearing Res.* **47**: 235–256.

Montgomery, N. M. and Fite, K. V., 1991. Organization of ascending projections from the optic tectum and mesencephalic pretectal gray in *Rana pipiens*. *Vis. Neurosci.* **7**: 459–478.

Morrell, J. I., Kelley, D. B. and Pfaff, D. W., 1975. Autoradiographic localization of hormone-concentrating cells in the brain of an amphibian, *Xenopus laevis*. II. Estradiol. *J. Comp. Neurol.* **164**: 63–78.

Mudry, K. M. and Capranica, R. R., 1980. Evoked auditory activity within the telencephalon of the bullfrog (*Rana catesbeiana*). *Brain Res.* **182**: 303–311.

Mudry, K. M. and Capranica, R. R., 1987a. Correlation between auditory evoked responses in the thalamus and species-specific call characteristics. I. *Rana catesbeiana* (Anura: Ranidae). *J. Comp. Physiol.* **A 160**: 477–489.

Mudry, K. M. and Capranica, R. R., 1987b. Correlation between auditory thalamic area evoked responses and species-specific call characteristics. II. *Hyla cinerea* (Anura: Hylidae). *J. Comp. Physiol.* **A 161**: 407–416.

Muñoz, A., Muñoz, M., González, A. and ten Donkelaar, H. J., 1994. Spinothalamic projections in amphibians as revealed with anterograde tracing techniques. *Neurosci. Lett.* **171**: 81–84.

Münz, H., 1989. Functional organization of the lateral line periphery. Pp. 99–115 *in* "The Mechanosensory Lateral Line: Neurobiology and Evolution", ed by S. Coombs, P. Görner and H. Münz. Springer, New York.

Münz, H. and Claas, B., 1991. Activity of lateral line efferents in the axolotl (*Ambystoma mexicanum*). *J. Comp. Physiol.* **169**: 461–469.

Myers, S. F. and Lewis, E. R., 1990. Hair-cell tufts and afferent innervation of the bullfrog crista ampullaris. *Brain Res.* **534**: 15–24.

Narins, P. M., 1992. Biological constraints on anuran acoustic communication: auditory capabilities of naturally behaving animals. Pp. 439–454 *in* "The Evolutionary Biology of Hearing", ed by D. Webster, A. Popper and R. R. Fay. Springer, New York.

Narins, P. M., Ehret, G. and Tautz, J., 1988. Accessory pathway for sound transfer in a neotropical frog. *Proc. Natl. Acad. Sci.* **85**: 1508–1512.

Naujoks-Manteuffel, C. and Manteuffel, G., 1988. Origins of descending projections to the medulla oblongata and rostral medulla spinalis in the urodele *Salamandra salamandra*. *J. Comp. Neurol.* **273**: 187–206.

Neary, T. J., 1988. Forebrain auditory pathways in ranid frogs. Pp. 233–252 *in* "The Evolution of the Amphibian Auditory System", ed by B. Fritzsch, M. J. Ryan, W. Wilczynski, T. E. Hetherington and W. Walkowiak. Wiley and Sons, New York.

Neary, T. J., 1990. The pallium of anuran amphibians. Pp. 107–138 *in* "Cerebral Cortex, Vol V/8A", ed by E. G. Jones and A. Peters. Plenum, New York.

Neary, T. J., 1995. Afferent projections to the hypothalamus in ranid frogs. *Brain Behav. Evol.* **46**: 1–13.

Neary, T. J. and Fritzsch, B., 1992. Stage and concentration specific effects of retinoic acid on the differentiation of *Xenopus* hindbrain and ear. *Soc. Neurosci. Abstr.* **18**: 955.

Neary, T. J. and Northcutt, R. G., 1983. Nuclear organization of the bullfrog diencephalon. *J. Comp. Neurol.* **213**: 262–278.

Neary, T. J. and Northcutt, R. G., 1990. Septal area connections in ranid frogs. *Soc. Neurosci. Abstr.* **16**: 129.

Neary, T. J. and Wilczynski, W., 1977a. Ascending thalamic projections from the obex region in ranid frogs. *Brain Res.* **138**: 529–533.

Neary, T. J. and Wilczynski, W., 1977b. Autoradiographic demonstration of hypothalamic efferents in the bullfrog, *Rana catesbeiana*. *Anat. Rec.* **187**: 665.

Neary, T. J. and Wilczynski, W., 1980. Descending inputs to the optic tectum in ranid frogs. *Soc. Neurosci. Abstr.* **6**: 629.

Neary, T. J. and Wilczynski, W., 1986. Auditory pathways to the hypothalamus in ranid frogs. *Neurosci. Lett.* **71**: 142–146.

Neary, T. J. and Wilczynski, W., 1994. Midbrain roof-thalamic connections in ranid frogs. *Soc. Neurosci. Abstr.* **20**: 1418.

Noden, D. M. and van de Water, T. R., 1986. The developing ear: tissue origins and interactions. Pp. 15–46 *in* "The Biology of Change in Otolaryngology", ed by R. J. Ruben, T. R. van de Water and E. W. Rubel. Elsevier, Amsterdam.

Noden, D. M. and van de Water, T. R., 1992. Genetic analyses of mammalian ear development. *Trends. Neurosci.* **15**: 235–237.

Norris, H. W., 1892. Studies on the development of the ear in Amblystoma. I. Development of the auditory vesicle. *J. Morphol.* **7**: 23–34.

Northcutt, R. G., 1970. Pallial projection of sciatic, ulnar, and mandibular branch of the trigeminal nerve afferents in the frog (*Rana catesbeiana*). *Anat. Rec.* **166**: 356.

Northcutt, R. G., 1974. Some histochemical observations on the telencephalon of the bullfrog, *Rana catesbeiana* Shaw. *J. Comp. Neurol.* **157**: 379–390.

Northcutt, R. G., 1989. The phylogenetic distribution and innervation of craniate mechanoreceptive lateral lines. Pp. 17–78 *in* "The Mechanosensory Lateral Line: Neurobiology and Evolution", ed by S. Coombs, P. Görner and H. Münz. Springer, New York.

Northcutt, R. G., 1992a. The phylogeny of octavolateralis ontogenies: A reaffirmation of Garstang's phylogenetic hypothesis. Pp. 21–47 *in* "The Evolutionary Biology of Hearing", ed by D. Webster, A. N. Popper and R. R. Fay. Springer, New York.

Northcutt, R. G., 1992b. Distribution and innervation of lateral line organs in the axolotl. *J. Comp. Neurol.* **325**: 95–123.

Northcutt, R. G. and Gans, C., 1983. The genesis of neural crest and epidermal placodes: a reinterpretation of vertebrate origin. *Q. Rev. Biol.* **58**: 1–28.

Northcutt, R. G. and Plassmann, W., 1989. Electrosensory activity in the telencephalon of the axolotl. *Neurosci. Lett.* **99**: 79–84.

Northcutt, R. G. and Ronan, M., 1992. Afferent and efferent connections of the bullfrog medial pallium. *Brain Behav. Evol.* **40**: 1–16.

Northcutt, R. G. and Royce, G. J., 1975. Olfactory bulb projections in the bullfrog *Rana catesbeiana*. *J. Morphol.* **145**: 251–268.

Northcutt, R. G., Catania, K. C. and Criley, B. B., 1994. Development of lateral line organs in the axolotl. *J. Comp. Neurol.* **340**: 480–514.

Northcutt, R. G., Brändle, K. and Fritzsch, B., 1995. Electroreceptors and mechanosensory lateral line organs arise from single placodes in axolotls. *Devel. Biol.* **168**: 358–373.

Popper, A. N., Platt, C. and Edds, P. L., 1992. Evolution of the vertebrate ear: an overview of ideas. Pp. 49–57 *in* "The Evolutionary Biology of Hearing", ed by D. B. Webster, A. N. Popper and R. R. Fay. Springer, New York.

Rand, A. S., 1988. On overview of anuran acoustic communication. Pp. 415–432 *in* "The Evolution of the Amphibian Auditory System", ed by B. Fritzsch, M. Ryan, W. Wilczynski, T. Hetherington and W. Walkowiak. Wiley and Sons, New York.

Represa, J., Leon, Y., Miner, C. and Giraldez, F., 1991. The int-2 proto-oncogene is responsible for induction of the inner ear. *Nature* **353**: 561–563.

Retzius, G., 1881. Pp. 286 *in* "Das Gehörorgan der Wirbeltiere: I. Das Gehörorgan der Fische und Amphibien". Samson und Wallin, Stockholm.

Roberts, B. L. and Meredith, G. E., 1992. The efferent innervation of the ear: Variations on an enigma. Pp. 182–210 *in* "The Evolutionary Biology of Hearing", ed by D. B. Webster, A. N. Popper and R. R. Fay. Springer, New York.

Roper, S., 1989. The cell biology of vertebrate taste receptors. *Annu. Rev. Neurosci.* **12**: 329–353.

Roper, S., 1992. The microphysiology of peripheral taste organs. *J. Neurosci.* **12**: 1127–1134.

Rubinson, K. and Skiles, M. P., 1975. Efferent projections of the superior olivary nucleus in the frog, *Rana catesbeiana*. *Brain Behav. Evol.* **12**: 151–160.

Russel, I. J., 1976. Amphibian lateral line receptors. Pp. 513–549 *in* "Frog Neurobiology", ed by R. Llinás and W. Precht. Springer, Berlin.

Ryan, M., 1988. Constraints and patterns in the evolution of anuran accoustic communication. Pp. 637–678 *in* "The Evolution of the Amphibian Auditory System", ed by B. Fritzsch, M. Ryan, W. Wilczynski, T. Hetherington and W. Walkowiak. Wiley and Sons, New York.

Sarasin, P. and Sarasin, F., 1888. Zur Entwicklungsgeschichte und Anatomie der ceylonesischen Blindwühle *Ichthyophis glutinosus*. Das Gehörorgan. *Ergebnisse Naturwiss Forsch auf Ceylon* **2**: 207–222. Kreidels Verlag, Wiesbaden

Sarasin, P. and Sarasin, F., 1892. Über das Gehörorgan der Caeciliiden. *Anat. Anz.* **7**: 812–815.

Scalia, F., 1976. The optic pathway of the frog: Nuclear organization and connections. Pp. 386–406 *in* "Frog Neurobiology", ed by R. Llinás and W. Precht. Springer, Berlin.

Scalia, F. and Gregory, K., 1970. Retinofugal projections in the frog: Location of the postsynaptic neurons. *Brain Behav. Evol.* **3**: 16–29.

Scalia, F., Gallousis, G. and Roca, S., 1991. Differential projections of the main and accessory olfactory bulb in the frog. *J. Comp. Neurol.* **305**: 443–461.

Schellart, N. A. M. and Popper, A. N., 1992. Functional aspects of the evolution of the auditory system of actinopterygian fish. Pp. 295–322 *in* "The Evolutionary Biology of Hearing", ed by D. Webster, A. Popper and R. R. Fay. Springer, New York.

Schmitz, B., White, T. D. and Narins, P. M., 1992. Directionality of phase locking in auditory nerve fibers of the leopard frog *Rana pipiens pipiens*. *J. Comp. Physiol.* **170**: 589–604.

Sive, H. L. and Cheng, P. F., 1991. Retinoic acid perturbs the expression of Xhox.lab genes and alters mesodermal determination in *Xenopus laevis*. *Genes and Develop.* **5**: 1321–1332.

Smith, S. C., Lannoo, M. J. and Armstrong, J. B., 1990. Development of the mechanoreceptive lateral-line system in the axolotl: placode specification, guidance of migration, and the origin of neuromast polarity. *Anat. Embryol.* **182**: 171–180.

Swanson, G. J., Howard, M. and Lewis, J., 1990. Epithelial autonomy in the development of the inner ear of a bird embryo. *Dev. Biol.* **137**: 243–257.

ten Donkelaar, H. J., de Boer-vanHuizen, R., Schouten, F. T. M. and Eggen, S. J. H., 1981. Cells of origin of descending pathways to the spinal cord in the clawed toad *(Xenopus laevis)*. *Neurosci.* **6**: 2297–2312.

Tóth, P., Csank, G. and Lázár, G., 1985. Morphology of the cells of origin of descending pathways to the spinal cord in *Rana esculenta*. A tracing study using cobaltic-lysine complex. *J. Hirnforsch.* **26**: 365–383.

Urano, A. and Gorbman, A., 1981. Effects of pituitary hormonal treatment on responsiveness of anterior preoptic neurons in male leopard frogs, *Rana pipiens*. *J. Comp. Physiol.* **141**: 163–171.

van Bergijk, W. A., 1967. Evolution of the sense of hearing in vertebrates. *Am. Zool.* **6**: 371–377.

van Stokkum, I. H. M., 1987. Sensitivity of neurons in the dorsal medullary nucleus of the grassfrog to spectral and temporal characteristics of sound. *Hearing Res.* **29**: 223–235.

van Stokkum, I. H. M. and Melssen, W. J., 1991. Measuring and modelling the response of auditory midbrain neurons in the grassfrog to temporally structured binaural stimuli. *Hearing Res.* **52**: 113–132.

von Düring, M. and Andres, K. H., 1993. Electrosensory organs of the giant salamander *(Andrias davidianus)*. *J. Comp. Physiol.* **173**: 745.

von Kupffer, C., 1895. Studien zur vergleichenden Entwicklungsgeschichte des Kopfes der Kranioten. Heft 3. Die Entwicklung der Kopfnerven von Ammocoetes planeri. *München, Lehmann.* Pp. 1–80.

Vesselkin, N. P., Ermakova, T. V., Kenigfest, N. B. and Goikovic, M., 1980. The striatal connections in frog *Rana temporaria*: an HRP study. *J. Hirnforsch.* **21**: 381–392.

Wahnschaffe, U., Fritzsch, B. and Himstedt, W., 1985. The fine structure of the lateral-line organs of larval *Ichthyophis* (Amphibia: Gymnophiona). *J. Morphol.* **186**: 369–377.

Wahnschaffe, U., Bartsch, U. and Fritzsch, B., 1987. Metamorphic changes within the lateral-line system of Anura. *Anat. Embryol.* **175**: 431–442.

Wake, D. B., Nishikawa, K. C. and Roth, G., 1987. The fate of the lateral line system in plethodontid salamanders. *Am. Zool.* **27**: 166a.

Wake, D. B., Nishikawa, K. C., Dicke, U. and Roth, G., 1988. Organization of the motor nuclei in the cervical spinal cord of salamanders. *J. Comp. Neurol.* **278**: 195–208.

Walkowiak, W., 1988. Central temporal encoding. Pp. 275–294 *in* "The Evolution of the Amphibian Auditory System", ed by B. Fritzsch, M. J. Ryan, W. Wilczynski, T. E. Hetherington and W. Walkowiak. John Wiley and Sons, New York.

Walkowiak, W. and Luksch, H., 1994. Sensory motor interfacing in acoustic behavior of anurans. *Amer. Zool.* **34**: 685–695.

Webb, J. F. and Noden, D. M., 1993. Ectodermal placodes: contributions to the development of the vertebrate head. *Amer. Zool.* **33**: 434–447.

Wells, K. D., 1988. The effects of social interaction on anuran vocal behavior. Pp. 433–453 *in* "The Evolution of the Amphibian Auditory System", ed by B. Fritzsch, M. Ryan, W. Wilczynski, T. Hetherington and W. Walkowiak. Wiley and Sons, New York.

Wetzel, D. M., Haerter, U. L. and Kelley, D. B., 1985. A proposed neural pathway for vocalization in South African clawed frogs, *Xenopus laevis*. *J. Comp. Physiol.* **A 157**: 749–761.

Wever, E. G., 1974. The evolution of vertebrate hearing. Pp. 423–454 *in* "Handbook of Sensory Physiology, V/1: Auditory System", ed by W. D. Keidel and W. D. Neff. Springer, Berlin.

Wever, E. G., 1985. Pp. 405 *in* "The Amphibian Ear". Princeton University Press, New Jersey.

White, J. S. and Baird, I. L., 1982. Comparative morphological features of the apodan inner ear with comments on the evolution of amphibian auditory structures. *Scann. Electron. Micr.* **3**: 1301–1312.

Wicht, H. and Himstedt, W., 1988. Topologic and connectional analysis of the dorsal thalamus of *Triturus alpestris* (amphibia, urodela, salamandridae). *J. Comp. Neurol.* **267**: 545–561.

Wiederhold, M. L., Yamashita, M. and Asashima, M., 1992. Development of the gravity-sensing organs in the Japanese red-bellied newt, *Cynops pyrrogaster*. *Proc. 18th. Int. Symp. Space Technol. Sci.* **18**: 2103–2108.

Wilczynski, W., 1981. Afferents to the midbrain auditory center in the bullfrog, *Rana catesbeiana*. *J. Comp. Neurol.* **198**: 421–433.

Wilczynski, W., 1988. Brainstem auditory pathways in anuran amphibians. Pp. 209–231 *in* "The Evolution of the Amphibian Auditory System", ed by B. Fritzsch, M. J. Ryan, W. Wilczynski, T. E. Hetherington and W. Walkowiak. Wiley and Sons, New York.

Wilczynski, W., 1992. Auditory and endocrine inputs to forebrain centers in anuran amphibians. *Ethol. Ecol. Evol.* **4**: 75–87.

Wilczynski, W. and Allison, J. D., 1989. Acoustic modulation of neural activity in the hypothalamus of the leopard frog. *Brain Behav. Evol.* **33**: 317–324.

Wilczynski, W. and Neary, T. J., 1986. Terminal field organization in the bullfrog torus semicircularis. *Soc. Neurosci. Abstr.* **12**: 105.

Wilczynski, W. and Northcutt, R. G., 1983a. Connections of the bullfrog striatum: Afferent organization. *J. Comp. Neurol.* **214**: 321–332.

Wilczynski, W. and Northcutt, R. G., 1983b. Connections of the bullfrog striatum: Efferent projections. *J. Comp. Neurol.* **214**: 333–343.

Wilczynski, W., Allison, J. D. and Marler, C. A., 1993. Sensory pathways linking social and environmental cues to endocrine control regions of amphibian forebrains. *Brain Behav. Evol.* **42**: 252–264.

Will, U., Luhede, G. and Görner, P., 1985. The area octavo-lateralis in *Xenopus laevis*. I. The primary afferent projections. *Cell Tissue Res.* **239**: 147–161.

Will, U., 1988. Organization and projections of the area octavolateralis in amphibians. Pp. 185–208 *in* "The Evolution of the Amphibian Auditory System", ed by B. Fritzsch, M. J. Ryan, W. Wilczynski, T. E. Hetherington and W. Walkowiak. Wiley and Sons, New York.

Will, U., 1989. Central mechanosensory lateral line system in amphibians. Pp. 17–78 *in* "The Mechanosensory Lateral Line: Neurobiology and Evolution" ed by S. Coombs, P. Görner and H. Münz. Springer, New York.

Will, U. and Fritzsch, B., 1988. The eighth nerve of amphibians: Peripheral and central distribution. Pp. 159–184 *in* "The Evolution of the Amphibian Auditory System", ed by B. Fritzsch, M. Ryan, W. Wilczynski, T. Hetherington and W. Walkowiak. Wiley and Sons, New York.

Winklbauer, R., 1989. Development of the lateral line system of *Xenopus*. *Progr. Neurobiol.* **32**: 181–206.

Yntema, C. L., 1939. Self-differentiation of heterotopic ear ectoderm in the embryo of *Amblystoma punctatum*. *J. Exp. Zool.* **80**: 1–17.

Yntema, C. L., 1955. Ear and nose. Pp. 415–428 *in* "Analysis of Development", ed by B. H. Willier, P. A. Weiss and V. Hamburger. Saunders, Philadelphia.

Zakon, H. H. and Wilczynski, W., 1988. The physiology of the anuran eighth nerve. Pp. 125–155 *in* "The Evolution of the Amphibian Auditory System", ed by B. Fritzsch, M. Ryan, W. Wilczynski, T. Hetherington and W. Walkowiak. Wiley and Sons, New York.

Zittlau, K. E., Claas, B. and Münz, H., 1988. Horseradish peroxidase study of tectal afferents in *Xenopus laevis* with special emphasis on their relationship to the lateral-line system. *Brain Behav. Evol.* **32**: 208–219.

V. ABBREVIATIONS USED IN TEXT

AM = amplitude modulation
AP = amphibian papilla
AVC = anterior vertical canal
BP = basilar papilla
DLN = dorsolateral nucleus
DTAM = dorsal tegmental nucleus of the medulla
HC = horizontal canal
HRP = horseradish peroxidase
IID = interaural intensity difference
ITD = interaural time difference
PVC = posterior vertical canal
SO = superior olivary nucleus

CHAPTER 5

Skin Sensory Systems of Amphibian Embryos and Young Larvae

Alan Roberts

I. Introduction
II. Skin Structure
III. Trunk Touch System
IV. Trigeminal Sensory System
V. Skin Impulses That Detect Noxious Stimuli
VI. Conclusion
VII. Acknowledgements
VIII. References

I. INTRODUCTION

MANY amphibians hatch from the egg at an early developmental stage. As free-living animals they have sensory systems that inform them about stimuli impinging on their body surface and allow them to respond appropriately, for example by swimming away. During these first days of larval life the skin has a very simple structure but like most adult animals (Burgess and Perl 1973) it is provided, even before hatching, with separate sensory pathways that detect three types of mechanical stimuli: "noxious", which are strong or damaging; "touch", which cause local, rapid and transient indentation; and "pressure", where the indentation is over a larger area and occurs more slowly (reviewed by Roberts 1987). At least two of these pathways are transient and only function during the earliest few days of larval life. They degenerate when the dorsal root ganglia develop and provide a new, more adult replacement innervation of the trunk skin (Hughes 1957). The mechanoreceptor sensory system of the skin of the hatchling tadpole is based entirely on innervation by free nerve-endings and is of special interest for a number of reasons. Firstly, it is so simple that it allows the anatomy of the nerve endings for different modalities to be distinguished. Secondly, the behavioural roles of the different sensory modalities of the skin are clear. Thirdly, the "noxious" modality is mediated by a propagated impulse in the skin, a system otherwise found only in certain invertebrate groups, like jellyfish and sea-anemones, and in some primitive marine chordates (Mackie 1970).

It is the simplicity of the behaviour of the young larvae that makes an understanding of the roles of the skin sensory modalities possible. The development of this behaviour was studied in the first decades of this century by Coghill (1914, 1924) and Hooker (1911) in the tiger salamander *(Ambystoma tigrinum)* and by Wintrebert (1904, 1905, 1909, 1920) in a variety of European amphibians (e.g., *Alytes, Rana* and *Salamandra*). Later studies have largely confirmed the earlier picture (Abu Gidieri 1971; Boothby and Roberts 1995; Soffe 1991a).

Reactions to touch start shortly after the closure of the neural tube and consist initially of a simple flexion of the body on the side opposite to the stimulus. These become stronger as the muscles develop and eventually they can throw the body into a C-flexure or coil. The next step is the alternation of flexions on each side of the body following the stimulus; finally, these flexions become strong enough to move the tadpole through the water in what can be called swimming. Swimming is usually initiated by touch or stronger mechanical stimulation to any part of the body. In the *Xenopus* embryo, swimming can be stopped abruptly when the animal bumps into either the surface meniscus or any other solid surface (Roberts and Blight 1975; Boothby and Roberts 1992a,b). Finally, if attempts are made to hold young tadpoles, they wriggle vigorously and these movements are much slower than those involved in swimming (Soffe 1991b). When analysed in *Xenopus* embryos using high-speed movies, it became clear that the waves of bending during such struggling travelled from the tail towards the head (Kahn and Roberts 1982), which is in the opposite direction to the head-to-tail waves that occur during swimming. Struggling is a very effective method of escape.

II. SKIN STRUCTURE

The skin in embryos and young tadpoles is very simple and laid down at the time of neurulation; it consists of two layers of cells lying on an extracellular basal lamina (Fig. 1; Kelly 1966; Schroeder 1970; Roberts and Stirling 1971; Warburg *et al.* 1994). In the *Xenopus* embryo, the deep cells are flattened and appear to form a homogeneous class. The superficial layer consists mainly of broadly cuboidal cells whose outer surface is lined with small mucus vesicles. Interspersed among these mucus cells are two other types: (1) multiply ciliated cells with numerous mitochondria and (2) flask-shaped cells with fewer, but larger, mucus vesicles at their outer surface. At their outer surfaces all these cells are connected by a terminal bar

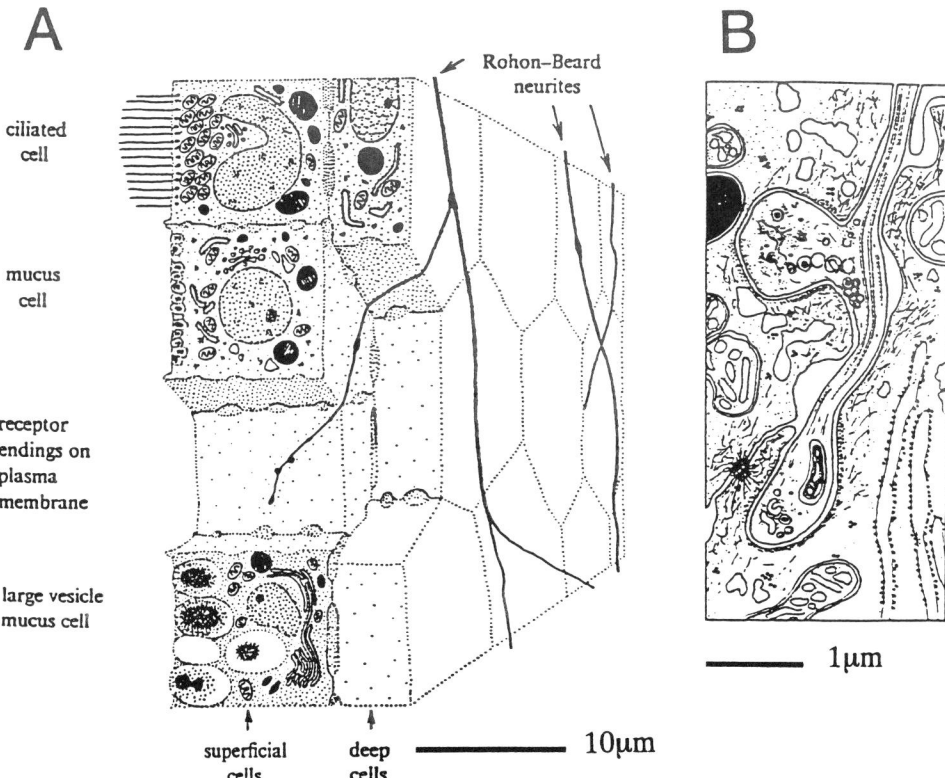

Fig. 1. Trunk skin anatomy in hatchling *Xenopus* embryo. **A:** Diagram showing two-layered structure with one class of deep cells and three classes of superficial cells. Sensory neurites from Rohon-Beard neurons in the spinal cord run over the inner skin surface and free nerve-endings penetrate between the skin cells. **B:** Diagram based on transmission electron microscopy of the terminal varicose endings of Rohon-Beard neurites invaginated into the plasma membranes of two superficial skin cells. From Roberts and Hayes (1977).

consisting of an uninterrupted belt of "tight" junctions which seals the extracellular space; there are frequent desmosomes. Deep to the surface, the cells are occasionally connected by small punctate "gap" junctions which would permit ionic currents and small molecules to pass from cell to cell. As the tadpoles develop, the skin becomes stretched much thinner and the basal lamina becomes thicker by the deposition of a layer of collagen fibrils.

III. TRUNK TOUCH SYSTEM

The trunk skin in amphibian embryos is initially innervated by touch-sensitive free nerve endings from Rohon-Beard neurons lying in the dorsal spinal cord (*Ambystoma*: Coghill 1914; *Xenopus*: Hughes 1957). Rohon-Beard neurons are amongst the earliest neurons to differentiate. In *Xenopus*, they undergo their last cell division at the gastrula stage (Lamborghini 1980) and begin to grow neurites to the skin shortly after the closure of the neural tube (Taylor and Roberts 1983). By the time of hatching (stage 37/38 of Nieuwkoop and Faber 1956) there are 200–300 Rohon-Beard neurons forming a nearly continuous column along the whole dorsal suface of the spinal cord which extends rostrally into the brain stem (Roberts and Clarke 1982). These neurons innervate the whole surface of the body from the gills to the tip of the tail. Rohon-Beard neuron anatomy is summarized in Figure 2. Within the spinal cord each Rohon-Beard neuron has a longitudinal central axon which ascends and descends in the dorsal sensory tracts (Nordlander 1984). The peripheral neurites leave the cord and grow between the somites to reach the inside surface of the skin where they form a plexus of fine, naked, unmyelinated neurites 0.1–0.8 μm in diameter (Roberts and Taylor 1982). Neurites from this plexus penetrate the skin to innervate small areas (0.04–0.25 mm^2) with varicose, free, nerve endings lying underneath and between the outer layer of skin cells (Fig. 1B; Roberts and Hayes 1977). No specialized junctions are seen between the endings or varicosities and the skin cells but the varicosities are often invaginated into the plasma membrane of the cells that they innervate; these invaginations are wrapped by intracellular tonofilaments. Such structures suggest a firm mechanical coupling between the sensory neurites and the skin cells. The innervation of the trunk skin by Rohon-Beard neurons is very similar in the newt *Triturus* (Roberts and Clarke 1983). Both Coghill (1914) studying *Ambystoma* and Hughes (1957) studying *Xenopus* concluded that Rohon-Beard neurons innervated the trunk muscles as they pass through on their way to the skin. However, evidence based on scanning electron microscopy and horseradish peroxidase backfilling suggests that this is not the case and that the earlier conclusion resulted from incomplete staining which gave the impression that neurites were ending in the muscles when in reality they pass through to the skin (Roberts and Hayes 1977).

The receptive fields and function of Rohon-Beard neurons have been studied in *Xenopus* embryos and tadpoles immobilized by neuromuscular blocking agents such as tubocurarine and α-bungarotoxin. Both extracellular and intracellular recordings have shown that Rohon-Beard neurons respond to local and quick distortion of the skin within their receptive field (Fig. 3; Roberts and Hayes 1977; Clarke *et al.* 1984). They are classic rapidly adapting "touch" or "rapid transient" detectors (Burgess and Perl 1973), fire very few impulses during stimulation, and show no maintained firing to sustained distortion. They are also very unresponsive to repeated mechanical stimuli unless about a minute is left for recovery. Extracellular recordings allowed exploration of the receptive fields and the range of effective stimuli, while intracellular recordings followed by dye injection permitted the Rohon-Beard neurons to be identified unambiguously. Together these recordings indicate that Rohon-Beard neurons form a single population of sensory neurons with similar response characteristics. These neurons provide the only innervation of the trunk skin in the embryo and young tadpole before the development of the dorsal root ganglia. At comparable developmental stages Rohon-Beard neurons in the newt *Triturus* show similar responses but tend to be stimulated by slower and broader mechanical distortion and fire more frequently in response to repeated stimuli (Fig. 3; Roberts and Clarke 1983).

Hughes (1957) and Lamborghini (1980) reported neurons in *Xenopus* embryos, which were named "extramedullary neurons", lying just outside the spinal cord with peripheral and central axons like Rohon-Beard neurons. These neurons are assumed to be sensory although no recordings have been made to establish their function. During the development of the

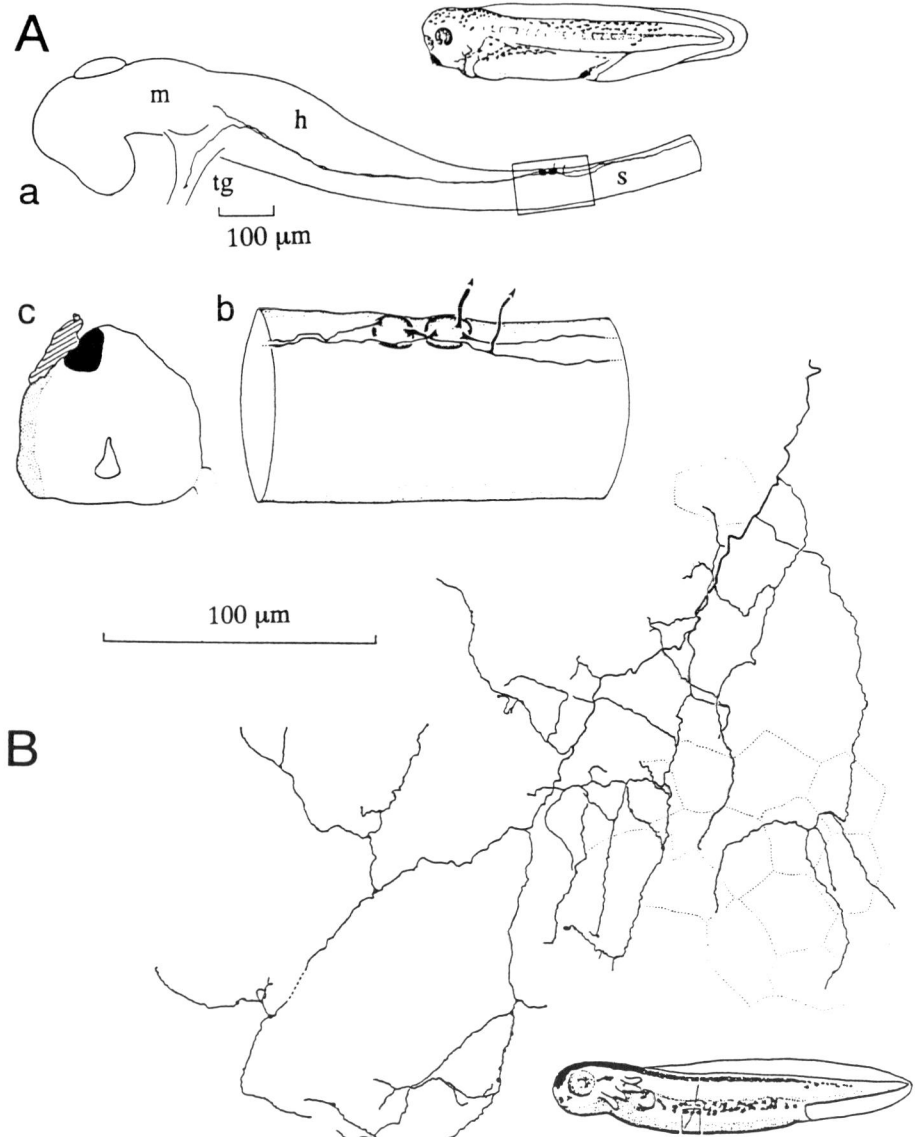

Fig. 2. Anatomy of touch-sensitive Rohon-Beard neurons innervating trunk skin. (**A**) In the hatchling *Xenopus* tadpole (diagram at stage 37/38 top right), (a) is a lateral view of the left side of the brain (m, midbrain; h, hindbrain) and spinal cord (s). It shows the ascending and descending axons of two Rohon-Beard sensory neurons that have been filled with horseradish peroxidase. The ascending axons pass through the hindbrain (h) and one runs into the trigeminal ganglion (tg). The somata of these same neurons are shown enlarged (b) and arrowheads mark the cut ends of their peripheral neurites. A transverse section of the spinal cord (c) shows dorsal intraspinal location of Rohon-Beard neuron soma (black). From Roberts and Clarke (1982). (**B**) Peripheral neurites of a touch-sensitive Rohon-Beard neuron filled with horseradish peroxidase and innervating ventral trunk skin in the hatchling *Triturus* larva (diagram bottom right shows location of neurities in area bounded by rectangle). From Roberts and Clarke (1983). 100 μm scale applies to Ac, Ab and B.

peripheral neurites of Rohon-Beard neurons some neuronal somata were noted bulging out from the spinal cord at the point of neurite exit and others were seen lying peripherally along the neurites (Taylor and Roberts 1983). A simple interpretation is that these "extramedullary neurons" are simply Rohon-Beard neurons that migrate from the spinal cord if their somata lie at the point of neurite exit.

Touching the skin usually initiates larval swimming, so a role for Rohon-Beard neurons in initiation of swimming seems likely for most amphibian embryos and young larvae. In *Xenopus* embryos, it has been possible to stimulate single Rohon-Beard neurons by injecting

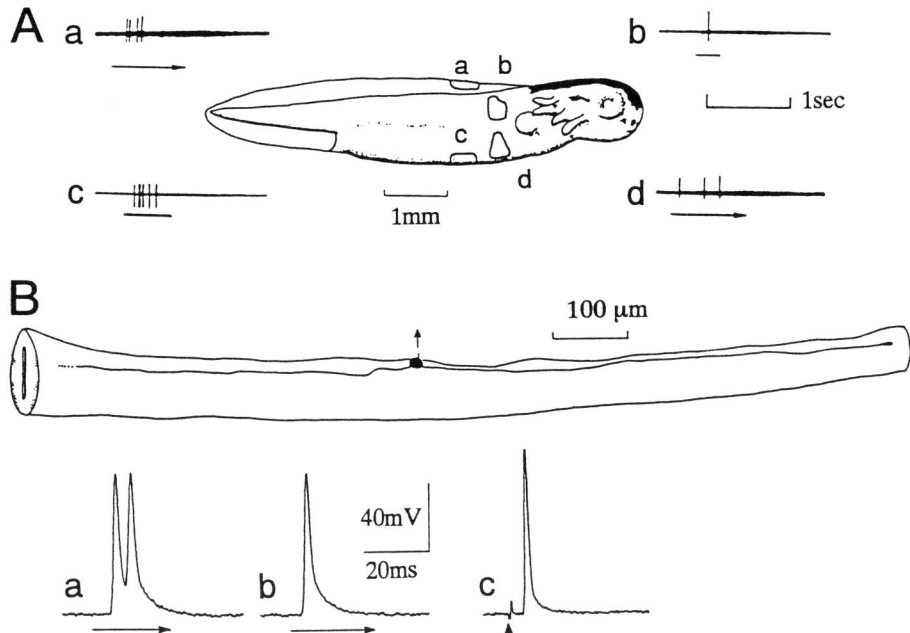

Fig. 3. Physiology of touch-sensitive Rohon-Beard neurons innervating trunk skin. (**A**) Responses from four Rohon-Beard neurons whose receptive fields are shown (a to d) in drawings of the European newt, *Triturus vulgaris*, at stages 33 to 36 of Gallien and Bidaud (1959). Extracellular recordings show impulses evoked by a stroke through the receptive field with a fine hair (at arrows) or pressure (during line). Based on data from Roberts and Clarke (1983). (**B**) Recordings from a single Rohon-Beard neuron made with a dye-filled microelectrode from a *Xenopus laevis* tadpole at stage 37/38 of Nieuwkoop and Faber (1956). The diagram shows the soma of the recorded neuron with its ascending and descending axons in the dorsal spinal cord viewed from the left side (brain to the left). (a) and (b) show impulses evoked by a stroke through the receptive field with a fine hair (at arrows). (c) shows impulse response to electrical pulse within receptive field. From Roberts *et al.* (1983).

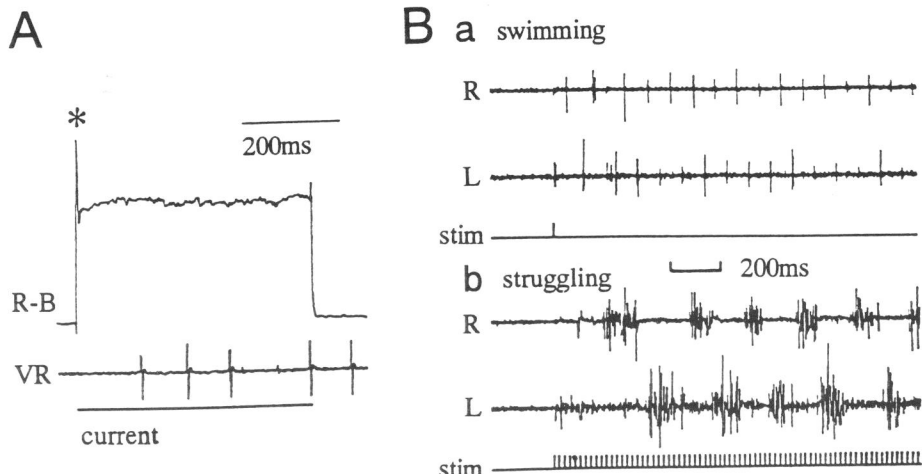

Fig. 4. Role of Rohon-Beard neurons in the behaviour of the *Xenopus* embryo. (**A**) A single impulse (at asterisk) evoked by current injection from a microelectrode into a single intracellularly recorded Rohon-Beard neuron can initiate swimming, as seen in a recording of motor ventral root activity (VR). From Clarke *et al.* (1984). (**B**) (a) a single current pulse to the skin producing single impulses in Rohon-Beard neurons initiates swimming as recorded from the left (L) and right (R) ventral roots, while (b) repetitive current pulses which produce repeated impulses in Rohon-Beard neurons evoke struggling as recorded in the ventral roots. From Soffe (1992).

current through a microelectrode. In about half such cases, a single current-evoked impulse in a single Rohon-Beard neuron is sufficient to start larval swimming, so the role of these neurons is conclusively established (Fig. 4; Clarke *et al.* 1984). Although some Rohon-Beard neurons have substance-P-like immunoreactivity (Clarke *et al.* 1984; Gallagher and Moody 1987), the role of any such peptide is unclear because Rohon-Beard neurons appear to release an excitatory amino acid, like glutamate, at their central synapses in the dorsal spinal cord (Sillar and Roberts 1988). This excitatory transmitter allows them to stimulate many sensory interneurons along the length of their central axon in the spinal cord; these interneurons amplify and distribute excitation that initiates swimming (Clarke and Roberts 1984; Roberts and Sillar 1990).

Sustained local pressure on the skin in the rostral trunk of *Xenopus* embryos evokes struggling (see section I above) and Soffe (1991b, 1992) has shown that this depends on repetitive firing of Rohon-Beard neurons. In spinal embryos, a single electrical pulse that stimulates the peripheral neurites of the Rohon-Beard neurons in the skin, equivalent to a brief touch, will evoke swimming (Fig. 4 Ba). However, repetitive pulses at 25 Hz, in the same place and at the same strength, equivalent to pressure on the skin, will evoke struggling for as long as the stimulation continues. These observations provide an extremely clear example of the pattern of activity in a single sensory pathway determining a behavioural response. A single impulse in one or more Rohon-Beard neurons results in swimming, while repetitive impulses in the same neurons result in struggling.

IV. TRIGEMINAL SENSORY SYSTEM

The skin of the head in amphibian embryos, as in the adults, is innervated by sensory neurons in the trigeminal ganglia. The anatomy of these neurons has been revealed in *Xenopus* embryos by electron microscopy, silver staining and horseradish peroxidase filling (Fig. 5; Hayes and Roberts 1983). Their central axons run caudally through the hindbrain (brainstem) and a few reach as far as the spinal cord. Of the 170 peripheral axons leaving the trigeminal ganglia on each side, approximately 40 innervate the cement gland (see below) leaving 130 to innervate the skin. These are divided roughly equally between the ophthalmic nerve, innervating the dorsal part of the head, and the maxillary nerve which innervates more ventral regions. The responses of the different types of trigeminal sensory neurons have been studied by making extracellular recordings of their impulse activity.

The pressure-sensitive nerve endings in the ventral part of the cement gland of *Xenopus* embryos have the simplest structure. Trigeminal neurites, 0.3 to 1 μm in diameter, penetrate the inside surface of the gland and run parallel to the long axis of the deep columnar epithelium formed by the mucus secreting cells (Fig. 5Aa and b; Roberts and Blight 1975). The neurites do not branch but near the secretory surface they form swollen terminal dendrites up to 3 μm across; transmission electron microscopy shows these dendrites to contain a dense mass of microfilaments (Fig. 5Ac). Recordings from trigeminal neurons innervating the cement gland show that they fire repetitively at up to about 50 Hz in response to any slow distortion of the gland (Fig. 5B). This effect can be produced by pressing on the gland with a blunt probe, or by pulling on the strand of mucus secreted by the gland. Since these receptors respond to pressure on the cement gland it is simplest to call them pressure-receptors by analogy with similar receptors elsewhere (Burgess and Perl 1973) but in fact they adapt within one second and show no sustained firing to maintained pressure.

The whole of the head skin in tadpoles at the tailbud stage is innervated by other trigeminal neurons which respond to broad pressure in a similar way to those innervating the cement gland (Roberts 1980). In the *Xenopus* embryo these neurons have receptive fields from 0.006 to 0.028 mm^2 in area (mean 0.017 ± 0.007) and fire repetitively when a probe pushes into the skin at speeds as low as 5 mm.s^{-1} (Fig. 6). Frequency of firing depends on the degree and velocity of indentation and can be up to 120 Hz. Just as in the cement gland, these neurons adapt within one second and give no sustained response to continued pressure. Trigeminal neurons with very similar response characteristics were also found innervating head skin in the tailbud stage of tadpoles of the anuran *Rana temporaria* (stages 24 to 27 of Cambar and Marrot (1954)) and in larvae of the urodele *Triturus helvetica* (stages 32–33 of Gallien and Bidaud (1959)). The anatomy of the nerve endings responsible for these responses are discussed below.

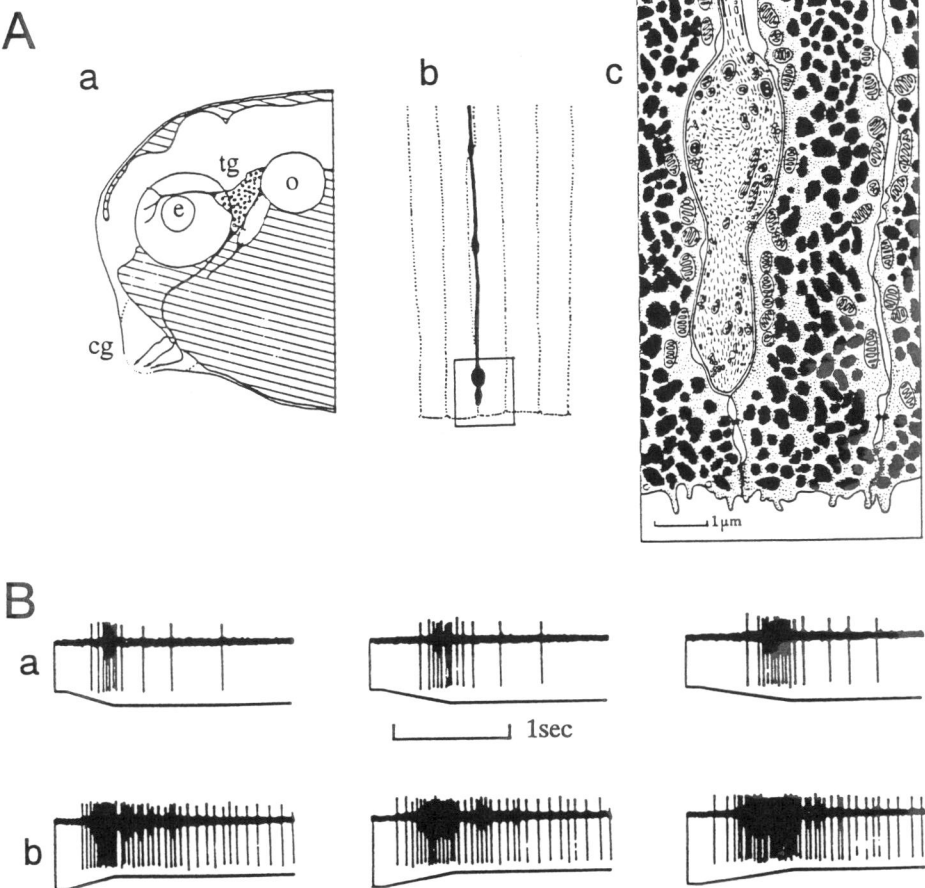

Fig. 5. Anatomy and physiology of pressure receptors in the *Xenopus* embryo cement gland. (**A**) (a) General diagram of the left side of the head showing trigeminal ganglion (tg) between the eye cup (e) and the otic capsule (o) with the ophthalmic nerve running behind the eye to innervate the dorsal head skin and the maxillary nerve running ventrally to innervate the cement gland (cg) and ventral head skin. (b) Neurites in the cement gland run between the columnar mucus-secreting cells to form varicose endings near the secretory surface (in square at bottom). (c) Diagram based on transmission electron microscopy of a varicose nerve ending in the cement gland. (**B**) Recordings of impulse discharge (top traces) from a trigeminal pressure receptor innervating the cement gland and responding (a) to 3 velocities of pressure into the gland (bottom trace moves down) and (b) to pull on the mucus strand secreted by the gland (bottom trace moves up). Modified from Roberts and Blight (1975).

A second overlapping innervation of the head skin allows sensitivity to touch in tailbud stages of *Xenopus*, *Triturus* and *Rana* (Roberts 1980). Recordings from the trigeminal ganglion show neurons which only respond to rapid local distortion of the skin surface, just like the response of Rohon-Beard neurons innervating the trunk skin (Fig. 7). In *Xenopus* and *Triturus* these neurons usually fire only once or twice following a local indentation but can fire more impulses to a stroke through the receptive field. In *Rana* the touch neurons fire slightly more repetitively. The receptive field areas of touch neurons in *Xenopus* were similar to the pressure sensitive neurons and ranged from 0.005 to 0.031 mm^2 (mean 0.015 ± 0.008).

In the *Xenopus* tadpole the anatomy of the nerve endings responsible for touch and pressure sensitivity was revealed by horseradish peroxidase backfilling and resolved after the discovery of a curious difference between the two types (Hayes and Roberts 1983). Recordings from trigeminal neurons showed that the neurites of touch-sensitive neurons from the trigeminal ganglion on one side sometimes strayed over the dorsal midline to innervate skin on the opposite side of the body, whereas the pressure-sensitive neurites never did this and always stopped abruptly at the dorsal midline (Kitson and Roberts 1983). The touch endings lie mainly between the two layers of cells in the skin, have occasional elongated varicosities, and

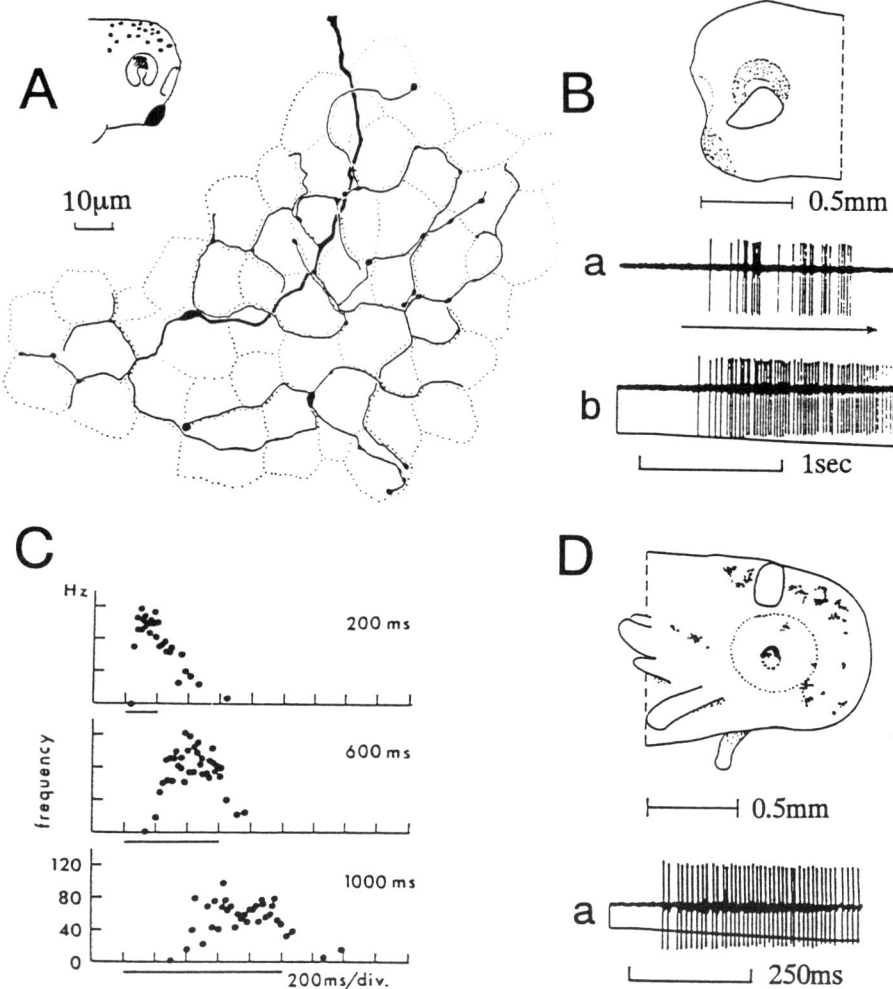

Fig. 6. Anatomy and physiology of pressure receptors in head skin of hatchling *Xenopus* and *Triturus* larvae. (A) to (C) in *Xenopus* tadpoles. (**A**) An example of horseradish peroxidase filled pressure-sensitive trigeminal nerve endings in skin over the dorsal part of the eye (shaded in the diagram of the head). The neurites run around the edges of the skin cells (shown as dotted lines). (**B**) Receptive field over lower part of the eye (outlined on diagram of head), and impulse responses of a trigeminal neuron to a hair stroking through the receptive field (a) and a small probe pushing into the field (b, lower trace moves down). (**C**) Plots of the impulse frequency as a function of time of a trigeminal neuron in *Xenopus* responding to a probe pushing into the skin with different rise-times (marked by bars). (**D**) In hatchling *Triturus helveticus* larva, the receptive field of a trigeminal neuron innervating skin on the dorsal head (outlined on diagram of head), and its response to a small probe pushing slowly into the skin in the receptive field (a). (A) from Hayes and Roberts (1983); (B to D) from Roberts (1980).

form few branches (Fig. 7). Their anatomy is similar to that of the Rohon-Beard neurites in the trunk skin (Fig. 2). On the other hand, the pressure-sensitive endings are more superficial and run between and often around the cells in the superficial outer layer of the skin where they branch quite extensively and have numerous large varicosities containing mitochondria (Fig. 6). Transmission electron microscopy showed that neither type of neurite made any specialized junctions, such as synapses or gap junctions, with the skin cells.

What is the behavioural significance of the dual innervation of the head skin and cement gland by touch-sensitive and pressure-sensitive trigeminal neurons? Simple behavioural tests established that slow and broad pressure stimuli applied to the head skin or cement gland never evoked swimming. However, if a restrained tadpole was already swimming, such stimuli were effective in stopping it (Boothby and Roberts 1992a;

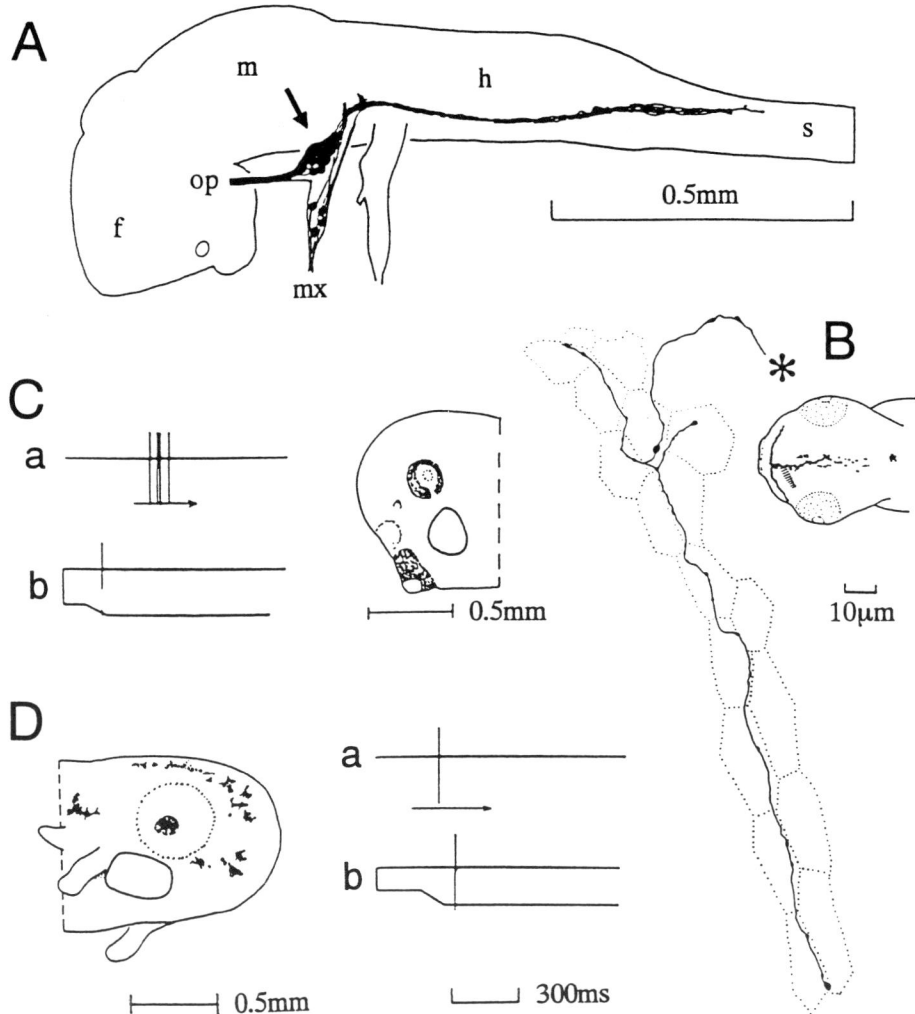

Fig. 7. Anatomy and physiology of trigeminal touch receptors in head skin of hatchling *Xenopus* and *Triturus* larvae. (**A**) to (**C**) Stage 37/38 *Xenopus* tadpole. (**A**) View of the left side of the brain showing sensory neurons in the left trigeminal ganglion (at arrow) with central axons projecting caudally into the hindbrain (h), seen after horseradish peroxidase filling in a lateral view of the brain (f, forebrain; m, midbrain; mx, maxillary nerve; op, ophthalmic nerve; s, spinal cord). (**B**) Touch-sensitive nerve endings in the dorsal head skin (in hatched area in inset dorsal view of head) arising from a trigeminal neurite under the skin (at asterisk). Outlines of superficial cells are dotted. From Hayes and Roberts (1983). (**C**) and (**D**) Responses of trigeminal neurons in *Xenopus* (**C**) and *Triturus* (**D**) larvae with receptive fields just ventral to the eye and outlined in diagrams of the head in side view. In each case (a) shows response to a stroke with a fine hair (during arrow) and (b) shows response to local indentation (when lower trace moves down). From Roberts (1980).

Roberts and Blight 1975). The current hypothesis is that the pressure-sensitive trigeminal neurons project into the hindbrain; there they excite inhibitory reticulospinal neurons that project to the spinal cord where they release the inhibitory transmitter GABA onto spinal neurons. In turn, these neurons stop the rhythm-generator for swimming (Boothby and Roberts 1992b). The pressure-sensitive trigeminal neurons therefore have entirely inhibitory effects on behaviour. The touch-sensitive trigeminal neurons are like the Rohon-Beard neurons in the spinal cord; stimuli which excite them will usually initiate swimming.

V. SKIN IMPULSES THAT DETECT NOXIOUS STIMULI

Most animals, from leeches (Nicholls and Baylor 1968) to mammals (Iggo 1976) have sensory mechanisms in their skin that allow the detection of noxious or damaging stimulation. The mechanism in amphibian larvae is unconventional. In the hatchling *Xenopus* embryo, a strong mechanical stimulus applied to any point on the body surface will evoke an all-or-none

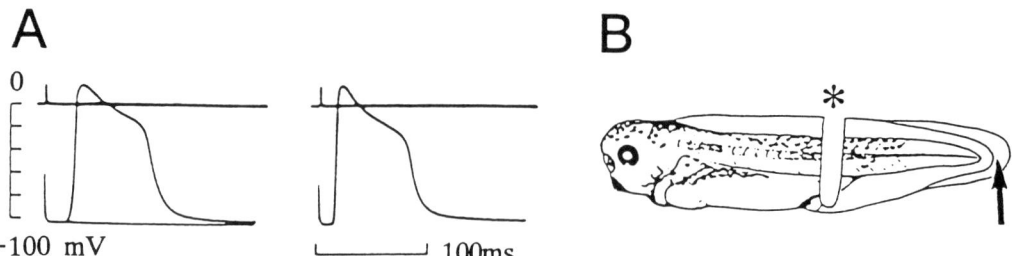

Fig. 8. The skin impulse in *Xenopus* embryos and tadpoles. (**A**) Intracellular recording from two skin cells (left and right records) in *Xenopus* embryos at stages 33 to 37 showing overshooting, all-or-none impulse evoked by a brief current pulse to the skin at the start of the trace. (**B**) Diagram of Wintrebert's 1904 experiment repeated on a stage 37/38 *Xenopus* tadpole. A cut is made (at asterisk) so the only tissue connecting the tail to the rostral trunk is a narrow, ventral bridge of skin. If the tail is then pricked (at arrow), a skin impulse is initiated, spreads across the bridge and the front part of the tadpole starts swimming movements.

propagated impulse which spreads in all directions from the point of stimulation at 5–11 cm.s^{-1} (mean 7.7 cm.s^{-1}) and can be recorded by a microelectrode inserted into any randomly chosen skin cell (Fig. 8A; Roberts 1969; Roberts and Stirling 1971). In many ways, the skin impulse is like the cardiac impulse which propagates through the heart and initiates contraction. It is primarily sodium-dependent, overshoots zero and is blocked by tetrodotoxin. It is generated across the inwardly facing membranes of the outer skin cells and probably invades all types of cells in both layers of the skin. It propagates as current spreads to neighbouring skin cells via gap junctions. The skin impulse in *Xenopus* embryos first appears at the time when the neural tube is closing and persists after hatching for the first days of larval life (Roberts and Smyth 1974). Its history later in development is difficult to assess as it becomes hard to record electrically.

What is the function of the skin impulse in the behaviour of the hatchling tadpole? The impulse is evoked by stronger mechanical stimuli and when it occurs it leads to vigorous swimming but does this depend on the coactivity of the Rohon-Beard or trigeminal touch sensory systems? This can be tested by denervating part of the tail and then using electrical stimulation to evoke a skin impulse on its own. The result is that swimming is still evoked. So, over the whole surface of the body of the young tadpole there are two sensory mechanisms sensitive to mechanical distortion. Light, local distortion excites only the free nerve endings from Rohon-Beard or trigeminal sensory neurons which have a lower threshold. Stronger stimuli will excite these endings but will also excite the skin cells directly and evoke a skin impulse (Roberts and Smyth 1974). How does the skin impulse gain access to the nervous system and initiate swimming? The mechanism and details of the neurons involved is still not clear but lesion experiments have shown that in *Xenopus* embryos there is no direct access to the spinal cord. The main pathway is via the trigeminal nerves to the brainstem (Roberts 1996).

The discovery of the excitability of amphibian tadpole skin is a fascinating story. It was first proposed by Wintrebert (1904, 1905) who showed that the rostral part of a *Rana* or *Ambystoma* larva would still contract when the tail was pricked, even if the only connection between the tail and the rostral part was a narrow bridge of skin (Fig. 8B). Hooker (1911) repeated Wintrebert's experiments on *Rana* tadpoles but did not accept Wintrebert's conclusion that the skin was excitable; rather, he suggested that his results depended on transmission of vibrations from the tail to the head. Wintrebert presented further evidence for his interpretation (1920) but neither Coghill (1924) nor Dushane (1938), working on *Ambystoma*, were convinced and it was not until direct electrical recordings were made from skin cells in *Xenopus* embryos (Roberts 1969) that Wintrebert's conclusions were finally confirmed for western scientists. The presence of a skin impulse in the newt, *Cynops orientalis*, actually had been confirmed by Fan and Dai (1962) in China using electrical recording but this work was not rediscovered until after the Cultural Revolution.

Skin impulses have now been found in a wide range of amphibian embryos and young larvae (Table 1). In most cases, the evidence is behavioural but in some the impulse has been recorded intracellularly. Impulses have been searched for but not found in the newt *Triturus*

Table 1. Distribution of skin impulses in some amphibian larvae as determined by direct electrical recording (E) or behavioural tests (B). + = present; — = not present; 0 = not tested.

Taxon	E	B	Authority
Anurans			
Alytes obstetricans	0	—	Wintrebert (1909, 1920)
Alytes cisternasi	0	—	Wintrebert (1920)
Bufo bufo	+	+	Roberts (1971)
Bufo vulgaris* (bufo)	0	+	Wintrebert (1920)
Discoglossus pictus	0	+	Wintrebert (1920)
Hyla arborea	0	+	Wintrebert (1920)
Rana esculenta	0	+	Wintrebert (1904, 1909)
Rana temporaria	0	+	Wintrebert (1909)
Rana temporaria	+	+	Roberts (1971)
Xenopus laevis	+	+	Roberts (1969)
Salamanders			
Ambystoma mexicana	+	0	Blight (pers. comm.)
Ambystoma punctatum	0	+	Wintrebert (1920)
Ambystoma tigrinum	0	+	Wintrebert (1909, 1920)
Cynops orientalis	+	+	Fan and Dai (1962)
Cynops pyrogaster	+	0	Sato *et al.* (1981)
Salamandra maculosa* (salamandra)	0	—	Wintrebert (1920)
Siredon pisciformis	0	+	Wintrebert (1904)
Triturus alpestris	0	+	Wintrebert (1909, 1920)
Triturus marmorata	0	+	Wintrebert (1920)
Triturus palmatus* (helveticus)	0	+	Wintrebert (1920)
Triturus vulgaris	0	+	Wintrebert (1909, 1920)

*Name change has occurred since original publication (current name in brackets when ambiguous).

vulgaris (Roberts and Clarke 1983). There is at present no evidence that similar skin impulses are found in the embryos or young larvae of fishes (Wintrebert 1920). However, propagated epithelial impulses or "neuroid" conduction has been described in hydrozoan jellyfish and various pelagic lower chordates (Mackie 1970). These impulses are initiated by external stimuli, propagate to the nervous system and evoke behavioural responses, so their function is analogous to that in amphibian tadpoles.

VI. CONCLUSIONS

The simplicity of the organization of the somatosensory system of the young amphibian larva has allowed a very complete picture of its three mechanosensory modalities to be established. Free nerve-endings from unmyelinated neurites provide touch and pressure sensitivity. There are small but clear differences in the anatomy of the nerve endings serving these two modalities. The third noxious modality is served by a skin impulse, evoked by stronger stimuli and propagated from cell to cell in the skin to reach the head and excite trigeminal sensory neurons (still to be identified). It appears that all cells in the skin are excitable and are invaded by the skin impulse (Roberts 1971; Roberts and Stirling 1971). Despite this large depolarization passing through the skin cells that they innervate, the majority of sensory neurons are not excited by the skin impulse. One can conclude that the sensory neurites must respond directly to mechanical stimulation rather than by being excited indirectly by a transmitter released when skin cells are depolarized by stimulation (Roberts 1975). The effects of other stimuli, such as temperature, pH, and salinity, on the skin's sensory neurons have not been investigated yet.

Again, as a result of the simplicity of the system, the behavioural roles of the different sensory modalities of the skin are distinct. Excitation of touch or noxious sensory modalities leads to movements, swimming if the stimulus is brief, and struggling if the stimulus is prolonged or if the tadpole is grasped by a predator. Excitation of the pressure modality does not lead to movement but inhibits any ongoing movements such as swimming. These obvious differences in the roles of different sensory modalities of the skin provide a clear example of the primitive function of sensory modalities. These modalities distinguish between stimuli and by making suitable central connections can lead to distinct and adaptive behaviour.

VII. ACKNOWLEDGEMENTS

I would like to thank my colleagues in Bristol, particularly Dr Steve Soffe, for discussions and I am grateful to the Wellcome Trust, MRC and SERC for financial support.

VIII. REFERENCES

Abu Gideiri, Y. R., 1971. The development of locomotory mechanisms in *Bufo regularis. Behaviour* **38**: 121–132.

Boothby, K. M. and Roberts, A., 1992a. The stopping response of *Xenopus laevis* embryos: behaviour, development and physiology. *J. Comp. Physiol.* **170**: 171–180.

Boothby, K. M. and Roberts, A., 1992b. The stopping response of *Xenopus laevis* embryos: Pharmacology and intracellular physiology of rhythmic spinal neurons and hindbrain neurons. *J. Exp. Biol.* **169**: 65–86.

Boothby, K. M. and Roberts, A., 1995. Effects of site of tactile stimulation on the escape response of *Xenopus laevis* embryos. *J. Zool., Lond.* **235**: 113–125.

Burgess, P. R. and Perl, E. R., 1973. Cutaneous mechanoreceptors and nociceptors. Pp. 29–78 in "Handbook of Sensory Physiology", Vol. 2, ed by A. Iggo. Springer Verlag, Berlin.

Cambar, R. and Marrot, B. R., 1954. Table chronologique du developpement de la grenouille agile (*Rana dalmatina* Bon.). *Bull. Biol. Fra. Belg.* **88**: 168–177.

Clarke, J. D. W. and Roberts, A., 1984. Interneurones in the *Xenopus* embryo spinal cord: sensory excitation and activity during swimming. *J. Physiol.* **354**: 345–362.

Clarke, J. D. W., Hayes, B. P., Hunt, S. P. and Roberts, A., 1984. Sensory physiology, anatomy and immuno-histochemistry of Rohon-Beard neurones in embryos of *Xenopus laevis. J. Physiol.* **348**: 511–525.

Coghill, G. E., 1914. The afferent system of the trunk of Amblystoma. *J. Comp. Neurol.* **24**: 161–233.

Coghill, G. E., 1924. Correlated anatomical and physiological studies of the growth of the nervous system in Amphibia. III. The floor plate of *Amblystoma. J. Comp. Neurol.* **37**: 37–69.

Dushane, G. P., 1938. Neural fold derivatives in the Amphibia: pigment cells, spinal ganglia and Rohon-Beard cells. *J. Exp. Zool.* **78**: 485–503.

Fan, S. and Dai, R., 1962. Electric activity of embryonic epithelium in Urodeles. *Kexuw Tongboa* **10**: 38–39.

Gallagher, B. C. and Moody, S. A., 1987. Development of substance P-like immunoreactivity in *Xenopus* embryos. *J. Comp. Neurol.* **260**: 175–185.

Gallien, L. and Bidaud, O., 1959. Table chronologique du developpement chez *Triturus helveticus* Razoumowski. *Bull. Soc. Zool. Fr.* **84**: 22–32.

Hayes, B. P. and Roberts, A., 1983. The anatomy of two functional types of mechanoreceptive "free" nerve-endings in the head skin of *Xenopus* embryos. *Proc. R. Soc. Lond. (B)* **218**: 61–76.

Hooker, D., 1911. The development and function of voluntary and cardiac muscle in embryos without nerves. *J. Exp. Zool.* **11**: 59–186.

Hughes, A. F. W., 1957. The development of the primary sensory system in *Xenopus laevis. J. Anat.* **91**: 323–338.

Iggo, A., 1976. Is the physiology of cutaneous receptors determined by morphology? *Prog. Brain Res.* **43**: 15–31.

Kahn, J. A. and Roberts, A., 1982. The neuromuscular basis of rhythmic struggling movements in embryos of *Xenopus laevis. J. Exp. Biol.* **99**: 197–206.

Kelly, D. E., 1966. Fine structure of desmosomes, hemidesmosomes and an adepidermal globular layer in developing newt epidermis. *J. Cell Biol.* **28**: 51–72.

Kitson, D. L. and Roberts, A., 1983. Competition during the innervation of embryonic head skin by trigeminal sensory neurites. *Proc. R. Soc. Lond. (B)* **218**: 49–59.

Lamborghini, J. E., 1980. Rohon-Beard cells and other large neurones in *Xenopus* embryos originate during gastrulation. *J. Comp. Neurol.* **189**: 323–334.

Mackie, G. O., 1970. Neuroid conduction and the evolution of conducting tissues. *Quart. Rev. Biol.* **45**: 319–332.

Nicholls, J. G. and Baylor, D. A., 1968. Specific modalities and receptive fields of sensory neurons in CNS of the leech. *J. Neurophysiol.* **31**: 740–756.

Nieuwkoop, P. D. and Faber, J., 1956. Normal table of *Xenopus laevis* (Daudin). North-Holland Publishing Co., Amsterdam.

Nordlander, R. H., 1984. Developing descending neurons of the early *Xenopus* tail spinal cord in the caudal spinal cord of early *Xenopus. J. Comp. Neurol.* **228**: 117–128.

Roberts, A., 1969. Conducted impulses in the skin of young tadpoles. *Nature* **222**: 1265–1266.

Roberts, A., 1971. The role of propagated skin impulses in the sensory system of young tadpoles. *Z. Vergl. Physiol.* **75**: 388–401.

Roberts, A., 1975. Mechanisms for the excitation of "free nerve endings". *Nature (Lond.)* **253**: 737–738.

Roberts, A., 1980. The function and role of two types of mechanoreceptive "free" nerve endings in the head skin of amphibian embryos. *J. Comp. Physiol.* **135**: 341–348.

Roberts, A., 1987. Skin sensory modalities, free nerve-endings and behaviour: a reappraisal based on studies of amphibian embryos. Pp. 80–103 in "Aims and Methods in Neuroethology", ed by D. M. Guthrie. Manchester University Press, Manchester.

Roberts, A., 1996. Trigeminal pathway for the skin impulse to initiate swimming in hatchling *Xenopus* embryos. *J. Physiol.* **493**: 40-41P.

Roberts, A. and Blight, A. R., 1975. Anatomy, physiology and behavioural role of sensory nerve endings in the cement gland of embryonic *Xenopus*. *Proc. R. Soc. Lond. B.* **192**: 111–127.

Roberts, A. and Clarke, J. D. W., 1982. The neuroanatomy of an amphibian embryo spinal cord. *Phil. Trans. Roy. Soc.* **296**: 195–212.

Roberts, A. and Clarke, J. D. W., 1983. The sensory systems of embryos of the newt: *Triturus vulgaris. J. comp. Physiol.* **152**: 529–534.

Roberts, A. and Hayes, B. P., 1977. The anatomy and function of "free" nerve endings in an amphibian skin sensory system. *Proc. R. Soc. Lond. B.* **196**: 415–429.

Roberts, A. and Sillar, K. T., 1990. Characterization and function of spinal excitatory interneurons with commissural projections in *Xenopus laevis* embryos. *Europ. J. Neurosci.* **2**: 1051–1062.

Roberts, A. and Smyth, D., 1974. The development of a dual touch sensory system in embryos of the amphibian, *Xenopus laevis. J. Comp. Physiol.* **88**: 31–42.

Roberts, A., Soffe, S. R., Clarke, J. W. D. and Dale, N., 1983. Initiation and control of swimming in amphibian embryos. Pp. 261–284 *in* "Neural origin of rhythmic movements", ed by A. Roberts and B. Roberts. Cambridge University Press, Cambridge.

Roberts, A. and Stirling, C. A., 1971. The properties and propagation of a cardiac-like impulse in the skin of young tadpoles. *Z. Vergl. Physiol.* **71**: 295–310.

Roberts, A. and Taylor, J. S. H., 1982. A scanning electron microscope study on the development of a peripheral sensory neurite network. *J. Embryol. Exp. Morph.* **69**: 237–250.

Sato, E., Adachi, K. S. and Itô, S., 1981. The genesis and transmission of epidermal potentials in an amphibian embryo. *Dev. Biol.* **88**: 137–146.

Schroeder, T. E., 1970. Neurulation in *Xenopus laevis*. An analysis and model based upon light and electron microscopy. *J. Embryol. Exp. Morph.* **23**: 427–462.

Sillar, K. T. and Roberts, A., 1988. Unmyelinated cutaneous afferent neurons activate two types of excitatory amino acid receptors in the spinal cord of *Xenopus laevis. J. Neurosci.* **8**: 1350–1360.

Soffe, S. R., 1991a. Centrally generated rhythmic and non-rhythmic behavioural responses in *Rana temporaria* embryos. *J. Exp. Biol.* **156**: 81– 99.

Soffe, S. R., 1991b. Triggering and gating of motor responses by sensory stimulation: behavioural selection in *Xenopus* embryos. *Proc. Roy. Soc. B* **246**: 197–203.

Soffe, S. R., 1992. To flex, swim or struggle? Behavioural selection in *Xenopus* embryos. Pp. 73–87 *in* "Neurobiology of Motor Programme Selection: New Approaches to the Study of Behavioural Choice", ed by J. Kein, C. R. McCrohan and W. Winlow. Manchester University Press, Manchester.

Taylor, J. and Roberts, A., 1983. The early development of primary sensory neurites in an amphibian embryo: an SEM study. *J. Embryol. Exp. Morph.* **75**: 49–66.

Warburg, M. R., Lewinson, D. and Rosenberg, M., 1994. Ontogenesis of amphibian epidermis. Pp. 33–63 *in* "The Integument", Vol. 1, "Amphibian Biology", ed by H. Heatwole, G. T. Barthalmus and A. Heatwole. Surrey Beatty & Sons, Chipping Norton.

Wintrebert, M. P., 1904. Sur l'existence d'une irritabilite excito-motrice primitive, independent des voies nerveuse chez les embryons cilies de Batraciens. *Comptes Rendus de la Societe de Biologie* **57**: 645–647.

Wintrebert, M. P., 1905. Nouvelles recherches sur la sensibilite primitive des Batraciens. *Comptes Rendus de la Societe de Biologie* **59**: 58–59.

Wintrebert, M. P., 1909. Sur la sensibilite primitive et la conductibilite centripete anerveuse du tegument chez les larves de Batracienns. *Ass. Français Avanc. Science* **38**: 127–128.

Wintrebert, M. P., 1920. L'epoque d'apparition et le mode d'extension de la sensibilite a la surface du tegument chez les vertebres anamniotes. *Comptes Rendus Acad. Sci.* **171**: 408–410.

CHAPTER 6

Diffuse Cutaneous and Muscular Sensory Systems: Mechanoreception, Thermoreception, Nociception, Chemoreception and Kinesthetic Sense

Harold Heatwole

I. Introduction
II. Touch
   A. Function of Merkel Cells
      1. Transduction
      2. Modulation of Response
      3. Developmental Guidance
      4. Mediation and Positional Information
      5. Other Functions
III. Cutaneous Pressure/Stretch Receptors
IV. Thermoreception
V. Nociception
VI. Chemoreception
VII. Kinesthetic Sense
VIII. Receptive Fields
IX. Acknowledgement
X. References

I. INTRODUCTION

IN addition to the modalities localized in discrete sensory organs, there are various senses more diffusely distributed in the amphibian body, especially in the skin. Frog skin is innervated by dorsal cutaneous nerves branching from sensory roots of the spinal cord (Spray 1976). The peripheral part of this system arises as myelinated and non-myelinated axons entering the subepidermal layers from two plexuses, one deep and the other superficial (Catton 1958; Whitear 1974).

Spray (1976), Catton (1976) and Fox *et al.* (1980) reviewed the skin receptors of amphibians. Collectively they noted four major types, designated here as types a—d:

(Type a) Free nerve endings with a relatively large axonal diameter (9–10 μm), located in the epidermis. These have relatively rapid conduction velocity (14–18 m sec^{-1}), respond with a phasic discharge to mechanical stimuli, and adapt rapidly. They occur in at least two Rapidly Adapting subtypes (RA I and RA II) based on discharge characteristics (Ogawa *et al.* 1981).

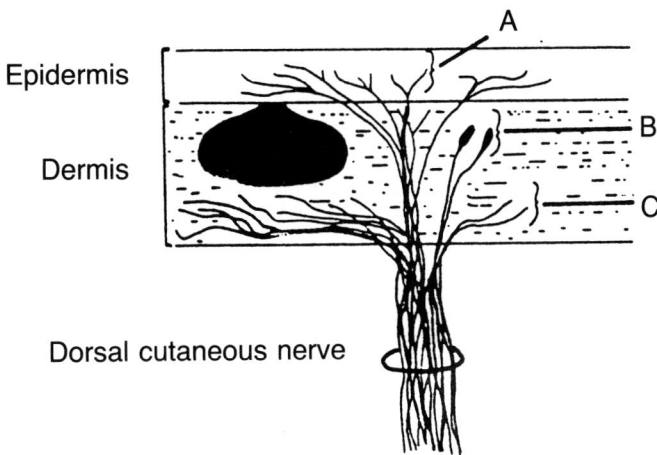

Fig. 1. Diagrammatic representation of the distribution of sensory endings of the dorsal cutaneous nerve in frog skin. A = endings in the epidermis. B = expanded tip endings in the upper dermis. C = free nerve endings in the lower dermis. Dark area is mucous gland. Modified from Spray (1976).

(Type b) Encapsulated nerve endings at the interface between the epidermis and dermis or in the outer dermis. These have slightly smaller axons (5–6 μm), conduct somewhat more slowly (7–9 m sec^{-1}) and respond with a sustained discharge to mechanical stimuli such as stretch, and adapt slowly.

(Type c) Small (2–3 μm in axon diameter) myelinated fibers with free nerve endings within the lower dermal layers. These conduct still more slowly (3.5–5 m sec^{-1}) and respond to thermal stimulation and perhaps to vibration.

(Type d) Very small fibers of less than 1 μm in axonal diameter, which are non-myelinated and end as free nerve endings in the lower dermis. They have very slow conduction velocities (0.8–1.8 m sec^{-1}) and respond to painful mechanical or chemical stimulation.

To these types can be added two kinds (here designated as types e and f) that were discovered later. These have been studied extensively and their neurophysiological characteristics described by Ogawa *et al.* (1981, 1984a,b), Taniguchi *et al.* (1984), Ogawa and Yamashita (1988) and Yamashita and Ogawa (1991).

(Type e) Slowly adapting, irregularly discharging afferents that respond to mechanical stimuli. These, designated as Slowly Adapting Frog type 1 (SA Ft–1), innervate both "warty" and "non-warty" skin.

(Type f) Slowly adapting, regularly discharging, mechanosensitive afferents, called Slowly Adapting Frog type 2 (SA Ft–2). These innervate only "non-warty" skin.

Types e and f respond differently to stimulation by direct current, concentrations of cations and application of pharmacological agents than do types a–d, suggesting that different mechanisms of mechano-electric transduction are employed.

Thus, it seems that types a, e, and f are receptors for touch whereas b, c and d are receptors for stretch or pressure, temperature and/or vibration, and pain, respectively. In addition, there are several kinds of pain receptors with larger axonal diameters than type d (see section on Nociception) and small, pressure-detecting fibers that are infrequently encountered (Maruhashi *et al.* 1952). Locations of cutaneous receptors are shown diagrammatically in Figure 1 and their discharge characteristics are portrayed in Figure 2. The various skin senses will now be discussed individually.

II. TOUCH

The sense of very light touch in early developmental stages of amphibians is mediated by sensory cell bodies located in the spinal cord and known as Rohon-Beard cells (see Chapter 5 this volume; Roberts 1996). These have naked peripheral neurites forming a loose network

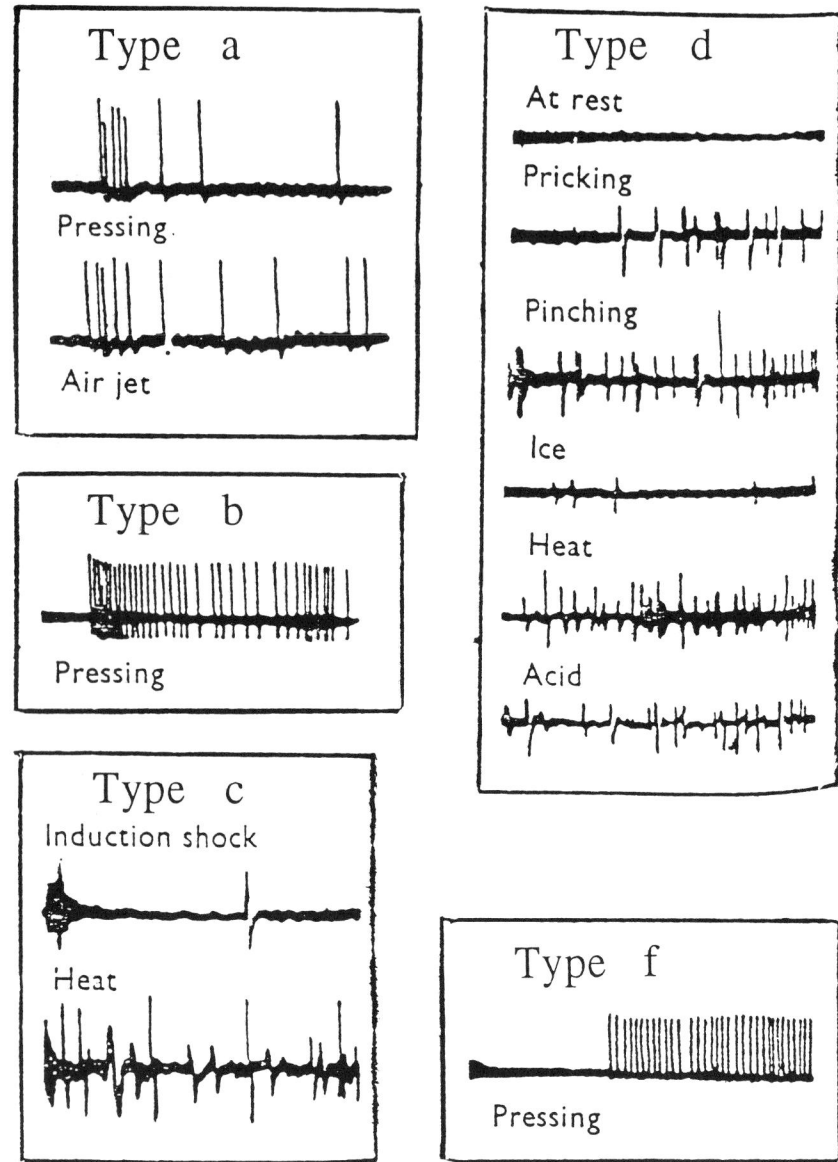

Fig. 2. Discharge characteristics of different types of afferent nerve fibers of toads to tactile, pressure, thermal and noxious stimuli. Modified from Maruhashi *et al.* (1952).

under the skin. Discharges from these cells adapt rapidly and, consequently, they detect transient stimuli and fatigue quickly. There is also a sensory system that monitors stronger tactile stimulation. Between these two systems, the larva is informed about transient stimuli from gentle strokes to strong mechanical distortions that produce tissue damage. The cement gland also seems to be an important organ of touch in the embryos and larvae of some frogs and is involved in inducing quiescence of tadpoles when they become attached to a substrate. In the adult amphibian the Rohon-Beard cells are superseded by sensory neurons whose cell bodies are located in dorsal root ganglia and which take over the function of mechanoreception (Hughes 1957).

The skin of adult amphibians is extremely sensitive to touch, the lowest minimum mechanical threshold measured being only 2 μ; frog skin is five times more sensitive than toad skin (Catton 1976).

A variety of reflexive behaviours is elicited by tactile stimulation, even in frogs with the spinal cord transected at its anterior end (Franzisket 1963). The particular response is determined by the area of the skin touched, although such responses can be modified by

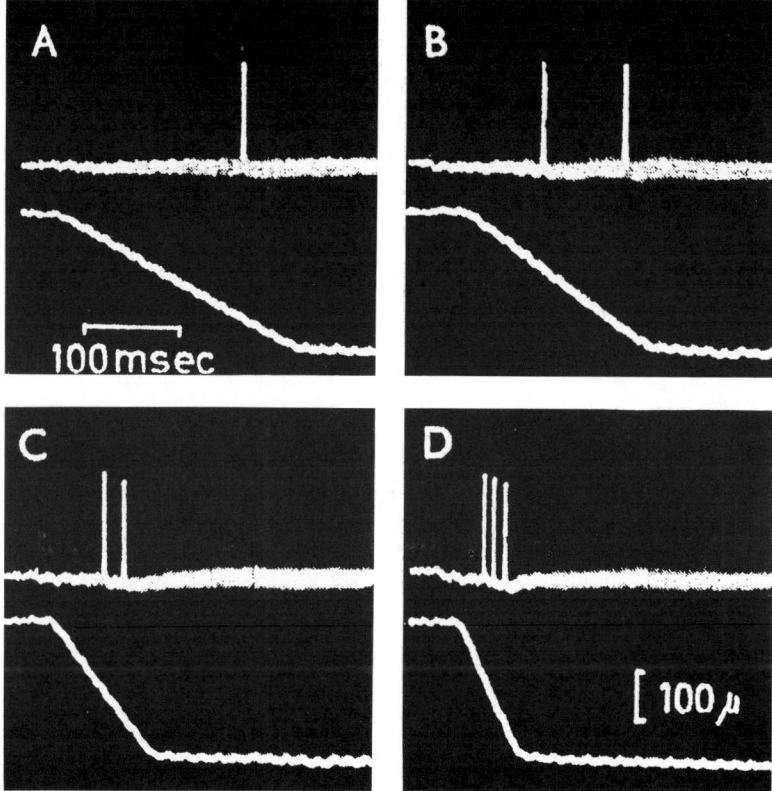

Fig. 3. Recordings from very rapidly adapting touch receptors of the lateral plantar pad of the foot of *Bufo bufo*. Upper trace indicates impulse discharge and lower trace shows the extent of deformation of the skin (as measured by capacitance). A = 1.0 mm sec^{-1} (critical slope); B = 1.3 mm sec^{-1}; C = 2.3 mm sec^{-1}; D = 3.8 mm sec^{-1}. Latency of first impulse in A–D 190, 85, 50 and 25 msec, respectively, including calculated time (5 msec) of impulse condution from receptor to recording site. Plateau amplitude of deformation 260 μm in all records. Rheobase value 20–30 μm (fluctuating). From Lindblom (1963).

training. There are three major dorsal reflexogenous areas in *Rana esculenta*: (1) The anterior zone includes the head and medial surfaces of the forelimb; when this area is touched the forelimb is flexed at the elbow and lifted so that the hand wipes the head from the occiput to the nose. This is a stereotyped response and occurs regardless of which part of the anterior zone is stimulated. (2) The major zone includes the skin of the back, the lateral surfaces of the forelimbs and the upper surfaces of the anterior half of the hindlimb. A stimulus to any part of this area results in a wiping movement of the ipsilateral hind foot over the stimulated area. (3) The caudal zone includes the rump and the heels. Tactile stimulation results in a wiping movement by the ipsilateral heel.

The sense of touch seems to be mediated by the most superficial of the skin receptors, i.e., those located in the epidermis (Fox and Whitear 1978) (Fig. 1). Type a receptors are characterized by large spike amplitude (about 400 μV), high velocity of conduction (14–25 m/sec), rapid adaptation, and axons of large diameter (Catton 1958, 1976; Spray 1976). When the skin is touched briefly, these receptors respond with only a few spikes. If the tactile stimulus is sustained, the receptors respond initially but then cease responding until the stimulus is removed, at which time there is again a brief response. These receptors are stimulated primarily by dynamic stimulation but are less responsive to static deformation of the skin; the latency of the response decreases and the number of spikes increases with increasing rate of skin deformation (Lindblom 1963) (Fig. 3). Noxious stimuli such as acid do not activate these receptors (Maruhashi *et al.* 1952).

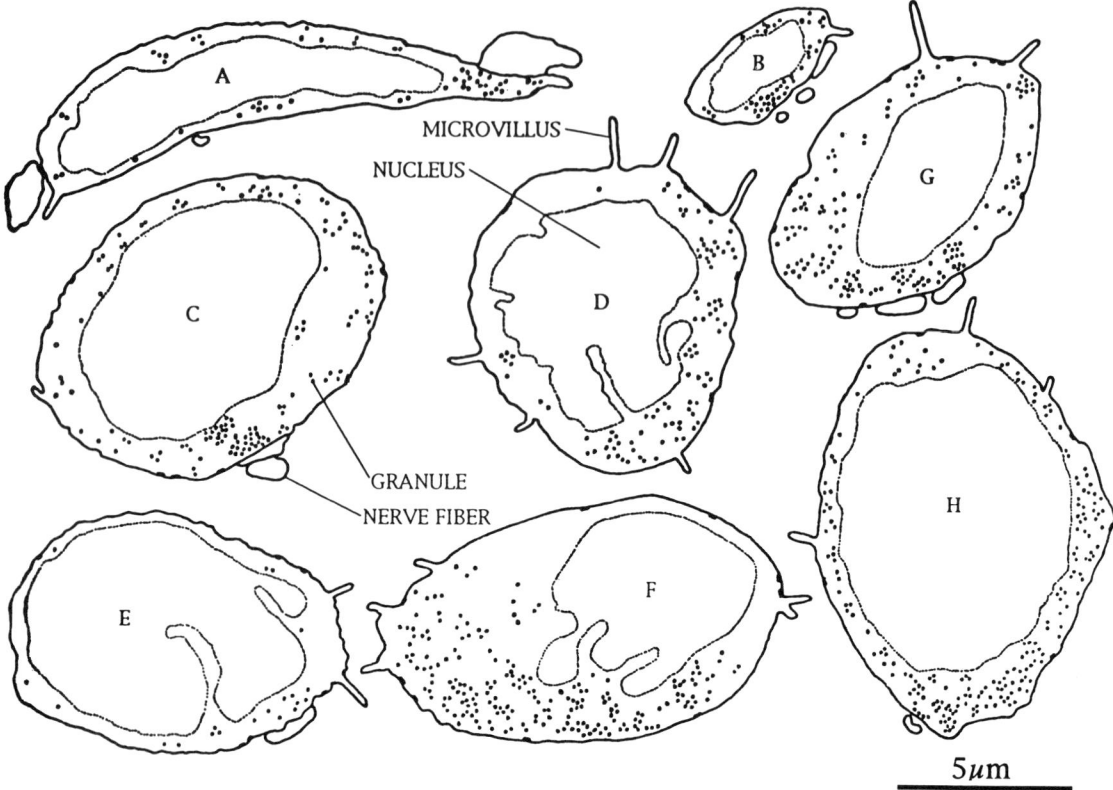

Fig. 4. Outline drawings of Merkel cells from a selection of amphibian species and body locations. Modified from Fox and Whitear (1978). A. Adult *Xenopus laevis* skin of digit; B. Adult *Xenopus laevis*, skin of back; C. Metamorphosing *Xenopus laevis*, skin of back; D. Adult *Rana temporaria*, skin of thumb; E. Larva of *Rana temporaria*, skin of tail; F. Larva of *Rana temporaria*, skin of lip; G. Adult *Bufo bufo*, skin of belly; H. Larva of *Bufo bufo*, skin of digit.

These tactile sensory nerve fibers are associated with epidermal Merkel cells (see review by Fox 1994), with which they form a neurite-Merkel cell complex (Cooper *et al.* 1975a). Merkel cells were discovered by Friedrich Sigmund Merkel (1880) and appropriately called touch-cells ("Tastzellen") by him; later, their name was changed in his honour. These cells have small, finger-like processes extending from the surface membrane. The cytoplasm contains, among other inclusions, numerous membrane-bounded granules (Fig. 4) of varying density and probably originating from the Golgi apparatus (Whitear 1977; Fox and Whitear 1978; Fox 1994). These Merkel granules are unevenly distributed within the cytoplam; they are concentrated on the dermal side of the cell (Parducz *et al.* 1977) and sometimes are closely packed near synapses with nerve cells (e.g., Fig. 4c).

The Merkel cell-neurite complex differs functionally among taxa. Salamanders have only rapidly adapting complexes (Cooper and Diamond 1977), whereas mammals have only slowly adapting ones (Diamond *et al.* 1986). Yamashita and Ogawa (1991), using the quinacrine technique, found that in *Rana catesbeiana* the distribution of Merkel cells correlated with the receptive fields of slowly adapting FT–1 units (Type e) around dermal warts. Thus, in frogs there are two kinds of afferents associated with Merkel cells, a rapidly adapting one (Type a) like that of salamanders and a slowly adapting one (Type e, or FT–1) like that of mammals. Lindblom (1963) noted that in toad *(Bufo bufo)* skin, there are various intermediary and transitional forms between the very rapidly adapting receptors and those that adapt less rapidly.

The two subtypes (or ends of a continuous spectrum) may play somewhat different roles in tactile sensation (Lindblom 1962, 1963; Ogawa *et al.* 1981; Taniguchi *et al.* 1984). RA I units discharge spikes phasically in response to ramp stimulation, and the maximum instantaneous frequency of discharge is expressed by monotonically increasing functions of indentation velocity. Thus, they probably detect the speed with which the skin is indented.

By contrast, RA II units respond with only a single phasic spike and the number of impulses is related neither to the velocity nor magnitude of indentation. These cells probably detect a higher derivative of the stimulus, such as acceleration of skin indentation. In type e (SA) units, the frequency of impulses is linearly related to the magnitude of indentation and thus encodes the strength of the mechanical stimulus. Together, these three kinds of receptors allow for very precise characterization of tactile stimuli, one indicating how strong the stimulus is, another how fast the skin is being indented, and the third whether the speed of indentation is accelerating or not. Of course, very strong stimuli bring into play additional sensory receptors, including those dealing with pressure (see following section). Calof et al. (1981), for example, classified SA receptors in amphibians subjected to high stimulus forces into two categories, compression receptors and stroke receptors.

Merkel cells have been found in various life history stages from larvae to adults, and in species representing all three orders of amphibians (Table 1); it is likely that they form an integral feature of the epidermis in all amphibians. Touch spots and Merkel cells seem to be randomly distributed over the general body surface in salamanders (Cooper et al. 1977). In

Table 1. Known occurrences of Merkel cells in amphibians.

Taxon	Stage; Location in skin	Authority
APODANS		
Ichthyophis kohtaoensis	Adult; back, belly	Fox and Whitear (1978)
Ichythyophis orthoplicatus	Adult; back	Fox and Whitear (1978)
SALAMANDERS		
Ambystoma mexicanum	Young larva; digits	Fox and Whitear (1978)
	Neotenous adult; digits	Fox and Whitear (1978)
	Metamorphosed adult; tail, digits	Fox and Whitear (1978)
Ambystoma punctatum	Adult; hind limb	Alguilar et al. (1973)
Ambystoma tigrinum	Adult; hind limb	Alguilar et al. (1973)
	Adult; dorsal hindlimb	Diamond et al. (1986); Nurse et al. (1983); Parducz et al. (1977); Scott et al. (1981)
	Adult; dorsal and ventral hind limbs	Cooper et al. (1977b)
Aneides lugubris	Adult; digits	Fox and Whitear (1978)
Pleurodeles waltl	Young larvae; digits	Fox and Whitear (1978)
Proteus anguinus	Neotenous juvenile; tail, digits	Fox and Whitear (1978)
Salamandra salamandra	Larva; digits	Fox and Whitear (1978)
	Juvenile; throat	Fox and Whitear (1978)
Triturus cristatus	Adult; back, belly, head, tail	Fox and Whitear (1978)
Triturus vulgaris	Adult; digit	Fox and Whitear (1978)
ANURANS		
Bufo bufo	Adult; belly	Budtz and Larsen (1975)
	Prometamorphic larva; digits	Fox and Whitear (1978)
	Adult; belly, leg	
Rana catesbeiana	Adult; around warts and skin glands	Yamashita and Ogawa (1991)
Rana esculenta	Adult; taste organ	Düring and Andres (1976)
Rana pipiens	Post-metamorphic juveniles, back, belly	Nafstad and Baker (1973)
Rana temporaria	Premetamorphic larva; body, tail	Fox and Whitear (1978)
	Prometamorphic larva; digits, back, lip	Fox and Whitear (1978)
	Metamorphic climax; back, tail, digits	Fox and Whitear (1978)
	Adult; back, belly, head, leg, thumb-pad, palate	Fox and Whitear (1978)
	Adult; taste organ	Düring and Andres (1976)
Xenopus laevis	Larva; tentacles	Nurse et al. (1983); Ovalle (1976)
	Premetamorphic larva	
	Prometamorphic larva; back, tail, tentacle, digits	Fox and Whitear (1978)
	Metamorphic climax; back, tail, digits	Fox and Whitear (1978)
	Juvenile; back, belly, digits	Fox and Whitear (1978)
	Juvenile; around ducts of skin glands	Crowe and Whitear (1978); Nurse et al. (1983)
	Adult; digits	Fox and Whitear (1978)
	Adult; around ducts of skin glands	Nurse et al. (1983)
	Adult; dorsal thigh	Mearow and Diamond (1988)

frogs, they also occur widely in the skin, but also appear in groups on the lips and toes of tadpoles, at least in some species. They are also concentrated near the ducts of skin glands in larval and adult *Xenopus laevis* (Crowe and Whitear 1978; Nurse *et al.* 1983) and in association with skin warts in *Rana catesbeiana* (Yamashita and Ogawa 1991) where the cutaneous sense of touch is concentrated (Maruhashi *et al.* 1952). They occur at the base of the taste organs in frogs, where they account for the mechanoreceptive part of the dual sensory function of these organs (Düring and Andres 1976). Surprisingly, Merkel cells are absent from the dermal papillae of the "tactile spots" in *Xenopus laevis* and *Rana temporaria* (Fox and Whitear 1978). The neurite-Merkel cell complex of amphibians differs from that of mammals in an important way: in mammals, the synapses are structurally polarized in the direction from Merkel cell to neurite whereas, in amphibians, the synapses can be reciprocal (Mearow and Diamond 1988; Diamond *et al.* 1988).

The neurite-Merkel cell complex seems to be the functional touch receptor. In amphibian skin there are low-threshold mechanosensitive areas ("touch spots") corresponding to the distribution of Merkel cells. Maps of areas of salamander and *Xenopus* skin sensitive to touch at low thresholds, as determined electrophysiologically, correspond closely to the distribution of Merkel cells within those same patches of skin when later examined by electron microscopy (Pardusz *et al.* 1977; Cooper *et al.* 1977; Mearow and Diamond 1988). Furthermore, in regenerating skin, sensitivity to mechanical stimuli is restored at the time the nerves make contact with Merkel cells (Scott *et al.* 1981).

A. Function of Merkel Cells

There is evidence that Merkel cells play a variety of roles in mediating the sense of touch. These will be reviewed in turn.

1. Transduction

The initial function ascribed to the Merkel cell was that of a mechanosensory transducer, i.e., that it detects a primary energy change in the environment and converts it to a form appropriate to excite an impulse in the sensory nerve cell (see Mills and Diamond 1995 for a review). Electron microscopy clearly indicates that reciprocal synapses, as defined morphologically, occur between the sensory nerve endings and the Merkel cell in amphibians (Diamond *et al.* 1986). The "straightened" regions of synapse-like contact between nerve endings and Merkel cells are characterized by an increased density of the "post-synaptic" membrane and by cytoplasmic densities abutting the "presynaptic" membrane; in association with these structures is an almost invariable occurrence of one or more membrane-bound clear vesicles within the nerve ending, and densely cored granules within the Merkel cell. Omega figures are often in close relation to specific contact sites. Morphological polarization of the contact sites can be directed either from neurite to Merkel cell or in the reverse direction, the oppositely polarized contact sites often lying adjacent to one another at the same junctional region.

The physiological characteristics of these synaptic regions, however, do not always conform to expectations for a transducer. Failure to detect monoamines in the Merkel cells of frog skin (Crowe and Whitear 1978) paralleled the finding that bath application of noradrenaline, 5-hydroxytryptamine or octopamine had negligible effect on the mechanosensory process. Furthermore, varying the concentrations of divalent cations to levels that depress or completely block transmitter release at conventional chemical synapses did not affect thresholds of mechanosensitivity, except where the effect was attributable to direct action upon the nerve itself (Diamond *et al.* 1986, 1988). Finally, irradiation of quinacrine-treated Merkel cells in *Xenopus* skin selectively killed the Merkel cells but left mechanosensory function unimpaired (Diamond *et al.* 1988). These data suggest that synapses do not represent a chemical link transducing mechanical stimuli to nerve transmissions and led Diamond *et al.* (1988) to conclude that the function of transduction appears to reside in the nerve ending itself.

It is likely that there is no clear resolution to the dichotomy of whether the Merkel cell or the nerve itself is the transducer. Rather, it appears that both theories may be correct. Yamashita and Ogawa (1991) suggested that Merkel cells are involved in perception of touch by FT–1 (type e) units, but that in FT–2 (type f) ones discharges are probably the result of direct stimulation of afferent nerve terminals not associated with Merkel cells. The former units are more sensitive to calcium blockers than are the latter (Yamashita et al. 1986).

In a recent review of the Merkel cell complexes of amphibians and mammals, Ogawa (1996) elaborated further on the hypothesis of two transduction sites by suggesting that Merkel cells may transduce maintained mechanical stimulation to provide ionic responses whereas the nerve terminals may transduce transient stimuli. He also presented a model of Merkel cell transduction involving mechanically gated ionic channels in the Merkel cell membrane. More recently Ogawa (pers. comm., July 1996) suggested that there probably are various functional types of Merkel cells or Merkel-like cells that are not distinguishable morphologically or immunohistochemically.

2. Modulation of responses

Merkel cells may harbour synaptic mechanisms involved in bringing about subtle modulation of a response generated directly in the nerve ending by mechanical stimulation (Diamond et al. 1986). Especially notable in this regard is the elevating of sensitivity threshold in nerves deprived of their Merkel cells. Mearow and Diamond (1988) suggested that low thresholds may depend on the Merkel cells providing a continuous secretion that lowers the mechanical threshold of the nerve ending. An alternate possibility is that the Merkel cells may induce the characteristic low threshold sensitivity to develop in the nerve ending (Diamond et al. 1988).

3. Developmental Guidance

Merkel cells may exert an influence over the maturation or differentiation of the nerve endings so that the latter are able to express their intrinsic mechanosensitivity (Nafstad and Baker 1973). Merkel cells act as developmental targets for nerve fibers. In development, a single nerve fiber sprouts to innervate a number of Merkel cells but individual Merkel cells do not ordinarily accept contact with more than one nerve fiber; in regenerating skin, once the Merkel cells are supplied with nerve fibers, sprouting ceases (Scott et al. 1981). Merkel cells do not degenerate in denervated skin and rapidly form in regenerating skin, prior to the appearance of nerve fibers (Cooper et al. 1977). Consequently, their development is not induced by nerves, nor is the presence of Merkel cells dependent upon nerves. Rather, preexisting Merkel cells serve as targets toward which nerve fibers grow and make contact. Once contact is established between a nerve and a Merkel cell, the latter loses its attractiveness and no further neurites are recruited. There seems to be some selectivity as given nerve fibers will expand only so far into the territory of other nerve fibers, even when the latter are degenerating (Cooper et al. 1975b; Diamond et al. 1976; Macintyre and Diamond 1981). Merkel cells thus play a key role in determining the peripheral organization of low-threshold mechanosensation (Nurse et al. 1983; Mearow and Diamond 1988). Mills and Diamond (1995) suggested that the Merkel cell may release an attractant for the ingrowing nerve fibers, or provide a "stop" signal for them.

Sprouting of nerves appears to be mediated by axoplasmic flow of trophic factors (Aguilar et al. 1973). When the supply of these factors is reduced, either through transection of the nerve, or by treatment with colchicine, collateral nerve fibers sprout and invade the territory of the damaged nerve fiber. In addition, the ability of nerve fibers to sprout is itself dependent upon the maintenance of axoplasmic flow. In the case of transected nerves, the number of new mechanosensory endings that sprout exactly match the number lost by denervation. Although nerves treated with colchicine may become unresponsive, they also may remain functional but at increased threshold, even though replaced. Accordingly, there is sometimes overcompensation in that the total number of endings exceed the number originally present (Cooper et al. 1977).

In *Ambystoma tigrinum* determination of distribution of nerve fibers relies at least partly on factors other than those residing in the skin. Axons sprout readily within portions of their own "domain" (the territory of their parent spinal nerve, roughly corresponding to their

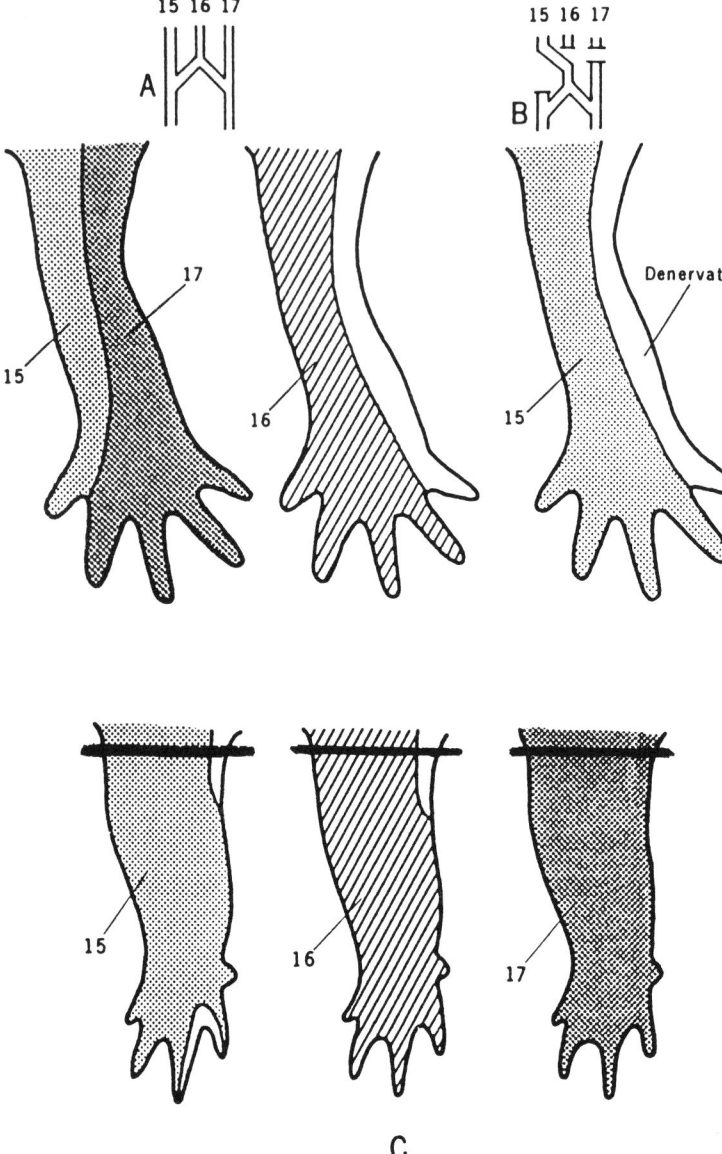

Fig. 5. Redirection of regenerating nerves and regeneration of amputated limbs in salamanders. A: Two views of a control limb showing receptive fields of nerves 15, 16 and 17. Note that the field of spinal nerve 16 overlaps with fields of both nerves 16 and 17 but that there is little overlap between 15 and 17. B: Expansion of the receptive field of nerve 15 following transection of nerves 16 and 17 and redirection of nerve 15 into the common stump of nerves 16 and 17. C: Receptive areas of spinal nerves 15, 16 and 17 (shown separately) all broadly overlap in a limb regenerating after amputation. Site of amputation denoted by heavy line. From Diamond *et al.* (1976).

dermatome) but are hindered from crossing the boundary into other domains (Macintyre and Diamond 1981). At least for certain spinal nerves, the mechanism responsible for this phenomenon selectively alters with time and is lost after about two months.

The stimulus that initiates regeneration in a cut nerve is different from that causing collateral sprouting of intact nerves at a target tissue. Regenerating fibers readily innervate areas of skin whose borders are not crossed by collateral sprouts of intact fibers of the same nerve. In newly regenerating limbs of salamanders, invading nerves also are less rigourously governed by spatial limits (Diamond *et al.* 1976; Macintyre and Diamond 1981) (Fig. 5).

4. Mediation of Positional Information

Merkel cells may be involved in the mediation of positional information. The development of the somatosensory nervous system, and especially the organization of the somatosensory cortex, is critically influenced by information arriving along afferent nerves. Thus, sensory nerves may convey positional information to the developing central nervous system that relates to the region they supply, and which eventually might influence the somatotopic configuration of the appropriate central circuitry. Diamond *et al.* (1988) pointed out that all of the peripheral nerves implicated in influencing the development of the somatosensory nervous system in vertebrates contain mechanosensory axons associated with Merkel cells, and they suggested that Merkel cells could be involved in "positional labelling". Thus, Merkel cells might not only regulate the exact location of the nerve endings in the skin, but also "instruct" nerve fibers as to their location in the body.

Early experiments of various kinds on frogs clearly indicated that part of the determination of positional identification resides directly in the skin (Miner 1956). Excision of the three spinal ganglia of the hindlimb segments in tadpoles of *Rana clamitans* resulted in innervation of the affected leg by trunk fibers from nerves on either side of those normally innervating that limb. Subsequent cutaneous localization was normal in the affected leg, indicating that positional information was not mediated to the central nervous system by identification of the spinal nerve transmitting the impulse, but rather arose within the skin itself. In *Rana pipiens*, limbs transplanted to the center of the back early in development became innervated by afferent fibers of the dorsal rami of the trunk nerves, rather than by those normally innervating the limbs. After metamorphosis these aberrant animals exhibited site-specific responses. When the transplanted limb was stimulated, the normal reflexive acts of a backward kick of the foot in response to light touch on the toe, or a sideward kick in response to touching the heel occurred, not by the stimulated transplanted foot, but by the unstimulated ipsilateral limb in normal position. Strips of skin of *Rana pipiens* tadpoles that were excised and rotated 180° subsequently developed the colour pattern characteristic of their original location (white "belly" skin developed on the dorsum, and spotted "back" skin developed on the belly). These areas also maintained their locational identity and when the dorsal skin on the belly was stimulated by touch, the frogs scratched their backs in a misdirected fashion, i.e., they scratched the former location of the patch of skin, rather than its current location. The reverse was true of belly skin on the back; its stimulation resulted in the frog scratching the belly where that part of the skin had been formerly located. In all of these experiments the diverted fibers innervating new areas acquired local properties suited to the type of skin in which they terminated, rather than those characteristic of the areas normally innervated by those fibers. It was concluded that (1) the cutaneous fibers of the spinal nerves establish their peripheral terminations largely at random within their respective dermatomes, (2) the skin undergoes highly specific local differentiation, (3) the differentiated integument induces a parallel specification in the primary sensory neurons via their local contacts, and (4) this local specification of the cutaneous fiber determines the patterning of central functional relations (Miner 1956). Immediately after metamorphosis the reflexive responses are normal, i.e., directed to the area of skin stimulated. Later, provided that sufficiently large patches of skin are used and they overlap the boundary between the reflexogenous zones of the forelimb and hindlimb, misdirected responses are elicited from small areas in the graft. This area widens until most of the graft gives misdirected responses (Baker and Jacobson 1970). Baker (1972) investigated the mechanisms whereby developing neurones, or those invading new areas such as grafted skin, establish central connections mediating reflexes characteristic of that dermal region. There was no peripheral regrowth of nerves from the donor region to the new site but rather local nerves innervated the graft. Given the fact that peripheral nerves can convey chemicals from their periphery to their cell bodies and that epidermal products have been found within cutaneous nerves and cell bodies, he suggested that retrograde transport of skin products to cutaneous nerve cell bodies via their axons might be a means of establishing site identity and organizing reflexes. Merkel cells were associated with grafted as well as normal skin from both the belly and back, and there were no structural differences in nerve morphology between grafted and normal skin (Nafstad and Baker 1973).

The situation may be somewhat different in salamanders. In *Ambystoma tigrinum* the original boundaries of domains and the reluctance of nerves to sprout across those boundaries are re-established in skin patches that have been excised and rotated 180° and regrafted (Macintyre and Diamond 1981). Thus, the skin alters in conformity to previously established patterns.

5. Other Functions

Merkel cells may influence fatiguability and/or provide metabolic support that allows the neurites to function (Diamond *et al.* 1988; Mills and Diamond 1995).

The above functions are not mutually exclusive and several may be operative. Whatever the suite of roles played by Merkel cells, it is now clear that in addition to their probable role as mechanosensory transducers, they may also have a trophic role, their secretions modulating neural function and development and controlling distribution of the tactile sense over the surface of the body. The chemical signals involved in such trophic interactions may be contained either within the granules of the Merkel cells and/or in the vesicles of the nerve endings (Diamond *et al.* 1986).

III. CUTANEOUS PRESSURE/STRETCH RECEPTORS

In addition to receptors that respond to light touch, there are others (type b) that have a higher threshold of stimulation and respond only to stronger mechanical distortions. They are encapsulated and are located in the upper dermis or at the interface of dermis and epidermis (Fig. 1). They are characterized by spike amplitudes of 200–300 μV, conducting velocities of 7–15 m/sec, and moderately large axons (Spray 1976; Catton 1976). Unlike the previously discussed light touch receptors, they adapt slowly, and continue to discharge during sustained pressure. Thus, they might be termed pressure receptors, although it is difficult to discern whether they detect pressure or stretch. The actual response may be to stretching of the receptor when the skin is indented rather than to pressure *per se*. It is likely that the croak reflex, released by blunt pressure on the back even in spinal frogs (Franzisket 1963), is mediated by these receptors.

A given stimulus may excite more than one kind of receptor. For example, Murahashi *et al.* (1952) noted that when afferent impulses were recorded from a number of cutaneous fibers in response to constant pressure, there was an initial short discharge from a tactile fiber followed by a more sustained one from a pressure fiber, i.e., the senses of both touch and pressure were affected by contact with a single object.

It is possible that the type b units also are capable of detecting changes in water content of the skin. Ogawa *et al.* (1981) found that slowly adapting units in the skin of *Rana catesbeiana* did not show "off" responses to ramp stimulation when fresh, but did so after desiccation. Addition of water sometimes caused the "off" responses to disappear. Perhaps frog skin warps on drying in ways that alters its mechanosensory perception. If so, this could have important implications for such moisture-sensitive animals as amphibians. This topic requires further investigation.

IV. THERMORECEPTION

Neither type a nor type b receptors detect temperature. For example, Ogawa *et al.* (1981) found that slowly adapting receptors in *Rana catesbeiana* produced no response to application of ice water. Although temperature changes may influence thresholds and rates of recovery, they do not evoke discharges in the absence of mechanical stimulation.

There are, however, specific thermoreceptors that are relatively insensitive to most mechanical stimuli although they have been characterized as "vibration receptors" by some workers. They are type c fibers (Catton 1976; Spray 1976), having small myelinated axons with free nerve endings in the lower dermis (Fig. 1). They are slow conductors (3.5 m/sec) and have spike amplitudes of 100–150 μV. Spray (1975) showed that their neurophysiological attributes are subject to thermal acclimation. Some of the pain receptors (type d) also respond to heat and cold (Fig. 2).

Spray (1974, 1976) found thermoreceptors to function in two different ways in ranid frogs; there are (1) static receptors that respond to maintained temperature and (2) dynamic receptors that respond to sudden thermal displacements.

Afferents within the dorsal cutaneous nerve displayed firing rates that were related to the static temperature of the skin. These receptors were maximally active at temperatures of 24°–25°C and 27°–28°C in *Rana pipiens* and *Rana catesbeiana*, respectively.

Heating the skin decreases the discharge of the dynamic receptors whereas cooling accelerates it. It is this pattern that characterizes these receptors as cold-sensitive rather than warm-sensitive and they are called "cold receptors".

Stimulation of the sympathetic nervous system markedly and progressively increases the sensitivity of the cold receptors, an action that can be mimicked by application of adrenergic agonists (Spray 1974). Cholinergic agonists increase sensitivity at low concentrations but decrease it at high ones (Spray and Galansky 1975).

V. NOCICEPTION

Early developmental stages of amphibians have limited touch perception. *Xenopus* embryos have only systems sensitive to transient tactile stimulation of various strengths, and they probably have no sensation of pain or sustained pressure (Roberts and Hayes 1979).

In adult amphibians, however, there are nociceptors (pain receptors). They are type d fibers (Spray 1976; Catton 1976) with free nerve endings in the lower dermis (Fig. 1). They have very small axons, are slow conductors (0.1–1.8 m/sec) and have spike amplitudes of less than 100 μV. They are unique among the sensory receptors in amphibian skin in that they are not myelinated.

They are not stimulated by light touch, but slow impulses are evoked by injurious stimuli including strong heating, intense mechanical stimulation, or application of acid (Murahashi *et al.* 1952) (Fig. 2). Thus, they are not specific to a particular kind of stimulus, but respond to various stimuli of sufficient intensity to be damaging to the organism.

Detection of pain and touch are not mediated by a single type of receptor responding differently to different intensities of stimulation; rather, the two senses have distinct receptors. In addition to the differences already mentioned, various anesthetics selectively abolish the slow pain impulses, but leave the fast fibers unaffected. The pain receptors have slightly larger receptive fields than do touch receptors. Scraping the skin eliminates the superficial touch receptors but not the pain receptors (Adrian *et al.* 1931). Pain receptors differ from all other skin receptors in their long impulse duration (15–70 msec for noxious stimuli; 3–4 msec for touch) (Spray 1976).

Several other kinds of pain receptors have been identified in the leg skin of toads and in the tongue of frogs (Maruhashi *et al.* 1952; Spray 1976) (Fig. 6). Some of these are myelinated and some have rather large axons and thus differ from type d nociceptors.

VI. CHEMORECEPTION

A general sensitivity to chemical stimulation resides in the skin and inner mucous surfaces of amphibians. It is mediated by sensory terminals of the Vth and Xth cranial nerves (Oksche and Ueck 1976). Also, strong chemicals, like acids, stimulate pain receptors (Adrian *et al.* 1931).

VII. KINESTHETIC SENSE

In common with other vertebrates, amphibians have a kinesthetic sense, i.e., a sense of the position of the various muscles and tendons of the body and of the degree to which muscles are stretched in their different states of relaxation or tension. For example, passsive movement of the limbs of salamanders evokes responses in the spinal nerves even when the skin is stripped away to eliminate the superficial touch receptors (Cooper and Diamond 1977). This sense is essential for the maintenance of posture, balance and muscle tone.

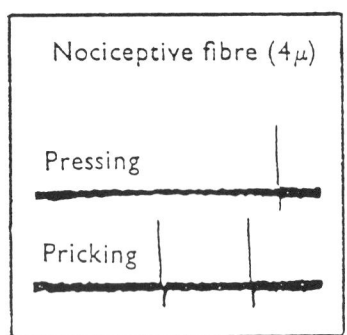 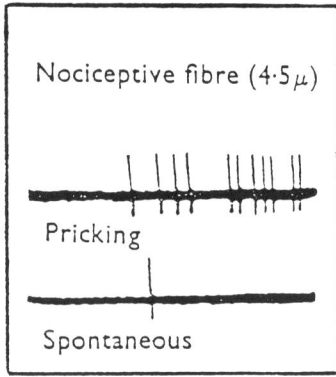 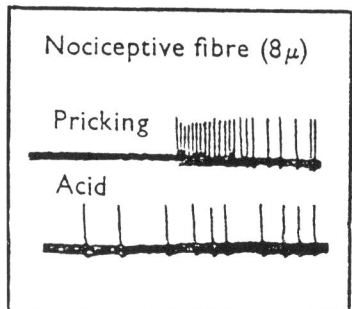

Fig. 6. Discharge characteristics of several kinds of nociceptive nerve fibers with larger diameters than those of type d. From Maruhashi *et al.* (1952).

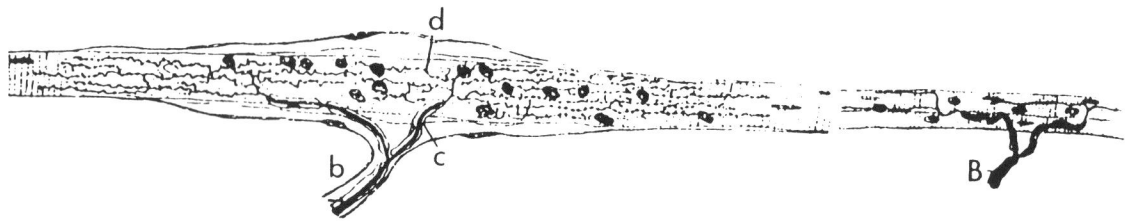

Fig. 7. The muscle spindle of a frog. The myelinated afferent fiber (b) branches (c) and gives rise to numerous beaded chains of unmyelinated endings (d) coiling around the intrafusal muscle fibers. B = motor terminal. From Ottoson (1976).

The sense organ that provides kinesthetic information is the muscle spindle (Fig. 7). Its structure and function in amphibians have been reviewed by Ottoson (1976). It is a fluid-filled, fusiform capsule of connective tissue containing slender muscle fibers (called intrafusal fibers) innervated by afferent sensory nerve fibers, and activated by motor nerve fibers. There are also satellite cells within the capsule. Each spindle unit consists of an equatorial sensory region, called a reticular zone, in which the muscle fibers are nonstriated and are interspersed with a network of connective tissue fibrils; there is a dense aggregation of nuclei in this region. At each end, the spindle narrows to a polar compact zone containing striated muscle fibers. A single sensory nerve fiber passes through the capsule, branches repeatedly and coils around the intrafusal muscle fibers (Fig. 7). The motor fibers are branches of the nerves leading to the extrafusal muscles in which the spindle is embedded.

Amphibian spindles differ from those of mammals in several ways. They have one sensory fiber per spindle unit, rather than two or more; they are of equal length to the surrounding extrafusal muscle fibers, rather than shorter and extending throughout the muscle from tendon to tendon; they show less clear distinction between different types of intrafusal fibers; finally, they have a number of spindles connected in series in a single encapsulation as a compound spindle system, rather than each unit occurring as an individual spindle.

The spindle is stimulated by stretch imposed by the changes in length of the muscle in which it is embedded, and is essentially a transducer measuring the relative length and rate of change of length of the extrafusal muscle fibers. It is highly sensitive to small changes in length or rate of stretch. The threshold for stimulation changes with length but may be as small as 1% of resting length.

A brief stretch produces in the afferent fiber a response characterized by a fast-rising dynamic phase and an exponential fall toward baseline. When stretch is maintained, the response declines from its dynamic peak to a lower static level, characterized by a regular discharge with slowly decreasing frequency. The spindle potential is graded and increases in amplitude to a maximum with progressive lengthening of a spindle. With step-like stretches the dynamic peak reaches a maximum amplitude when the spindle is extended by 20%–30%

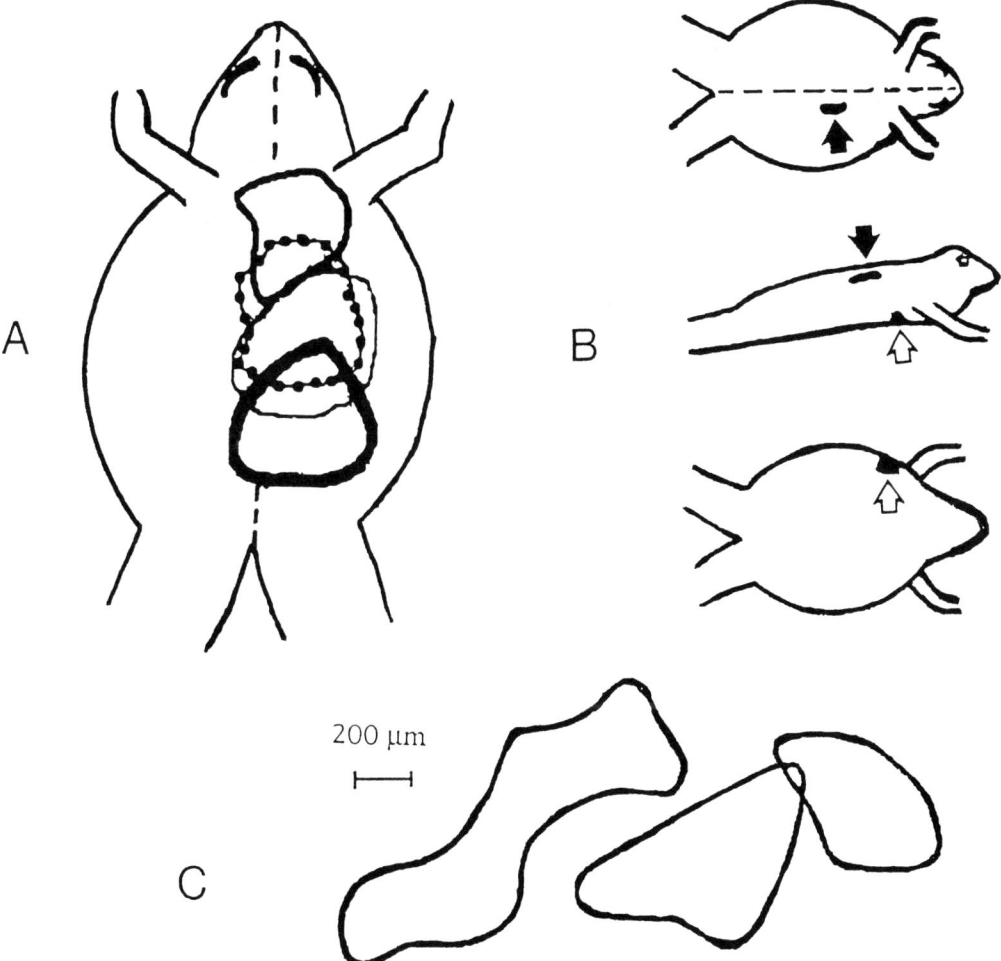

Fig. 8. Tactile receptive fields at the level of individual nerves, nerve fibers and axons in amphibians. A: Overlapping of sensory receptive fields of four dorsal cutaneous nerves (right side) in a frog. Each field outlined by a different width or type of line. Dashed line indicates dorsal midline. B: Area of tactile endings of single nerve fibers in a frog. Dark arrows point to the receptive field of a dorsolateral fiber; white arrows point to the receptive field of a ventrolateral fiber. Dashed line indicates dorsal midline. C: Map of three adjacent receptive fields of individual free nerve endings in the salamander *Ambystoma tigrinum*. A and B modified from Adrian *et al.* (1931) and C modified from Cooper and Diamond (1977).

of its resting length. For a given amount of stretching, the amplitude of the dynamic peak is a function of the velocity of stretching. Thus, the spindle conveys information about (1) onset of individual, brief stretches, (2) state of sustained stretching and (3) the magnitude and velocity of stretches. Furthermore, there is a pause between the dynamic and static phases of the impulse that may provide information about the transition from initial lengthening to the maintained statically stretched condition.

There are complex interactions of excitability changes of the transducer and its associated nerve fiber during stretch which determine the patterning of the impulse in response to a given stretch. A corresponding series of events takes place in the aftermath of a stretch that influences the nature of response to a subsequent one; thus, repeated stretches produce somewhat different responses.

It would appear that the muscle spindle is a transducer with a high degree of precision, and the firing at any moment during muscular stretch provides the central nervous system with exact information about the parameters of the stimulus.

Central processing of the input from the muscle spindles takes place largely in the cerebellum (Herrick 1948). Information from the spindle system is used to initiate appropriate proprioceptive responses in the maintenance of posture and balance.

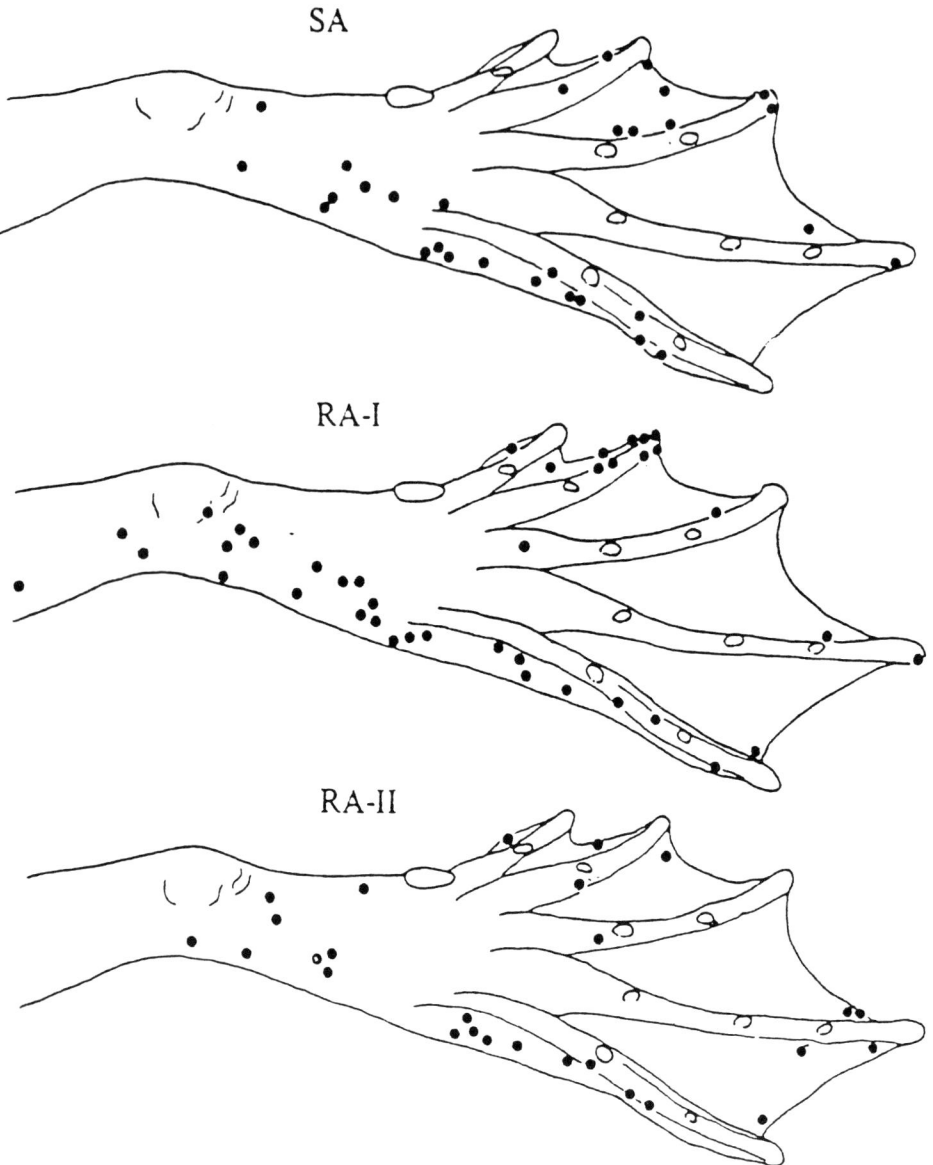

Fig. 9. Location of slowly adapting (SA) and two types of rapidly adapting (RA I and RA II) tactile receptors on the plantar and lateral surfaces of the foot of *Rana catesbeiana*. Each dot represents the centre of a receptive field. From Ogawa *et al.* (1981).

VIII. RECEPTIVE FIELDS

Catton (1976) reviewed somatosensory receptive fields. Each dorsal root fiber branches into various free endings. In salamanders it has been calculated that each axon branches from as few as 4 to as many as 100 times (Cooper *et al.* 1977). The receptive fields of individual nerve endings have approximate radii of 50–75 μm in *Ambystoma tigrinum* and do not overlap, but rather are spaced 200–250 μm apart (Cooper and Diamond 1977) (Fig. 8C). In general, the estimates of axonal receptive fields of anurans are much larger (1–30 mm^2) than those for salamanders (0.05–2.5 mm^2); however, this difference may be an artifact of different methodologies (Cooper and Diamond 1977).

The degree of overlap among receptive fields depends on the scale of measurement. In contrast to receptive fields of individual free nerve endings, those of nerve fibers and nerves do overlap (Fig. 8). When responses of an individual afferent fiber were recorded during point stimulation of the skin it was found that the receptive fields were irregular in shape,

2–35 mm² in area (Lindblom 1958; Catton 1976) and overlapped extensively (Fig. 8B). Thus, the skin is organized as a mosaic of discrete fields of individual nerve endings arising from different fibers. Aguilar *et al.* (1973) found that the average sizes for the tactile receptive fields of spinal nerves were 170 mm² (right side):168 mm² (left side) for nerve 15, 507R:502L for nerve 16 and 385R:394L for nerve 17 in the salamander, *Ambystoma tigrinum*. When the 16th nerve was sectioned, the denervated sensory field was invaded by sprouting collaterals from nerves 15 and 16, both of which increased their respective sensory fields. In *Rana pipiens*, the receptive fields on the belly are larger and of a different shape than those on lateral or dorsal surfaces of the trunk (Baker and Jacobson 1970).

Ogawa *et al.* (1981) mapped the distribution of mechanoreceptive fields of three kinds of receptive units on the hind feet of *Rana catesbeiana*. They found that all types were distributed throughout the various parts of the foot (Fig. 9) except that there were fewer units of any type in the web between the toes. In warty skin, the receptive field of slowly adapting units and one kind of rapidly adapting units (RA II) were usually located beside warts and were small (SA: mean = 7.4 mm², range = 0.4–25.9 mm²; RA II: mean = 2.8 mm², range = 0.2–9.4 mm²). By contrast, the other rapidly adapting unit (RA I) had larger receptive fields (mean = 13.3 mm², range = 0.2–60.5 mm²) not restricted to warty areas. Lindblom (1958) noted that the mechanoreceptive units located on the tips of the toes and on plantar pads of *Bufo bufo* have very small receptive fields (1 mm²) and diffuse sensitivity; he considered them a distinctive type which he termed "apical units".

IX. ACKNOWLEDGEMENTS

I am indebted to Drs Hisashi Ogawa and Robert Grossfeld whose constructive criticism greatly improved this chapter.

X. REFERENCES

Adrian, E. D., Cattell, M. and Hoagland, H., 1931. Sensory discharges in single cutaneous nerve fibers. *J. Physiol.* **72**: 377–391.

Aguilar, C. E., Bisby, M. A., Cooper, E. and Diamond, J., 1973. Evidence that axoplasmic transport of trophic factors is involved in the regulation of peripheral nerve fields in salamanders. *J. Physiol.* **234**: 449–464.

Baker, R. E., 1972. Biochemical specification versus specific regrowth in the innervation of skin grafts in anurans. *Nature* **236**: 235–237.

Baker, R. E. and Jacobson, M., 1970. Development of reflexes from skin grafts in *Rana pipiens*: influence of size and position of grafts. *Developmental Biol.* **22**: 476–494.

Budtz, P. E. and Larsen, L. O., 1975. Structure of the toad epidermis during the moulting cycle. II. Electron microscopic observations on *Bufo bufo* (L.). *Cell Tiss. Res.* **159**: 459–483.

Calof, A. L., Jones, R. B. and Roberts, W. J., 1981. Sympathetic modulation of mechanoreceptor sensitivity in frog skin. *J. Physiol.* **310**: 481–499.

Catton, W. T., 1958. Some properties of frog skin mechanoreceptors. *J. Physiol.* **141**: 305–322.

Catton, W. T., 1976. Cutaneous mechanoreceptors. Chapter 21, pp. 630–642 in "Frog Neurobiology, a Handbook" ed by R. Llinás and W. Precht. Springer-Verlag, Berlin.

Cooper, E. and Diamond, J., 1977. A quantitative study of the mechanosensory innervation of the salamander skin. *J. Physiol.* **264**: 695–723.

Cooper, E., Diamond, J. and Turner, C., 1977a. The effects of nerve section and of colchicine treatment on the density of mechanosensory nerve endings. *J. Physiol.* **264**: 725–749.

Cooper, E., Scott, S. A. and Diamond, J., 1977b. Control of mechanosensory nerve sprouting in salamander skin. *Neuroscience Symposia* **2**: 120–138.

Cooper, E., Diamond, J., Leslie, R., Parducz, A. and Turner, C., 1975a. Touch receptors of the salamander skin. *Proc. Physiol. Soc.* **256**: 117P–118P.

Cooper, E., Diamond, J., Macintyre, L. and Turner, C., 1975b. Control of collateral sprouting in mechanosensory nerves of salamander skin. *Proc. Physiol. Soc.* **252**: 20P–21P.

Crowe, R. and Whitear, M., 1978. Quinacrine fluorescence of Merkel cells in *Xenopus laevis*. *Cell Tiss. Res.* **190**: 273–283.

Diamond, J., Holmes, M. and Nurse, C. A., 1986. Are Merkel cell-neurite reciprocal synapses involved in the initiation of tactile responses in salamander skin? *J. Physiol.* **376**: 101–120.

Diamond, J., Mills, L. R. and Mearow, K. M., 1988. Evidence that the Merkel cell is not the transducer in the mechanosensory Merkel cell-neurite complex. *Progress Brain Res.* **74**: 51–56.

Diamond, J., Cooper, E., Turner, C. and Macintyre, L., 1976. Trophic regulation of nerve sprouting. *Science* **193**: 371–377.

Düring, M. and Andres, K. H., 1976. The ultrastructure of taste and touch receptors of the frog's taste organ. *Cell Tiss. Res.* **165**: 185–198.

Fox, H., 1994. The structure of the Integument. Pp. 1–32 in "The Integument", Vol. 1 of "Amphibian Biology", ed by H. Heatwole, G. T. Barthalmus and A. Y. Heatwole. Surrey Beatty & Sons, Chipping Norton.

Fox, H. and Whitear, M., 1978. Observations on Merkel cells in amphibians. *Biologie cellulaire* **32**: 223–231.

Fox, H., Lane, E. B. and Whitear, M., 1980. Sensory nerve endings and receptors in fish and amphibians. *Linn. Soc. Symp. Ser.* **9**: 271–281.

Franzisket, L., 1963. Characteristics of instinctive behaviour and learning in reflex activity of the frog. *Anim. Behav.* **11**: 318–324.

Herrick, C. J., 1948. "The Brain of the Tiger Salamander *Ambystoma tigrinum*". University of Chicago Press, Chicago.

Hughes, A., 1957. The development of the primary sensory system in *Xenopus laevis* (Daudin). *J. Anatomy* **91**: 323–338.

Lindblom, U. F., 1958. Excitability and functional organization within a peripheral tactile unit. *Acta Physiologica Scandinavica (Suppl.)* **44(153)**: 1–84.

Lindblom, U., 1962. The relation between stimulus and discharge in a rapidly adapting touch receptor. *Acta Physiologica Scandinavica* **56**: 349–361.

Lindblom, U., 1963. Phasic and static excitability of touch receptors in toad skin. *Acta Physiologica Scandinavica* **59**: 410–423.

Maruhashi, J., Mizuguchi, K. and Tasaki, I., 1952. Action currents in single afferent nerve fibres elicited by stimulation of the skin of the toad and the cat. *J. Physiol.* **117**: 129–151.

Macintyre, L. and Diamond, J., 1981. Domains and mechanosensory nerve fields in salamander skin. *Proc. Roy. Soc. Lond., B* **211**: 471–499.

Mearow, K. M. and Diamond, J., 1988. Merkel cells and the mechanosensitivity of normal and regenerating nerves in *Xenopus* skin. *Neuroscience* **26**: 695–708.

Merkel, F., 1880. Uber die Eindigungen der sensiblen Nerven in der Haut der Wirbelthiere. H. Schmidt, Rostock.

Mills, L. R. and Diamond, J., 1995. Merkel cells are not the mechanosensory transducers in the touch dome of the rat. *J. Neurocytol.* **24**: 117–134.

Miner, N., 1956. Integumental specification of sensory fibers in the development of cutaneous local sign. *J. Compar. Neurol.* **105**: 161–170.

Nafstad, P. H. J. and Baker, R. E., 1973. Comparative ultrastructural study of normal and grafted skin in the frog, *Rana pipiens*, with special rererence to neuroepithelial connections. *Z. Zellforsch.* **139**: 451–462.

Nurse, C. A., Mearow, K. M., Holmes, M., Visheau, B. and Diamond, J., 1983. Merkel cell distribution in the epidermis as determined by quinacrine fluorescence. *Cell Tiss. Res.* **228**: 511–524.

Ogawa, H., 1996. The Merkel cell as a possible mechanoreceptor cell. *Progress in Neurobiol.* **156**: in press.

Ogawa, H. and Yamashita, Y., 1988. Mechano-electric transduction in the slowly adapting cutaneous afferent units of frogs. Chapter, pp. 63–68 in "Progress in Brain Research", ed by W. Hamann and A. Iggo. Elsevier Science Publishers B. V., New York.

Ogawa, H., Morimoto, K. and Yamashita, Y., 1981. Physiological characteristics of low threshold mechanoreceptor afferent units innervating frog skin. *Quart. J. Experimen. Physiol.* **66**: 105–116.

Ogawa, H., Yamashita, Y., Nomura, T. and Taniguchi, K., 1984a. Discharge patterns of the slowly adapting mechanoreceptor afferent units innervating the non-warty skin of the frog. *Japanese J. Physiol.* **34**: 255–267.

Ogawa, H., Yamashita, Y., Nomura, T. and Taniguchi, K., 1984b. Functional properties of mechanoreceptors in frogs. Pp. 169–177 in "Sensory Receptor Mechanisms", ed by W. Hamann and A. Iggo. World Scientific Publ. Co., Singapore.

Oksche, A. and Ueck, M., 1976. The nervous system. Chapter 7, pp. 313–419 in "Frog Neurobiology, a Handbook", ed by R. Llinás and W. Precht. Springer-Verlag, Berlin.

Ottosen, D., 1976. Morphology and physiology of muscle spindles. Chapter 22, pp. 643–675 in "Frog Neurobiology, a Handbook", ed by R. Llinás and W. Precht. Springer-Verlag, Berlin.

Ovalle, W. K., 1976. The larval tentacle of an anuran amphibian tadpole: muscle fiber ultrastructure and innervation. *Anatomical Record* **184**: 494.

Pardusz, A., Leslie, R. A., Cooper, E., Turner, C. J. and Diamond, J., 1977. The Merkel cells and the rapidly adapting mechanoreceptors of the salamander skin. *Neuroscience* **2**: 511–521.

Roberts, A., 1996. Skin Sensory Systems of Amphibian Embryos and Young Larvae. Chapter 5, Pp. 923–935 in "Amphibian Biology", Vol. 3, "Sensory Perception", ed by H. Heatwole and E. Dawley. Surrey Beatty & Sons, Chipping Norton.

Roberts, A. and Blight, A. R., 1975. Anatomy, physiology and behavioural role of sensory nerve endings in the cement gland of embryonic *Xenopus*. *Proc. Roy. Soc. Lond. B* **192**: 111–127.

Roberts, A. and Hayes, B. P., 1979. The anatomy and function of "free" nerve endings in an amphibian skin sensory system. *Proc. Roy. Soc. Lond. B* **196**: 415–429.

Roberts, A. and Smyth, D., 1974. The development of a dual touch sensory system in embryos of the amphibian *Xenopus laevis*. *J. Comp. Physiol.* **88**: 31–42.

Scott, S. A., Cooper, E. and Diamond, J., 1981. Merkel cells as targets of the nechanosensory nerves in salamander skin. *Proc. Roy. Soc. Lond. B* **211**: 455–470.

Spray, D. C., 1974. Characteristics, specificity, and efferent control of frog cutaneous cold receptors. *J. Physiol.* **237**: 15–38.

Spray, D. C., 1975. Effect of reduced acclimation temperature on responses of frog cold receptors. *Comp. Biochem. Physiol.* **50A**: 391–395.

Spray, D. C., 1976. Pain and temperature receptors in anurans. Chapter 20, pp. 607–628 in "Frog Neurobiology, a Handbook", ed by R. Llinás and W. Precht. Springer-Verlag, Berlin.

Spray, D. C. and Galansky, S. H., 1975. Effects of cholinergic agonists and antagonists on frog cutaneous cold receptors. *Comp. Biochem. Physiol.* **50C**: 97–103.

Taniguchi, K., Yamashita, Y. and Ogawa, H., 1984. Threshold response phase to sinusoidal stimulation of frog cutaneous mechanoreceptor afferent units. *Japanese J. Physiol.* **34:** 1065–1075.

Whitear, M., 1974. The nerves in frog skin. *J. Zool.* **172:** 503–529.

Whitear, M., 1977. A functional comparison between the epidermis of fish and amphibians. *Symp. Zool. Soc. Lond.* **39:** 291–313.

Yamashita, Y. and Ogawa, H., 1991. Slowly adapting cutaneous mechanoreceptor afferent units associated with Merkel cells in frogs and effects of direct currents. *Somatosensory Motor Res.* **8:** 87–95.

Yamashita, Y., Ogawa, H. and Taniguchi, K., 1986. Differential effects of manganese and magnesium on two types of slowly adapting cutaneous mechanoreceptor afferent units in frogs. *Pflügers Arch.* **406:** 218–224.

CHAPTER 7

Magnetoreception

John B. Phillips

I. Introduction
II. Evidence for the Use of the Geomagnetic Field in Spatial Orientation
   A. Introduction
   B. Compass Orientation
   C. Homing
III. Potential Transduction Mechanisms
   A. Electroreceptors
   B. Magnetite-Based Receptors
   C. Homing
IV. Characterization of Magnetoreception Mechanisms in Newts
   A. Light-Dependent Magnetic Compass
   B. Mechanism(s) of Magnetoreception Involved in Homing
V. Conclusions
VI. Acknowledgements
VII. References

I. INTRODUCTION

SENSITIVITY to the geomagnetic field appears in widely separated taxa of amphibians (Phillips 1977; Phillips and Borland 1992a; Sinsch 1990), as well as among vertebrates in general (e.g., Wiltschko 1983; Burda *et al.* 1990; Light *et al.* 1993). Magnetoreception is unique among the major sensory systems, however, in that the nature of the transduction mechanism(s) and the location of the receptor(s) are yet to be determined. Recent evidence from amphibians (Phillips 1986a; Phillips and Borland 1994) and other vertebrates (Semm and Demaine 1986; Semm and Beason 1990), suggests that at least some vertebrate species may possess two distinct magnetoreception mechanisms, one detecting magnetic field direction and a second detecting magnetic field intensity or inclination (see below). This chapter focuses on: (1) evidence for the use of the geomagnetic field in compass orientation and homing by amphibians, and (2) behavioural and neurophysiological evidence concerning the nature of the underlying receptor mechanism(s).

The use of the geomagnetic field in orientation by amphibians appears to involve sensory mechanisms and behavioural strategies that are similar to those found in other vertebrate groups (Rodda and Phillips 1992). The types of behavioural responses exhibited by amphibians, however, make them particularly well-suited as experimental subjects for investigating the detection and use of the geomagnetic field in spatial orientation. The advantages of amphibians include: (1) the short distances over which they move under natural conditions (Sinsch 1990), (2) the relative ease with which some species can be trained to orient in a particular compass direction relative to the magnetic field (Phillips 1986b), (3) the strength and consistency of the magnetic orientation that can be elicited by appropriate manipulation of factors such as temperature and humidity that influence the motivation of amphibians to orient (Phillips 1986b, 1987), and (4) the ability to study long-distance homing under controlled laboratory conditions (Phillips 1987; Phillips and Borland 1994).

II. EVIDENCE FOR THE USE OF THE GEOMAGNETIC FIELD IN SPATIAL ORIENTATION

A. Introduction

The use of the geomagnetic field in compass orientation has been demonstrated in two species of urodele amphibians (Phillips 1977, 1986a,b; Phillips and Borland 1992a). The geomagnetic field also appears to be involved in homing by both urodeles and anurans, although the nature of this involvement remains unclear (Phillips 1987; Phillips and Borland 1994; Sinsch 1990).

Homing, or true navigation, requires the use of both map (i.e., geographic positional) information, and compass (i.e., directional) information. Map information may be derived by path integration (i.e., by keeping track of the direction and distance of each segment of the displacement path) which, in principle, could make use of directional information derived from a magnetic compass. Alternatively, bico-ordinate map information may be derived at unfamiliar sites from geophysical gradients that are extrapolated beyond the organism's area of familiarity.

The geomagnetic field has been suggested to play a role in a bico-ordinate map, as well as a source of compass information (Gould 1980; Moore 1980; Walcott 1980). This hypothesis postulates that one or both co-ordinates of a bico-ordinate map are derived from learned spatial gradients of magnetic field parameters such as inclination and total intensity. By extrapolating these gradients beyond its normal range of movements, an organism in unfamiliar territory could use the values of the two map co-ordinates at the site to "fix" its position relative to home. Having derived its geographic position relative to home using the bico-ordinate map, the organism could then use one of its compass systems (e.g., the magnetic compass) to orient in the homeward direction.

B. Compass Orientation

Evidence for the use of an earth-strength magnetic field for compass orientation (i.e., orientation along a fixed compass bearing that does not rely on map information) has been obtained for two species of salamanders (the cave salamander *Eurycea lucifuga*, Phillips 1977, and the eastern red-spotted newt *Notophthalmus viridescens* Phillips 1986a,b; Phillips and Borland 1992a–c). The magnetic compass response of the eastern newt has been characterized in some detail. Eastern newts readily learn the magnetic compass direction of an artificial shoreline in an outdoor training tank. In this respect, newts appear similar to other pond-dwelling amphibians that orient perpendicular to a natural or artificial shoreline using any of a variety of compass systems (Ferguson 1971).

The magnetic compass of shoreward-orienting newts (Phillips 1986a), like that of nocturnally migrating birds (Wiltschko and Wiltschko 1972), is affected by the axial sensitivity, but not the polar sensitivity, of the magnetic field (Fig. 1). The inclination or dip angle of the magnetic field is used to distinguish between the two ends of the magnetic axis. As in the earlier study by Wiltschko and Wiltschko (1972), inversion of the vertical component of the magnetic field (which reverses the inclination, but does not alter the horizontal polarity of the magnetic field) caused shoreward-orienting newts to reverse their direction of orientation relative to the magnetic field (Phillips 1986a).

C. Homing

The geomagnetic field also has been implicated in the homing orientation of both anuran and urodele amphibians (Phillips 1987; Phillips and Borland 1994; Sinsch 1990). Sinsch (1987, 1988, 1990, 1992) displaced three species of toads *(Bufo)* from their breeding ponds, and observed both their initial orientation away from the release point (by means of a mechanical tracking device; Fig. 2A), and their homing success. He investigated the relative importance of magnetic, olfactory and visual cues by reversibly blocking or swamping input from each of these sensory modalities.

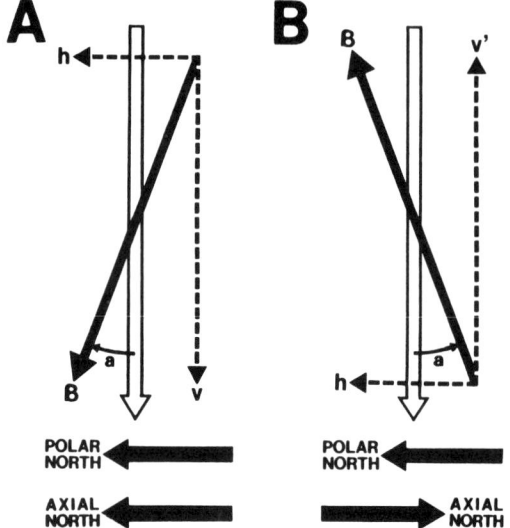

Fig. 1. Axial and polar magnetic responses (from Phillips 1986a). Newts that are homing respond to the horizontal polarity (h) of the magnetic field (β), which defines "polar" north. In contrast, newts exhibiting shoreward magnetic compass orientation use the slope or inclination of the magnetic field lines to distinguish "axial" north. Axial north is defined as the direction in which the magnetic field lines form the smallest angle (a) with the gravity vector (open arrow). **A.** For any given horizontal component (h), when the vertical component of the magnetic field (v) points downward, polar and axial norths coincide. **B.** Inversion of the vertical component (v') reverses the inclination but does not change the horizontal polarity of the magnetic field. Thus, when the vertical component is inverted, newts exhibiting shoreward magnetic compass orientation reverse their direction of orientation, while newts that are homing are unaffected.

An inability to detect the geomagnetic field disrupted initial homeward orientation in all three species of toads. Individuals carrying strong bar magnets that prevented detection of the geomagnetic field exhibited initial orientation that was either randomly distributed (*Bufo bufo*, Sinsch 1987 and *Bufo calamita*[1], Sinsch 1992) or oriented in a non-home direction (*Bufo spinulosus*, Sinsch 1988) (Fig. 2B). Controls carrying brass bars of comparable size and weight oriented in the homeward direction.

In contrast to magnetic deprivation, the effects of other types of sensory deprivation (e.g., olfactory and visual) on the initial homing orientation of toads were less consistent (Fig. 2C, D). Olfactory deprivation eliminated initial homing orientation in *Bufo bufo* which migrates over relatively long distances (up to 3 km) to permanent bodies of water. In *B. calamita* and *B. spinulosus* which migrate over relatively short distances (less than 1 km) to temporary ponds, individuals deprived of olfactory information were oriented in the home direction (Fig. 2C; and see Sinsch 1992 for additional data on *B. calamita*). In subsequent experiments with *B. calamita*, however, anosmia was found to produce an increase in the scatter of initial orientation (Sinsch 1992). Conversely, visual deprivation disrupted initial orientation in the two short distance migrants (*B. calamita* and *B. spinulosus*), but had no effect on initial orientation in *B. bufo* (Fig. 2D, and see Sinsch 1992 for additional data from *B. calamita*).

Sinsch's findings indicate that toads utilize multiple sources of directional information for homing (reviewed by Sinsch 1990), as also has been found to be the case in other vertebrates (e.g., birds, Benvenuti and Ioale 1988; Wiltschko *et al.* 1989). Furthermore, the toad studies suggest that the relative importance of different sensory modalities may vary depending upon the type of habitat (i.e., temporary versus permanent ponds), or distance of migration. Of the three sensory modalities studied by Sinsch, however, only geomagnetic deprivation eliminated initial homeward orientation in all three species. These findings, therefore, provide compelling evidence that the geomagnetic field plays an integral role in the homing orientation of these amphibians under natural conditions.

In contrast to the effects observed on initial homing orientation, the homing success of toads (proportion of individuals returning to the original capture site within a three day period) was unaffected by magnetic or olfactory deprivation, although individuals in both treatments took longer to home than did controls. These findings suggest that neither magnetic nor olfactory information plays an essential role in the map component of homing. The toads deprived of map information derived from one or both of these sensory modalities may, however, have been able to switch to an alternative source of map information or to home without relying on map information, i.e., using familiar landmarks and/or home-emanating

---

[1] In *Bufo calamita*, only males exhibited initial orientation that was directed toward the pond from which they were collected and was disrupted by magnetic deprivation. Females oriented towards nearby choruses of calling males.

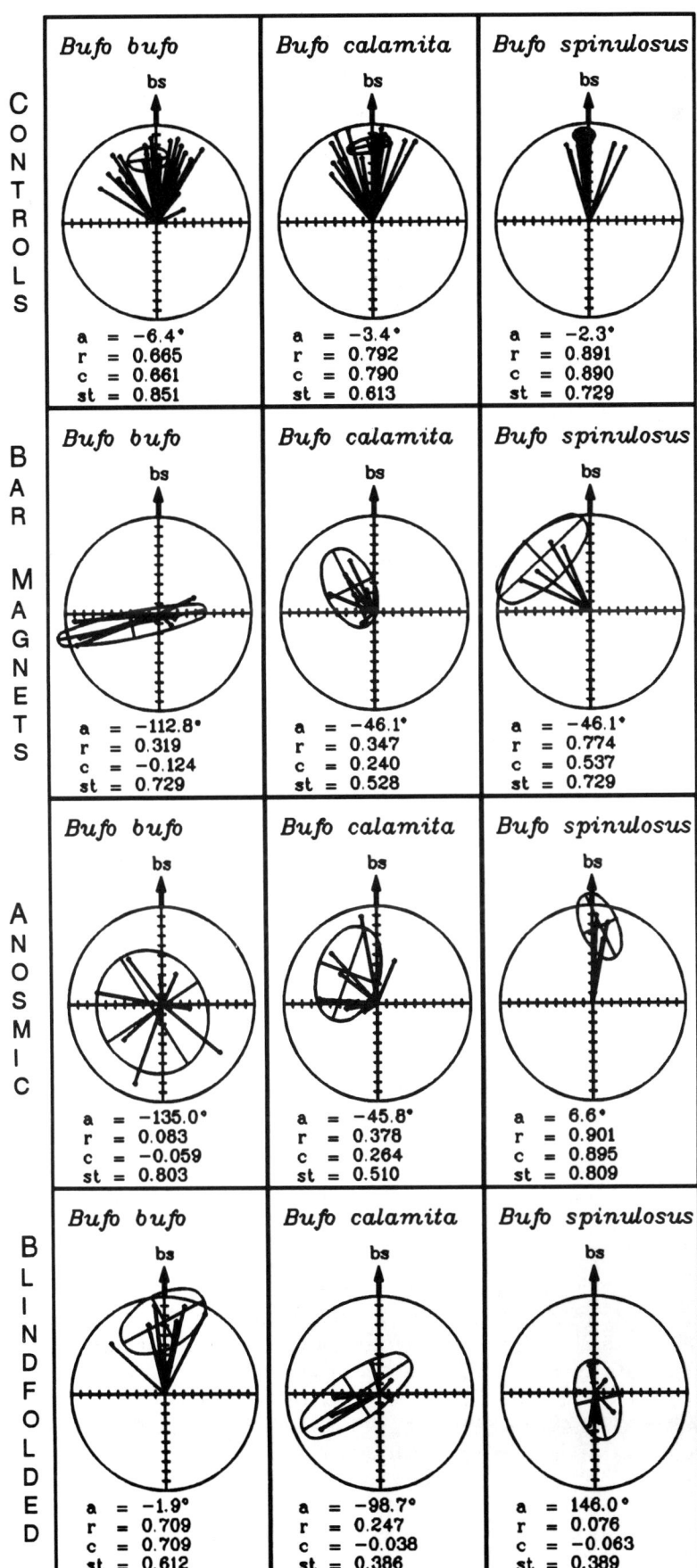

Fig. 2. Effects of sensory deprivation on initial homing orientation of three species of toads (data from Sinsch 1990). Initial orientation was obtained using a mechanical tracking device after displacement from the site of capture (see Sinsch 1987, 1988, 1992 for detailed descriptions of methods). **A.** Orientation of toads tested without sensory deprivation. **B.** Orientation of toads carrying bar magnets of sufficient strength to swamp the geomagnetic field. **C.** Orientation of toads rendered anosmic by blocking the external nares. **D.** Orientation of blindfolded toads. The individual vectors in each distribution indicate the mean vector bearings for groups of toads released simultaneously at one testing site (length of each vector is proportional to the mean vector length, r, with the radius of the circle corresponding to $r = 1$). An ellipse indicates the 95% confidence interval for the second order mean vector calculated from the distribution of group mean vector bearings. Distributions for which the 95% confidence ellipse does not include the origin differ significantly from a random distribution. Confidence ellipses for non-random distributions that do not include the home direction (line from the top of each distribution) indicate that the observed orientation differs significantly from the home direction.

cues since experimental manipulations involved displacements of less than 200 m. Thus, while Sinsch's studies clearly establish that the geomagnetic field plays an important role in homing, his findings leave unanswered the question of whether the geomagnetic field is involved in the map and/or compass components.

Homing orientation has been demonstrated in laboratory studies of the eastern red-spotted newt after displacements of from 10 to more than 40 km (approximately 20 times their normal range of movement; Phillips 1987; Phillips and Borland 1994). Deprivation of olfactory, magnetic, visual and inertial cues during displacement from the home pond was found to have no effect on the strength or accuracy of homing orientation, suggesting that newts are able to utilize a bico-ordinate map for homing (Phillips *et al.* 1995). To date, the eastern newt is the only organism studied that has been shown to exhibit map-based homing under controlled laboratory conditions where the involvement of the geomagnetic field in the map and compass components of homing can be rigorously investigated.

Interestingly, newts exhibiting shoreward compass orientation and the compass component of homing were found to respond differently to the magnetic field (Phillips 1986a). As discussed previously (Fig. 1), shoreward orienting newts responded to an inversion of the vertical component of the magnetic field by reversing their direction of orientation, indicating that they are sensitive to the axis, but not the polarity, of the magnetic field ("axial" sensitivity). In contrast, newts that were homing were unaffected by an inversion of the magnetic field's vertical component, indicating that they are sensitive to the polarity of the magnetic field ("polar" sensitivity). More recent experiments suggest that the shoreward orientation and homing responses of newts also are affected differently by changes in the wavelength of light under which they are tested (Phillips and Borland 1994, and see below). These differences provide evidence of two distinct magnetoreception systems (Phillips 1986a; Phillips and Borland 1994). In birds, neurophysiological evidence also suggests that two distinct magnetoreception mechanisms may be present (Semm and Demaine 1986; Beason and Semm 1987; Semm and Beason 1990).

The involvement of a second magnetoreception system in homing by newts suggests that the geomagnetic field may be involved in the map, as well as the compass, component of homing. A second magnetoreception system is unlikely to be present merely if it provides a redundant source of compass information (Phillips 1986a), but may be required if map information is derived from the geomagnetic field. It has been suggested on theoretical grounds that the type of magnetoreception mechanism that would function effectively as a compass is unlikely to provide the high level of sensitivity necessary to detect the subtle geographic variation in the geomagnetic field (Yorke 1979; Kirschvink *et al.* 1985). While there are many theoretical arguments both for and against the hypothesis that a short-distance migrant like the newt might derive map information from the geomagnetic field (Phillips and Borland 1994), the ability to study map-based homing under controlled laboratory conditions makes it possible for the first time to carry out critical empirical tests of hypotheses (Phillips 1996).

III. POTENTIAL TRANSDUCTION MECHANISMS

A. Electroreceptors

Many aquatic amphibians have well-developed electroreception organs (Fritzsch and Neary 1993). In marine elasmobranchs, electroreception organs have been suggested to be responsible for magnetic field sensitivity by detecting the induced electrical fields generated by movement through the geomagnetic field (Kalmijn 1982). This type of "induction" mechanism would be expected to exhibit a polar response (i.e., a magnetic compass response that is sensitive to the horizontal polarity of the magnetic field), which would be consistent with the polar responses exhibited by eastern newts when homing (see earlier discussion). Detection of the geomagnetic field by this type of induction is unlikely for organisms living in fresh water or terrestrial environments, however, due to the requirement of a low resistance pathway for electrical current that must be provided by the surrounding medium (Kalmijn 1982). Fresh water and air have much lower conductivities than does sea water. Hence, in these media electroreceptors are less effective in detecting induced electrical fields resulting from movement through the geomagnetic field. An induction mechanism, therefore, is unlikely to account for the magnetic field sensitivity observed in studies of amphibians, which, to date, have been carried out under terrestrial conditions.

B. Magnetite-Based Receptors

Particles of the biogenic mineral magnetite (Fe_3O_4) have been shown to mediate the passive magnetic orientation of some single-celled organisms (Kalmijin and Blakemore 1982). The discovery of single domain (SD) and superparamagnetic (SPM) particles of magnetite in a wide variety of organisms (reviewed by Kirschvink et al. 1985), including salamander species in which magnetic compass orientation has been demonstrated (J. Kirschvink, pers. comm.), has led to speculation that magnetite-based receptors may be responsible for magnetic field sensitivity in higher animals. Because the strength of interaction of a SD or SPM particle of magnetite with the geomagnetic field is considerably above the thermal randomization energy (kT), this type of receptor could transduce magnetic stimuli by detecting changes in particle alignment or torque caused by the geomagnetic field.

Although, in theory, a magnetite-based receptor could exhibit either axial or polar sensitivity (Kirschvink and Gould 1981; Kirschvink et al. 1985), it is unlikely that the axial magnetic compass used by newts for shoreward orientation is mediated by a magnetite-based receptor. Phillips and Borland (1992a,b) have shown that the magnetic compass used for shoreward orientation is light-dependent (see below), while the interaction of a magnetite particle with an external magnetic field is expected to be independent of light. Although it is possible that a magnetite-based receptor associated with a visual neuron would exhibit both magnetic field and light sensitivity, it is unlikely that the dependence on light would be maintained by natural selection, because light-dependence reduces the range of environmental conditions under which the magnetic compass can provide accurate directional information (Phillips and Borland 1992 a,b). The evidence from newts, as well as evidence for the presence of a magnetic compass with similar light-dependent properties in *Drosophila melanogaster* (Phillips and Sayeed 1993), suggests, therefore, that light-dependence is an intrinsic property of the underlying magnetoreception mechanism.

In contrast to the light-dependent axial magnetic compass used by newts for shoreward orientation, the magnetoreception system used for homing exhibits polar sensitivity (Phillips 1986a). Polar sensitivity is consistent with input from a magnetite-based, rather than a photoreceptor-based mechanism (see below). Moreover, a magnetite-based receptor is theoretically capable of the high level of sensitivity that would be necessary for newts to derive map information from subtle spatial gradients in the geomagnetic field (Yorke 1979; Kirschvink et al. 1985; and see below). In birds, Semm and Beason (1990) have reported single-unit responses to magnetic stimuli in the trigeminal nerve that are independent of visual input, and are sensitive to small (i.e., <1%) changes in magnetic field intensity. While additional work is needed to identify conditions under which these responses can be obtained reliably, it is intriguing to note that the trigeminal nerve innervates the anterior region of the head where particles of the mineral magnetite have been localized in a number of different vertebrates (Kirschvink et al. 1985), raising the possibility that a magnetite-based receptor may be present in vertebrates, possibly including amphibians.

C. Specialized Photoreceptors

Schulten (1982) and Schulten and Windemuth (1986) proposed specialized photoreceptors detecting the earth's magnetic field as an explanation of the axial sensitivity observed in behavioural studies of nocturnally-migrating birds (Wiltschko and Wiltschko 1972; see also Phillips 1986a; Beason 1989; Light et al. 1983). The "radical pair" mechanism proposed by Schulten (1982) would amplify the weak interaction of the geomagnetic field with a single electron spin to the level of photon detection, resulting in a magnetic-field-dependent modulation of a photoreceptor's response to light. The electron spin-resonance interactions that underlie this general class of models are inherently insensitive to the polarity of the magnetic field (Schulten 1982) and, therefore, could mediate responses exhibiting axial, but not polar, sensitivity.

An earlier model implicating a specialized photoreceptor in magnetoreception, the so-called "optical-pumping" model (Leask 1977), has been criticized on several grounds: (1) that the proposed energy of interaction with the geomagnetic field is considerably below the thermal randomization energy (kT), and (2) that the requirement of an independent source of energy in the range of 3 MHz is unlikely to be met in a biological system. Despite these criticisms, Leask's model played an important role in calling attention to the possible involvement of the visual system in magnetoreception, and stimulated research in this area.

The "radical pair" model proposed by Schulten (1982) is not limited in the same way by thermal noise because the proposed interactions have a much shorter time-course than do thermal relaxation processes (Schulten, pers. comm.). In addition, an independent source of relaxation energy in the MHz range is not a requirement of these models.

Although the presence of photoreceptors that exhibit magnetic field sensitivity has yet to be convincingly demonstrated in any organism, behavioural (Phillips and Borland 1992a,b, and see below) and neurophysiological (Semm et al. 1984; Semm and Demaine 1986; Olcese et al. 1988) evidence for the presence of light-dependent magnetoreception mechanisms in vertebrates provide strong support for this class of models.

IV. CHARACTERIZATION OF MAGNETORECEPTION MECHANISMS IN NEWTS

A. Light-Dependent Magnetic Compass

Behavioural studies of shoreward magnetic compass orientation in eastern newts (Phillips 1986a; Phillips and Borland 1992a,b) provided important clues to the identity of the underlying magnetoreception mechanism. As noted previously, shoreward orienting newts exhibit an axial, or inclination, magnetic compass response, which is insensitive to the polarity of the magnetic field (Phillips 1986a). More recent experiments have shown that shoreward magnetic compass orientation by newts is eliminated in the absence of visible light (i.e., in tests carried out under near-infrared light; Phillips and Borland 1992b). Although the elimination of behavioural responses to the magnetic field in total darkness is compatible with a photoreceptor-based magnetoreception mechanism (Wiltschko and Wiltschko 1981), this type of finding does not provide critical evidence for or against such a mechanism. Elimination of magnetic orientation in the absence of light could result from a change in the newts' motivation to orient, rather than from an effect on the underlying magnetoreception mechanism. Conversely, an ability to orient in the absence of light would not rule out a photoreceptor-based magnetoreceptor, because the presence of a biochemical ("dark") source of energy might enable this type of magnetoreception mechanism to operate in the absence of environmental light.

More compelling evidence for the presence of a light-dependent magnetoreception mechanism in newts has been obtained in studies of the influence of changes in the wavelength of light on shoreward magnetic compass orientation (Phillips and Borland 1992a; Fig. 3). After training to an artificial shore under natural (i.e., full-spectrum) light, newts tested under short-wavelength light exhibited magnetic orientation in the appropriate shoreward direction. When tested under long wavelength (i.e., >500 nm) light, however, the newts' responses were rotated 90° counterclockwise of the shoreward direction. Experiments in which newts were trained under long-wavelength light demonstrated that this 90° rotation was due to a direct effect of light on the underlying magnetoreception mechanism, i.e., under long-wavelength light, the directional information obtained by the newts from the magnetic compass was rotated by 90° (Phillips and Borland 1992a). These findings indicate that the magnetoreception mechanism used by newts in shoreward orientation is light-dependent, and involves at least two distinct spectral mechanisms.

Phillips and Borland (1992a) proposed that the 90° rotation of the newt's shoreward magnetic compass orientation was due to the magnetic field producing complementary, bimodal patterns of response to light absorbed by the short-wavelength and long-wavelength spectral mechanisms (Fig. 4). If so, equal excitation of these two spectral mechanisms should cause the complementary patterns to cancel each other and prevent the light-dependent magnetic compass from operating. Consistent with this prediction, newts tested under wavelengths of light around 475 nm, i.e., intermediate between the spectral regions that

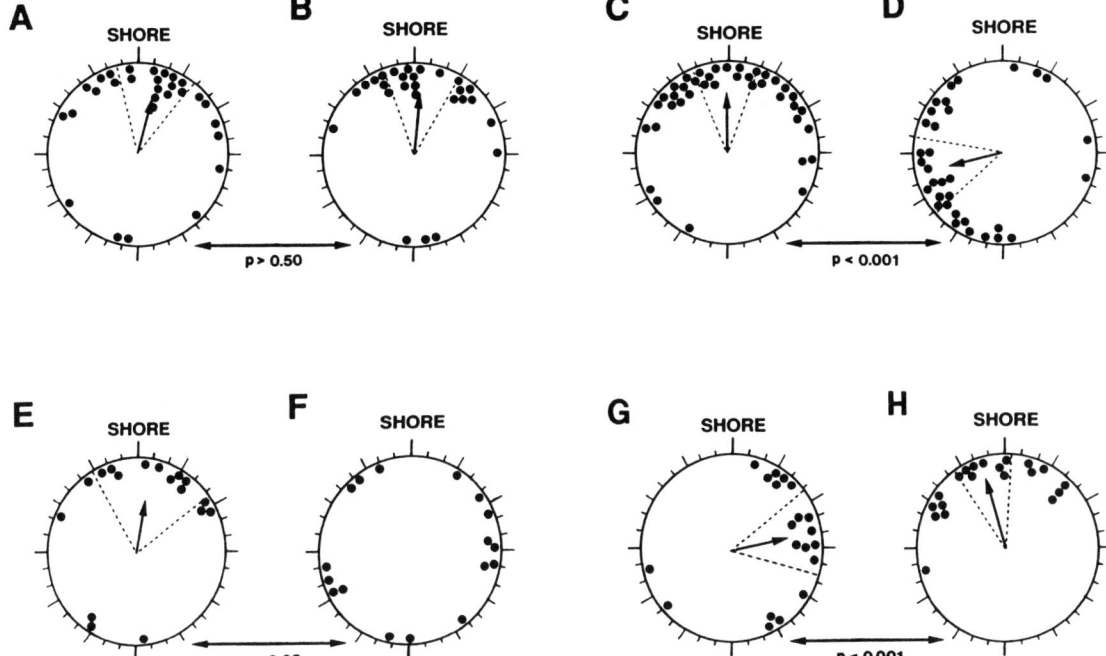

Fig. 3. Dependence of the shoreward magnetic compass orientation of newts on the wavelength of light (from Phillips and Borland 1992a). Each pair of distributions show data from experiments in which newts were tested in an indoor arena under either full spectrum light (left) or under wavelengths from a specific region of the spectrum adjusted to produce equal quantal flux (right). Shoreward magnetic orientation of newts trained in outdoor tanks under natural skylight: **A.** Controls tested under full spectrum light oriented towards shore. **B.** Newts tested under 400 and 450 nm light also oriented towards shore, and were indistinguishable from controls. **C.** Full spectrum controls were oriented towards shore. **D.** Newts tested under 500, 550 and 600 nm light oriented approximately 90° counterclockwise of shore, and differed significantly from full spectrum controls. **E.** Full spectrum controls oriented toward shore. **F.** Newts tested under 475 nm were randomly distributed with respect to the magnetic field and differed significantly from full spectrum controls. Shoreward magnetic orientation of newts trained in outdoor tanks under long-wavelength light: **G.** Newts tested under full spectrum light oriented approximately 90° clockwise of shore. **H.** Newts tested under wavelengths greater than 500 nm were oriented in the shoreward magnetic direction, and differed significantly from newts tested under full spectrum light (G). The two distributions were significantly different. The newts tested under full spectrum light after long-wavelength training (G) also differed from newts tested under full spectrum light after full spectrum training (A, C, E). Each data point is the magnetic bearing of an individual new, tested only once. Distributions are the pooled magnetic bearings from an approximately equal number of newts tested in each of four symmetrical magnetic field alignments, i.e., with magnetic north aligned along each of the four cardinal compass directions. As a consequence of this symmetrical testing format, the distributions retain only the component of orientation that was a consistent response to the magnetic fields (see Phillips and Borland 1992a,c for more detailed descriptions of methods). Arrows at the centre of distributions indicate significant mean vector bearings; lengths of arrows are proportional to mean vector lengths "r" (radius of circle corresponds to r = 1). Dashed lines indicate the 95% confidence interval for the mean vector bearing.

produced normal (400–450 nm) and shifted (500–600 nm) magnetic orientation, were randomly distributed with respect to the magnetic field (Phillips and Borland 1992a).

Final confirmation that the visual system of amphibians is involved in magnetoreception will require the use of neurophysiological techniques to directly sample the responses of specific classes of photoreceptors to magnetic stimuli. Because photoreceptors are present in both the retina and pineal complex of amphibians, however, current research is focusing on localizing the site of the light-dependent magnetic compass. Once this has been accomplished, the relative ease of neurophysiological recording from amphibian preparations should facilitate comparisons of behavioural and neurophysiological responses to the magnetic field.

B. Mechanism(s) of Magnetoreception Involved in Homing

Evidence that newts using the magnetic field for shoreward compass orientation and for the compass component of homing are affected differently by changes in the vertical component of the magnetic field (Phillips 1986a) and by changes in the wavelength of light (Phillips and

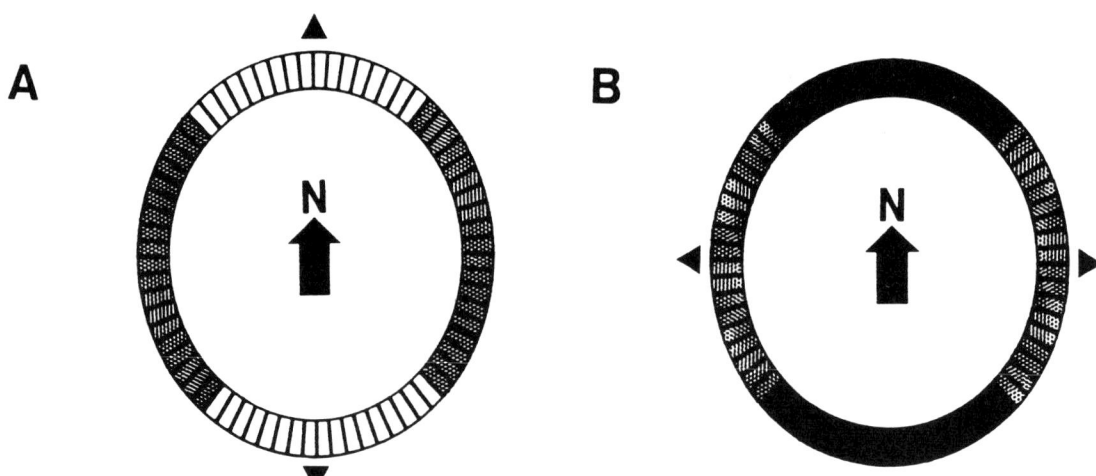

Fig. 4. Hypothetical magnetoreception system consisting of a circular array of receptors (from Phillips and Borland 1992a). **A.** Under short-wavelength light, receptors aligned within a certain angle of either end of the magnetic axis exhibit an increase in response (light-coloured rectangles) relative to receptors in alignments that are not affected by the magnetic field (hatched rectangles). Arrows at the edge of the circular array indicate the axis that will appear to have the highest level of response. **B.** Under long-wavelength light, receptors in alignments that are affected by the magnetic field exhibit a decrease in response (dark rectangles) relative to receptors in alignments that are unaffected by the magnetic field (hatched rectangles). The axes with the highest level of response (indicated by black triangles) differ by 90° in (A) and (B). This hypothesis does not predict the direction along the megnatic axis that will be interpreted by a given organism as magnetic north.

Borland 1994) provide support for the presence of two magnetoreception mechanisms. While on theoretical grounds the presence of a second magnetoreception mechanism suggests that the geomagnetic field is involved in the map component of homing (Phillips 1986a, and see earlier discussion), the distinct functional properties exhibited by newts engaged in the compass component of homing suggests that the second magnetoreception mechanism may also be involved in deriving compass information. This second magnetoreception mechanism probably does not function as an autonomous compass system, however, because natural selection would be unlikely to have maintained two such mechanisms if they provided redundant sources of directional information (Phillips and Borland 1994).

Phillips and Borland (1994) proposed that the second magnetoreception mechanism inferred to be present in newts is a light-independent intensity or inclination detector that is sensitive to the polarity of the magnetic field. The principle function of the light-independent mechanism may be detection of subtle spatial variation in magnetic field intensity. Phillips and Borland propose that the map and compass components of homing may be mediated by a "hybrid" magnetoreception system that receives inputs both from the hypothesized light-independent mechanism and from the light-dependent magnetic compass. Integration of information from these two magnetoreception mechanisms may be required to measure magnetic field intensity or inclination with the precision and reproducibility necessary to detect geographic variation, which averages only about 0.01% per km in total intensity and 0.005° per km in inclination.

Phillips and Borland (1994) argued that newts use the light-dependent magnetic compass to align the intensity or inclination detector in a fixed relationship to the magnetic field (e.g., along the magnetic north-south axis) when taking measurements of magnetic field intensity. Precise alignment of an intensity detector relative to the magnetic field, for example, would increase measurement accuracy if the response of the intensity detector were affected by magnetic field alignment. An intensity detector that measures the variance in the alignment of magnetite particles (Yorke 1979) is likely to involve a transduction mechanism (e.g., hair cell, stretch receptor) that exhibits some degree of directional sensitivity. By aligning the intensity detector with an independent magnetic compass system each time that a measurement is taken, the directional component of the intensity detector's response would be held constant.

Because directional information from the light-dependent magnetic compass of newts is wavelength-dependent (Phillips and Borland 1992a), changes in the wavelength of light under which newts are tested can be used to manipulate the input from the light-dependent magnetic compass to provide a test for the involvement of the proposed hybrid magnetoreception system in homing. Under long-wavelength light, the directional information from the light-dependent magnetic compass is shifted by 90° (Phillips and Borland 1992a). A 90° shift in the input from light-dependent magnetic compass to the hybrid magnetoreception system should cause the apparent position of the north-south magnetic axis to be at right angles to the horizontal polarity of the magnetic field measured by the intensity detector. When newts are tested under long wavelength light, therefore, the polarity of the magnetic field could not be used to distinguish between the two ends of the magnetic axis. If newts that are homing are unable to use input from the light-dependent magnetoreception mechanism as an inclination compass, this inability to distinguish between the two ends of the magnetic axis under long-wavelength light should result in either bimodal or random (Wiltschko and Wiltschko 1972; Light et al. 1993) orientation.

In experiments investigating the effects of different wavelengths of light on the use of the magnetic field for the compass component of homing, Phillips and Borland (1994) found that orientation of newts tested under short-wavelength (i.e., 400 and 450 nm) light was indistinguishable from controls tested under full spectrum light. However, as predicted, under long-wavelength (i.e., 550 and 600 nm) light, newts exhibiting the compass component of homing were randomly distributed with respect to the magnetic field. These findings confirm the need for a more detailed investigation of the proposed hybrid magnetoreception system.

V. CONCLUSIONS

Investigations of magnetic field sensitivity in amphibians are clearly in their infancy. Nevertheless, studies of this vertebrate group have already made important contributions to an understanding of the role of the geomagnetic field in orientation behaviour, and to characterizing the underlying receptor mechanism(s). The ease with which the magnetic responses of amphibians can be studied under both laboratory and field conditions insures that these animals will continue to play an important role in attempts to identify and characterize the receptor(s) responsible for detection of the geomagnetic field.

VI. ACKNOWLEDGEMENTS

Studies of magnetic compass orientation in newts were supported by the National Science Foundation. I would like to thank Chris Borland, Mark Deutschlander, Omer Sayeed and Wendy Wente for comments on earlier drafts of this manuscript.

VII. REFERENCES

Beason, R. C., 1989. Use of an inclination compass during migratory orientation by the bobolink (*Dolichonyx oryzivorus*). *Ethology* **81**: 291–299.

Beason, R. C. and Semm, P., 1987. Magnetic responses of the trigeminal nerve system of the bobolink (*Dolichonyx oryzivorus*). *Neurosci. Lett.* **80**: 229–234.

Benvenuti, S. and Ioale, P., 1985. Initial orientation of pigeons: different sensitivity to altered magnetic fields in birds of different countries. *Experientia* **44**: 358–359.

Burda, H., Marhold, S., Westenberger, Wiltschko, R. and Wiltschko, W., 1990. Magnetic compass orientation in the subterranean rodent *Chryptomys hottentotus*. *Experientia* **46**: 528–530.

Ferguson, D. E., 1971. The sensory basis of orientation in amphibians. *Ann. N.Y. Acad. Sci.* **188**: 30–36.

Fritzch, B. and Neary, T., 1997. The octavolateralis system of mechanosensory and electrosensory organs. Chapter 4. Pp. 878–922 *in* "Amphibian Biology", Vol. 3, "Sensory Perception", ed by H. Heatwole and E. M. Dawley. Surrey Beatty and Sons, Chipping Norton.

Gould, J. L., 1980. The case for magnetic sensitivity in birds and bees (such as it is). *Amer. Sci.* **68**: 256–267.

Kalmijn, A., 1982. Electric and magnetic field detection in elasmobranch fishes. *Science* **218**: 916–918.

Kalmijn, A. and Blakemore, K., 1982. Magnetic behavior of mud bacteria. Pp. 354–355 *in* "Animal Migration, Navigation and Homing", ed by K. Schmidt-Koenig and W. T. Keeton. Springer Verlag, Berlin.

Kirschvink, J. L. and Gould, J. L., 1981. Biogenic magnetite as a basis for magnetic field detection in animals. *Biosystems* **13**: 181–201.

Kirschvink, J. L. and Walker, M. M., 1985. Particle-size considerations for magnetite-based magnetoreceptors. Pp. 243–254 in "Magnetite Biomineralization and Magnetoreception: A New Biomagnetism", ed by J. L. Kirschvink, D. S. Jones and B. J. MacFadden. Plenum Press, New York.

Kirschvink, J. L., Jones, D. S. and Walker, M. M., 1985. "Magnetite Biomineralization and Magnetoreception in Organisms: A New Biomagnetism", ed by J. L. Kirschvink, D. S. Jones and B. J. MacFadden. Plenum Press, New York.

Leask, M. J. M., 1977. A physicochemical mechanism for magnetic field detection by migrating birds and homing pigeons. *Nature* **267**: 144–145.

Leask, M. J. M., 1978. Primitive models of magnetoreception. Pp. 318–324 in "Animal Migration, Navigation and Homing", ed by K. Schmidt-Koenig and W. T. Keeton. Springer Verlag, Berlin.

Light, P., Salmon, M. and Lohmann, K. J., 1993. Geomagnetic orientation of loggerhead sea turtles: evidence for an inclination compass. *J. Exp. Biol.* **182**: 1–10.

Moore, B. R., 1980. Is the homing pigeon's map geomagnetic? *Nature* **285**: 69–70.

Olcese, J., Reuss, S., Stehle, J., Steinlechner, S. and Vollrath, L., 1988. Responses of the mammalian retina to experimental alteration of the ambient magnetic field. *Brain Res.* **448**: 325–330.

Phillips, J. B., 1977. Use of the earth's magnetic field by orienting cave salamanders *(Eurycea lucifuga)*. *J. Comp. Physiol.* **121**: 273–288.

Phillips, J. B., 1986a. Two magnetoreception pathways in a migratory salamander. *Science* **233**: 765–767.

Phillips, J. B., 1986b. Magnetic compass orientation in the Eastern redspotted newt *(Notophthalmus viridescens)*. *J. Comp. Physiol.* **158**: 103–109.

Phillips, J. B., 1987. Homing orientation in the Eastern red-spotted newt *(Notophthalmus viridescens)*. *J. Exp. Biol.* **131**: 215–229.

Phillips, J. B., 1996. Magnetic navigation. *J. Theoret. Biol.* (in press).

Phillips, J. B. and Borland, S. C., 1992a. Behavioural evidence for use of a light-dependent magnetoreception mechanism by a vertebrate. *Nature* **359**: 142–144.

Phillips, J. B. and Borland, S. C., 1992b. Magnetic compass orientation is eliminated under near-infrared light in the eastern red-spotted newt *Notophthalmus viridescens*. *Anim. Behav.* **44**: 796–797.

Phillips, J. B. and Borland, S. C., 1992c. Wavelength specific effects of light on magnetic compass orientation of the eastern red-spotted newt *Notophthalmus viridescens*. *Ethol. Ecol. Evol.* **4**: 33–42.

Phillips, J. B. and Borland, S. C., 1994. Use of a specialized magnetoreception system for homing by the eastern red-spotted newt *Notophthalmus viridescens*. *J. Exp. Biol.* **188**: 275–291.

Phillips, J. B., Adler, K. and Borland S. C., 1995. True navigation by an amphibian. *Anim. Behav.* **50**: 855–858.

Phillips, J. B. and Sayeed, O., 1993. Wavelength-dependent effects of light on magnetic compass orientation in Drosophila melanogaster. *J. Comp. Physiol.* **172**: 303–308.

Rodda, G. H. and Phillips, J. B., 1992. Navigational systems develop along similar lines in amphibians, reptiles and birds. *Ethol. Ecol. Evol.* **4**: 43–51.

Schulten, K., 1982. Magnetic field effects in chemistry and biology. *Advan. Solid State Phys.* **22**: 61–83.

Schulten, K. and Windemuth, A., 1986. Model for a physiological magnetic compass. Pp. 99–106 in "Biophysical Effects of Steady Magnetic Fields", ed by G. Maret. Springer-Verlag, Berlin.

Semm, P. and Beason, C., 1990. Responses to small magnetic variations by the trigeminal system of the bobolink. *Brain Res. Bull.* **25**: 735–740.

Semm, P. and Demaine, C., 1986. Neurophysiological properties of magnetic cells in the pigeon's visual system. *J. Comp. Physiol.* **159**: 619–625.

Semm, P., Nohr, D., Demaine, C. and Wiltschko. W., 1984. Neural basis of the magnetic compass: Interactions of visual, magnetic and vestibular inputs in the pigeon's brain. *J. Comp. Physiol.* **155**: 283–288.

Sinsch, U., 1987. Orientation behaviour of toads *(Bufo bufo)* displaced from the breeding site. *J. Comp. Physiol.* **161**: 715–727.

Sinsch, U., 1988. El sapo andino *Bufo spinulosus*: análisis preliminar de su orientación hacia sus lugares de reproducción. *Boletin de Lima* **57**: 83–91.

Sinsch, U., 1990. The orientation behaviour of three toad species (genus *Bufo*) displaced from the breeding site. *Fortschr. Zool.* **38**: 75–83.

Sinsch, U., 1992. Sex-biased site fidelity and orientation behavior in reproductive natterjack toads *(Bufo calamita)*. *Ethol. Ecol. Evol.* **4**: 15–32.

Walcott, C., 1980. Magnetic orientation in homing pigeons. *I.E.E.E. Trans. Mag.* **16**: 1008–1013.

Wiltschko, R., Schops, M. and Kowalski, U., 1989. Pigeon homing: wind exposition determines importance of olfactory input. *Naturwissenschaften* **76**: 229–231.

Wiltschko, W., 1983. Compasses used by birds. *Comp. Biochem. Physiol.* **76**: 709–713.

Wiltschko, W. and Wiltschko, R., 1972. Magnetic compass of European robins. *Science* **176**: 62–64.

Wiltschko, W. and Wiltschko, R., 1981. Disorientation of unexperienced young pigeons after transportation in total darkness. *Nature* **291**: 433–434.

Yorke, E. D., 1979. A possible magnetic transducer in birds. *J. Theoret. Biol.* **77**: 101–105.

Indexes to Amphibian Biology

Volume 3 — Sensory Perception

Subject Index   page 966

Taxonomic Index   page 970

Chemical, Enzyme and Hormonal Index   page 971

SUBJECT INDEX

Bold face indicates pages on which an item is illustrated or diagrammed, or on which the legend to the figure is located.

A

abducens nerve (cranial nerve VI) 815
AC (amacrine cells) 841
accessory olfactory bulb 723–725, 726, 728, **809**
accessory olfactory organ 879
accessory olfactory tract 726
accomodation 799
acoustic nerve/*nervus stato-acusticus* (cranial nerve VIII) **812**, 886, 889, 901, 904
action potential 721, 766, 772
alarm substances 730
alar plate 882, 889, 914, 916
amacrine cells **801**, 803, 804, 841
ambush strategy 784, 787, 788
amphibian papilla **886**, 887, 890, 892, **895**, 896, 897, 898, 899, 909, 910
amphibian recess 899
ampullary cristae 886, 887
ampullary electroreceptors 879, **880**, 884, 910, **913**
ampullary organs 879, 880, 882, **883**, 884, **911**, 912, **913**, 916
amygdala 726, 727, 808, **809**, 812, 831, 860, 907
amygdaloid nucleus 725, 726
anatomical resolution 806
anterior commissure 726
anterior olfactory habenular tract 727
anterior preoptic area 904, 906, 907, 908, 909
anterior thalamic nucleus **812**, 906, 907, 908
anterior vertical canal **886**
anterodorsal tegmental nucleus 901, 903, 909
anteroventral tegmental nucleus 909
AP (amphibian papilla) **886, 893**, 901
aposematic 729
area uncinata **816**
Arkansas 734
auricula cerbelli **809**
autostylic skull 900
AVC (anterior vertical canal) **886**, 887, 889, 891

B

BC (bipolar cells) 841
basal cells 712, **713**, 715–716, 720, 722, 748, 749, 750, 751, 755, 756, 757, 759, 761, 767, 770
basal lamina 716, 720
basal optic tract 812, 815
basal optic neuropil 815, **816**, 865
basal stem cells 758
basement membrane 712, **713**, 720, 986
basilar papilla 885, **886**, 887, 892, **893**, 894, **895**, 896–899, 900, 909, 910
basilar recess 886, 890, 891, 899
bed nucleus 909
bico-ordinate map 955, 958
binocular triangulation/binocular vision 786, 798, 799, 845, 846, 854, 855
bipolar neurons/bipolar cells 712, **801**, 803, 804, 841, 843, 860
BOD (basal optic tract) 815, **816**
BON (basal optic nucleus) **812**, 815, 817, 858, 865

Bowman's glands 712, **713**, 715, 716, 720, 721, 722, 729
BP (basilar papilla) 885, **886, 893**, 901
branchial baskets 763
bulbus olfactorius **809**

C

camera anterior bulbi 797
camera posterior bulbi 797
cannibalism 744–746, 747
caudal tectum 798
caudomedial foramen 886
cement gland 928, **929**, 930
Central America 732
central amygdalar nucleus 909
central thalamic nucleus **902**, 903, 904, 905, 906, 907, 908, 909
cerebellum 807, **808, 809, 810**, 812–815, 859, 882, 889, 890, 892
cerebral hemispheres 725
CGT (*corpus geniculatum thalamicum*) 815, **816**, 817
cheek glands 736
chiasma opticum 806, 815
China 932
chin glands 736
chin-touching 732
choanae 711
chromatin 712, **713**
cilia 712, 715, 716, 721, 722
Class O cells 841
Class 1 cells 804, **805**, 841, 843, 862
Class 2 cells 804, **805**, 806, 842, 843, 862, 863, 868
Class 3 cells 806, 842, 843, 844, 861, 862, 863, 868, 869
Class 4 cells **805**, 842, 843, 844, 861, 863, 869
Class 5 cells 806
Class 6 cells 806
Class 7 cells 806
cloacal glands 736
cloacal papillae 736
CNS (central nervous system) 714
cochlea 896
cochlear nuclei 903
colliculus inferior 811
colouration 747
command elements 787, 863, 869
commissura anterior **809**
commissura habenulae **809**, 810
commissura posterior **809**, 810, 811, 857, 858
commissura postoptica 829, **838**
commissura tecti mesencephali 832
commissura tuberculi posterioris **809**
compass orientation 861, 954, 955, 958, 960–961, 962, 963
competition 733, 734, 735
cone **802**–803, 839, 841, 843
corpus cerebelli 815
corpus geniculatum thalamicum **812**, 815, **816**, 855
cortex 725, 726, 727
corticohabenular tract 725, 726
courtship 727, 733, 735, 736
cranial nerve 0 728
cranial nerve III (see oculomotor nerve)
cranial nerve IV (see troahlear nerve)

cranial nerve V (see trigeminal nerve)
cranial nerve VII (see facial nerve)
cranial nerve VIII (see acoustic nerve/stato-acoustic nerve)
cranial nerve IX (see glossopharyngeal nerve)
cranial nerve X (see vagal nerve)
cranial nerve XII (see hypoglossal nerve)
cranial nerves **808, 835**, 836, 879, 882
cristae 888, 889, 898
croak reflex 946
crus 888
cupula 888

D

defense 735
depth perception 786, 799, 807, 854
dermal papillae 942
dermatome 944, 945
desmosomes 758, 925
development 764–766, 915, 943–944, 945–946
diencephalon 725, 807, **808, 809, 810**, 811, 812, 815, 817, **824**, 830, 833, **834**, 841, 859, 905–908, 909
dioptric apparatus 797, 807
direct development 884
DLN (dorsolateral nucleus) 901, **902**, 903, 904
dorsal auditory nucleus **883**
dorsal cutaneous nerves 936, **937**, 947
dorsal hypothalamic nucleus 908
dorsal root ganglion 923, 925, 938
dorsal tegmental nucleus of the medulla 908
dorsolateral (auditory) nucleus 897, 901, **902**, 910
dorsolateral placodes 915
dorsomedial foramen 886
dorsolateral nucleus **895**
dorsolateral placodes 879–880
DTAM (dorsal tegmental nucleus of the medulla) 908, 909

E

eardrum 910
ear placodes 915
ectopic transplantation 764
edge length 842, 854
EL (edge length) 842, 854
electro-olfactograms 720, 722
electron spin-resonance 959
electroreceptive organs 884
embryogenesis 765, 766
eminentia olfactoria 717
endolymph 882, 886, 888
endolymphatic duct 886
endoplasmic reticulum 750, 758
endosome 714
entopeduncular nuclei 904, 909
EOG (electro-olfactogram) 722
epibranchial placodes 879
epiphysis 859
epiphysis cerebri 860
epithalamus 808, 810
ERF (excitatory receptive field) 841, 842, 843, 846, 847

INDEX

eustachian tube 899
excitatory receptive field 841
exocytosis 714
external mandibular nerve 912
external nares 711, 718, **911**
external plexiform layer **809**
extramedullary neurons 925
eye cup **929**

F

facial nerve/*nervus facialis* (cranial nerve VIIth) 748, 752, 760, 761, 773, 815, 834, **835, 840,** 912
facial nucleus 885
faeces 734
fasciculus medialis telencephali 808
fasciculus lateralis telencephali 808
fasciculus longitudinalis lateralis **809**
fasciculus longitudinalis medialis 858
fasciculus postolfactorius **809**
FAP (fixed action pattern) 787
feature detectors 787
feeding 784–795, 861–869
feeding reflex 863
filia olfactoria **809**
filiform papillae 756, 762
fixed action pattern 787
foraging 731
foramen interventriculare 808
fovea centralis 801, 807
frontal nerve 861
frontal organ 859, 860, 861
FT-1 units 940, 943
FT-2 units 943
fungiform papillae 756, 762
fusiform cells 749, 750, 751, 758, 759, 767, 773

G

ganglion cells **801,** 804, **805,** 806, 807, 817, 843, 879, 882, 883, 899, 916
gap junctions 839, 925, 930, 932
gated channels 881, 882, 943
genes 721
geomagnetic field 954, 955–960, 961, 963
glial cells 713, **714,** 759, 806, 879
glomeruli 716, 721, 722, 723, 724, 725, **809**
glossopharyngeal nerve/*nervus glossopharyngeus* (cranial nerve IX) 748, 752, 760, 761, 770, 771, 772, 773, **812,** 815, **835,** 912
glossopharyngeal ganglion 761
GnRH-ir fibers 728, 729
goblet cells 720
Golgi cells/apparatus 812, 940
granule cells 723, 724, 725
green rod **802,** 803, 843
growth regulators 730

H

H1 cells 803
H2 cells 803
habenula 727, 808, **809**
habenular commissure 725, 727, 859, **860**
habenular nucleus **812, 860**
habitat selection 730
hair-cell **881,** 882, 885, 886, 887, 888, 890, 891, 892, 894, 896, 897, 898, 899, 910, 911, 912, 916

HC (horizontal canal) [Chapter 4] **886,** 887, 889, 891
HC (horizontal cells) [Chapter 3] 803, 841
hemispheric sulcus 725, 726
hervbivores 913
hippocampus 808, 908
home range 732
homesite 733
homing 732, 954, 955–958, 961
horizontal canal **886,** 888, 898
horizontal cells 799, **801,** 803, 804, 841
horopter 849, 854, 869
*Hox*A1 gene 880
hunter strategy 784, 787, 788
hybridomas 722
hybrid zone 733
hyperplasy 716
hyperpolarization 839, 841
hypoglossal motor nuclei 869
hypoglossal nerve/*nervus hypoglossus* (cranial nerve XII) **812,** 815
hypophysis **809**
hypothalamic nucleus **812**
hypothalamus 728, 729, 808, **809,** 811, 869, 906, 907, 908

I

implantation cone 723, 724, 725, 728
infundibulum 729
inhibitory receptive field 841
INL (inner nuclear layer) 799, **800,** 801, 803, 804
inner nuclear layer 799, 801
inner plexiform layer 799, **805**
inner segment 801, 802
Int-2 gene 880
internal nares 711, 715, 718
IPL (inner plexiform layer) 799, **800,** 804, **805,** 806
IRF (inhibitory receptive field) 841, 842
IS (inner segment) 801
isthmal nucleus 907
isthmic neuropil **837**
isthmo-tectal tract **837**
Italy 755

J

Japan 755
Julesz pattern 789

K

kinocilia **881,** 882, 887, 888, 890, 891, 892, 897, 912
kin recognition 730
kT 960

L

lagena 885, **886,** 887, 889, 891–892, **895,** 898, 899, 901, 903, 904
lagenar papilla 894
lagenar recess 886, 890, 892, 894
lamina propria 712, **713,** 715, 716, 720, 728, 729
laminar (toral) nucleus 901, 903, 904, 905, 907, 908
lamina terminalis 808
Landolt club 804
lateral diverticulum 718, 720, 722
lateral foramen 886
lateral forebrain bundle 904

lateral geniculate body 816
lateral hypothalamic nucleus 908, 909
lateral lemniscus 904
lateral line 880, 882, 884, 910–916
lateral line organs 879, 882, **883**
lateral line placodes 915
lateral neuropil **836**
lateral recess 718
lateral septal nucleus 904, 907, 908
lateral thalamic nucleus 905, 906, 907
Layer-1 units 842, 843
Layer-2 units 843, 844
Layer-3 units 843, 844
learning 866
lemniscal tract **837**
lemniscus spinalis **809, 812,** 836
lens placode 879
lesions 855–856, 932
levator bulbi 799
LGB (lateral geniculate body) 816
limbic tissue 886, 896

M

maculae 886, 890, 892, 898
magnocellular toral nucleus 901, 903, 904, 905, 907
mating 783, 865
mating calls 905, 907, 908, 909
maxillary nerve **929, 931**
M cells 844
medial cortex 908
medial diverticulum 718, 722
medial foramen 886
medial forebrain bundle 904, 907, 908
medial nucleus **895**
medial olfactory tract 727
medial recess 718
medial septal nucleus 908
medial superior olivary nucleus 903
medulla oblongata 807, **808, 809, 812,** 815, **816,** 818, 820, 821, **822,** 823, **824,** 825, 826, 827, **828,** 831, **832, 833, 834, 835,** 836, 837, **840,** 869, 897, 901, 909, 916
medulla spinalis **808,** 820
Merkel cells 750, 751, 757, 758, 759, **940,** 941, 942, 943, 945, 946
Merkel granules **940**
Merkel-like cells 758, 759, 760, 763, 766, 767, 771, 943
mesencephalon 807, **808, 809,** 811–812, 841
microvilli 712, 714, 721, 749, 750, 755, 757, 758, 767, 770, **881,** 882, **911,** 912
middle nerve 912
migration 731–733
mitochondria 750, 802, 924, 930
mitosis 715, 716
mitral cells 723, 724, 725, 726, **809**
monoclonal antibodies 722
motivation 869
mucus cells 757, 758, 760, **924**
mucus gland **937**
Müller cells 806
muscle spindle **948**
musculus rectus cervicis profundus **835**
musculus subarcualis rectus **835**
myeloma cells 722
myoid 802, 803

N

nares constrictor muscle 729
nasal organ 718

nasolabial grooves 720, 733, 734
nasolacrimal duct 718
NBl (*pars lateralis* of the neuropil of Bellonci) 815, 817
Nbm (*pars medialis* of the neuropil of Bellonci) 815
nBON (nucleus of the basal optic neuropil) 855
nBOT (nucleus of the basal optic root) 855
NCAM (neural cell adhesion molecule) 715
neglected papilla 885, **886,** 887, 892, 897, 898
neocortex 808, 822
neural ectoderm 716
neural tube 924, 925, 932
neurogenesis 716, 721, 727, 879
neuromasts 879, **880,** 910, **911,** 912, **913,** 916
neuromodulatory cells 759
neuropil of Bellonci **812, 816,** 855, 907
neuropil posterior thalami 816
neurotransmitters 728, 751, 770
North America 734
nose-tapping 732, 733, 734
NTS (nucleus of the solitary tract) 752, 761, 772
nucleus accumbens **812**
nucleus caudalis 903, 904
nucleus Darkschewitsch 810, 858
nucleus dorsalis 832
nucleus dorsalis tegmenti 811
nucleus dorsomedialis anterior 907
nucleus dorsolateralis anterior 907
nucleus isthmi 821, 826, 828, 832, 836–839, **837,** 844, **845,** 854, 855, 859, 904
nucleus laminaris 811, **812**
nucleus lemnisci spinalis 832
nucleus lentiformis mesencephali 811
nucleus magnocellularis 811, **812**
nucleus of Bellonci 811, **812, 816,** 855, 904, 906, 907
nucleus of the lateral lemniscus 904
nucleus of the solitary tract 752
nucleus opticus tegmenti 811
nucleus praeopticus 808, **809,** 810
nucleus praetectalis 810, 831, **834,** 858
nucleus reticularis **812,** 832
nucleus ruber 811
nucleus saccularis 891
nuclei septi 808
nucleus tegmenti mesencephali **812**
nucleus tuberculi posterioris 811
nucleus ventralis tegmenti 811
nucleus vestibularis 832
nystagmus 796, 889

O

obex region 836, 889, 904, 906, 907
OBP (odourant-binding protein) 716, 721
octaval nerve 890, 891, 892
octaval placode 879, 880
octval root 890
octavolateral efferent nucleus 885
octavolateral organs **880**
oculomotor nerve/*nervus oculomotorius* (cranial nerve III) 799, 811
oculomotor nucleus **812**
Oklahoma 734

olfactohabenular tract 726, 727
olfactometer 732, 733
olfactory buds 720
olfactory bulb 714, 721, 722, 723–725, 726, 727, 728, 736, 808, **809, 810,** 909
olfactory chamber 715, 719, 722
olfactory epithelium 711, 712, 718, 720, 721, 722, 729
olfactory glands 711
olfactory knobs 712, 714, 715, 716, 722
olfactory mucosa **713,** 714, 720
olfactory nerve 714, 717, 723, 725, 726, 728, **810**
olfactory placodes 716, 879
olfactory tract 724, 725, 726, 727, **809**
oliva superior **812**
omega figures 942
ONL (outer nuclear layer) 799, **800**
oophagy 747
opercularis muscle 899
operculum 886, 899, 900, 910
ophthalmic ganglia 882
ophthalmic nerve 928, **929, 931**
OPL (outer plexiform layer) 799, **800,** 803, 804, 806
optic chiasma 812
optic nerve 714, 801, 806, **812,** 815, **840,** 857, 868
optic nucleus 812
optic tectum 726, 807, **810, 812,** 817–831, **828,** 842, 904, 906, 907
optic tract/*tractus opticus* **812, 816**
optical resolution 806
optokinetic responses 795, 796, 855
optomotor nerve 858
optomotor responses 795, 855
oral cavity 711
orientation 731–733
oropharynx 748, 749, 753, 754, 760, 763, 765, 772, 773
OS (outer segment) 801, 802
osmotic balance 772
ossicles 909, 916
otic capsule 885, 886, 890, 892, **893,** 900, 929
otic cup 879
otic duct 886
otic labyrinth 885, **886**
otic placode 916
optoconia 890, 892
otocyst 879, 916
otoliths 889, 891
outer nuclear layer 799, 801
outer plexiform layer 799
outer segment 801
oval window 886, 892, 899, 900

P

P (pretectal neuropil) 815, 817
paedomorphosis 807, 871
palatine ganglia 729
pallial commissure **812**
pallium 808, **809, 812,** 859, 907, 909
paracrine cells 759
parallax 785
P cells 843
pedicel 801, 802
peptide-immunoreactive fibers 751, 752
periamygdaloid region 725
periglomerular cells 724
perilymph 886, 899

perilymphatic tissue/space/system 886, 892, 893
perilymphatic sac **893**
periotic canal **893**
periotic duct **893,** 894, 896
periotic labyrinth 892, **893,** 894, 896, 897, 898, 899, 909, 916
periotic sac 892
periotic tissue/space/system 885, 886, 890, 892–893, 894
periventricular grey matter 817, 818, **819,** 820, 828, **831**
pheromones 729, 733, 734, 735
philopatric 730
photoisomerization 839
phototaxis 861
pineal complex 859–861, 961
pineal organ 859–861
pineal tract **860**
pit organ **911,** 912, **913**
placodes 879, 881, 882, 898, 913, 915, 916
plexiform layer 723, 724, 725
poison glands 747
pontine gustatory nucleus 752
post cloacal press 734
posterior commissure 859, **860**
posterior thalamic nucleus 905, 906
posterior vertical canal **886**
posteroventral thalamic nucleus 906, 909
postolfactory eminence 725
postoptic commissure 820, 821, 828, 830, 832
postotic ganglionic complex 752, 760
praetectum [see pretectum]
predator saturation 730
preoptic nucleus **812**
preoptic region 726, 728, 729, 860, 907
pretectal neuropil 815, **816,** 819, 830, 839, 857
pretectal nuclei 904
pretectum **809,** 811, 816, 820, 821, 823, **824,** 826, **828,** 830, 844, 855–858, 859, 860, 861, 865, 905, 907
prey recognition 789–795, 861–869
principal (toral) nucleus 901, 903, 904, 905
profundus placode 879
protractor lentis 785, 799
pulvinar 832
Purkinje cells 812, 815
PVC (posterior vertical canal) **886,** 887, 888, 889, 892, 894, 897
PVG (periventricular grey) 828, 829, **831**

Q (nil)

R

R2 cells 842, 843
R3 cells 842, 844
R4 cells 842, 844
RA I (rapidly adapting nerve type I) 936, 950, 951
RA II (rapidly adapting nerve type II) 936, 941, 950, 951
RCP (*musculus rectus cervicis profundus*) **835**
receptive field 944, 949, 950–951
receptor cells 712–714, 802
receptor terminals **801**

INDEX

red rod **802**, 803
reflex 945
reflex arc 912
reflexogenous zones 945
regeneration 764, 944
Reissner's fibre **860**
resting potential 881
reticular formation 870
retina 839–845, **845**
retinal ganglion cells 799, 841, 867
retinotopic map 844
retractor bulbi 799
RF (receptive field) 845, 846, 854, 858, 866
RGC (retinal ganglion cells) 799, **800**, 801, 804, 806, 820, 841, 842, 843, 844, 861, 862, 863, 866, 868, 869
RGCL (retinal ganglion cell layer) **805**
rhombencephalon 909
rods 801–802, 843
Rohon-Beard neurites **924**, 925, **930**
Rohon-Beard neurons/cells **924**, 925, **926**, 927, 928, 929, 931, 937, 938
round window 886, 892

S

saccule **886**, 887, 889, 890–891, 892, 893, 897, 899, 901, 903, 904
SA units (slowly adapting nerves) 941, 950, 951
SA Ft-1 (slowly adapting nerve, frog type 1) 937
SA Ft-2 (slowly adapting nerve, frog type 2) 937
SAR (*musculus subarcualis rectus*) **835**
satellite support cells 757, 758
scavengers 913
schooling 730
scotoma 859
SD (single domain particles) 959
secondary isthmal nucleus 904, 906, 908, 909
secretory vesicles 750
selector gene 880
SEM (scanning electron microscopy) 721
semicircular canals 886, 887–889, 890
septal nucleus 725, 726, 727, **812**, 908
septum 728, **809**, 904
sexual dimorphism 727, 896
sibling recognition 729, 730
sign stimuli 787
single domain particles 959
SO (superior olivary nucleus) 901, **902**, 903, 904, 906
somatotopic map 772
spatial facilitation 869
spermatophore 736
spherule 801, 802
sphincter pupillae 803
spinal cord 906, 907, 909, 914, 923, 925, **926**, 927, 928, **931**, 932, 936
spinal ganglia 945
spinal nerve 945
SPM (superparamagnetic particles) 959
stapes 899, 900, 910
stapes foot plane 886
stato-acoustic nerve/*nervus stato-acusticus* (cranial nerve VIII) 815, 889, 901, 904
stem cells 748

stereocilia 750, 881, 887, 888
stereovilli **881**, 882, 887, 888, 890, 896
Stirnorgan 859, **860**
stitches 911, 912, 913, 915
stratum album centrale 818
stratum griseum centrale 818
stratum griseum periventriculare 817
stria medullaris 726, 727
stria medullaris thalami 725
stria terminalis 909
striatum 726, 808, **812**, 901, **902**, 904, 906, 908, 909
striola 886, 889, 890, 892
subcommisural organ **860**, 861
subepidermal nerve plexus 751
subepithelial plexus 760
sulcus dorsalis 810
sulcus hypothalamicus 810
sulcus limitans **812**
sulcus medialis 810, **812**
sulcus ventralis 810
superior olivary nucleus 901, **902**, 903
superficial reticular nucleus 901, 903, 904, 909
superparamagnetic particles 959
suprachiasmatic nucleus **812**, 904, 907, 909
supratemporal nerve 912
sustentacular cells 712, **713**, **714**, 714–715, 716, 720, 757, 758, 759
swimbladder 900
synapses 716, 724, 725, 750, 751, 758, 760, 761, 770, 802, 803, 804, 822, 839, 860, 928, 930, 942
synaptic terminal 803
synaptic vesicles 750

T

T1, T2, T3, T4, T5 (tectal cell types) 820–829, **822**, **824**, 837, 846–853, 862, 863, 865, 866, 868, 869, 870
taste bud 748, 749, 750, 751, 752, 753, 754, 755, 757, 761, 763, 764, 765, 766, 767, 770, 773, 879, 882
taste disks 754, 755, 756, 757, 759, 760, 762–763, 764, 771, 772, 773
taste papilla 761
taste pore 767
taste receptor cell 758, 760, 766, 766, 767, 768, 769, 773
Tastzellen 940
tectal cell types (T-1 to T-4) 820–829
tectobulbar tract 859, 870
tecto-bulbo-spinal tract 820, 826, 828, 829, **835**, **836**, **837**
tectorium 887
tectum 811, 817–839, **819**, **827**, **840**, 841, 844–854, **845**, 865, 866, 907, 914
tectum mesencephali 807
tectum opticum **809**, 811, **816**, 817–831, **845**
tegmental nuclei 906
tegmentum **809**, 810, 811, **818**, 820, 832, 833, **837**, 859, 909
telencephalon 715, 716, 723, 725, 728, 729, 807, **808**, **809**, **810**, **812**, **816**, **832**, 859, 869, 901, 908–909
telencephalon impar 808
telodendra 802, 819, 820, 837, **838**

temporal facilitation 860
tentacles 717
terminal nerve 728, 729
terminal nerve ganglion 879
territory 732, 733, 734, 735
TH 1-10 neurons 855, 862, 863
thalamus 752, 808, **809**, 810, 811, 815, **816**, 817, 820, 821, 826, **828**, 830, **832**, 836, 844, 855–858, 859, 904, 906, 907, 908–909, 914
thalamic eminence 811
thalamic nucleus/neuropil **812**, 817, 819, 837, 839, 904
thermal randomization energy (kT) 960
TN (terminal nerve) 728, 729
tongue-snapping 731
tonofilaments 925
tonotopy 905, 910
torus semicircularis 807, 811, **818**, 822, 901, **902**, 903, 904–905, 908, 914
toxicitiy 746–748
TP phenomena 858, 859
tractus opticus basalis 816
tractus opticus marginalis **812**, 816
tractus opticus medialis 816
transduction 721, 723, 743, 767–768, 769, 771, 881, 887, 942–943, 949, 954, 958–960
trigeminal ganglion 751, 752, 761, **926**, 928, **929**, **931**
trigeminal/facial nerve 729
trigeminal nerve (cranial nerve V) 815, 879, 882, 912, 928–931, 932, 947, 959
trochlear nerve (cranial nerve IV) 811, 812
tuberculum 811, **812**, 909
tufted cells 725
tympanum 899, 900, 909
Type I cells 758
Type II cells 758
Type III cells 759

U

UF (uncinate field) 815
uncinate field (*area uncinata*) 815, **816**, 817
United States 755
utricular papilla 894
utricle **886**, 887, 888, 889–890, 891
utriculo-saccular foramen 886, 887, 890, 897, 898

V

vagal nerve/*nervus vagus* (cranial nerve X) 748, 752, 760, 761, 773, 815, **835**, **836**, 912, 947
Vancouver Island 734
ventral hypothalamic neuropil 904
ventral hypothalamic nucleus **902**, 904, 907, 908, 909
ventral nucleus **895**
ventrolateral isthmus 901
ventrolateral thalamic nucleus 906
ventromedial thalamic nucleus 906, 907
vestibular nucleus 890, 891, **895**
vestibulo-ocular reflex 799
visual acuity 806–807
visual field 797–798
VNO (vomeronasal organ) 717, 718, 722, 729

vomeronasal mucosae 712

vomeronasal nerve 713, 723, 725, 727, 728

vomeronasal organ 711, 716, 717, 718, 719, 720, 722, 727, 728, 733

vomeronasal epithelium 712, 714, 716, 718, 719, 720, 729

W

wing cells 757, 758, 759, 760

X

X cells 843
Xhox 1.6 gene 880

Y

yawing 888
Y cells 844

Z

z-axis 844, 862
zona limitans 725

TAXONOMIC INDEX

Bold face indicates pages on which a photograph of the species appears.

A

Afrixalus fornasii 830
Alytes 923
Alytes obstetricans 790, 933
Alytes cisternsi 933
Ambystoma 713, 715, 719, 721, 724, 925, 932
Ambystoma annulatum 745
Ambystoma barbouri 730
Ambystoma jeffersonianum 735
Ambystoma macrodactylum croceum 745
Ambystoma maculatum 719, 730, 732, 735, 745
Ambystoma mexicanum 714, 719, 744, 748, 753, 818, 830, 933, 941
Ambystoma opacum 730, 830
Ambystoma platineum 735
Ambystoma punctatum 933, 941
Ambystoma texanum 730
Ambystoma tigrinum 711, 716, 721, 726, 727, 728, 731, 745, 746, 806, 860, 861, 923, 933, 941, 943, 946, 949, 950, 951
Amphiuma 720
Andrias 884
Andrias davidianus 911
Aneides ferreus 734
Aneides flavipunctatus 728
Aneides lugubris 941
Arenophryne 806
Arenophryne rotunda 829, 830
Ascaphus 891, 894, 897, 898
Ascaphus truei **886**

B

Batrachosep 854
Batrachoseps attenuatus 728, 788, 806, 807, 829, 830, 854
Bolitoglossa rufescens 728, 792
Bolitoglossa rostrata 728, 787
Bolitoglossa subpalmata 724, 726, 727, 728, **786,** 787, 788, 789, 797, 805, 809, 816, 830, 852
Bolitoglossini 787, 788, 792, 815, 816
Bombina 796, 797, 891, 896
Bombina bombina 792
Bombina orientalis 754, 756, 762, 763, 830
Bombina variegata 745, 747, 790, 796
Breviceps verrucosus 787
Bufo 784, 795, 797, 842, 847, 852, 891, 896, 955
Bufo americanus 729, 730, 731, 806, 855
Bufo boreas 730, 732
Bufo boreas halophilus 745

Bufo bufo 732, 762, **784,** 786, 788, 790, 792, 794, 795, 796, 798, 842, 846, 847, 850, 851, 855, 858, 860, 862, 863, 864, 933, 939, 940, 941, 951, 956, 957
Bufo calamita 731, 732, 745, 956, 957
Bufo cognatus 730
Bufo japonicus 732, 762, 869
Bufo marinus 724, 731, 801, 802
Bufo spinulosus 732, 956, 957
Bufo valliceps 731, 732
Bufo vulgaris 933

C

Caeciliidae 717
Calliphora erythrocephala 788
Calyptocephalella gayi 762
Ceratophrys dorsata 745
Ceratophrys ornata 787
Chacophrys pierottii 745, 746
Chauliognathus 788
Chiropterotriton 728
Chthonerpeton indistinctum 716
Cryptobranchus 720
Cryptobranchus alleganiensis 746
Cynops orientalis 932, 933
Cynops pyrrhogaster 728, 933

D

Dendrobates auratus 745
Dendrobates histrionicus 745
Dendrobates lehmanni 745
Dendrobates occultator 745
Dendrobates pumilio 732, 745, 747, 830
Dendrobates viridis 745
Dendrotriton 728
Dendrotriton bromeliacea 787
Dermophis 723
Desmognathinae 815
Desmognathus 733, 734, 787, 797, 859
Desmognathus aeneus 830
Desmognathus fuscus 732, 733, 787
Desmognathus fuscus fuscus 746
Desmognathus imitator 733
Desmognathus monticola 734, 735, 830
Desmognathus ochrophaeus 727, 728, 732, 733, 736, 746, 807, 830
Desmognathus quadramaculatus 728, 734, 735, 830
Desmognathus wrighti 728, 830
Dicamptodon copei 746
Dicamptodon ensatus 746, 787
Discoglossus 828
Discoglossus pictus 756, 792, 827, 830, 831, 834, 835, 933
Drosophila 788

E

Eleutherodactylus coqui 763, 827, 829, 830, 831, 835
Ensatina eschscholtzi 728
Epicrionops petersi 717
Eurycea 787, 788
Eurycea bislineata 728, 730, 731, 736, 807, 816, 830
Eurycea guttolineata 728
Eurycea longicauda 730, 787
Eurycea lucifuga 955
Eurycea wilderae 727, 728

F (nil)

G

Gastrotheca riobambae 830

H

Heleioporus eyrei 801
Hemidactylinii 788, 815
Hoplophryne rogersi 745
Hydromantes 788, 796, 797, 807, 818, 829, 845, 852, 854
Hydromantes genei 788, 792, 816, 829
Hydromantes italicus 728, 731, **785,** 786, 788, 789, 792, 793, 796, 797, 798, 807, 808, 818, 822, 823, 829, 830, 844, 852, 853, 854, 855, 869
Hydromantes shastae 800
Hyla arborea 933
Hyla brunnea 745
Hyla chrysoscelis 730
Hyla cinerea 724, 788, 792, 903, 906
Hyla crucifer 730
Hyla pseudopuma 745
Hyla raniceps 801
Hyla regilla 728, 730
Hyla septentrionalis 830, 846
Hyla zeteki 745
Hymenochirus boettgeri 745
Hynobius retardatus 753
Hyperolius quinquevittatus 830
Hypogeophis rostratus 717

I

Ichthyophis 723, 754
Ichthyophis glutinosus **886**
Ichthyophis kohtaoensis 755, 941
Ichthyophis orthoplicatus 941

J (nil)

K (nil)

INDEX

L

Latimeria 891, 893, 894, 896, 913
Limnodynastes 818
Limnodynastes tasmaniensis 818, 830
Lepidobatrachus laevis 744
Lepomis cyanellus 730
Leptodactylus pentadactylus 744, 745

M

Mantella aurantiaca 830
Mantella cowani 830
Musca 792
Musca domestica 788

N

Necturus 713, 720, 735, 750, 751, 802, 843
Necturus maculosus 720, 744, 746, 748
Necturus tigrinum 720
Notophthalmus viridescens 730, 731, 732, 735, 736, 746, 803, 806, 955

O (nil)

P

Parvimolge townsendi 830
Phyllobates aurotaenia 745
Phyllobates bicolor 745
Phyllobates lugubris 745
Phyllobates terribilis 745
Phyllobates vittatus 745
Pipa 722, 894
Pipidae 718, 912
Proteidae 735
Proteus 720, 735
Plethodon 733, 734, 788
Plethodon aureolus 733
Plethodon cinereus 713, 718, 719, 721, 727, 728, 729, 731, 733, 734, 746, 788, 789, 791, 802, 830
Plethodon caddoensis 734
Plethodon dunni 734, 746
Plethodon glutinosus 723, 733, 746
Plethodon jordani 724, 726, 727, 728, 732, 733, 789, 791, 792, 806, 819, 822, 823, 824, 829, 830, 831, 832, 833, 835, 838, 852, 854, 857
Plethodon kentucki 733
Plethodon ouachitae 734
Plethodon shenandoah 733
Plethodon teyahalee 733
Plethodontini 815, 816
Plethodontidae 733
Plethodon vandykei 734
Plethodon vehiculum 733, 734

Pleurodeles waltl 830, 941
Proteus anguinus 941
Pseudacris clarki 731
Pseudacris crucifer 730, 905
Pseudacris strecki 731
Pseudacris triseriata 730
Pyxicephalus adspersus 745, 787

Q (nil)

R

Rana 711, 723, 797, 817, 842, 891, 923, 929, 932
Rana arvalis 745
Rana cascadae 729
Rana catesbeiana 712, 713, 722, 724, 725, 727, 728, 730, 744, 745, 759, 763, 771, 787, 810, 811, 812, 906, 907, 940, 941, 942, 946, 947, 950, 951
Rana clamitans 724, 730, 945
Rana cyanophlictus 745
Rana esculenta 725, 726, 730, 745, 757, 789, 790, 792, 798, 812, 816, 825, 843, 846, 860, 933, 939, 941
Rana japonica 763
Rana lessonae 730
Rana palustris 745
Rana pipiens 712, 715, 722, 724, 725, 726, 727, 728, 730, 731, 732, 745, 756, 784, 786, 802, 804, 806, 811, 817, 818, 846, 858, 859, 862, 905, 906, 907, 941, 945, 94 7, 951
Rana pretiosa 730
Rana ridibunda 724, 725, 745
Rana sphenocephala 771
Rana sylvatica 729, 730, 732
Rana temporaria 730, 732, 745, 757, 763, 790, 796, 798, 799, 807, 830, 847, 849, 850, 860, 928, 933, 940, 941, 942
Rana tigrina 745
Rana utricularia 732
Rhacophorus leucomystax 830
Rhinophrynus dorsalis 745

S

Salamandra 791, 797, 804, 852, 855, 858, 859, 923
Salamandra maculosa 933
Salamandra salamandra 724, 726, 727, 728, 731, 746, 753, 785, 786, 788, 789, 790, 791, 792, 793, 794, 795, 796, 798, 799, 806, 807, 830, 842, 843, 858, **886,** 903, 933, 941
Salamandridae 732
Scaphiopus 746

Scaphiopus bombifrons 744, 745, 746
Scaphiopus couchii 787
Scaphiopus hammondi hammondi 745
Scaphiopus holbrooki 745
Scaphiopus multiplicatus 730
Scolecomorphus kirkii 717
Siredon pisciformis 933
Siren 720, 914
Sminthillus limbatus 830
Spodoptera 788

T

Taricha 732, 735, 736, 747
Taricha granulosa 728, 729
Taricha rivularis 732
Taricha torosa 746, 747
Thorius 797
Thorius macdougali 728
Thorius narisovalis 807, 830
Thorius pennatulus 830
Triton 722
Triturus 720, 736, 804, 843, 894, 926, 929, 930, 931
Triturus alpestris 727, 728, 933
Triturus cristatus 736, 941
Triturus helveticus 928, 930, 933
Triturus marmorata 933
Triturus palmatus 933
Triturus vulgaris 792, 927, 932, 933, 941
Typhlonectes 754
Typhlonectes compressicaudus 754
Typhlonectes natans 716, 717
Tylotriton 796, 797
Tylotriton verrucosus 796
Typhlonectidae 716

U

Ureaotyphlus narayani 717

V (nil)

W (nil)

X

Xenopus 713, 722, 843, 861, 880, 891, 895, 908, 909, 912, 914, 915, 916, 924, 925, 926, 927, 928, 929, 930, 931, 932, 942
Xenopus laevis 729, 754, 759, 762, 771, 804, 806, 830, 860, 880, 927, 933, 940, 941, 942

Y (nil)

Z (nil)

CHEMICAL, ENZYME, AND HORMONAL INDEX

Alphabetical index is followed by a numerical index of chemicals whose names begin with numbers.

A

acetylcholine 728, 885
acetylcholinesterase 728
actin 882
adenylate cyclase 721
alpha-bungarotoxin 925
aragonite 891
arginine 771
aspartic acid 771
atropine 799

B

biocytin 804, 819, 820, 822, 823, 826, 827, 828, 829, 830, 831, 833, 836, 837, 839, 852

C

calbindin 759
calcite 890
calcitonin gene-related peptide 751, 761
calmodulin 759

cAMP (cyclic adenosine monophosphate) ***, 769
calretinin 749, 751
cantharidin 744
carnosine 722
cGMP 769
CGRP (calcitonin gene-related peptide) 751, 761
cholinesterase 761
cobaltic lysine 839, 850
colchicine 943

D

denatoniuim 769

E (nil)

F

FMRFamide (molluscan caridoexcitatory peptide) 728

G

GABA (gamma-aminobutyric acid) 825, 841, 931
GABA-lir 815
gamma-aminobutyric acid 825
gastrin-releasing peptide 761
GDP (guanine diphosphate) 721
glutamate 751, 825, 928
glutamate-lir 825
glycine 825, 841
glycine-lir 825
glycoprotein 721, 760
gonadotropin 729, 732, 908
gonadotropin-releasing hormone 728, 729
GnRH (gonadotropin-releasing hormone) 728, 729
G protein 767, 768, 769
G-protein (G_{olf}) 721
GRP (gastrin-releasing peptide) 761
GTP (guanine triphosphate) 721

H

horseradish peroxidase 727, 817, 839, 925, 926, 929, 930, 931
HRP (horseradish peroxidase) 721, 817, 818, 819, 831, 858, 901

I

IP-3 769

J (nil)

K

kainic acid 858, 859

L

lanthanum 760
L-arginine 769
lectins 722
L-proline 769

M

magnetite 959
molluscan caridoexcitatory peptide 728

N

neuron specific enolase 759
neuropeptides 728
neurotoxins 747
noradrenaline 942
NSE (neuron specific enolase) 759

O

octopamine 942

P

parvalbum 759
peptide histadine isoleucine 761
PHI (peptide histadine isoleucine) 761
phospholipase C 769
porphyropsin 802, 803
protein kinase 769

Q

quinacrine 940

R

retinoic acid 880
rhodopsin 802, 803, 839, 860

S

SBA (soybean agglutinin) 722
serotonin 749, 750, 770, 823
S-100 protein 759
Spot 35 protein 759
substance P 751, 761

T

tetrodoxin 932
TH (tyrosine hydroxylase) 761
thyroxin 753
TTX 766, 768
tubocurarine 925
tyrosine hydroxylase 761

U (nil)

V

vasoactive intestinal peptide 761
VIP (vasoactive intestinal peptide) 751, 761

W (nil)

X (nil)

Y (nil)

Z (nil)

Numerical

3H-thymidine 721
5-hydroxytryptamine 942